LINEAR ALGEBRA

Theory and Applications

Yuma Ju, Wenxun Xing
Cuiqin Lin, Jinde Hu
Feiyan Wang
Tsinghua University, Beijing

CENGAGE Learning™

TSINGHUA UNIVERSITY PRESS

Australia • Brazil • Japan • Korea • Mexico • Singapore • Spain • United Kingdom • United States

CENGAGE Learning™

LINEAR ALGEBRA
Theory and Applications

Yuma Ju, Wenxun Xing, Cuiqin Lin, Jinde Hu and Feiyan Wang

Publishing Director:
Paul Tan

Senior Development Editor:
Yang Liping

Associate Development Editors:
Joe Ng
Tanmayee Ashok Bhatwadekar

Acquisition Editors:
Liu Ying
Zhou Kaiyan

Senior Product Director:
Janet Lim

Product Managers:
Kevin Joo
Tan Lee Hong

Assistant Publishing Manager:
Pauline Lim

Production Executive:
Cindy Chai

Copy Editor:
Parthasarathi Mukhopadhyay

Cover Designer:
Ong Lay Keng

Compositor:
Integra Software Services (P) Ltd

ISBN-13: 978-981-4232-00-5
ISBN-10: 981-4232-00-9

Cengage Learning Asia Pte Ltd
5 Shenton Way
#01-01 UIC Building
Singapore 068808

Printed in Singapore
1 2 3 4 5 14 13 12 11 10

Contents

Preface

This book was conceptualized and written to be used as a main text or as a reference for students who need to apply linear algebraic techniques for solving problems related to their specific disciplines. All the content in this book can be divided into four sections; basic theory, extended theory (marked with *), applications, and the appendices. There are seven chapters in the book;

- Determinants,
- Matrices,
- Systems of Linear Equations,
- Vector Spaces and Linear Transformations,
- Eigenvalues and Eigenvectors of a Matrix; Matrix Diagonalization,
- Quadratic Forms, and
- Applications of Linear Algebra.

Furthermore, a brief introduction to the concept of Inner Product Space, Hermitian Quadratic Forms, and the Jordan Canonical Form of a matrix is presented in the appendices.

As a comprehensive text, the book maintains clarity by explaining the profound ideas in simple terms. Theorems, corollaries, properties, definitions, proofs, and worked examples are presented in a brief and direct style.

The text has the following six major features:

1. **Target audience:** The book was written for students whose major is not mathematics. Hence, we have focused on the algebra of matrix, linear equations, eigenvalues and eigenvectors of a matrix, matrix diagonalization, and quadratic forms. We also provide a brief introduction to the structure of a linear space and the theory of linear transformations. In addition, Chapter 7 provides a useful introduction on the applications of linear algebra to students who will use linear algebra as a tool to solve various problems in their future courses.
2. **Teaching hours:** The book can be customized for courses ranging from 35 to 72 teaching hours.
3. **Proofs of theorems:** In the book, there are some key theorems that are difficult to prove. We have provided lucid yet brief proofs to allow students to better grasp the theory and concepts behind the theorems. For example, in Chapter 3, we describe the relationship among the highest order of a nonsingular matrix, the column rank and row rank of a matrix, that implies that the ranks of the columns and rows of a matrix are the same.
4. **Exercises:** This book also contains a large number of exercises, totaling 452 in the first six chapters. All exercises are categorized into three levels: basic, tough (marked with*), and supplementary exercises. Such

an arrangement allows instructors to select the exercises in each chapter according to the abilities and competencies of their classes.

5. **Range of application problems:** Chapter 7 presents seven application-based topics that may be used to introduce students to the widespread use of linear algebraic concepts in everyday life.
6. **Instructor resources:** Teaching tools, such as PowerPoint slides and the answers to the exercises of all seven chapters and two appendices, are available for instructors.

This book was translated from its Chinese edition, which sold more than 342,000 copies since it was first published. The Chinese edition was used as a textbook in more than 100 universities in Mainland China until 2009. Professor Wenxun XING led two PhD candidates–Qiang CHOU and Juanwei ZHANG–to work on the translation of the book, while Professor Cuiqin LIN prepared the PowerPoint slides for this book.

The authors wish to thank the PhD candidates, Qiang CHOU and Juanwei ZHANG, for assisting in the preparation and careful editing of the final manuscript. The authors are also grateful for the valuable comments on the manuscript from Mr. Ying LIU. Similarly, special appreciation is due to Dr. Parthasarathi MUKHOPADHYAY who refined the linguistic expressions of the translated text, proofread every word of the book carefully to clean up the errors, and also corrected the wrong answers to the exercises. Lastly, we would like to thank Liping YANG and Joe NG from Cengage Learning Asia Pte Ltd for their support in the successful publication of this book.

Last but not least, we wish to acknowledge the financial support from the Key Project No. 108005 of the Chinese Ministry of Education and Grant No. 10801087 of the China National Science Foundation.

Authors
April 2010

About the Authors

Yuma JU is Professor at the Department of Mathematical Sciences, Tsinghua University, Beijing, China. He was the Deputy Director of the Institute of Mathematics and Deputy Head of the Department of Applied Mathematics, responsible for undergraduate and postgraduate mathematics teaching in Tsinghua University. He was the Chief Editor or first author of 10 of his 12 published textbooks since 1980.

Wenxun XING is Professor at the Department of Mathematical Sciences, Tsinghua University. He received his BSc in Mathematics from Peking University in 1982 and a PhD in Operations Research in 1997 from Tsinghua University. His current research interests include production planning and scheduling, combinatorial optimization, global optimization, and metaheuristics. He has published over 40 journal papers and is a member of the Operations Research Society of China and the Mathematical Programming Society of China.

Cuiqin LIN is Professor at the Department of Mathematical Sciences, Tsinghua University. Her research areas include combinatorial theory, graph theory, and network optimization. She has published over 10 journal papers and five textbooks.

Jinde HU is Professor at the Department of Mathematical Sciences, Tsinghua University. He has rich experience in teaching courses such as calculus, linear algebra, analytic geometry, differential equations, and differential geometry. He has coauthored five textbooks with Professor Ju Yuma since 1986.

Feiyan WANG is Professor at the Department of Mathematical Sciences, Tsinghua University. She has taught courses on calculus, differential equations, linear algebra, analytic geometry, and mathematical modeling. Her research area is operations research. She has published four textbooks.

About the Copy Editor

Parthasarathi MUKHOPADHYAY was recipient of the medal for achieving first position in the first class in his MSc examination from the Department of Pure Mathematics, University of Calcutta, India. Later, he obtained his MPhil and PhD degrees from the same university department.

Currently, he is Associate Professor in Mathematics at the Ramakrishna Mission Residential College, Narendrapur, a highly reputed autonomous postgraduate college, affiliated to the University of Calcutta, where he teaches courses on abstract and linear algebra, real analysis, metric space, and differential equations. He is also a guest faculty at the Belur Vidyamandira College, where he teaches linear algebra at the postgraduate level. His research area is algebraic theory of semirings and history of mathematics, on which he has published 18 research papers and has presented some of those papers during various national and international conferences in India and abroad. Dr. Mukhopadhyay has also authored three textbooks (seven volumes altogether) on mathematics at the undergraduate and plus-two level in India.

Notations

$\mathbb{R}, \mathbb{R}^+, \mathbb{R}^n$	Real numbers, reals no less than zero, n-tuples of reals
$\mathbb{C}$	Complex numbers
$\{\cdots \mid \cdots\}$	Set of ... such that ...
Σ, Π	Summation of serial numbers, multiple times of serial numbers
$\in, \forall, \exists$	belongs to, for all, there exists
$\boldsymbol{A}, \boldsymbol{B}, \boldsymbol{C}, \boldsymbol{D}$	Matrices
$M_{m\times n}$	Set of $m \times n$ matrices
$\det\boldsymbol{A}, \lvert\boldsymbol{A}\rvert$	Determinant of the matrix $\boldsymbol{A}$
a_{ij}	Matrix entry in row i, column j
M_{ij}	Complementary minor
$\boldsymbol{A}_{ij}$	Algebraic complementary minor
$\mathbf{0}, \mathbf{0}_3$	Zero vector, zero vector of 3 dimensions
$\boldsymbol{I}, \boldsymbol{I}_n$	Unit matrix, unit matrix of order n
$\mathrm{diag}(d_1, d_2, \ldots, d_n), \boldsymbol{A}$	Diagonal matrix
$\boldsymbol{A}^T, \boldsymbol{A}^*, \boldsymbol{A}^{-1}$	Transpose, adjoint, inverse matrix of $\boldsymbol{A}$
$\mathrm{r}(\boldsymbol{A})$ or $\mathrm{rank}\,(\boldsymbol{A})$	Rank of $\boldsymbol{A}$
$Re(\mathrm{z})$	Real part of z
$\overline{\boldsymbol{A}}$	Conjugate of $\boldsymbol{A}$
$\boldsymbol{A} \cong \boldsymbol{B}$	Matrix $\boldsymbol{A}$ is equivalent to matrix $\boldsymbol{B}$
$\boldsymbol{A} \sim \boldsymbol{B}$	Matrix $\boldsymbol{A}$ is similar to matrix $\boldsymbol{B}$
$\boldsymbol{A} \simeq \boldsymbol{B}$	Matrix $\boldsymbol{A}$ is congruent to matrix $\boldsymbol{B}$
$\boldsymbol{\alpha}, \boldsymbol{\beta}$	Vectors
$V, V(F)$	Vector space, vector space over F
$W, W^{\perp}$	Vector subspace, orthogonal subspace
$(\boldsymbol{\alpha}, \boldsymbol{\beta})$ of $\boldsymbol{\alpha} \cdot \boldsymbol{\beta}$	Inner product of $\boldsymbol{\alpha}$ and $\boldsymbol{\beta}$
$<\boldsymbol{\alpha}, \boldsymbol{\beta}>$	Angle between $\boldsymbol{\alpha}$ and $\boldsymbol{\beta}$
$\perp$	Perpendicular or orthogonal
$\lvert x\rvert, \lVert\boldsymbol{\alpha}\rVert$	Absolute value of x, norm of $\boldsymbol{\alpha}$
$\boldsymbol{\varepsilon}_1, \boldsymbol{\varepsilon}_2, \ldots, \boldsymbol{\varepsilon}_{\mathrm{n}}$	Standard basis of $\mathbb{R}^{\mathrm{n}}$
$\mathrm{R}[x]_{\mathrm{n}}$	Set of $(n-1)$th degree polynomials over R
$\mathrm{L}(S), \mathrm{R}(S)$	Linear span of the set S
$M \oplus N$	Direct sum of subspaces M and N
$\mathrm{Ker}\sigma, \mathrm{Im}\sigma$	Kernel of σ, image of σ

Lower case Greek alphabet

name	character	name	character	name	character
alpha	α	iota	ι	rho	ρ
beta	β	kappa	κ	sigma	σ
gamma	γ	lambda	λ	tau	τ
delta	δ	mu	μ	upsilon	υ
epsilon	ε	nu	ν	phi	ϕ
zeta	ζ	xi	ξ	chi	χ
eta	η	omicron	o	psi	ψ
theta	θ	pi	π	omega	ω

List of Figures

List of Tables

List of Examples

1 Determinants

CHAPTER OVERVIEW

Determinants are among the most basic components of linear algebra. The theory of determinants is often used as a powerful tool in solving various real-life problems. Henceforth, we will introduce the definitions of the second-order and the third-order determinants and their expansion formulae along their respective first rows, and then discuss certain properties of the nth-order determinants. In this chapter, we

- introduce the definition of the nth-order determinants and their properties,
- show the different methods of calculation of determinants, including the use of Sarrus' rule,
- use Cramer's rule to solve a particular kind of nonhomogeneous systems of linear equations, and
- discuss the resultant necessary condition(s) under which a homogeneous system of linear equations, with an equal number of equations and unknowns, has a nonzero solution.

In Figure 1.1, we briefly describe the notations that will be used in this chapter.

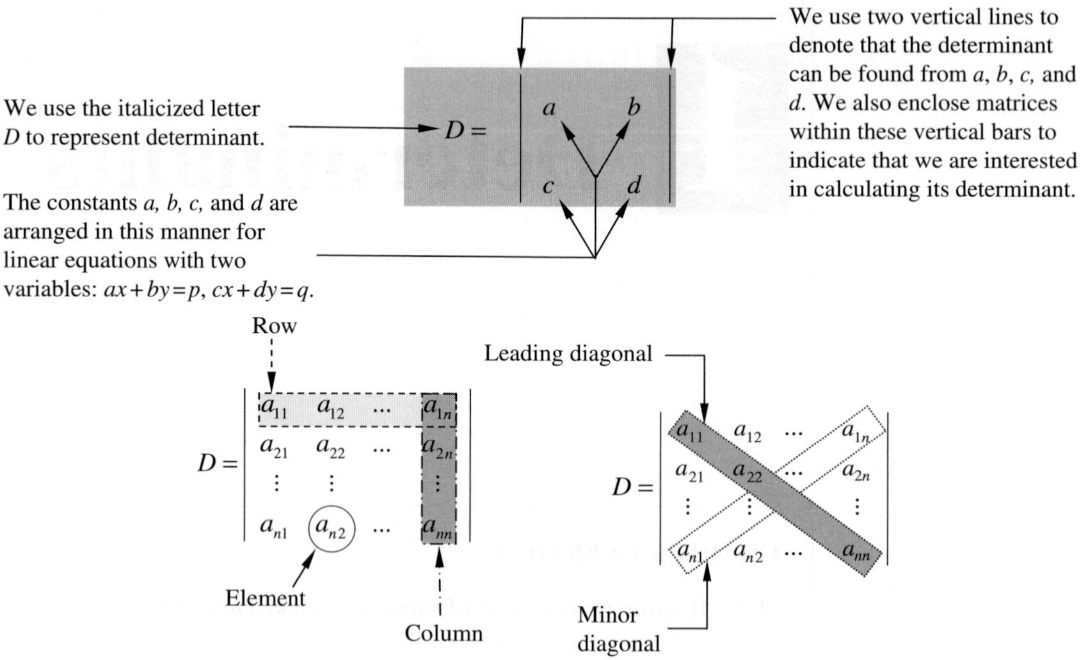

Figure 1.1 Notations of determinants

1.1 Definition of *n*th-order Determinants and Their Properties

We encounter the notion of a determinant for the first time while solving a system of simultaneous equations, in which the number of equations is the same as the number of unknowns (a system of simultaneous equations shall be referred to as a system of linear equations henceforth).

For instance, consider a system of linear equations in two variables,

$$\begin{cases} a_{11}x_1 + a_{12}x_2 = b_1, \\ a_{21}x_1 + a_{22}x_2 = b_2. \end{cases} \tag{1.1}$$

When $a_{11}a_{22} - a_{12}a_{21} \neq 0$, one can obtain its solution by the elimination method as shown below

$$x_1 = \frac{b_1a_{22} - a_{12}b_2}{a_{11}a_{22} - a_{12}a_{21}}, \quad x_2 = \frac{a_{11}b_2 - b_1a_{21}}{a_{11}a_{22} - a_{12}a_{21}}. \tag{1.2}$$

Now, if we denote

$$D = \begin{vmatrix} a & b \\ c & d \end{vmatrix} = ad - bc, \tag{1.3}$$

then consequently, we can represent the expressions (1.2) by

$$x_1 = \frac{\begin{vmatrix} b_1 & a_{12} \\ b_2 & a_{22} \end{vmatrix}}{\begin{vmatrix} a_{11} & a_{12} \\ a_{21} & a_{22} \end{vmatrix}}, \quad x_2 = \frac{\begin{vmatrix} a_{11} & b_1 \\ a_{21} & b_2 \end{vmatrix}}{\begin{vmatrix} a_{11} & a_{12} \\ a_{21} & a_{22} \end{vmatrix}}.$$

We call D in the formula (1.3) a **second-order determinant**.

For an expression ranged over three rows and three columns using nine elements a_{ij} $(i, j = 1, 2, 3)$ as below, we define

$$\begin{vmatrix} a_{11} & a_{12} & a_{13} \\ a_{21} & a_{22} & a_{23} \\ a_{31} & a_{32} & a_{33} \end{vmatrix} = \begin{array}{l} a_{11}a_{22}a_{33} + a_{12}a_{23}a_{31} + a_{13}a_{21}a_{32} \\ -a_{13}a_{22}a_{31} - a_{11}a_{23}a_{32} - a_{12}a_{21}a_{33}, \end{array} \tag{1.4}$$

which is called a **third-order determinant** (each of the three horizontal arrays is called a **row** and each of the vertical ones is called a **column**).

The six terms in the expression (1.4) may be obtained from the following schematic method (also known as **Sarrus' rule**), as shown below

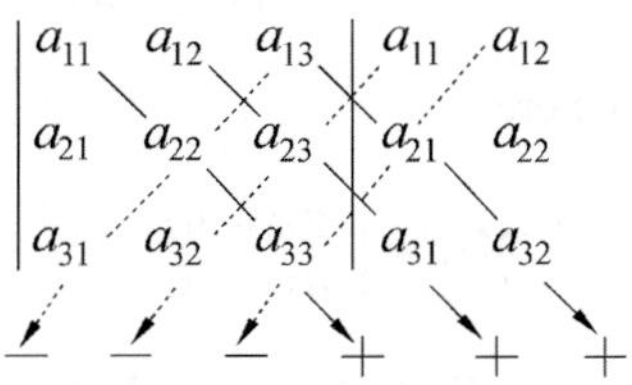

(1.5)

If

$$D = \begin{vmatrix} a_{11} & a_{12} & a_{13} \\ a_{21} & a_{22} & a_{23} \\ a_{31} & a_{32} & a_{33} \end{vmatrix} \neq 0,$$

happens to be the determinant made out of the coefficients of a linear equation system in three unknowns;

$$\begin{cases} a_{11}x_1 + a_{12}x_2 + a_{13}x_3 = b_1, \\ a_{21}x_1 + a_{22}x_2 + a_{23}x_3 = b_2, \\ a_{31}x_1 + a_{32}x_2 + a_{33}x_3 = b_3, \end{cases}$$

we solve the system of linear equations by the elimination method and may obtain

$$x_1 = \frac{D_1}{D}, \quad x_2 = \frac{D_2}{D}, \quad x_3 = \frac{D_3}{D}, \tag{1.6}$$

where D_j $(j = 1, 2, 3)$ is the third-order determinant formed out of D by replacing its jth column with the constant terms b_1, b_2, b_3; that is,

$$D_1 = \begin{vmatrix} b_1 & a_{12} & a_{13} \\ b_2 & a_{22} & a_{23} \\ b_3 & a_{32} & a_{33} \end{vmatrix}, \quad D_2 = \begin{vmatrix} a_{11} & b_1 & a_{13} \\ a_{21} & b_2 & a_{23} \\ a_{31} & b_3 & a_{33} \end{vmatrix}, \quad D_3 = \begin{vmatrix} a_{11} & a_{12} & b_1 \\ a_{21} & a_{22} & b_2 \\ a_{31} & a_{32} & b_3 \end{vmatrix}.$$

However, an nth-order determinant $(n > 3)$ in general cannot be defined by an analogical expression created out of a schematic diagram similar to (1.5) (**Sarrus' rule**). A general nth-order determinant needs to be defined by some other suitable method. Although it can be defined by three different methods, we will present the definition using a concise recursive method.

From the expansions of determinants of orders two or three, we find that they follow a common law—it can be expanded along the first row—that is,

$$\begin{aligned} D &= \begin{vmatrix} a_{11} & a_{12} & a_{13} \\ a_{21} & a_{22} & a_{23} \\ a_{31} & a_{32} & a_{33} \end{vmatrix} \\ &= a_{11} \begin{vmatrix} a_{22} & a_{23} \\ a_{32} & a_{33} \end{vmatrix} - a_{12} \begin{vmatrix} a_{21} & a_{23} \\ a_{31} & a_{33} \end{vmatrix} + a_{13} \begin{vmatrix} a_{21} & a_{22} \\ a_{31} & a_{32} \end{vmatrix}, \end{aligned} \tag{1.7}$$

where

$$M_{11} = \begin{vmatrix} a_{22} & a_{23} \\ a_{32} & a_{33} \end{vmatrix}, \quad M_{12} = \begin{vmatrix} a_{21} & a_{23} \\ a_{31} & a_{33} \end{vmatrix} \text{ and } M_{13} = \begin{vmatrix} a_{21} & a_{22} \\ a_{31} & a_{32} \end{vmatrix}$$

are called the **complementary minors** of elements a_{11}, a_{12}, and a_{13}, respectively.

We call $A_{11} = (-1)^{1+1} M_{11}$, $A_{12} = (-1)^{1+2} M_{12}$, and $A_{13} = (-1)^{1+3} M_{13}$ the **algebraic complements** of a_{11}, a_{12}, and a_{13}, respectively. So, the expression (1.7) may also be denoted as

$$D = a_{11}A_{11} + a_{12}A_{12} + a_{13}A_{13}. \tag{1.8}$$

Similarly,

$$D = \begin{vmatrix} a_{11} & a_{12} \\ a_{21} & a_{22} \end{vmatrix} = a_{11}A_{11} + a_{12}A_{12}, \tag{1.9}$$

where $A_{11} = (-1)^{1+1} |a_{22}| = a_{22}$, and $A_{12} = (-1)^{1+2} |a_{21}| = -a_{21}$.

Here, $|a_{22}|$ and $|a_{21}|$ stand for the first-order determinants (not to be confused with the notation for the absolute value of a number). We define $|a|$ to be a, for the first-order determinant $|a|$.

If (1.8) and (1.9) are considered as the definitions of the second- and third-order determinants, respectively, one can easily note a uniform pattern in their definition. They are both defined in terms of the determinants of lower order. Given this information, it is quite natural to seek a definition of the general nth-order determinant by the recursive method. The determinants of every order defined by this method will have uniform operational properties. We now provide the recursive definition of the nth-order determinant as follows.

1.1.1 Definition of nth-order determinants

Definition 1.1

An nth-order **determinant** composed of n^2 numbers a_{ij} $(i, j = 1, 2, \ldots, n)$,

$$D = \begin{vmatrix} a_{11} & a_{12} & \cdots & a_{1n} \\ a_{21} & a_{22} & \cdots & a_{2n} \\ \vdots & \vdots & & \vdots \\ a_{n1} & a_{n2} & \cdots & a_{nn} \end{vmatrix} \quad \text{(also denoted as } \left|a_{ij}\right|_1^n) \qquad (1.10)$$

is recursively defined for $n \geq 2$ by

$$D = a_{11}A_{11} + a_{12}A_{12} + \cdots + a_{1n}A_{1n} = \sum_{j=1}^{n} a_{1j}A_{1j}, \qquad (1.11)$$

where, $A_{1j} = (-1)^{1+j}M_{1j}$, M_{1j} being the $(n-1)$th-order determinant obtained from D by deleting the first row and the jth column without disturbing the remaining rows and columns, that is,

$$M_{1j} = \begin{vmatrix} a_{21} & \cdots & a_{2j-1} & a_{2j+1} & \cdots & a_{2n} \\ a_{31} & \cdots & a_{3j-1} & a_{3j+1} & \cdots & a_{3n} \\ \vdots & & \vdots & \vdots & & \vdots \\ a_{n1} & \cdots & a_{nj-1} & a_{nj+1} & \cdots & a_{nn} \end{vmatrix} \quad (j = 1, 2, \ldots, n),$$

and M_{1j} is called the **complementary minor** of element a_{1j}, whereas A_{1j} is called the **algebraic complement** of a_{1j}. For the case of $n = 1$, we define $D = |a_{11}| = a_{11}$.

In (1.10), the diagonal with $a_{11}, a_{22}, \ldots, a_{nn}$ is called the **leading diagonal** of the determinant, while $a_{11}, a_{22}, \ldots, a_{nn}$ are correspondingly called the **leading diagonal elements**. The other diagonal is called the **minor diagonal** of the determinant.

It is clear from the above definition that the expression of a determinant is a homogeneous polynomial of degree n (called **expansion**), which is formed by n^2 elements a_{ij} $(i, j = 1, 2, \ldots, n)$. There are 2! terms in the expansion of a second-order determinant, 3! terms in the expansion of a third-order determinant, and $n!$ terms in the expansion of an nth-order determinant, in which each term is a product of n elements from different rows and different columns. Half of the $n!$ terms are with positive sign, while the remaining half of the terms are with negative sign (this conclusion can be proved according to the definition by invoking mathematical induction). An nth-order determinant is a first-degree homogeneous polynomial in $x_1, x_2, \ldots, x_n$ if its elements in the first row are $x_1, x_2, \ldots, x_n$.

Example 1.1 Lower triangular nth-order determinant

Prove that a lower triangular nth-order determinant,

$$D_n = \begin{vmatrix} a_{11} & 0 & \cdots & 0 \\ a_{21} & a_{22} & \cdots & 0 \\ \vdots & \vdots & \ddots & \vdots \\ a_{n1} & a_{n2} & \cdots & a_{nn} \end{vmatrix} = a_{11}a_{22}\cdots a_{nn},$$

where $a_{ij} = 0$ as $i < j$, that is, elements above the leading diagonal of the determinant are all zeros.

Proof: Let's use mathematical induction on n. For $n = 2$, the equality is true. Assume that the conclusion is true for a lower triangular determinant of order $(n-1)$. Then from Definition 1.11, we have

$$D_n = \begin{vmatrix} a_{11} & 0 & \cdots & 0 \\ a_{21} & a_{22} & \cdots & 0 \\ \vdots & \vdots & \ddots & \vdots \\ a_{n1} & a_{n2} & \cdots & a_{nn} \end{vmatrix} = (-1)^{1+1}a_{11} \begin{vmatrix} a_{22} & 0 & \cdots & 0 \\ a_{32} & a_{33} & \cdots & 0 \\ \vdots & \vdots & \ddots & \vdots \\ a_{n2} & a_{n3} & \cdots & a_{nn} \end{vmatrix}.$$

As the determinant on the right-hand side (*RHS*) is a lower triangular determinant of order $(n-1)$, one obtains by induction hypothesis that

$$D_n = a_{11}(a_{22}a_{33}\cdots a_{nn}).$$

■

Similarly, we can prove that an nth-order diagonal determinant (all of whose elements off the leading diagonal are zeros):

$$\begin{vmatrix} a_{11} & 0 & \cdots & 0 \\ 0 & a_{22} & \cdots & 0 \\ \vdots & \vdots & \ddots & \vdots \\ 0 & 0 & \cdots & a_{nn} \end{vmatrix} = a_{11}a_{22}\cdots a_{nn}.$$

Example 1.2 Computing determinants using recursive formula

Compute the nth-order determinant (elements above the minor diagonal are all zeros)

$$D_n = \begin{vmatrix} 0 & 0 & \cdots & 0 & a_n \\ 0 & 0 & \cdots & a_{n-1} & * \\ \vdots & \vdots & ⋰ & \vdots & \vdots \\ 0 & a_2 & \cdots & * & * \\ a_1 & * & \cdots & * & * \end{vmatrix},$$

where $a_i \neq 0 \quad (i = 1, 2, \ldots, n)$ and "$*$" denotes that the element is an arbitrary number.

Solution: Note that for general n, the determinant is not equal to $-a_1a_2\ldots a_n$. Using the definition of determinant (see Definition 1.11), we obtain

$$D_n = (-1)^{n+1}a_n \begin{vmatrix} 0 & 0 & \cdots & a_{n-1} \\ \vdots & \vdots & ⋰ & \vdots \\ 0 & a_2 & \cdots & * \\ a_1 & * & \cdots & * \end{vmatrix} = (-1)^{n-1}a_nD_{n-1}.$$

By repeated use of the above relation (generally called a **recursion relation** or a **recursion formula**) between successive determinants, replacing n with $(n-1)$, $(n-2)$, ..., successively, we get

$$\begin{aligned} D_n &= (-1)^{n-1}a_nD_{n-1} \\ &= (-1)^{n-1}a_n(-1)^{n-2}a_{n-1}D_{n-2} \\ &\vdots \\ &= (-1)^{(n-1)+(n-2)+\cdots+2+1}a_na_{n-1}\cdots a_2a_1 \\ &= (-1)^{\frac{n(n-1)}{2}}a_na_{n-1}\cdots a_2a_1. \end{aligned}$$

For instance, $D_2 = -a_1a_2$, $D_3 = -a_1a_2a_3$ for $n = 2, 3$; $D_4 = a_1a_2a_3a_4$, $D_5 = a_1a_2a_3a_4a_5$ when $n = 4, 5$.

The conclusion is also true for a determinant whose elements below the minor diagonal are all zeros. We can also arrive at a similar recursion formula by expanding the determinant along the last column, taking into consideration Property 1.2, that is discussed in the next section. ■

1.1.2 Properties of *n*th-order determinants

Generally speaking, it is quite complicated to compute a determinant directly from its definition. Hence, we want to deduce some properties of the determinant from its definition to simplify the calculation of determinants.

Property 1.1

The value of a determinant does not change if its rows and columns are exchanged (in original order), that is,

$$\begin{vmatrix} a_{11} & a_{12} & \cdots & a_{1n} \\ a_{21} & a_{22} & \cdots & a_{2n} \\ \vdots & \vdots & & \vdots \\ a_{n1} & a_{n2} & \cdots & a_{nn} \end{vmatrix} = \begin{vmatrix} a_{11} & a_{21} & \cdots & a_{n1} \\ a_{12} & a_{22} & \cdots & a_{n2} \\ \vdots & \vdots & & \vdots \\ a_{1n} & a_{2n} & \cdots & a_{nn} \end{vmatrix}. \tag{1.12}$$

This property can be proved by mathematical induction. The body of the proof is loaded down with trivial details and is omitted. Readers who are interested in it can refer to §1.4.1.

From this property, it follows that all the properties of determinant stated in terms of rows are true for columns as well. Next, we will discuss properties of determinant in terms of rows only.

Property 1.2

Determinant (1.10) is expanded along any row by the following expression. Their values are equal, that is,

$$D = a_{i1}A_{i1} + a_{i2}A_{i2} + \cdots + a_{in}A_{in} = \sum_{j=1}^{n} a_{ij}A_{ij} \ (i = 1, 2, \ldots, n), \tag{1.13}$$

where $A_{ij} = (-1)^{i+j}M_{ij}$, M_{ij} being the $(n-1)$th-order determinant formed out of D with the original arrangement after deleting all elements on the ith row and the jth column. M_{ij} is called the **complementary minor** of a_{ij}, whereas A_{ij} is called the **algebraic complement** of a_{ij}.

The proof of Property 1.2 is similar to Property 1.1, and is also carried out by mathematical induction (see §1.4.2).

Property 1.3 (Linearity)

There are two results:

$$\text{(i)} \quad \begin{vmatrix} a_{11} & a_{12} & \cdots & a_{1n} \\ \vdots & \vdots & & \vdots \\ ka_{i1} & ka_{i2} & \cdots & ka_{in} \\ \vdots & \vdots & & \vdots \\ a_{n1} & a_{n2} & \cdots & a_{nn} \end{vmatrix} = k \begin{vmatrix} a_{11} & a_{12} & \cdots & a_{1n} \\ \vdots & \vdots & & \vdots \\ a_{i1} & a_{i2} & \cdots & a_{in} \\ \vdots & \vdots & & \vdots \\ a_{n1} & a_{n2} & \cdots & a_{nn} \end{vmatrix}. \tag{1.14}$$

$$\text{(ii)} \quad \begin{vmatrix} a_{11} & a_{12} & \cdots & a_{1n} \\ \vdots & \vdots & & \vdots \\ a_{i1}+b_{i1} & a_{i2}+b_{i2} & \cdots & a_{in}+b_{in} \\ \vdots & \vdots & & \vdots \\ a_{n1} & a_{n2} & \cdots & a_{nn} \end{vmatrix} = \begin{vmatrix} a_{11} & a_{12} & \cdots & a_{1n} \\ \vdots & \vdots & & \vdots \\ a_{i1} & a_{i2} & \cdots & a_{in} \\ \vdots & \vdots & & \vdots \\ a_{n1} & a_{n2} & \cdots & a_{nn} \end{vmatrix} + \begin{vmatrix} a_{11} & a_{12} & \cdots & a_{1n} \\ \vdots & \vdots & & \vdots \\ b_{i1} & b_{i2} & \cdots & b_{in} \\ \vdots & \vdots & & \vdots \\ a_{n1} & a_{n2} & \cdots & a_{nn} \end{vmatrix}. \tag{1.15}$$

Applying Property 1.2, expanding the determinants on the left-hand side (*LHS*) of the equalities in (1.14) and (1.15) along the ith row, we immediately obtain the results on the *RHS*.

From expression (1.14), we also get

COROLLARY 1.1

If a determinant has a row consisting entirely of zeros, then its value is zero.

Property 1.4

If a determinant has two identical rows, then its value is zero, that is,

$$D = \begin{vmatrix} a_{11} & a_{12} & \cdots & a_{1n} \\ \vdots & \vdots & & \vdots \\ a_{i1} & a_{i2} & \cdots & a_{in} \\ \vdots & \vdots & & \vdots \\ a_{j1} & a_{j2} & \cdots & a_{jn} \\ \vdots & \vdots & & \vdots \\ a_{n1} & a_{n2} & \cdots & a_{nn} \end{vmatrix} = 0, \tag{1.16}$$

where $a_{il} = a_{jl}$ $(i \neq j, l = 1, 2, \ldots, n)$.

Proof: This can be proved by mathematical induction on n. The conclusion is clearly true for a second-order determinant. Suppose the conclusion is true for an $(n-1)$th-order determinant $(n > 3)$. In the case of nth-order, expand the determinant along the kth row $(k \neq i, j)$, then

$$D = a_{k1}A_{k1} + a_{k2}A_{k2} + \cdots + a_{kn}A_{kn} = \sum_{l=1}^{n} a_{kl}A_{kl}.$$

Since the complementary minors M_{kl} $(l = 1, 2, \ldots, n)$ are $(n-1)$th-order determinants, and each of these has two rows with identical elements,

$$A_{kl} = (-1)^{k+l} M_{kl} = 0 \ (l = 1, 2, \ldots, n).$$

So,

$$D = 0.$$

From result (i) in Property 1.3 and Property 1.4, we immediately obtain:

COROLLARY 1.2

A determinant with corresponding elements of two rows proportional is equal to zero (i.e., $a_{jl} = ka_{il},\ i \neq j,\ l = 1, 2, \ldots, n;$ *k is a constant).*

Property 1.5

In a determinant, if every element in a row multiplied by a constant number k $(k \neq 0)$, is added to the corresponding element in another row, the value of the determinant remains the same (in short, when a determinant undergoes a row operation of "multiple addition," its value does not change), that is,

$$\begin{vmatrix} a_{11} & a_{12} & \cdots & a_{1n} \\ \vdots & \vdots & & \vdots \\ a_{i1} & a_{i2} & \cdots & a_{in} \\ \vdots & \vdots & & \vdots \\ a_{j1} & a_{j2} & \cdots & a_{jn} \\ \vdots & \vdots & & \vdots \\ a_{n1} & a_{n2} & \cdots & a_{nn} \end{vmatrix} = \begin{vmatrix} a_{11} & a_{12} & \cdots & a_{1n} \\ \vdots & \vdots & & \vdots \\ a_{i1} & a_{i2} & \cdots & a_{in} \\ \vdots & \vdots & & \vdots \\ ka_{i1}+a_{j1} & ka_{i2}+a_{j2} & \cdots & ka_{in}+a_{jn} \\ \vdots & \vdots & & \vdots \\ a_{n1} & a_{n2} & \cdots & a_{nn} \end{vmatrix}. \quad (1.17)$$

The proof of expression (1.17) follows from the result (ii) in Property 1.3 and Corollary 1.2.

Property 1.6 (Antisymmetry of a determinant)

The value of the determinant changes its sign if any two of its rows are interchanged.

Proof: Repeatedly applying Property 1.5, and then using the result (i) in Property 1.3, we have

$$\begin{vmatrix} a_{11} & a_{12} & \cdots & a_{1n} \\ \vdots & \vdots & & \vdots \\ a_{i1} & a_{i2} & \cdots & a_{in} \\ \vdots & \vdots & & \vdots \\ a_{j1} & a_{j2} & \cdots & a_{jn} \\ \vdots & \vdots & & \vdots \\ a_{n1} & a_{n2} & \cdots & a_{nn} \end{vmatrix} = \begin{vmatrix} a_{11} & a_{12} & \cdots & a_{1n} \\ \vdots & \vdots & & \vdots \\ a_{i1}+a_{j1} & a_{i2}+a_{j2} & \cdots & a_{in}+a_{jn} \\ \vdots & \vdots & & \vdots \\ a_{j1} & a_{j2} & \cdots & a_{jn} \\ \vdots & \vdots & & \vdots \\ a_{n1} & a_{n2} & \cdots & a_{nn} \end{vmatrix}$$

$$= \begin{vmatrix} a_{11} & a_{12} & \cdots & a_{1n} \\ \vdots & \vdots & & \vdots \\ a_{i1}+a_{j1} & a_{i2}+a_{j2} & \cdots & a_{in}+a_{jn} \\ \vdots & \vdots & & \vdots \\ -a_{i1} & -a_{i2} & \cdots & -a_{in} \\ \vdots & \vdots & & \vdots \\ a_{n1} & a_{n2} & \cdots & a_{nn} \end{vmatrix}$$

$$
= \begin{vmatrix} a_{11} & a_{12} & \cdots & a_{1n} \\ \vdots & \vdots & & \vdots \\ a_{j1} & a_{j2} & \cdots & a_{jn} \\ \vdots & \vdots & & \vdots \\ -a_{i1} & -a_{i2} & \cdots & -a_{in} \\ \vdots & \vdots & & \vdots \\ a_{n1} & a_{n2} & \cdots & a_{nn} \end{vmatrix}
= - \begin{vmatrix} a_{11} & a_{12} & \cdots & a_{1n} \\ \vdots & \vdots & & \vdots \\ a_{j1} & a_{j2} & \cdots & a_{jn} \\ \vdots & \vdots & & \vdots \\ a_{i1} & a_{i2} & \cdots & a_{in} \\ \vdots & \vdots & & \vdots \\ a_{n1} & a_{n2} & \cdots & a_{nn} \end{vmatrix}.
$$

Property 1.7

The sum of products, with elements of one row and the algebraic complements of elements in a different row, is equal to zero, that is,

$$\sum_{k=1}^{n} a_{ik}A_{jk} = a_{i1}A_{j1} + a_{i2}A_{j2} + \cdots + a_{in}A_{jn} = 0 \quad (i \neq j). \tag{1.18}$$

Proof: Applying Property 1.2, expanding the determinant (1.10) along the row j yields

$$D = \sum_{k=1}^{n} a_{jk}A_{jk}.$$

If the elements $a_{j1}, a_{j2}, \ldots, a_{jn}$ in the jth row of (1.10) are replaced by $a_{i1}, a_{i2}, \ldots, a_{in}$, the expansion of the obtained determinant is $\sum_{k=1}^{n} a_{ik}A_{jk}$, that is,

$$
\sum_{k=1}^{n} a_{ik}A_{jk} = \begin{vmatrix} a_{11} & a_{12} & \cdots & a_{1n} \\ \vdots & \vdots & & \vdots \\ a_{i1} & a_{i2} & \cdots & a_{in} \\ \vdots & \vdots & & \vdots \\ a_{i1} & a_{i2} & \cdots & a_{in} \\ \vdots & \vdots & & \vdots \\ a_{n1} & a_{n2} & \cdots & a_{nn} \end{vmatrix} \begin{matrix} \\ \\ i\text{th row} \\ \\ j\text{th row} \\ \\ \\ \end{matrix}.
$$

Consequently, since the corresponding elements in the ith and jth rows of the above determinant are equal,

$$\sum_{k=1}^{n} a_{ik} A_{jk} = 0.$$

Rewriting (1.11), (1.13), and (1.18) in a compact form, we get

$$\sum_{k=1}^{n} a_{ik} A_{jk} = \delta_{ij} D,$$

where

$$\delta_{ij} = \begin{cases} 1, & \text{when} \quad i = j, \\ 0, & \text{when} \quad i \neq j. \end{cases} \tag{1.19}$$

Similarly, expanding (1.10) along its column also yields

$$\sum_{k=1}^{n} a_{ki} A_{kj} = \delta_{ij} D. \tag{1.20}$$

1.2 Calculation of *n*th-order Determinants

In this section, we will introduce common methods of computing the nth-order determinant, applying the definition of determinants and their properties by means of the following examples.

Example 1.3 Calculating an *n*th-order determinant

For an upper triangular determinant ($a_{ij} = 0$, as $i > j$), we have

$$D = \begin{vmatrix} a_{11} & a_{12} & \cdots & a_{1n} \\ 0 & a_{22} & \cdots & a_{2n} \\ \vdots & \vdots & \ddots & \vdots \\ 0 & 0 & \cdots & a_{nn} \end{vmatrix} = a_{11} a_{22} \cdots a_{nn}.$$

Solution: Let D' be a determinant obtained from exchanging rows and columns (in original orders) of the determinant D. Then, by applying Property 1.1 and the conclusion of Example 1.1 in §1.1, we immediately obtain

$$D = D' = a_{11} a_{22} \cdots a_{nn}.$$ ■

Example 1.4 Computing a fourth-order determinant (Basic method)

Compute the fourth-order determinant

$$D=\begin{vmatrix} 1 & 1 & -1 & 2 \\ -1 & -1 & -4 & 1 \\ 2 & 4 & -6 & 1 \\ 1 & 2 & 4 & 2 \end{vmatrix}. \quad \begin{matrix} ① \\ ② \\ ③ \\ ④ \end{matrix}$$

Solution: Transform the determinant into an upper triangular determinant by row operations of multiple addition and interchanging two rows.

Row Operations

Step 1: With respect to Property 1.5, we multiply row ① by 1, −2, −1 and add the results to rows ②, ③, and ④, respectively.

$$D=\begin{vmatrix} 1 & 1 & -1 & 2 \\ 0 & 0 & -5 & 3 \\ 0 & 2 & -4 & -3 \\ 0 & 1 & 5 & 0 \end{vmatrix} \qquad \begin{matrix} \\ ②'=①\times 1+② \\ ③'=①\times(-2)+③ \\ ④'=①\times(-1)+④ \end{matrix}$$

Step 2: Interchange rows ② and ④.

$$=-\begin{vmatrix} 1 & 1 & -1 & 2 \\ 0 & 1 & 5 & 0 \\ 0 & 2 & -4 & -3 \\ 0 & 0 & -5 & 3 \end{vmatrix} \qquad \begin{matrix} \\ ②'\rightleftharpoons ④' \\ \\ ④'\rightleftharpoons ②' \end{matrix}$$

Step 3: Multiply row ② by (−2) and add to row ③.

$$=-\begin{vmatrix} 1 & 1 & -1 & 2 \\ 0 & 1 & 5 & 0 \\ 0 & 0 & -14 & -3 \\ 0 & 0 & -5 & 3 \end{vmatrix} \qquad ③''=②''\times(-2)+③'$$

Step 4: Multiply row ③ by (−5/14) and add to row ④.

$$=-\begin{vmatrix} 1 & 1 & -1 & 2 \\ 0 & 1 & 5 & 0 \\ 0 & 0 & -14 & -3 \\ 0 & 0 & 0 & \frac{57}{14} \end{vmatrix} \qquad ④'''=③''\times\left(-\tfrac{5}{14}\right)+④''$$

$$=(-1)\times 1\times 1\times(-14)\times\left(\frac{57}{14}\right)$$
$$=57.$$

■

This example applies Properties 1.5 and 1.6 to transform a determinant into an upper triangular determinant. This is a basic method of computing a numerical determinant. However, for a numerical determinant of order 3, it is much easier and quicker to expand it along diagonals by **Sarrus' rule** (computing the product of six terms).

Example 1.5 Computing a fourth-order determinant

Compute the fourth-order determinant

$$D = \begin{vmatrix} 1 & 4 & -1 & 4 \\ 2 & 1 & 4 & 3 \\ 4 & 2 & 3 & 11 \\ 3 & 0 & 9 & 2 \end{vmatrix}.$$

Solution: Applying Property 1.5, transform a row (column) of the determinant into zero except for one nonzero element. Then, expand the new determinant along that row (or column) by using Property 1.2 to calculate it by reduced order. This is another basic method of expanding a determinant.

Note that there is already one zero in the second column of D. Transform the other elements of this column into zeros by using $a_{22} = 1$. Multiply the second row by (-4) and add it to the first row,

$$D = \begin{vmatrix} -7 & 0 & -17 & -8 \\ 2 & 1 & 4 & 3 \\ 4 & 2 & 3 & 11 \\ 3 & 0 & 9 & 2 \end{vmatrix}.$$

Then, multiply the second row by (-2), before adding to the third row,

$$D = \begin{vmatrix} -7 & 0 & -17 & -8 \\ 2 & 1 & 4 & 3 \\ 0 & 0 & -5 & 5 \\ 3 & 0 & 9 & 2 \end{vmatrix}.$$

Now expanding it along the second column, we obtain

$$D = (-1)^{2+2} \times 1 \times \begin{vmatrix} -7 & -17 & -8 \\ 0 & -5 & 5 \\ 3 & 9 & 2 \end{vmatrix}$$

$$= \begin{vmatrix} -7 & -17 & -8 \\ 0 & -5 & 5 \\ 3 & 9 & 2 \end{vmatrix}.$$

Add the third column to the second column, and expand it along the second row,

$$D=\begin{vmatrix} -7 & -17 & -8 \\ 0 & -5 & 5 \\ 3 & 9 & 2 \end{vmatrix} = \begin{vmatrix} -7 & -25 & -8 \\ 0 & 0 & 5 \\ 3 & 11 & 2 \end{vmatrix}$$

$$=(-1)^{2+3}\times 5\times \begin{vmatrix} -7 & -25 \\ 3 & 11 \end{vmatrix}$$

$$=-5\times(-77+75)=10.$$

■

Example 1.6 The antisymmetric determinant of odd order

A determinant D is called an **antisymmetric determinant** if $a_{ij}=-a_{ji}$ $(i, j=1, 2, \ldots, n)$ for $D=\left|a_{ij}\right|_1^n$. Prove that an odd order antisymmetric determinant is zero.

Proof: Suppose

$$D=\begin{vmatrix} 0 & a_{12} & \cdots & a_{1n} \\ -a_{12} & 0 & \cdots & a_{2n} \\ \vdots & \vdots & \ddots & \vdots \\ -a_{1n} & -a_{2n} & \cdots & 0 \end{vmatrix}.$$

Applying Property 1.1, we have

$$D=\begin{vmatrix} 0 & -a_{12} & -a_{13} & \cdots & -a_{1n} \\ a_{12} & 0 & -a_{23} & \cdots & -a_{2n} \\ \vdots & \vdots & \vdots & \ddots & \vdots \\ a_{1n} & a_{2n} & a_{3n} & \cdots & 0 \end{vmatrix}.$$

Then applying Property 1.3(i), taking out the common factor (-1) from every row, we get

$$D=(-1)^n D.$$

Since n is an odd number, it yields $D=-D$. Hence, $D=0$. ■

Example 1.7 Manipulating a third-order determinant

Prove that

$$\begin{vmatrix} a_1+b_1 & b_1+c_1 & c_1+a_1 \\ a_2+b_2 & b_2+c_2 & c_2+a_2 \\ a_3+b_3 & b_3+c_3 & c_3+a_3 \end{vmatrix} = 2\begin{vmatrix} a_1 & b_1 & c_1 \\ a_2 & b_2 & c_2 \\ a_3 & b_3 & c_3 \end{vmatrix}.$$

Proof:
Method 1: Add the second and third columns to the first column of the determinant on the *LHS*. Factor out the common number 2.

$$2\begin{vmatrix} a_1+b_1+c_1 & b_1+c_1 & c_1+a_1 \\ a_2+b_2+c_2 & b_2+c_2 & c_2+a_2 \\ a_3+b_3+c_3 & b_3+c_3 & c_3+a_3 \end{vmatrix}.$$

Then, the new first column multiplied by (-1) is added to the second and the third columns separately. Now, the determinant on the *LHS* is

$$2\begin{vmatrix} a_1+b_1+c_1 & -a_1 & -b_1 \\ a_2+b_2+c_2 & -a_2 & -b_2 \\ a_3+b_3+c_3 & -a_3 & -b_3 \end{vmatrix}.$$

Then, adding the second and the third columns to the first column,

$$2\begin{vmatrix} c_1 & -a_1 & -b_1 \\ c_2 & -a_2 & -b_2 \\ c_3 & -a_3 & -b_3 \end{vmatrix}.$$

Finally, after factoring out the common number (-1) from the second and third columns, followed by interchanging the columns twice, we can prove the required equality.

Method 2: Applying Property 1.3(ii) for every column of the *LHS* determinant in turn, we express the *LHS* determinant as a summation of 2^3 determinants, among which there are six determinants having the same two columns individually, that is,

$$= \begin{vmatrix} a_1 & b_1+c_1 & c_1+a_1 \\ a_2 & b_2+c_2 & c_2+a_2 \\ a_3 & b_3+c_3 & c_3+a_3 \end{vmatrix} + \begin{vmatrix} b_1 & b_1+c_1 & c_1+a_1 \\ b_2 & b_2+c_2 & c_2+a_2 \\ b_3 & b_3+c_3 & c_3+a_3 \end{vmatrix}$$

$$= \begin{vmatrix} a_1 & b_1 & c_1+a_1 \\ a_2 & b_2 & c_2+a_2 \\ a_3 & b_3 & c_3+a_3 \end{vmatrix} + \begin{vmatrix} a_1 & c_1 & c_1+a_1 \\ a_2 & c_2 & c_2+a_2 \\ a_3 & c_3 & c_3+a_3 \end{vmatrix} +$$

$$+ \begin{vmatrix} b_1 & b_1 & c_1+a_1 \\ b_2 & b_2 & c_2+a_2 \\ b_3 & b_3 & c_3+a_3 \end{vmatrix} + \begin{vmatrix} b_1 & c_1 & c_1+a_1 \\ b_2 & c_2 & c_2+a_2 \\ b_3 & c_3 & c_3+a_3 \end{vmatrix}$$

$$= \begin{vmatrix} a_1 & b_1 & c_1 \\ a_2 & b_2 & c_2 \\ a_3 & b_3 & c_3 \end{vmatrix} +0+0+0+0+0+0+ \begin{vmatrix} b_1 & c_1 & a_1 \\ b_2 & c_2 & a_2 \\ b_3 & c_3 & a_3 \end{vmatrix}$$

$=RHS.$ ■

Example 1.8 Computing a determinant of order *n*

Compute the determinant of order n given by

$$D = \begin{vmatrix} x & a & a & \cdots & a \\ a & x & a & \cdots & a \\ a & a & x & \cdots & a \\ \vdots & \vdots & \vdots & \ddots & \vdots \\ a & a & a & \cdots & x \end{vmatrix}.$$

Solution: The sum of elements in every row is the same. In this case, we add all other columns to the first column, and take out the common factor $[x+(n-1)a]$.

$$D = [x+(n-1)a] \begin{vmatrix} 1 & a & a & \cdots & a \\ 1 & x & a & \cdots & a \\ 1 & a & x & \cdots & a \\ \vdots & \vdots & \vdots & \ddots & \vdots \\ 1 & a & a & \cdots & x \end{vmatrix}.$$

Then we add the first row multiplied by (-1) individually to each of the other rows.

$$D = [x+(n-1)a] \begin{vmatrix} 1 & a & a & \cdots & a \\ 0 & x-a & 0 & \cdots & 0 \\ 0 & 0 & x-a & \cdots & 0 \\ \vdots & \vdots & \vdots & \ddots & \vdots \\ 0 & 0 & 0 & \cdots & x-a \end{vmatrix}$$

Finally, we transform D into an upper triangular determinant, that is,

$$D = [x + (n-1)a]\,(x-a)^{n-1}. \qquad \blacksquare$$

Example 1.9 Computing a third-order determinant

If $xyz \neq 0$, compute the third-order determinant

$$D = \begin{vmatrix} 1+x & 2 & 3 \\ 1 & 2+y & 3 \\ 1 & 2 & 3+z \end{vmatrix}.$$

Solution:
Method 1: Add the first row multiplied by (-1) to the second and the third row as well.

$$D = \begin{vmatrix} 1+x & 2 & 3 \\ -x & y & 0 \\ -x & 0 & z \end{vmatrix}.$$

Next, add the second column multiplied by x/y, the third column multiplied by x/z to the first column respectively.

$$D = \begin{vmatrix} 1+x+\frac{2x}{y}+\frac{3x}{z} & 2 & 3 \\ 0 & y & 0 \\ 0 & 0 & z \end{vmatrix}.$$

We transform the determinant into an upper triangular determinant,

$$D = \left(1 + x + \frac{2x}{y} + \frac{3x}{z}\right) yz = yz + xyz + 2zx + 3xy.$$

Method 2: Elements 1, 2, 3 in D are individually described as $(1+0)$, $(2+0)$, $(3+0)$, respectively. Applying Property 1.3(ii), D can then be transformed into 2^3 determinants, out of which four determinants are zero. Indeed,

$$D = \begin{vmatrix} 1 & 0 & 0 \\ 1 & y & 0 \\ 1 & 0 & z \end{vmatrix} + \begin{vmatrix} x & 2 & 0 \\ 0 & 2 & 0 \\ 0 & 2 & z \end{vmatrix} + \begin{vmatrix} x & 0 & 3 \\ 0 & y & 3 \\ 0 & 0 & 3 \end{vmatrix} + \begin{vmatrix} x & 0 & 0 \\ 0 & y & 0 \\ 0 & 0 & z \end{vmatrix}$$

$$= yz + 2zx + 3xy + xyz. \qquad \blacksquare$$

Example 1.10 The Vandermonde determinant

Prove that the Vandermonde determinant

$$V_n = \begin{vmatrix} 1 & 1 & 1 & \cdots & 1 \\ x_1 & x_2 & x_3 & \cdots & x_n \\ x_1^2 & x_2^2 & x_3^2 & \cdots & x_n^2 \\ \vdots & \vdots & \vdots & & \vdots \\ x_1^{n-1} & x_2^{n-1} & x_3^{n-1} & \cdots & x_n^{n-1} \end{vmatrix} = \prod_{1\le j<i\le n} (x_i - x_j),$$

where the continued multiplication,

$$\prod_{1\le j<i\le n} (x_i - x_j) = (x_2 - x_1)(x_3 - x_1)\cdots(x_n - x_1)(x_3 - x_2)\cdots$$
$$(x_n - x_2)\cdots(x_{n-1} - x_{n-2})(x_n - x_{n-2})(x_n - x_{n-1})$$

is the product of all factors $(x_i - x_j)$ satisfying $1 \le j < i \le n$.

Proof: The proof is conducted by mathematical induction on n.

For $n = 2$,

$$V_2 = \begin{vmatrix} 1 & 1 \\ x_1 & x_2 \end{vmatrix} = x_2 - x_1 = \prod_{1\le j<i\le 2} (x_i - x_j),$$

the conclusion is true. Suppose the conclusion is true for any $(n-1)$th-order Vandermonde determinant. We shall prove it to be true for the nth-order Vandermonde determinant.

In V_n, beginning from the nth row, adding the former row multiplied by $(-x_1)$ to the next row and continuing it till the second row, we obtain

$$V_n = \begin{vmatrix} 1 & 1 & 1 & \cdots & 1 \\ 0 & x_2 - x_1 & x_3 - x_1 & \cdots & x_n - x_1 \\ 0 & x_2(x_2 - x_1) & x_3(x_3 - x_1) & \cdots & x_n(x_n - x_1) \\ \vdots & \vdots & \vdots & & \vdots \\ 0 & x_2^{n-2}(x_2 - x_1) & x_3^{n-2}(x_3 - x_1) & \cdots & x_n^{n-2}(x_n - x_1) \end{vmatrix}.$$

Expanding along the first column and taking out the common factors, we get

$$V_n = (x_2 - x_1)(x_3 - x_1)\cdots(x_n - x_1) \begin{vmatrix} 1 & 1 & \cdots & 1 \\ x_2 & x_3 & \cdots & x_n \\ x_2^2 & x_3^2 & \cdots & x_n^2 \\ \vdots & \vdots & & \vdots \\ x_2^{n-2} & x_3^{n-2} & \cdots & x_n^{n-2} \end{vmatrix}.$$

Clearly, the determinant in the above expression is an $(n-1)$th-order Vandermonde determinant. The induction hypothesis implies that

$$V_n = (x_2 - x_1)(x_3 - x_1)\cdots(x_n - x_1) \prod_{2\le j<i\le n} (x_i - x_j),$$

where

$$V_n = \prod_{1\le j<i\le n} (x_i - x_j).$$

■

Example 1.11 Laplace expansion of a determinant

Prove that,

$$D = \begin{vmatrix} a_{11} & a_{12} & \cdots & a_{1k} & 0 & \cdots & 0 \\ a_{21} & a_{22} & \cdots & a_{2k} & 0 & \cdots & 0 \\ \vdots & \vdots & & \vdots & \vdots & & \vdots \\ a_{k1} & a_{k2} & \cdots & a_{kk} & 0 & \cdots & 0 \\ c_{11} & c_{12} & \cdots & c_{1k} & b_{11} & \cdots & b_{1m} \\ \vdots & \vdots & & \vdots & \vdots & & \vdots \\ c_{m1} & c_{m2} & \cdots & c_{mk} & b_{m1} & \cdots & b_{mm} \end{vmatrix}$$

$$= \begin{vmatrix} a_{11} & a_{12} & \cdots & a_{1k} \\ a_{21} & a_{22} & \cdots & a_{2k} \\ \vdots & \vdots & & \vdots \\ a_{k1} & a_{k2} & \cdots & a_{kk} \end{vmatrix} \bullet \begin{vmatrix} b_{11} & \cdots & b_{1m} \\ \vdots & & \vdots \\ b_{m1} & \cdots & b_{mm} \end{vmatrix}. \quad (1.21)$$

Proof: Denote

$$|\boldsymbol{A}| = \left|a_{ij}\right|_1^k, \ |\boldsymbol{B}| = \left|b_{ij}\right|_1^m.$$

The proof is carried out by mathematical induction on order k of $|\boldsymbol{A}|$.

For $k = 1$, expanding D along the first row yields $D = a_{11}|\boldsymbol{B}| = |a_{11}| \bullet |\boldsymbol{B}|$ (here $|a_{11}|$ is a first-order determinant), and the expression (1.21) is true.

Suppose (1.21) is true when $|\boldsymbol{A}|$ is a $(k-1)$th-order determinant. Now consider the case when $|\boldsymbol{A}|$ is a kth-order determinant: expanding D along the first row yields

$$D = a_{11}(-1)^{1+1}M_{11}^D + a_{12}(-1)^{1+2}M_{12}^D + \cdots + a_{1k}(-1)^{1+k}M_{1k}^D, \quad ①$$

where M_{1j}^D is the complementary minor of a_{1j} $(j = 1, 2, \ldots, k)$ in D. Explicitly, M_{1j}^D is also a determinant in the form of the *LHS* of expression

(1.21), and its left upper corner block made of a_{ij} is of $(k-1)$th-order. The induction hypothesis implies that

$$M_{1j}^{D} = M_{1j}^{|A|}\,|\boldsymbol{B}|, \quad j = 1, 2, \ldots, k, \qquad ②$$

where $M_{1j}^{|A|}$ is the complementary minor of a_{1j} in $|\boldsymbol{A}|$.

Substituting the expression ② into ① yields

$$D = [a_{11}(-1)^{1+1}M_{11}^{|A|} + a_{12}(-1)^{1+2}M_{12}^{|A|} + \cdots + a_{1k}(-1)^{1+k}M_{1k}^{|A|}]\,|\boldsymbol{B}| = |\boldsymbol{A}| \cdot |\boldsymbol{B}|.$$

Hence, the expression (1.21) is true when $|\boldsymbol{A}|$ is of order k. So, based on induction, we conclude that (1.21) is true when $|\boldsymbol{A}|$ is of any order. ■

The expression (1.21) can be denoted as

$$D = \begin{vmatrix} \boldsymbol{A} & \boldsymbol{0} \\ * & \boldsymbol{B} \end{vmatrix} = |\boldsymbol{A}| \cdot |\boldsymbol{B}|.$$

If $|\boldsymbol{A}|$ and $|\boldsymbol{B}|$ have the same form as above, similarly it yields

$$D = \begin{vmatrix} \boldsymbol{A} & * \\ \boldsymbol{0} & \boldsymbol{B} \end{vmatrix} = |\boldsymbol{A}| \cdot |\boldsymbol{B}|. \tag{1.22}$$

But note that,

$$D = \begin{vmatrix} \boldsymbol{0} & \boldsymbol{A} \\ \boldsymbol{B} & * \end{vmatrix} \neq -|\boldsymbol{A}| \cdot |\boldsymbol{B}|.$$

Indeed, by interchanging two adjacent columns once at a time (interchange $k \times m$ times in all), we can transform D into the form of (1.21). Hence,

$$D = \begin{vmatrix} \boldsymbol{0} & \boldsymbol{A} \\ \boldsymbol{B} & * \end{vmatrix} = (-1)^{k\times m} \begin{vmatrix} \boldsymbol{A} & \boldsymbol{0} \\ * & \boldsymbol{B} \end{vmatrix} = (-1)^{k\times m}\,|\boldsymbol{A}| \cdot |\boldsymbol{B}|. \tag{1.23}$$

Example 1.12 Roots of equation $D(x) = 0$

Find the roots of the equation $D(x) = 0$, where

$$D(x) = \begin{vmatrix} x-1 & x-2 & x-1 & x \\ x-2 & x-4 & x-2 & x \\ x-3 & x-6 & x-4 & x-1 \\ x-4 & x-8 & 2x-5 & x-2 \end{vmatrix}.$$

Solution: Indeed $x=0$ is a root, as the first and the second column are proportional when $x=0$, hence $D(0)=0$. In order to find other solutions, the determinant needs to be expanded. Adding the first column multiplied by (-1) to the second, third, and fourth columns, respectively, we have

$$D(x)=\begin{vmatrix} x-1 & -1 & 0 & 1 \\ x-2 & -2 & 0 & 2 \\ x-3 & 3 & 1 & 2 \\ x-4 & -4 & x-1 & 2 \end{vmatrix}.$$

Then, we add the new second column to the new fourth column, and apply expression (1.21), we get

$$\begin{aligned} D(x) &= \begin{vmatrix} x-1 & -1 & 0 & 0 \\ x-2 & -2 & 0 & 0 \\ x-3 & -3 & -1 & -1 \\ x-4 & -4 & x-1 & -2 \end{vmatrix} \\ &= \begin{vmatrix} x-1 & -1 \\ x-2 & -2 \end{vmatrix} \cdot \begin{vmatrix} -1 & -1 \\ x-1 & -2 \end{vmatrix} \\ &= -x(x+1). \end{aligned}$$

Hence, the equation $D(x)=0$ has two roots: 0 and -1. ■

***Example 1.13 Tri-diagonal determinant of order n**

Compute the tri-diagonal determinant of order n,

$$D_n=\begin{vmatrix} a & b & & & & \\ c & a & b & & & \\ & c & a & b & & \\ & & \ddots & \ddots & \ddots & \\ & & & c & a & b \\ & & & & c & a \end{vmatrix}.$$

Solution: Expanding D_n along the first row and then expanding the determinant in the second term of the expansion along the first column, we obtain

$$\begin{aligned} D_n &= aD_{n-1}+(-1)^{1+2}b \begin{vmatrix} c & b & & & & \\ 0 & a & b & & & \\ 0 & c & a & b & & \\ \vdots & \vdots & \ddots & \ddots & \ddots & \\ 0 & 0 & \cdots & c & a & b \\ 0 & 0 & \cdots & 0 & c & a \end{vmatrix}_{(n-1)\text{th-order}} \quad ① \\ &= aD_{n-1}-bcD_{n-2}. \end{aligned}$$

From expression ① (recursion formula), D_3 can be obtained from D_1 and D_2, D_4 can be obtained from D_2 and D_3, and so on. In order to obtain the recursion formula of D_n from D_1 and D_2, rewrite ① as

$$D_n - kD_{n-1} = l(D_{n-1} - kD_{n-2}), \tag{②}$$

where

$$k + l = a, \quad kl = bc. \tag{③}$$

Denote $\Delta_n = D_n - kD_{n-1}$, in ②. Then ② becomes

$$\Delta_n = l\Delta_{n-1}. \tag{②'}$$

From this recursion formula, we easily obtain

$$\Delta_n = l\Delta_{n-1} = l(l\Delta_{n-2}) = \cdots = l^{n-2}\Delta_2,$$

that is,

$$D_n - kD_{n-1} = l^{n-2}(D_2 - kD_1), \tag{④}$$

where

$$D_2 = \begin{vmatrix} a & b \\ c & a \end{vmatrix} = a^2 - bc, \quad D_1 = |a| = a.$$

Substituting them into expression ④ and utilizing expression ③, we easily get $D_2 - kD_1 = l^2$, hence

$$D_n = l^n + kD_{n-1}. \tag{⑤}$$

Then applying the recursion formula ⑤, we recursively obtain

$$\begin{aligned} D_n &= l^n + k(l^{n-1} + kD_{n-2}) = l^n + kl^{n-1} + k^2 D_{n-2} \\ &= l^n + kl^{n-1} + k^2(l^{n-2} + kD_{n-3}) \\ &= l^n + kl^{n-1} + k^2 l^{n-2} + k^3 D_{n-3} \\ &\vdots \\ &= l^n + kl^{n-1} + k^2 l^{n-2} + \cdots + k^{n-2} l^2 + k^{n-1} D_1, \end{aligned}$$

where $D_1 = |a| = a = k + l$, hence

$$D_n = l^n + kl^{n-1} + k^2 l^{n-2} + \cdots + k^{n-2} l^2 + k^{n-1} l + k^n. \tag{⑥}$$

For instance, when $a = 3$, $b = 2$, $c = 1$, we get $k = 1$, $l = 2$, or $k = 2$, $l = 1$ from ③. In this case,

$$D_n = \frac{l^{n+1} - k^{n+1}}{l - k} = 2^{n+1} - 1.$$

■

1.3 Cramer's Rule

In this section, we will provide a determinant-based method to solve a linear system of n equations in n unknowns where the determinant of coefficients is nonzero. It is usually called **Cramer's rule**. Further, a necessary condition for a homogeneous system of n linear equations in n unknowns to have a nonzero solution is presented.

THEOREM 1.1 *(Cramer's rule)*

For a nonhomogeneous system of linear equations

$$\begin{cases} a_{11}x_1 + a_{12}x_2 + \cdots + a_{1n}x_n = b_1, \\ a_{21}x_1 + a_{22}x_2 + \cdots + a_{2n}x_n = b_2, \\ \cdots \\ a_{n1}x_1 + a_{n2}x_2 + \cdots + a_{nn}x_n = b_n, \end{cases} \tag{1.24}$$

or simply

$$\sum_{j=1}^{n} a_{ij}x_j = b_i, \quad i = 1, 2, \ldots, n, \tag{1.25}$$

the system (1.24) has a unique solution

$$x_j = \frac{D_j}{D}, \; j = 1, 2, \ldots, n, \tag{1.26}$$

if the determinant of its coefficients,

$$D = \begin{vmatrix} a_{11} & a_{12} & \cdots & a_{1n} \\ a_{21} & a_{22} & \cdots & a_{2n} \\ \vdots & \vdots & & \vdots \\ a_{n1} & a_{n2} & \cdots & a_{nn} \end{vmatrix} \neq 0,$$

and D_j is a determinant obtained from D by replacing column j with the constant terms $b_1, b_2, \ldots, b_n$, that is,

$$D_j = \begin{vmatrix} a_{11} & \cdots & a_{1j-1} & b_1 & a_{1j+1} & \cdots & a_{1n} \\ a_{21} & \cdots & a_{2j-1} & b_2 & a_{2j+1} & \cdots & a_{2n} \\ \vdots & & \vdots & \vdots & \vdots & & \vdots \\ a_{n1} & \cdots & a_{nj-1} & b_n & a_{nj+1} & \cdots & a_{nn} \end{vmatrix}. \tag{1.27}$$

Proof: First, we prove that the expression (1.26) is a solution of equations (1.24). By (1.27), we have

$$D_j = b_1 A_{1j} + b_2 A_{2j} + \cdots + b_n A_{nj} = \sum_{k=1}^{n} b_k A_{kj},$$

where A_{kj} is the algebraic complement of the element a_{kj} of the coefficient determinant.

$$x_j = \frac{1}{D} \sum_{k=1}^{n} b_k A_{kj} \qquad (j = 1, 2, \ldots, n)$$

is substituted into the *LHS* of expression (1.25), yielding

$$\sum_{j=1}^{n} a_{ij} \left(\frac{1}{D} \sum_{k=1}^{n} b_k A_{kj} \right) = \frac{1}{D} \sum_{j=1}^{n} \left(\sum_{k=1}^{n} a_{ij} A_{kj} b_k \right)$$

$$= \frac{1}{D} \sum_{k=1}^{n} \left(\sum_{j=1}^{n} a_{ij} A_{kj} b_k \right) \longleftarrow$$

This is true despite the switching of double summation signs. See §1.4.3 for reference.

$$= \frac{1}{D} \sum_{k=1}^{n} b_k \left(\sum_{j=1}^{n} a_{ij} A_{kj} \right)$$

$$= \frac{1}{D} \sum_{k=1}^{n} b_k \delta_{ik} D \, (\delta_{ik} = 1, \text{ for } k = i, \delta_{ik} = 0 \text{ for } k \neq i)$$

$$= \frac{1}{D} (b_i \bullet 1 \bullet D)$$

$$= b_i (i = 1, 2, \ldots, n).$$

Hence, $x_j = D_j / D (j = 1, 2, \ldots, n)$ in (1.26) satisfies each equation in system (1.24). It is therefore a solution of the system of equations (1.24).

To prove that the solution of the system of equations (1.24) must be in the form (1.26), let $c_1, c_2, \ldots, c_n$ be a solution. Then,

$$\begin{cases} a_{11}c_1 + a_{12}c_2 + \cdots + a_{1n}c_n = b_1, \\ a_{21}c_1 + a_{22}c_2 + \cdots + a_{2n}c_n = b_2, \\ \cdots \\ a_{n1}c_1 + a_{n2}c_2 + \cdots + a_{nn}c_n = b_n. \end{cases}$$

The two sides of the above n equations are individually multiplied by $A_{1j}, A_{2j}, \ldots, A_{nj}$, respectively, then we sum up all the n equations to get

$$\left(\sum_{i=1}^{n} a_{i1}A_{ij}\right)c_1+\cdots+\left(\sum_{i=1}^{n} a_{ij}A_{ij}\right)c_j+\cdots+\left(\sum_{i=1}^{n} a_{in}A_{ij}\right)c_n=\sum_{i=1}^{n} b_iA_{ij}.$$

Note that, on the *LHS* of the above expression, the coefficients of $c_1, c_2, \ldots, c_{j-1}, c_{j+1}, \ldots, c_n$ are all zeros, the coefficients of c_j is D, and $\sum_{i=1}^{n} b_iA_{ij} = D_j$. Then, $Dc_j = D_j$.

Hence,

$$c_j = \frac{D_j}{D}$$

for each $j = 1, 2, \ldots, n$. Thus, the uniqueness of the solution of (1.24) is proved. □

The next corollary follows immediately from Cramer's rule.

COROLLARY 1.3

If the determinant of coefficient $D = |a_{ij}|_1^n \neq 0$, the homogeneous system of linear equations

$$\sum_{j=1}^{n} a_{ij}x_j = 0 \quad (i = 1, 2, \ldots, n) \tag{1.28}$$

has only a zero solution $x_j = 0,\ j = 1, 2, \ldots, n$.

The proof is obvious since $D_j = 0,\ j = 1, 2, \ldots, n$.

An equivalent proposition to Corollary 1.3 (i.e., **contra-positive proposition**) is: if a homogeneous system of linear equations has a nonzero solution, then its coefficient determinant $D = 0$.

In Chapter 3, we will further prove that $D = 0$ is also a sufficient condition for a homogeneous system of linear equations (1.28) to have a nonzero solution.

By Cramer's rule, $(n + 1)$ determinants of order n are to be computed for a nonhomogeneous system of linear equations in n unknowns whose coefficient determinant is not equal to zero. The computation is very heavy. In fact, Gaussian elimination (discussed in Chapter 2) is generally used for solving a system of linear equations with numerical coefficients (including systems of linear equations with coefficient determinants zero, but the number of equations does not equal that of its unknowns). Cramer's rule is significant in theory, especially since it explicitly uncovers the relation between solutions and the coefficients of the system.

Example 1.14 Applying Cramer's rule

Given the values of cubic curve $y = f(x) = a_0 + a_1 x + a_2 x^2 + a_3 x^3$ at four points $x = \pm 1$, $x = \pm 2$: $f(1) = f(-1) = f(2) = 6$, $f(-2) = -6$. Find its coefficients a_0, a_1, a_2, and a_3.

Solution: Substituting the values of the cubic curve at the four points into its equation, we obtain a nonhomogeneous system of linear equations in a_0, a_1, a_2, a_3:

$$\begin{cases} a_0 + a_1 + a_2 + a_3 = 6, \\ a_0 + a_1(-1) + a_2(-1)^2 + a_3(-1)^3 = 6, \\ a_0 + a_1(2) + a_2(2)^2 + a_3(2)^3 = 6, \\ a_0 + a_1(-2) + a_2(-2)^2 + a_3(-2)^3 = -6. \end{cases}$$

The determinant of its coefficients is a Vandermonde determinant (exchange the rows and the columns of Example 1.10).

$$\begin{aligned} D &= \begin{vmatrix} 1 & 1 & 1 & 1 \\ 1 & -1 & (-1)^2 & (-1)^3 \\ 1 & 2 & 2^2 & 2^3 \\ 1 & -2 & (-2)^2 & (-2)^3 \end{vmatrix} \\ &= (-1-1)(2-1)(-2-1)(2+1)(-2+1)(-2-2) \\ &= 72. \end{aligned}$$

Thus, the coefficients of the equation of the given cubic curve can be obtained by Cramer's rule,

$$a_j = \frac{D_j}{D}, \quad j = 0, 1, 2, 3,$$

where

$$D_0 = \begin{vmatrix} 6 & 1 & 1 & 1 \\ 6 & -1 & 1 & -1 \\ 6 & 2 & 4 & 8 \\ -6 & -2 & 4 & -8 \end{vmatrix} = 576, \quad D_1 = \begin{vmatrix} 1 & 6 & 1 & 1 \\ 1 & 6 & 1 & -1 \\ 1 & 6 & 4 & 8 \\ 1 & -6 & 4 & -8 \end{vmatrix} = -72,$$

$$D_2 = \begin{vmatrix} 1 & 1 & 6 & 1 \\ 1 & -1 & 6 & -1 \\ 1 & 2 & 6 & 8 \\ 1 & -2 & -6 & -8 \end{vmatrix} = -144, \quad D_3 = \begin{vmatrix} 1 & 1 & 1 & 6 \\ 1 & -1 & 1 & 6 \\ 1 & 2 & 4 & 6 \\ 1 & -2 & 4 & -6 \end{vmatrix} = 72.$$

Hence, $a_0 = 8$, $a_1 = -1$, $a_2 = -2$, $a_3 = 1$ is the unique solution. The equation of the cubic curve through the above four points is determined uniquely as

$$f(x) = 8 - x - 2x^2 + x^3. \qquad \blacksquare$$

In general, through $(n+1)$ points (x_i, y_i), $i = 1, 2, \ldots, n+1$, whose x-coordinates are different, we can uniquely determine the equation of an nth-order curve, $y = a_0 + a_1 x + a_2 x^2 + \ldots + a_n x^n$.

Example 1.15 Point of intersection of four planes

Find a necessary and sufficient condition under which four planes $a_i x + b_i y + c_i z + d_i = 0 \quad (i = 1, 2, 3, 4)$ intersect at one point (x_0, y_0, z_0).

Solution: Suppose the four planes intersect at point $P_0(x_0, y_0, z_0)$. Then, there are at least three planes intersecting at the point P_0. Otherwise, the four planes cannot intersect at that point. Without loss of generality, suppose the first three planes intersect at the point P_0. Therefore, the nonhomogeneous system of linear equations made out of the equations of those three planes must have a unique solution.

Thus,

$$D_3 = \begin{vmatrix} a_1 & b_1 & c_1 \\ a_2 & b_2 & c_2 \\ a_3 & b_3 & c_3 \end{vmatrix} \neq 0.$$

Let $a_i x + b_i y + c_i z + d_i t = 0$ (where $t = 1, i = 1, 2, 3, 4$) be equations of the four planes. The intersection of the four planes at the point P_0 amounts to the existence of a unique nonzero solution $(x_0, y_0, z_0, 1)$ of the homogeneous linear system of four equations in x, y, z, and t. Thus, its coefficient determinant $D = 0$. Therefore, $D = 0$ and $D_3 \neq 0$ is a necessary condition under which the four given planes intersect at a point. It is also a sufficient condition. This is due to the fact that we can get the first three planes intersecting at a point P_0 from $D_3 \neq 0$. P_0 is the unique point of intersection of the four planes follows from $D = 0$, however, it can be justified only after studying Chapter 3. ■

*1.4 Proof of Property 1.1, Property 1.2, and Double Summation Signs

*1.4.1 Proof of Property 1.1

Prove that

$$\begin{vmatrix} a_{11} & a_{12} & \cdots & a_{1n} \\ a_{21} & a_{22} & \cdots & a_{2n} \\ \vdots & \vdots & & \vdots \\ a_{n1} & a_{n2} & \cdots & a_{nn} \end{vmatrix} = \begin{vmatrix} a_{11} & a_{21} & \cdots & a_{n1} \\ a_{12} & a_{22} & \cdots & a_{n2} \\ \vdots & \vdots & & \vdots \\ a_{1n} & a_{2n} & \cdots & a_{nn} \end{vmatrix}.$$

Proof: The proof is by mathematical induction on the order of the determinant. Denote the determinants on the *LHS* and *RHS* of the equality as D and D'.

It is obvious that $D = D'$ for $n = 2$. Assuming it to be true for the determinants of order less than n, we consider the case of nth-order.

By definition,

$$D = a_{11}A_{11} + a_{12}A_{12} + \cdots + a_{1n}A_{1n},$$
$$D' = a_{11}A'_{11} + a_{21}A'_{21} + \cdots + a_{n1}A'_{n1}.$$

The induction hypothesis implied that $A'_{11} = A_{11}$, thus

$$D' = a_{11}A_{11} + (-1)^{1+2}a_{21}\begin{vmatrix} a_{12} & a_{32} & \cdots & a_{n2} \\ a_{13} & a_{33} & \cdots & a_{n3} \\ \vdots & \vdots & & \vdots \\ a_{1n} & a_{3n} & \cdots & a_{nn} \end{vmatrix}$$

$$+ (-1)^{1+3}a_{31}\begin{vmatrix} a_{12} & a_{22} & a_{42} & \cdots & a_{n2} \\ a_{13} & a_{23} & a_{43} & \cdots & a_{n3} \\ \vdots & \vdots & \vdots & & \vdots \\ a_{1n} & a_{2n} & a_{4n} & \cdots & a_{nn} \end{vmatrix} + \cdots$$

$$+ (-1)^{1+n}a_{n1}\begin{vmatrix} a_{12} & a_{22} & \cdots & a_{n-1,2} \\ a_{13} & a_{23} & \cdots & a_{n-1,3} \\ \vdots & \vdots & & \vdots \\ a_{1n} & a_{2n} & \cdots & a_{n-1,n} \end{vmatrix}.$$

Each of the above $(n-1)$ determinants of order $(n-1)$ is expanded along the respective first column. If it is proved that the summation of terms containing a_{12} equals to $a_{12}A_{12}$, the summation of terms containing a_{13} equals to $a_{13}A_{13}, \ldots,$ the summation of terms containing a_{1n} equals to $a_{1n}A_{1n}$, then $D' = D$. Now, let us prove that the summation of terms containing a_{12} equals $a_{12}A_{12}$, other cases being similar.

$$(-1)^{1+2}a_{21}a_{12}\begin{vmatrix} a_{33} & \cdots & a_{n3} \\ \vdots & & \vdots \\ a_{3n} & \cdots & a_{nn} \end{vmatrix} + (-1)^{1+3}a_{31}a_{12}\begin{vmatrix} a_{23} & a_{43} & \cdots & a_{n3} \\ \vdots & \vdots & & \vdots \\ a_{2n} & a_{4n} & \cdots & a_{nn} \end{vmatrix} +$$

$$\cdots + (-1)^{1+n}a_{n1}a_{12}\begin{vmatrix} a_{23} & \cdots & a_{n-1,3} \\ \vdots & & \vdots \\ a_{2n} & \cdots & a_{n-1,n} \end{vmatrix}$$

$$=(-1)^{1+2}a_{12}\left\{\begin{vmatrix} a_{21} & 0 & \cdots & 0 \\ 0 & a_{33} & \cdots & a_{n3} \\ \vdots & \vdots & & \vdots \\ 0 & a_{3n} & \cdots & a_{nn} \end{vmatrix} + \begin{vmatrix} 0 & a_{31} & 0 & \cdots & 0 \\ a_{23} & 0 & a_{43} & \cdots & a_{n3} \\ \vdots & \vdots & \vdots & & \vdots \\ a_{2n} & 0 & a_{4n} & \cdots & a_{nn} \end{vmatrix} + \right.$$

$$\left. \cdots + \begin{vmatrix} 0 & \cdots & 0 & a_{n1} \\ a_{23} & \cdots & a_{n-1,\,3} & 0 \\ \vdots & & \vdots & \vdots \\ a_{2n} & \cdots & a_{n-1,\,n} & 0 \end{vmatrix}\right\}$$

$$=(-1)^{1+2}a_{12}\begin{vmatrix} a_{21} & a_{31} & \cdots & a_{n1} \\ a_{23} & a_{33} & \cdots & a_{n3} \\ \vdots & \vdots & & \vdots \\ a_{2n} & a_{3n} & \cdots & a_{nn} \end{vmatrix} = a_{12}(-1)^{1+2}M'_{12} = a_{12}(-1)^{1+2}M_{12}$$

$$= a_{12}A_{12},$$

where the complementary minor M'_{12} [a determinant of order $(n-1)$] is the determinant obtained from interchanging rows and columns of M_{12}. This ends our proof. □

*1.4.2 Proof of Property 1.2

Prove that

$$a_{i1}A_{i1} + a_{i2}A_{i2} + \cdots + a_{in}A_{in} = a_{11}A_{11} + a_{12}A_{12} + \cdots + a_{1n}A_{1n} \, (i = 2, 3, \ldots, n).$$

Proof: The proof is similar to that of Property 1.1. □

*1.4.3 About the double summation symbol "$\sum\sum$"

For the sum of n numbers

$$a_1 + a_2 + \cdots + a_n,$$

we use the symbol $\sum$ to denote it as $\sum_{i=1}^{n} a_i$ for short.

For instance,

$$\sum_{i=1}^{n} i^2 = 1^2 + 2^2 + \cdots + n^2,$$

where i is the summation index and takes natural numbers $1, 2, \ldots, n$ successively.

Sometimes, a summation is numbered by two indices. For example, suppose $m \times n$ numbers are arranged as m rows and n columns:

$$\begin{array}{c} a_{11}, a_{12}, \ldots, a_{1j}, \ldots, a_{1n} \\ a_{21}, a_{22}, \ldots, a_{2j}, \ldots, a_{2n} \\ \cdots \\ a_{i1}, a_{i2}, \ldots, a_{ij}, \ldots, a_{in} \\ \cdots \\ a_{m1}, a_{m2}, \ldots, a_{mj}, \ldots, a_{mn} \end{array} \tag{1.29}$$

To denote their summation S, we can add the n numbers in the ith row and denote it as $S_i = \sum_{j=1}^{n} a_{ij}$ (called summation over index j). Then add the m sums of the rows $S_1, S_2, \ldots, S_m$ and denote it as

$$S = \sum_{i=1}^{m} S_i = \sum_{i=1}^{m} \sum_{j=1}^{n} a_{ij}. \tag{1.30}$$

Similarly, we can also add the m numbers in the jth column first, denote it as $S'_j = \sum_{i=1}^{m} a_{ij}$ (called summation over index i). Then add the n sums of columns $S'_1, S'_2, \ldots, S'_n$ and denote it as

$$S = \sum_{j=1}^{n} S'_j = \sum_{j=1}^{n} \sum_{i=1}^{m} a_{ij}. \tag{1.31}$$

In (1.30), we are summing over j before summing over i, while in (1.31), we are summing over i before summing over j. Both expressions are equal explicitly. This shows that switching the order of summation over i and j can be done in this case.

Sometimes, a conditional summation index is used for summing a part of numbers numbered by two indices. For instance,

$$\begin{aligned} \sum_{j=1}^{n-1} \sum_{j<i\le n} a_{ij} = (a_{21} + a_{31} + a_{41} + \cdots + a_{n1}) \quad & j = 1 \\ +(a_{32} + a_{42} + \cdots + a_{n2}) \quad & j = 2 \\ +(a_{43} + \cdots + a_{n3}) \quad & j = 3 \\ + \cdots \quad & \\ +a_{nn-1}. \quad & j = n-1 \end{aligned}$$

Another example, for two polynomials

$$f(x) = a_n x^n + a_{n-1} x^{n-1} + \cdots + a_1 x + a_0,$$
$$g(x) = b_m x^m + b_{m-1} x^{m-1} + \cdots + b_1 x + b_0,$$

give the coefficient of x^k for $f(x)g(x)$. It can be denoted by

$$\sum_{i+j=k} a_i b_j.$$

For example, the coefficient of x^4 is $\sum_{i+j=4} a_i b_j = a_0 b_4 + a_1 b_3 + a_2 b_2 + a_3 b_1 + a_4 b_0$.

In addition, the double summation symbol may also be used to describe a product of two summations, that is,

$$(a_1 + a_2 + \cdots + a_n)(b_1 + b_2 + \cdots + b_m) = \sum_{i=1}^{n} a_i \sum_{j=1}^{m} b_j.$$

By distributive law, we obtain that the above expression

$$= \sum_{i=1}^{n} \left(\sum_{j=1}^{m} a_i b_j \right) = \sum_{i=1}^{n} \sum_{j=1}^{m} a_i b_j \overset{\text{or}}{=} \sum_{j=1}^{m} \left(\sum_{i=1}^{n} a_i b_j \right) = \sum_{j=1}^{m} \sum_{i=1}^{n} a_i b_j,$$

where $\sum_{j=1}^{m} a_i b_j$ is the summation over index j for a given i. In this case, the summation formula is

$$\sum_{j=1}^{m} a_i b_j = a_i \sum_{j=1}^{m} b_j.$$

Exercises and Supplementary Exercises

Exercises

Q1.1–1.8: Compute the following determinants of orders two or three (by Sarrus' rule and definition).

1.1 $\begin{vmatrix} a^2 & ab \\ ab & b^2 \end{vmatrix}$ **1.2** $\begin{vmatrix} \cos\alpha & -\sin\alpha \\ \sin\alpha & \cos\alpha \end{vmatrix}$

1.3 $\begin{vmatrix} a+bi & b \\ 2a & a-bi \end{vmatrix}$ **1.4** $\begin{vmatrix} 3 & 2 & -4 \\ 4 & 1 & -2 \\ 5 & 2 & -3 \end{vmatrix}$

1.5 $\begin{vmatrix} 1 & 2 & 3 \\ 4 & 5 & 6 \\ 7 & 8 & 9 \end{vmatrix}$ **1.6** $\begin{vmatrix} 2 & 2 & 1 \\ 4 & 1 & -1 \\ 202 & 199 & 101 \end{vmatrix}$

1.7 $\begin{vmatrix} 1 & \omega & \omega^2 \\ \omega^2 & 1 & \omega \\ \omega & \omega^2 & 1 \end{vmatrix}$, where $\omega = -\frac{1}{2} + \mathrm{i}\frac{\sqrt{3}}{2}$.

1.8 $\begin{vmatrix} 1 & x & x \\ x & 2 & x \\ x & x & 3 \end{vmatrix}$

Q1.9–1.18: Compute the following numerical determinants (expand them by properties of determinant).

1.9 $\begin{vmatrix} 0 & 0 & 0 & 4 \\ 0 & 0 & 4 & 3 \\ 0 & 4 & 3 & 2 \\ 4 & 3 & 2 & 1 \end{vmatrix}$

1.10 $\begin{vmatrix} 0 & 0 & \cdots & 0 & 1 & 0 \\ 0 & 0 & \cdots & 2 & 0 & 0 \\ \vdots & \vdots & ⋰ & \vdots & \vdots & \vdots \\ 0 & 8 & \cdots & 0 & 0 & 0 \\ 9 & 0 & \cdots & 0 & 0 & 0 \\ 0 & 0 & \cdots & 0 & 0 & 10 \end{vmatrix}$

1.11 $\begin{vmatrix} 1 & 1 & 1 & 1 \\ 1 & -1 & 1 & 1 \\ 1 & 1 & -1 & 1 \\ 1 & 1 & 1 & -1 \end{vmatrix}$ **1.12** $\begin{vmatrix} 1 & 2 & 3 & 4 \\ 2 & 3 & 4 & 1 \\ 3 & 4 & 1 & 2 \\ 4 & 1 & 2 & 3 \end{vmatrix}$

1.13 $\begin{vmatrix} 5 & 0 & 4 & 2 \\ 1 & -1 & 2 & 1 \\ 4 & 1 & 2 & 0 \\ 1 & 1 & 1 & 1 \end{vmatrix}$

1.14 $\begin{vmatrix} 3 & 6 & 5 & 6 & 4 \\ 2 & 5 & 4 & 5 & 3 \\ 3 & 6 & 3 & 4 & 2 \\ 2 & 5 & 4 & 6 & 5 \\ 1 & 1 & 1 & -1 & -1 \end{vmatrix}$ **1.15** $\begin{vmatrix} 1 & 2 & 0 & 0 \\ 3 & 4 & 0 & 0 \\ 0 & 0 & -1 & 3 \\ 0 & 0 & 5 & 1 \end{vmatrix}$

1.16 $\begin{vmatrix} 1 & 2 & 3 & 4 & 5 \\ 6 & 7 & 8 & 9 & 10 \\ 0 & 0 & 0 & 1 & 3 \\ 0 & 0 & 0 & 2 & 4 \\ 0 & 1 & 0 & 1 & 1 \end{vmatrix}$

1.17 $\begin{vmatrix} 0 & 0 & 1 & -1 & 2 \\ 0 & 0 & 3 & 0 & 2 \\ 0 & 0 & 2 & 4 & 0 \\ 1 & 2 & 4 & 0 & -1 \\ 3 & 1 & 2 & 5 & 8 \end{vmatrix}$

1.18 $\begin{vmatrix} * & \boldsymbol{A} \\ \boldsymbol{B} & \boldsymbol{0} \end{vmatrix}$, where $\boldsymbol{A} = \begin{pmatrix} 1 & 0 & 0 \\ 1 & 2 & 0 \\ 1 & 2 & 3 \end{pmatrix}$,

$$\boldsymbol{B} = \begin{pmatrix} 0 & 0 & 0 & 0 & -1 \\ 0 & 0 & 0 & -2 & 0 \\ 0 & 0 & -3 & 0 & 0 \\ 0 & -4 & 0 & 0 & 0 \\ -5 & 0 & 0 & 0 & 0 \end{pmatrix}.$$

Q1.19–1.22: Prove the following identities.

1.19 $$\begin{vmatrix} a_1+b_1x & a_1x+b_1 & c_1 \\ a_2+b_2x & a_2x+b_2 & c_2 \\ a_3+b_3x & a_3x+b_3 & c_3 \end{vmatrix} = (1-x^2)\begin{vmatrix} a_1 & b_1 & c_1 \\ a_2 & b_2 & c_2 \\ a_3 & b_3 & c_3 \end{vmatrix}$$

1.20 $$\begin{vmatrix} 1+x & 1 & 1 & 1 \\ 1 & 1-x & 1 & 1 \\ 1 & 1 & 1+y & 1 \\ 1 & 1 & 1 & 1-y \end{vmatrix} = x^2y^2$$

1.21 $$\begin{vmatrix} 1 & 1 & 1 \\ a & b & c \\ a^2 & b^2 & c^2 \end{vmatrix} = (a+b+c)(b-a)(c-a)(c-b)$$

1.22 $$\begin{vmatrix} 1 & a^2 & a^3 \\ 1 & b^2 & b^3 \\ 1 & c^2 & c^3 \end{vmatrix} = (ab+bc+ca)\begin{vmatrix} 1 & a & a^2 \\ 1 & b & b^2 \\ 1 & c & c^2 \end{vmatrix}$$

Q1.23–1.30: Calculate the following determinants.

1.23 $$\begin{vmatrix} 1 & 0 & 2 & a \\ 2 & 0 & b & 0 \\ 3 & c & 4 & 5 \\ d & 0 & 0 & 0 \end{vmatrix}$$

1.24 $$\begin{vmatrix} a & 1 & 0 & 0 \\ -1 & b & 1 & 0 \\ 0 & -1 & c & 1 \\ 0 & 0 & -1 & d \end{vmatrix}$$

1.25 $$\begin{vmatrix} a^2 & (a+1)^2 & (a+2)^2 & (a+3)^2 \\ b^2 & (b+1)^2 & (b+2)^2 & (b+3)^2 \\ c^2 & (c+1)^2 & (c+2)^2 & (c+3)^2 \\ d^2 & (d+1)^2 & (d+2)^2 & (d+3)^2 \end{vmatrix}$$

1.26 $$\begin{vmatrix} a & b & c & 1 \\ b & c & a & 1 \\ c & a & b & 1 \\ \frac{b+c}{2} & \frac{c+a}{2} & \frac{a+b}{2} & 1 \end{vmatrix}$$

1.27 $$\begin{vmatrix} a_1 & 0 & 0 & b_1 \\ 0 & a_2 & b_2 & 0 \\ 0 & b_3 & a_3 & 0 \\ b_4 & 0 & 0 & a_4 \end{vmatrix}$$

1.28 $$\begin{vmatrix} 1 & 2 & 2 & \cdots & 2 & 2 \\ 2 & 2 & 2 & \cdots & 2 & 2 \\ 2 & 2 & 3 & \cdots & 2 & 2 \\ \vdots & \vdots & \vdots & \ddots & \vdots & \vdots \\ 2 & 2 & 2 & \cdots & n-1 & 2 \\ 2 & 2 & 2 & \cdots & 2 & n \end{vmatrix}$$

1.29 $$\begin{vmatrix} 1 & 1 & 1 & \cdots & 1 \\ a & a-1 & a-2 & \cdots & a-n \\ a^2 & (a-1)^2 & (a-2)^2 & \cdots & (a-n)^2 \\ \vdots & \vdots & \vdots & & \vdots \\ a^n & (a-1)^n & (a-2)^n & \cdots & (a-n)^n \end{vmatrix}$$

1.30 $$\begin{vmatrix} a_1^n & a_1^{n-1}b_1 & a_1^{n-2}b_1^2 & \cdots & a_1b_1^{n-1} & b_1^n \\ a_2^n & a_2^{n-1}b_2 & a_2^{n-2}b_2^2 & \cdots & a_2b_2^{n-1} & b_2^n \\ \vdots & \vdots & \vdots & & \vdots & \vdots \\ a_{n+1}^n & a_{n+1}^{n-1}b_{n+1} & a_{n+1}^{n-2}b_{n+1}^2 & \cdots & a_{n+1}b_{n+1}^{n-1} & b_{n+1}^n \end{vmatrix}$$

Q1.31–1.32: Solve the following systems of linear equations by Cramer's rule.

1.31 $$\begin{cases} 5x_1 + 4x_3 + 2x_4 = 3 \\ x_1 - x_2 + 2x_3 + x_4 = 1 \\ 4x_1 + x_2 + 2x_3 = 1 \\ x_1 + x_2 + x_3 + x_4 = 0 \end{cases}$$

1.32 $$\begin{cases} x_2 + x_3 + x_4 + x_5 = 1 \\ x_1 + x_3 + x_4 + x_5 = 2 \\ x_1 + x_2 + x_4 + x_5 = 3 \\ x_1 + x_2 + x_3 + x_5 = 4 \\ x_1 + x_2 + x_3 + x_4 = 5 \end{cases}$$

1.33 What conditions do a and b have to satisfy when the homogeneous system of linear equations

$$\begin{cases} x_1 + x_2 + x_3 + ax_4 = 0 \\ x_1 + 2x_2 + x_3 + x_4 = 0 \\ x_1 + x_2 - 3x_3 + x_4 = 0 \\ x_1 + x_2 + ax_3 + bx_4 = 0 \end{cases}$$

has a nonzero solution?

1.34 Find the equation of a straight line in the plane passing through two points (x_1, y_1), (x_2, y_2) (expressed by determinant).

1.35 Find a cubic polynomial $f(x) = a_0 + a_1x + a_2x^2 + a_3x^3$, such that

$$f(-1) = 0,\ f(1) = 4,\ f(2) = 3,\ f(3) = 16.$$

Supplementary Exercises

Q1.36–1.39: Prove the following identities.

1.36
$$\begin{vmatrix} 1+a_1 & 1 & \cdots & 1 \\ 1 & 1+a_2 & \cdots & 1 \\ \vdots & \vdots & & \vdots \\ 1 & 1 & \cdots & 1+a_n \end{vmatrix} = \left(1+\sum_{i=1}^{n}\frac{1}{a_i}\right)\prod_{i=1}^{n}a_i$$

(a) Use mathematical induction;
(b) Compute the determinant by denoting the original determinant as a sum of 2^n determinants by linearity properties;
(c) Calculate the determinant by recursion formula.

1.37
$$\begin{vmatrix} x & -1 & 0 & \cdots & 0 & 0 \\ 0 & x & -1 & \cdots & 0 & 0 \\ \vdots & \vdots & \vdots & \ddots & \vdots & \vdots \\ 0 & 0 & 0 & \cdots & x & -1 \\ a_n & a_{n-1} & a_{n-2} & \cdots & a_2 & x+a_1 \end{vmatrix} = x^n + \sum_{k=1}^{n} a_k x^{n-k}$$

1.38
$$\begin{vmatrix} a_1 & -1 & 0 & \cdots & 0 & 0 \\ a_2 & x & -1 & \cdots & 0 & 0 \\ a_3 & 0 & x & \cdots & 0 & 0 \\ \vdots & \vdots & \vdots & \ddots & \vdots & \vdots \\ a_{n-1} & 0 & 0 & \cdots & x & -1 \\ a_n & 0 & 0 & \cdots & 0 & x \end{vmatrix} = \sum_{k=1}^{n} a_k x^{n-k}$$

1.39
$$\begin{vmatrix} \cos\theta & 1 & & & & \\ 1 & 2\cos\theta & 1 & & & \\ & \ddots & \ddots & \ddots & & \\ & & 1 & 2\cos\theta & 1 \\ & & & 1 & 2\cos\theta \end{vmatrix} = \cos n\theta$$

Q1.40–1.43: Calculate the following determinants:

1.40
$$\begin{vmatrix} \frac{1}{3} & -\frac{5}{2} & \frac{2}{5} & \frac{3}{2} \\ 3 & -12 & \frac{21}{5} & 15 \\ \frac{2}{3} & -\frac{9}{2} & \frac{4}{5} & \frac{5}{2} \\ -\frac{1}{7} & \frac{2}{7} & -\frac{1}{7} & \frac{3}{7} \end{vmatrix}$$

1.41
$$\begin{vmatrix} 1 & 1 & \cdots & 1 & -n \\ 1 & 1 & \cdots & -n & 1 \\ \vdots & \vdots & ⋰ & \vdots & \vdots \\ 1 & -n & \cdots & 1 & 1 \\ -n & 1 & \cdots & 1 & 1 \end{vmatrix}$$

1.42
$$\begin{vmatrix} a_1+\lambda_1 & a_2 & a_3 & \cdots & a_n \\ a_1 & a_2+\lambda_2 & a_3 & \cdots & a_n \\ a_1 & a_2 & a_3+\lambda_3 & \cdots & a_n \\ \vdots & \vdots & \vdots & \ddots & \vdots \\ a_1 & a_2 & a_3 & \cdots & a_n+\lambda_n \end{vmatrix}$$

1.43
$$\begin{vmatrix} 1 & 2 & 3 & \cdots & n-1 & n \\ 2 & 3 & 4 & \cdots & n & 1 \\ 3 & 4 & 5 & \cdots & 1 & 2 \\ \vdots & \vdots & \vdots & & \vdots & \vdots \\ n-1 & n & 1 & \cdots & n-3 & n-2 \\ n & 1 & 2 & \cdots & n-2 & n-1 \end{vmatrix}$$

1.44 Prove

$$\begin{vmatrix} 1 & 1 & 1 & \cdots & 1 \\ x_1 & x_2 & x_3 & \cdots & x_n \\ x_1^2 & x_2^2 & x_3^2 & \cdots & x_n^2 \\ \vdots & \vdots & \vdots & & \vdots \\ x_1^{n-2} & x_2^{n-2} & x_3^{n-2} & \cdots & x_n^{n-2} \\ x_1^n & x_2^n & x_3^n & \cdots & x_n^n \end{vmatrix} = \left(\sum_{i=1}^{n} x_i\right)\prod_{1\le j<i\le n}(x_i - x_j).$$

1.45 Prove (by using mathematical induction) the relational expression of derivative,

$$\frac{d}{dt}\begin{vmatrix} a_{11}(t) & a_{12}(t) & \cdots & a_{1n}(t) \\ a_{21}(t) & a_{22}(t) & \cdots & a_{2n}(t) \\ \vdots & \vdots & & \vdots \\ a_{n1}(t) & a_{n2}(t) & \cdots & a_{nn}(t) \end{vmatrix} = \sum_{j=1}^{n} \begin{vmatrix} a_{11}(t) & \cdots & \frac{d}{dt}a_{1j}(t) & \cdots & a_{1n}(t) \\ a_{21}(t) & \cdots & \frac{d}{dt}a_{2j}(t) & \cdots & a_{2n}(t) \\ \vdots & \vdots & \vdots & & \vdots \\ a_{n1}(t) & \cdots & \frac{d}{dt}a_{nj}(t) & \cdots & a_{nn}(t) \end{vmatrix}.$$

1.46 The three points $P_1(x_1, y_1)$, $P_2(x_2, y_2)$, and $P_3(x_3, y_3)$ do not fall on a straight line. Find the equation of the circle passing through P_1, P_2, and P_3.

1.47 Find a necessary and sufficient condition under which three points (x_1, y_1), (x_2, y_2), and (x_3, y_3) are on a straight line.

1.48 Write the equation of the plane passing through the three points (1,1,1), $(2, 3, -1)$, and $(3, -1, -1)$.

1.49 Write the equation of the sphere passing through the points (1,1,1), $(1, 1, -1)$, $(1, -1, 1)$, and $(-1, 0, 0)$ and find its center and radius.

1.50 Given $a^2 \neq b^2$, prove that the following system of equations has a unique solution, and solve it.

$$\begin{cases} ax_1 & & & & + bx_{2n} = 1 \\ & ax_2 & & + bx_{2n-1} & = 1 \\ & & \cdots\ \cdots\ \cdots & & \\ & & ax_n + bx_{n+1} & & = 1 \\ & & bx_n + ax_{n+1} & & = 1 \\ & & \cdots\ \cdots\ \cdots & & \\ & bx_2 & & + ax_{2n-1} & = 1 \\ bx_1 & & & & + ax_{2n} = 1 \end{cases}$$

2 Matrices

CHAPTER OVERVIEW

The matrix is an important and basic concept in mathematics. Some quantitative relations in many problems may be described using matrices. Readers will find numerous applications of matrices in many branches of mathematics and other subjects as well.

In this chapter, we start with Gaussian elimination, and subsequently introduce the concept of a matrix, followed by its basic operations—addition, scalar multiplication, multiplication, transposition, and the inversion of an invertible matrix. Furthermore, we present the elementary operations and finally show the partitioning of a matrix and the operation of block matrices.

Matrices are useful mathematical objects. Readers must study the special rules of matrix operations carefully and master the skills of manipulating matrices, such as its various basic operations, in order to solve fundamental problems of linear algebra.

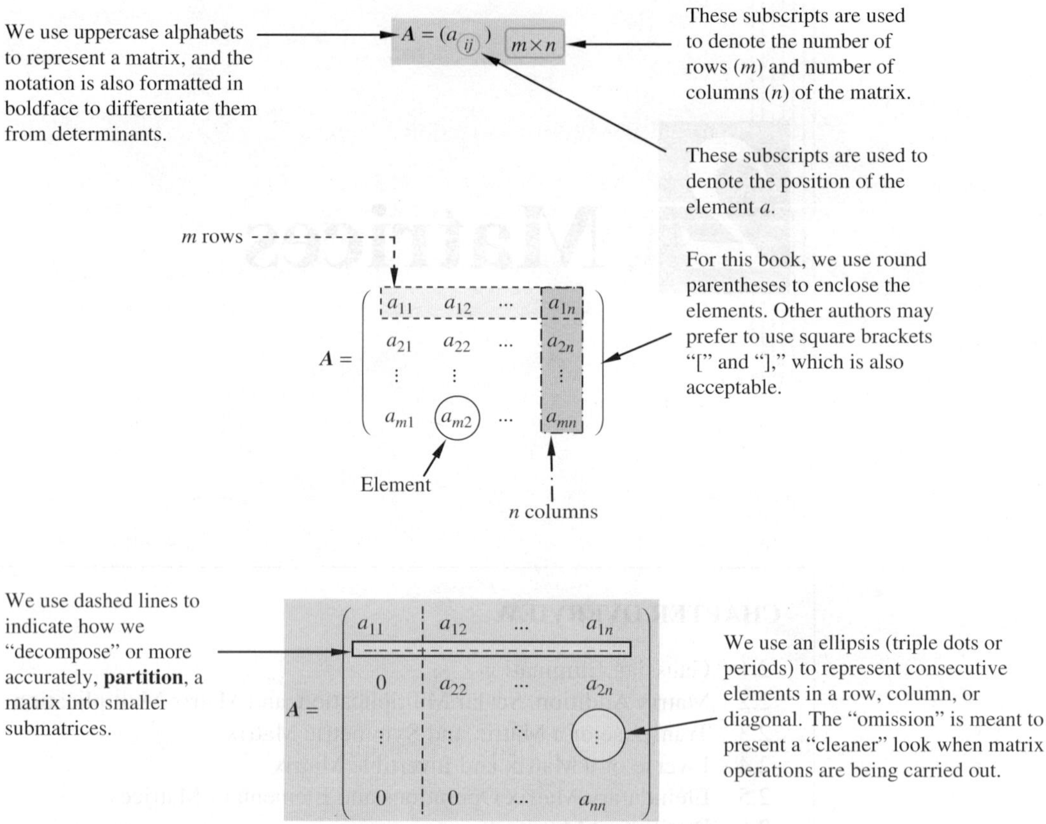

Figure 2.1 Notations of matrices

2.1 Gaussian Elimination

In Chapter 1, we introduced Cramer's rule for solving the linear system of n equations in n unknowns. Before starting this course on linear algebra, readers would have learnt how to solve systems of linear equations in two or three unknowns by the methods of substitution or elimination. Now we shall extend this idea to the general case of m equations in n unknowns, and from this perspective, introduce the concept of a matrix. In Chapter 3, we will present the general theory of solution of a system of linear equations.

The basic idea of elimination is to change a system of equations into an equivalent system which is easier to solve. When solving a system of equations in a large number of unknowns, we need to organize the elimination steps in a standard and simple way. In the following example, we illustrate the specific method of Gaussian elimination.

Example 2.1 Method of Gaussian elimination

Solve the system of linear equations

$$\begin{cases} 2x_1 - 2x_2 + 6x_4 = -2, \\ 2x_1 - x_2 + 2x_3 + 4x_4 = -2, \\ 3x_1 - x_2 + 4x_3 + 4x_4 = -3, \\ 5x_1 - 3x_2 + x_3 + 20x_4 = -2. \end{cases} \tag{2.1}$$

Solution: Multiplying the first equation by $\frac{1}{2}$, we get

$$\begin{cases} x_1 - x_2 + 3x_4 = -1, & ① \\ 2x_1 - x_2 + 2x_3 + 4x_4 = -2, & ② \\ 3x_1 - x_2 + 4x_3 + 4x_4 = -3, & ③ \\ 5x_1 - 3x_2 + x_3 + 20x_4 = -2. & ④ \end{cases}$$

Add the first equation multiplied by (-2), (-3), (-5) to equations ②, ③, and ④, respectively. Thus, eliminating the unknown x_1 from the equations ②, ③, and ④ we have

$$\begin{cases} x_1 - x_2 + 3x_4 = -1, \\ x_2 + 2x_3 - 2x_4 = 0, \\ 2x_2 + 4x_3 - 5x_4 = 0, \\ 2x_2 + x_3 + 5x_4 = 3. \end{cases}$$

It is easy to prove that this system of linear equations has the same solution as the original system. Now adding the second equation multiplied by (-2) to equations ③ and ④ individually, so that x_2 is eliminated from these two equations, we obtain

$$\begin{cases} x_1 - x_2 + 3x_4 = -1, \\ x_2 + 2x_3 - 2x_4 = 0, \\ - x_4 = 0, \\ - 3x_3 + 9x_4 = 3. \end{cases}$$

Then, taking equation ③ multiplied by (-1), multiply equation ④ by $-\frac{1}{3}$, and then interchanging the positions of equations ③ and ④, we have

$$\begin{cases} x_1 - x_2 + 3x_4 = -1, \\ x_2 + 2x_3 - 2x_4 = 0, \\ x_3 - 3x_4 = -1, \\ x_4 = 0. \end{cases} \tag{2.2}$$

The system of linear equations (2.2) has the same solution as (2.1). From (2.2), we can easily see that $x_4 = 0$, substituting this value into equation ③ we have $x_3 = -1$. Further substituting it into equations ② and ①, we have $x_2 = 2$ and $x_1 = 1$, respectively. So, (1, 2, − 1, 0) is a solution of the original system of equations (2.1). ■

The system of linear equations in the form (2.2) is called a **staircase system of linear equations**.

Any system of linear equations may be changed into a system of linear equations, which is easy to solve and is equivalent to the original system, by using the Gaussian elimination as shown in Example 2.1. The crux of the so-called elimination method lies in changing the coefficient of an unknown into zero. In order to set the elimination process in motion, we can write the corresponding coefficients of the linear equations as

$$\begin{cases} a_{11}x_1 & + & a_{12}x_2 & + & \cdots & + & a_{1n}x_n & = & b_1, \\ a_{21}x_1 & + & a_{22}x_2 & + & \cdots & + & a_{2n}x_n & = & b_2, \\ \cdots & & \cdots & & \cdots & & \cdots & & \cdots \\ a_{m1}x_1 & + & a_{m2}x_2 & + & \cdots & + & a_{mn}x_n & = & b_m. \end{cases} \tag{2.3}$$

in a rectangular table form

$$\begin{pmatrix} a_{11} & a_{12} & \cdots & a_{1n} & b_1 \\ a_{21} & a_{22} & \cdots & a_{2n} & b_2 \\ \vdots & \vdots & & \vdots & \vdots \\ a_{m1} & a_{m1} & \cdots & a_{mn} & b_m \end{pmatrix}, \tag{2.4}$$

where a_{ij} $(i = 1, 2, \ldots, m;\ j = 1, 2, \ldots, n)$ indicates the coefficient of the jth unknown in the ith equation. So, the necessary calculations pertaining to the process of Gaussian elimination may be done in this table and the table is called a matrix.

Definition 2.1

An array of numbers

$$\begin{pmatrix} a_{11} & a_{12} & \cdots & a_{1n} \\ a_{21} & a_{22} & \cdots & a_{2n} \\ \vdots & \vdots & & \vdots \\ a_{m1} & a_{m1} & \cdots & a_{mn} \end{pmatrix} \tag{2.5}$$

arranged in m rows and n columns with $m \times n$ numbers a_{ij} $(i = 1, 2, \ldots, m;$ $j = 1, 2, \cdots, n)$ taken from a number field F, and put in parentheses (or square brackets), is called a **matrix** over the number field F. The matrix is usually written as $\boldsymbol{A}$ or $\boldsymbol{A}_{m\times n}$ in upper case. Sometimes, it may also be written as

$$\boldsymbol{A} = (a_{ij})_{m\times n} \quad (i = 1, 2, \cdots, m;\ j = 1, 2, \ldots, n)$$

in which a_{ij} denotes the element placed at the intersection of row i and column j in the matrix $\boldsymbol{A}$. When $a_{ij} \in \mathbb{R}$ (field of real numbers), $\boldsymbol{A}$ is called a real matrix, when $a_{ij} \in \mathbb{C}$ (complex field), $\boldsymbol{A}$ is called a complex matrix.

The matrix with all $m \times n$ elements equal to zero is called the **zero matrix**, written as $\mathbf{0}$.

When $m = n$, we call $\boldsymbol{A}$ an nth-order **matrix** (or ***n*th-order square matrix**).

The set consisting of all $m \times n$ matrices over the number field F is written as $F^{m \times n}$ or $M_{m \times n}(F)$; the set consisting of all $n \times n$ real matrices (or nth-order real matrices) is written as $\mathbb{R}^{n \times n}$ or $M_n(\mathbb{R})$.

The matrix (2.4) which corresponds to the system of linear equations (2.3) is called the **augmented matrix** of the system (2.3), and is written as $(\boldsymbol{A}, \boldsymbol{b})$. The matrix $\boldsymbol{A}$ made with the coefficients of unknowns is called the **coefficient matrix** of the system of linear equations.

The elimination process to solve the system of linear equations can be carried out in its augmented matrix. An illustration is provided below.

Example 2.2 Gaussian elimination using the augmented matrix

Solve the system of linear equations

$$\begin{cases} x_1 - x_2 - x_3 + 3x_5 = -1, & ① \\ 2x_1 - 2x_2 - x_3 + 2x_4 + 4x_5 = -2, & ② \\ 3x_1 - 3x_2 - x_3 + 4x_4 + 5x_5 = -3, & ③ \\ x_1 - x_2 + x_3 + x_4 + 8x_5 = 2. & ④ \end{cases} \tag{2.6}$$

Solution: The augmented matrix of the system of linear equations is

$$(\boldsymbol{A}, \boldsymbol{b}) = \left(\begin{array}{ccccc|c} 1 & -1 & -1 & 0 & 3 & -1 \\ 2 & -2 & -1 & 2 & 4 & -2 \\ 3 & -3 & -1 & 4 & 5 & -3 \\ 1 & -1 & 1 & 1 & 8 & 2 \end{array}\right). \tag{2.6$'$}$$

We multiply equation ① by (-2), (-3), (-1) and add the result respectively to equations ②, ③, and ④; in (2.6) in turn, we eliminate x_1 from the last three equations (meanwhile x_2 is also eliminated, in this particular problem). That is, adding row ① of the matrix multiplied respectively by (-2), (-3), (-1) to rows ②, ③, and ④, in turn in (2.6)$'$. We obtain

$$(\boldsymbol{A}, \boldsymbol{b}) \xrightarrow[④+①\times(-1)]{\substack{②+①\times(-2) \\ ③+①\times(-3)}} \left(\begin{array}{ccccc|c} 1 & -1 & -1 & 0 & 3 & -1 \\ 0 & 0 & 1 & 2 & -2 & 0 \\ 0 & 0 & 2 & 4 & -4 & 0 \\ 0 & 0 & 2 & 1 & 5 & 3 \end{array}\right). \begin{array}{c} ① \\ ② \\ ③ \\ ④ \end{array}$$

The last three elements of columns 1 and 2 of the matrix are all zeros. This indicates that x_1 and x_2 in the equations ②, ③, and ④ of system (2.6) have been eliminated. Adding the row ② multiplied by (-2) to rows ③ and ④ yields

$$\xrightarrow[④+②\times(-2)]{③+②\times(-2)} \left(\begin{array}{ccccc|c} 1 & -1 & -1 & 0 & 3 & -1 \\ 0 & 0 & 1 & 2 & -2 & 0 \\ 0 & 0 & 0 & 0 & 0 & 0 \\ 0 & 0 & 0 & -3 & 9 & 3 \end{array}\right) \begin{array}{c} ① \\ ② \\ ③ \\ ④ \end{array}$$

$$\xrightarrow[③\rightleftharpoons④]{④\times(-\frac{1}{3})} \left(\begin{array}{ccccc|c} 1 & -1 & -1 & 0 & 3 & -1 \\ 0 & 0 & 1 & 2 & -2 & 0 \\ 0 & 0 & 0 & 1 & -3 & -1 \\ 0 & 0 & 0 & 0 & 0 & 0 \end{array}\right), \begin{array}{c} ① \\ ② \\ ③ \\ ④ \end{array} \tag{2.7}$$

where ④ $\times (-\frac{1}{3})$ indicates that row ④ is multiplied by $(-\frac{1}{3})$; ③ $\rightleftharpoons$ ④ indicates interchange in the positions of rows ③ and ④.

A system of linear equations corresponding to the staircase augmented matrix (2.7) is equivalent to the original system of linear equations. In order to omit the process of back substitution, we eliminate all elements in the column above the first nonzero element of each row into zeros, that is,

$$\xrightarrow{②+③\times(-2)} \left(\begin{array}{ccccc|c} 1 & -1 & -1 & 0 & 3 & -1 \\ 0 & 0 & 1 & 0 & 4 & 2 \\ 0 & 0 & 0 & 1 & -3 & -1 \\ 0 & 0 & 0 & 0 & 0 & 0 \end{array}\right) \begin{array}{c} ① \\ ② \\ ③ \\ ④ \end{array}$$

$$\xrightarrow{①+②} \left(\begin{array}{ccccc|c} 1 & -1 & 0 & 0 & 7 & 1 \\ 0 & 0 & 1 & 0 & 4 & 2 \\ 0 & 0 & 0 & 1 & -3 & -1 \\ 0 & 0 & 0 & 0 & 0 & 0 \end{array}\right). \begin{array}{c} ① \\ ② \\ ③ \\ ④ \end{array} \tag{2.8}$$

We call the matrix (2.8) a **row-reduced echelon matrix,** the corresponding system of linear equations being

$$\begin{cases} x_1 - x_2 + 7x_5 = 1, \\ x_3 + 4x_5 = 2, \\ x_4 - 3x_5 = -1. \end{cases} \tag{2.8$'$}$$

It is equivalent to the original system of linear equations. Here, we have three equations and five unknowns. On inspection, we see that the columns corresponding to the first, third, and fourth variables resemble columns of an identity matrix. (See Definition 2.6 for a formal definition of an identity matrix.) This implies that we have to assign arbitrary values to the remaining two variables. Hence, selecting $x_2 = k_1, x_5 = k_2$ and substituting them into the system of linear equations (2.8)′, we can get the unique solution for x_1, x_3, x_4 corresponding to k_1, k_2. Thus, we can get all the solutions of the given system of linear equations, $x_1 = 1 + k_1 - 7k_2$, $x_2 = k_1, x_3 = 2 - 4k_2, x_4 = -1 + 3k_2, x_5 = k_2$, in which k_1 and k_2 are arbitrary constants. Finally, we represent the solution of linear equations in the following form

$$(x_1, x_2, x_3, x_4, x_5) = (1 + k_1 - 7k_2, k_1, 2 - 4k_2, -1 + 3k_2, k_2). \quad \blacksquare$$

When all the constant terms $b_1 = b_2 = \cdots = b_m = 0$ in the system of linear equations (2.3), we call it a **homogeneous system of linear equations,** otherwise we call it a **nonhomogeneous system of linear equations**.

The solution procedure of the homogeneous system of linear equations is the same as shown in Example 2.2. If the constant terms of the four equations in Example 2.2 were all zeros, its solution would have been

$$(x_1, x_2, x_3, x_4, x_5) = (k_1 - 7k_2, k_1, -4k_2, 3k_2, k_2).$$

Example 2.3 Inconsistent system of linear equations

Solve the system of linear equations

$$\begin{cases} x_1 + x_2 + x_3 = 1, \\ x_1 + 2x_2 - 5x_3 = 2, \\ 2x_1 + 3x_2 - 4x_3 = 5. \end{cases}$$

Solution:

$$(\boldsymbol{A}, \boldsymbol{b}) = \left(\begin{array}{ccc|c} 1 & 1 & 1 & 1 \\ 1 & 2 & -5 & 2 \\ 2 & 3 & -4 & 5 \end{array}\right) \xrightarrow[③+①\times(-2)]{②+②\times(-1)} \left(\begin{array}{ccc|c} 1 & 1 & 1 & 1 \\ 0 & 1 & -6 & 1 \\ 0 & 1 & -6 & 3 \end{array}\right)$$

$$\xrightarrow[③+②\times(-1)]{} \left(\begin{array}{ccc|c} 1 & 1 & 1 & 1 \\ 0 & 1 & -6 & 1 \\ 0 & 0 & 0 & 2 \end{array}\right). \qquad (2.9)$$

Where the equation $0x_1 + 0x_2 + 0x_3 = 2$ indicated by the third row 0, 0, 0, 2 obviously has no solution, so the original system of linear equation cannot have any solution as well. $\blacksquare$

This is because the *LHS* of the third equation is equal to the sum of the *LHS* of the first two equations, while the *RHS* of the third equation is not equal to the sum of the *RHS* of the first two equations. This indicates that the third equation is contradictory to the first two, that is, the solution satisfying the first two equations does not satisfy the third one. Such a system of linear equations that contains contradictory equations is called an **inconsistent system of equations**. Otherwise, we call it a **consistent system of equations**. Examples 2.1 and 2.2 are consistent systems of equations. Note that, in the elimination process of Example 2.2, a complete zero row appears in row ③ of the augmented matrix because equation ③ in (2.6) is equal to {(equation ② × 2) + [equation ① × (−1)]}. This indicates that all the solutions of equations ① and ② satisfy equation ③, and thus equation ③ becomes redundant with respect to the solving of the system (2.6). We call ③ a **redundant equation**. During the Gaussian elimination process, the augmented matrix can single out the redundant and contradictory equations.

We can see from Examples 2.2 and 2.3, that for a general system of linear equations (2.3), the augmented matrix (2.4) can be changed into the row-reduced echelon matrix form as shown in (2.8) through the elimination process. In brief, we may transform (2.4) into the following reduced row echelon matrix with certain column interchanges:

$$(\boldsymbol{A}, \boldsymbol{b}) \Rightarrow \begin{pmatrix} c_{11} & 0 & \cdots & 0 & c_{1,r+1} & \cdots & c_{1n} & d_1 \\ 0 & c_{22} & \cdots & 0 & c_{2,r+1} & \cdots & c_{2n} & d_2 \\ \vdots & \vdots & \ddots & \vdots & \vdots & & \vdots & \vdots \\ 0 & 0 & \cdots & c_{rr} & c_{r,r+1} & \cdots & c_{rn} & d_r \\ 0 & 0 & \cdots & 0 & 0 & \cdots & 0 & d_{r+1} \\ 0 & 0 & \cdots & 0 & 0 & \cdots & 0 & 0 \\ \vdots & \vdots & & \vdots & \vdots & & \vdots & \vdots \\ 0 & 0 & \cdots & 0 & 0 & \cdots & 0 & 0 \end{pmatrix}, \tag{2.10}$$

where $c_{ii} = 1, i = 1, 2, \ldots, r$.

The nonhomogeneous systems of linear equations corresponding to (2.10) and (2.3) are equivalent. We can obviously see that a necessary and sufficient condition of a system of linear equations to have a solution is $d_{r+1} = 0$. Under the conditions where a solution can be computed:

(i) when $r = n$, (2.10) has a unique solution $x_1 = d_1, x_2 = d_2, \ldots, x_n = d_n$;

(ii) when $r < n$, (2.10) has infinitely many solutions. In order to solve it, we take unknowns $(x_1, x_2, \ldots, x_r)$ corresponding to the columns containing $c_{ii} (i = 1, 2, \cdots r)$ as basic variables, where c_{ii} is the first nonzero element of each row in (2.10), while the other unknowns $(x_{r+1}, x_{r+2}, \ldots, x_n)$

are taken as free variables. Assigning arbitrary constants $k_1, k_2, \ldots, k_{n-r}$ to the free unknown variables in turn, and substituting them into the equations corresponding to expression (2.10), we can get

$$\begin{cases} x_1 = d_1 - c_{1,r+1}k_1 - \cdots - c_{1n}k_{n-r}, \\ x_2 = d_2 - c_{2,r+1}k_1 - \cdots - c_{2n}k_{n-r}, \\ \qquad\cdots \\ x_r = d_r - c_{1,r+1}k_1 - \cdots - c_{rn}k_{n-r}, \\ x_{r+1} = k_1, \\ \qquad\cdots \\ x_n = k_{n-r}. \end{cases} \tag{2.11}$$

Here, $k_1, k_2, \ldots, k_{n-r}$ are mutually independent arbitrary constants. These are all solutions of the given system of linear equations.

A homogeneous system of linear equations has $b_1, b_2, \ldots, b_m = 0$ in (2.3); such a system always has a solution because this leads to $d_1 = \cdots = d_r = d_{r+1} = 0$ in (2.10). When $r = n$, it has zero solution only, that is, $x_1 = x_2 = \cdots = x_n = 0$. When $r < n$, it has infinitely many solutions. These solutions are described by (2.11) with $d_1 = d_2 = \cdots = d_r = 0$.

If the number of m equations in a homogeneous system of linear equations is less than the number of n unknowns, the system must have infinitely many nonzero solutions.

Finally, we need to point out an important fact. The form of echelon matrices that one may reach through different elimination processes, starting from a given matrix, is not unique. However, the number of nonzero rows in the echelon matrix is unique. When the system of linear equations has a solution, it indicates that the number of arbitrary constants in solutions are the same, even when one uses different elimination processes. The expressions of solutions may not be unique, while the solution sets are identical. However, these important conclusions cannot be formally proven until we study the theorems about the relationship between the rank of a matrix and the concept of linear independence in Chapter 3.

2.2 Matrix Addition, Scalar Multiplication, and Matrix Multiplication

In the previous section, we learnt that it is easy to solve a system of linear equations, denoted by matrices, using the method of elimination. In this section, we will elaborate on certain theorems on the system of equations by further studies on matrices.

Matrices are not only important for studying a system of linear equations, but also for studying various basic problems of linear algebra. Furthermore, the matrix as a tool is needed for studying many practical problems. Readers can find some applications of matrices in Chapter 7.

Matrix addition, scalar multiplication, and matrix multiplication are the most basic operations involving matrices. In order to discuss these matrix operations, let us first define the equality of two matrices.

Definition 2.2

Two matrices $\boldsymbol{A} = (a_{ij})$ and $\boldsymbol{B} = (b_{ij})$ are said to be equal, denoted as $\boldsymbol{A} = \boldsymbol{B}$, if the two matrices have the same number of rows and same number of columns with identical corresponding elements, that is, $a_{ij} = b_{ij} (i = 1, 2, \ldots, m;\ j = 1, 2, \cdots, n)$.

From the definition, if two $m \times n$ matrices are equal, then elements of the two matrices have $m \times n$ equalities. For example,

$$\begin{pmatrix} x & -1 & -8 \\ 0 & y & 4 \end{pmatrix} = \begin{pmatrix} 3 & -1 & z \\ 0 & 2 & 4 \end{pmatrix},$$

we can immediately obtain $x = 3$, $y = 2$, and $z = -8$.

Readers must be aware of the essential difference between a matrix and a determinant. A determinant is a formula and one can get its numerical value by calculation. But, a matrix is a table of numbers and the numbers of its rows and columns may be different. For an nth-order matrix $\boldsymbol{A}$, its determinant (written as $|\boldsymbol{A}|$ or $\det \boldsymbol{A}$) sometimes is calculated; however, the matrix $\boldsymbol{A}$ and its determinant are two completely different concepts. When $\det \boldsymbol{A} = 0$ ($\boldsymbol{A}$ may or may not be the zero matrix), we call $\boldsymbol{A}$ a **singular matrix**, when $\det \boldsymbol{A} \neq 0$, we call $\boldsymbol{A}$ a **nonsingular matrix**.

2.2.1 Matrix addition

Definition 2.3

Let $\boldsymbol{A} = (a_{ij})$ and $\boldsymbol{B} = (b_{ij}) \in F^{m \times n}$, set

$$\boldsymbol{A} + \boldsymbol{B} = (a_{ij} + b_{ij}) = \begin{pmatrix} a_{11} + b_{11} & a_{12} + b_{12} & \cdots & a_{1n} + b_{1n} \\ a_{21} + b_{21} & a_{22} + b_{22} & \cdots & a_{2n} + b_{2n} \\ \vdots & \vdots & & \vdots \\ a_{m1} + b_{m1} & a_{m2} + b_{m2} & \cdots & a_{mn} + b_{mn} \end{pmatrix}, \tag{2.12}$$

and call $\boldsymbol{A} + \boldsymbol{B}$ the sum of $\boldsymbol{A}$ and $\boldsymbol{B}$.

It is noteworthy that only matrices with the same number of rows and the same number of columns (i.e., matrices of the same size) can be added, and the sum of matrices of the same size results in a matrix of that size again.

According to Definition 2.3, the following operational rules can easily be verified:

(i) Commutative law: $\boldsymbol{A}+\boldsymbol{B}=\boldsymbol{B}+\boldsymbol{A}$;
(ii) Associative law: $(\boldsymbol{A}+\boldsymbol{B})+\boldsymbol{C}=\boldsymbol{A}+(\boldsymbol{B}+\boldsymbol{C})$;
(iii) Zero matrix satisfies: $\boldsymbol{A}+\boldsymbol{0}=\boldsymbol{A}$, in which $\boldsymbol{0}$ and $\boldsymbol{A}$ are matrices of the same size;
(iv) For a matrix $\boldsymbol{A}=(a_{ij})_{m\times n}$, the matrix $(-\boldsymbol{A})$ exists if it satisfies $\boldsymbol{A}+(-\boldsymbol{A})=0$, where $-\boldsymbol{A}=(-a_{ij})_{m\times n}$ [i.e., each element of $\boldsymbol{A}$ is multiplied by (-1)]. We also call $(-\boldsymbol{A})$ the **negative matrix** of $\boldsymbol{A}$.

Furthermore, we define the subtraction of matrices of the same size as follows:

$$\boldsymbol{A}-\boldsymbol{B}=\boldsymbol{A}+(-\boldsymbol{B}).$$

2.2.2 Multiplication of matrix by a number (scalar multiplication)

Definition 2.4

Let k be an arbitrary element of a number field F, $\boldsymbol{A}=(a_{ij})\in F^{m\times n}$, set

$$k\boldsymbol{A}=(ka_{ij})=\begin{pmatrix} ka_{11} & ka_{12} & \cdots & ka_{1n} \\ ka_{21} & ka_{22} & \cdots & ka_{2n} \\ \vdots & \vdots & & \vdots \\ ka_{m1} & ka_{m1} & \cdots & ka_{mn} \end{pmatrix}, \tag{2.13}$$

and call this matrix the **scalar multiplication** of $\boldsymbol{A}$ by k.

Notice that when a matrix $\boldsymbol{A}$ is multiplied by a number k, we multiply k to each element of that matrix $\boldsymbol{A}$. This is different from Property 1.3(i) of a determinant.

Suppose that 1, k, and l are numbers of F, then scalar multiplication of a matrix satisfies the following operational rules:

(i) $1\boldsymbol{A}=\boldsymbol{A}$;
(ii) $(kl)\boldsymbol{A}=k(l\boldsymbol{A})$;
(iii) $(k+l)\boldsymbol{A}=k\boldsymbol{A}+l\boldsymbol{A}$;
(iv) $k(\boldsymbol{A}+\boldsymbol{B})=k\boldsymbol{A}+k\boldsymbol{B}$.

2.2.3 Matrix multiplication

Definition 2.5

Let $\boldsymbol{A}$ be an $m \times n$ matrix and $\boldsymbol{B}$ an $n \times s$ matrix, that is,

$$\boldsymbol{A} = \begin{pmatrix} a_{11} & a_{12} & \cdots & a_{1n} \\ a_{21} & a_{22} & \cdots & a_{2n} \\ \vdots & \vdots & & \vdots \\ a_{m1} & a_{m2} & \cdots & a_{mn} \end{pmatrix}, \qquad \boldsymbol{B} = \begin{pmatrix} b_{11} & b_{12} & \cdots & b_{1s} \\ b_{21} & b_{22} & \cdots & b_{2s} \\ \vdots & \vdots & & \vdots \\ b_{n1} & b_{n2} & \cdots & b_{ns} \end{pmatrix}.$$

Then, the product $\boldsymbol{AB}$ [written as $\boldsymbol{C} = (c_{ij})$] of $\boldsymbol{A}$ and $\boldsymbol{B}$ is an $m \times s$ matrix, and

$$c_{ij} = a_{i1}b_{1j} + a_{i2}b_{2j} + \cdots + a_{in}b_{nj} = \sum_{k=1}^{n} a_{ik}b_{kj}. \tag{2.14}$$

The (i, j)th entry c_{ij} of the matrix $\boldsymbol{C} = \boldsymbol{AB}$ is the sum of products of n elements of row i in $\boldsymbol{A}$ respectively by n elements of column j in $\boldsymbol{B}$.

Readers will notice that the product $\boldsymbol{AB}$ of two matrices $\boldsymbol{A}$ and $\boldsymbol{B}$ makes sense (or we say they are multiplicable) when the number of columns of $\boldsymbol{A}$ equals the number of rows of $\boldsymbol{B}$, otherwise $\boldsymbol{A}$ and $\boldsymbol{B}$ are not multiplicable.

Example 2.4 Multiplication of two matrices

Suppose

$$\boldsymbol{A} = \begin{pmatrix} 1 & 2 & -1 \\ -1 & 3 & 4 \\ 1 & 1 & 1 \end{pmatrix}, \qquad \boldsymbol{B} = \begin{pmatrix} 5 & 6 \\ -5 & -6 \\ 6 & 0 \end{pmatrix}.$$

Compute $\boldsymbol{AB}$.

Solution: $\boldsymbol{AB}$ is a 3×2 matrix, that is,

$$\boldsymbol{AB} = \begin{pmatrix} 1 \times 5 + 2 \times (-5) + (-1) \times 6 & 1 \times 6 + 2 \times (-6) + 0 \\ (-1) \times 5 + 3 \times (-5) + 4 \times 6 & (-1) \times 6 + 3 \times (-6) + 0 \\ 5 + (-5) + 6 & 6 + (-6) + 0 \end{pmatrix}$$

$$= \begin{pmatrix} -11 & -6 \\ 4 & -24 \\ 6 & 0 \end{pmatrix}.$$

■

Example 2.5 Matrix multiplication without commutative law

Let $\boldsymbol{A}$ and $\boldsymbol{B}$ be two $n \times 1$ and $1 \times n$ matrices, respectively, given by

$$\boldsymbol{A}=\begin{pmatrix} a_1 \\ a_2 \\ \vdots \\ a_n \end{pmatrix}, \qquad \boldsymbol{B}=\begin{pmatrix} b_1, & b_2, & \cdots, & b_n \end{pmatrix}.$$

Compute $\boldsymbol{AB}$ and $\boldsymbol{BA}$.

Solution:

$$\boldsymbol{AB}=\begin{pmatrix} a_1 \\ a_2 \\ \vdots \\ a_n \end{pmatrix}\begin{pmatrix} b_1, & b_2, & \cdots, & b_n \end{pmatrix}=\begin{pmatrix} a_1b_1 & a_1b_2 & \cdots & a_1b_n \\ a_2b_1 & a_2b_2 & \cdots & a_2b_n \\ \vdots & \vdots & & \vdots \\ a_nb_1 & a_nb_2 & \cdots & a_nb_n \end{pmatrix}.$$

$$\boldsymbol{BA}=\begin{pmatrix} b_1, & b_2, & \cdots, & b_n \end{pmatrix}\begin{pmatrix} a_1 \\ a_2 \\ \vdots \\ a_n \end{pmatrix}=b_1a_1+b_2a_2+\cdots+b_na_n.$$

$\boldsymbol{AB}$ is an nth-order matrix, $\boldsymbol{BA}$ is a first-order matrix.

Hint: When the final result of the computation is a first-order matrix, we may consider it as a number and need not use matrix notation. But, during computations, we generally do not treat a first-order matrix as a number. ■

Example 2.6 Matrix multiplication without cancellation law

Suppose

$$\boldsymbol{A}=\begin{pmatrix} a & a \\ -a & -a \end{pmatrix}, \quad \boldsymbol{B}=\begin{pmatrix} b & -b \\ -b & b \end{pmatrix}, \quad \boldsymbol{C}=\begin{pmatrix} -1 & 1 \\ 1 & -1 \end{pmatrix}.$$

Compute $\boldsymbol{AB}$, $\boldsymbol{AC}$, and $\boldsymbol{BA}$.

Solution:

$$\boldsymbol{AB}=\boldsymbol{AC}=\begin{pmatrix} 0 & 0 \\ 0 & 0 \end{pmatrix}, \qquad \boldsymbol{BA}=\begin{pmatrix} 2ab & 2ab \\ -2ab & -2ab \end{pmatrix}.$$ ■

From Examples 2.4, 2.5, and 2.6, we have the following:

Conclusion (i) Multiplication of matrices does not satisfy commutative law.

In Example 2.4, $\boldsymbol{B}$ and $\boldsymbol{A}$ are not multiplicable, although $\boldsymbol{A}$ and $\boldsymbol{B}$ are; in Example 2.5, both $\boldsymbol{AB}$ and $\boldsymbol{BA}$ can be computed but they are not of the same size; in Example 2.6, $\boldsymbol{AB}$ and $\boldsymbol{BA}$ are not identical, although they are both second-order matrices.

Multiplication of matrices does not satisfy commutative law, however, this does not mean $\boldsymbol{AB} \neq \boldsymbol{BA}$ for two arbitrary matrices $\boldsymbol{A}$ and $\boldsymbol{B}$. For example, if

$$\boldsymbol{A} = \begin{pmatrix} 2 & 0 \\ 0 & 2 \end{pmatrix}, \qquad \boldsymbol{B} = \begin{pmatrix} a & b \\ c & d \end{pmatrix},$$

we have

$$\boldsymbol{AB} = \boldsymbol{BA} = \begin{pmatrix} 2a & 2b \\ 2c & 2d \end{pmatrix} = 2\boldsymbol{B}.$$

When $\boldsymbol{AB} \neq \boldsymbol{BA}$, we say $\boldsymbol{AB}$ is not commutative (or $\boldsymbol{A}$ and $\boldsymbol{B}$ are not commutative). When $\boldsymbol{AB} = \boldsymbol{BA}$, we say $\boldsymbol{AB}$ is commutative (or $\boldsymbol{A}$ and $\boldsymbol{B}$ are commutative). It is easy for readers to prove that if $\boldsymbol{AB} = \boldsymbol{BA}$, then $\boldsymbol{A}$ and $\boldsymbol{B}$ must be square matrices of the same order.

Conclusion (ii) $\boldsymbol{AB} = \boldsymbol{0}$ (zero matrix) does not necessarily imply that $\boldsymbol{A} = \boldsymbol{0}$ or $\boldsymbol{B} = \boldsymbol{0}$. Equivalently, even if $\boldsymbol{A} \neq \boldsymbol{0}$ and $\boldsymbol{B} \neq \boldsymbol{0}$, it is possible that $\boldsymbol{AB} = \boldsymbol{0}$.

It is a peculiar phenomenon if one looks at it in terms of the multiplication among numbers. However, it is easy for readers to comprehend. For example, if the homogeneous system of linear equations with coefficient matrix $\boldsymbol{A}$ (which is not a zero matrix) has a nonzero solution and such a nonzero solution is arranged as a column matrix $\boldsymbol{B}$, then $\boldsymbol{AB} = \boldsymbol{0}$.

In the case of $\boldsymbol{A} \neq \boldsymbol{0}$, $\boldsymbol{B} \neq \boldsymbol{0}$, but $\boldsymbol{AB} = \boldsymbol{0}$, we call $\boldsymbol{B}$ a right zero divisor of $\boldsymbol{A}$ and $\boldsymbol{A}$ a left zero divisor of $\boldsymbol{B}$. For some nonzero matrices, there do exist certain nonzero matrices as their left or right zero divisors. In Example 2.6, $\boldsymbol{B}$ and $\boldsymbol{C}$ are both right zero divisors of $\boldsymbol{A}$. Generally speaking, if $\boldsymbol{A}$ has a zero divisor, then its zero divisor is not unique.

Conclusion (iii) Matrix multiplication does not satisfy the cancellation law.

When $\boldsymbol{A} \neq \boldsymbol{0}$, we cannot deduce that $\boldsymbol{B} = \boldsymbol{C}$ from $\boldsymbol{AB} = \boldsymbol{AC}$. This is implied by conclusion **(ii)**. Indeed,

$$\boldsymbol{AB} = \boldsymbol{AC} \Rightarrow \boldsymbol{AB} - \boldsymbol{AC} = \boldsymbol{0} \overset{*}{\Rightarrow} \boldsymbol{A}(\boldsymbol{B} - \boldsymbol{C}) = \boldsymbol{0},$$

which does not necessarily imply that $\boldsymbol{B} - \boldsymbol{C} = \boldsymbol{0}$, that is, $\boldsymbol{B} = \boldsymbol{C}$ (the left distributive rule, which we are going to introduce shortly, is used after mark *), as $(\boldsymbol{B} - \boldsymbol{C})$ may be a right divisor of zero.

For instance in Example 2.6, $A \neq 0$, $AB = AC$, but $B \neq C$ (when $b \neq -1$).

Readers will learn in the following sections that $A \neq 0$, $AB = AC$ implies that $B = C$ and $B = 0$ if $AB = 0$ holds true when A is a nonsingular matrix, that is, determinant $|A| \neq 0$. That is to say, when A is nonsingular, matrix multiplication by A does not enjoy the peculiar phenomena described in conclusions **(ii)** and **(iii)** above, which are different from number multiplication.

Matrix multiplication does not satisfy commutative law and cancellation law. This is an important characteristic difference with number multiplication. However, matrix multiplication and number multiplication still share a few same or similar operational rules. In fact, matrix multiplication satisfies the following operational rules:

(i) Associative rule: $(AB)C = A(BC)$;
(ii) Associative law of scalar multiplication: $k(AB) = (kA)B = A(kB)$, where k is a number;
(iii) Left distributive law: $A(B + C) = AB + AC$;
Right distributive law: $(B + C)A = BA + CA$.

Proof: (i) Let A be an $m \times n$ matrix, B an $n \times p$ matrix, C a $p \times s$ matrix, then $(AB)C$ and $A(BC)$ are both $m \times s$ matrices. In the following, we will show that their (i, j)th entries ($i = 1, 2, \cdots, m$; $j = 1, 2, \ldots, s$) are the same.

$$\begin{aligned}[(AB)C]_{ij} &= \sum_{k=1}^{p} (AB)_{ik} C_{kj} = \sum_{k=1}^{p} (\sum_{l=1}^{n} a_{il} b_{lk}) c_{kj} \\ &= \sum_{k=1}^{p} (\sum_{l=1}^{n} a_{il} b_{lk} c_{kj}) = \sum_{l=1}^{n} (\sum_{k=1}^{p} a_{il} b_{lk} c_{kj}) \\ &= \sum_{i=1}^{n} a_{il} (\sum_{k=1}^{p} b_{lk} c_{kj}) = \sum_{l=1}^{n} a_{il} (BC)_{lj} \\ &= [A(BC)]_{ij}.\end{aligned}$$

□

Proofs of (ii) and (iii) are left for readers as exercises.

In the following sections, we will introduce several important special matrices and their multiplication operations.

Definition 2.6

An nth-order matrix whose leading diagonal elements are all ones, while other elements are all zeros, is called an nth-order **identity matrix** (or **unit matrix**), written as I_n or I; an nth-order matrix whose leading diagonal elements are all number k, while the other elements are all zeros is called an nth-order **scalar matrix**, written as $k\boldsymbol{I}_n$ or $k\boldsymbol{I}$. That is,

$$\boldsymbol{I}_n = \begin{pmatrix} 1 & & & \\ & 1 & & \\ & & \ddots & \\ & & & 1 \end{pmatrix}_{n\times n}, \quad k\boldsymbol{I}_n = \begin{pmatrix} k & & & \\ & k & & \\ & & \ddots & \\ & & & k \end{pmatrix}_{n\times n}.$$

The role of the identity matrix in matrix multiplication is similar to that of the number 1 in number multiplication; indeed, $\boldsymbol{I}_m\boldsymbol{A}_{m\times n} = \boldsymbol{A}_{m\times n}$, $\boldsymbol{A}_{m\times n}\boldsymbol{I}_n = \boldsymbol{A}_{m\times n}$.

Moreover, $(k\boldsymbol{I})\boldsymbol{A} = k(\boldsymbol{I}\boldsymbol{A}) = k\boldsymbol{A}$; $\boldsymbol{A}(k\boldsymbol{I}) = k(\boldsymbol{A}\boldsymbol{I}) = k\boldsymbol{A}$, so multiplying scalar matrix $k\boldsymbol{I}$ with matrix $\boldsymbol{A}$ is the same as the scalar multiplication of the number k with matrix $\boldsymbol{A}$, and nth-order scalar matrix $k\boldsymbol{I}$ commutes with arbitrary nth-order matrix $\boldsymbol{A}$ under multiplication. We can also prove: the nth-order matrix which commutes with arbitrary nth-order matrix under multiplication must be an nth-order scalar matrix (left for readers to practice).

Definition 2.7

An nth-order matrix in which all elements off the leading diagonal are zeros is called an nth-order **diagonal matrix**, written as $\boldsymbol{\Lambda}$, that is,

$$\boldsymbol{\Lambda} = \begin{pmatrix} a_1 & & & \\ & a_2 & & \\ & & \ddots & \\ & & & a_n \end{pmatrix},$$

or written as $\operatorname{diag}(a_1, a_2, \ldots, a_n)$.

Matrix $\boldsymbol{A}$ multiplied by diagonal matrix $\boldsymbol{\Lambda}$ from the left is equal to multiplying each element in row i of $\boldsymbol{A}$ by $a_i (i = 1, \cdots, n)$. While $\boldsymbol{A}$ multiplied by diagonal matrix $\boldsymbol{\Lambda}$ from the right is equal to multiplying each element in column i of $\boldsymbol{A}$ by $a_i (i = 1, \cdots, n)$. That is,

$$\begin{pmatrix} a_1 & & & \\ & a_2 & & \\ & & \ddots & \\ & & & a_n \end{pmatrix}\begin{pmatrix} a_{11} & a_{12} & \cdots & a_{1s} \\ a_{21} & a_{22} & \cdots & a_{2s} \\ \vdots & \vdots & & \vdots \\ a_{n1} & a_{n2} & \cdots & a_{ns} \end{pmatrix} = \begin{pmatrix} a_1a_{11} & a_1a_{12} & \cdots & a_1a_{1s} \\ a_2a_{21} & a_2a_{22} & \cdots & a_2a_{2s} \\ \vdots & \vdots & & \vdots \\ a_na_{n1} & a_na_{n2} & \cdots & a_na_{ns} \end{pmatrix},$$

$$\begin{pmatrix} a_{11} & a_{12} & \cdots & a_{1n} \\ a_{21} & a_{22} & \cdots & a_{2n} \\ \vdots & \vdots & & \vdots \\ a_{m1} & a_{m2} & \cdots & a_{mn} \end{pmatrix} \begin{pmatrix} a_1 & & & \\ & a_2 & & \\ & & \ddots & \\ & & & a_n \end{pmatrix} = \begin{pmatrix} a_1a_{11} & a_2a_{12} & \cdots & a_na_{1n} \\ a_1a_{21} & a_2a_{22} & \cdots & a_na_{2n} \\ \vdots & \vdots & & \vdots \\ a_1a_{m1} & a_2a_{m2} & \cdots & a_na_{mn} \end{pmatrix},$$

$$\begin{pmatrix} a_1 & & & \\ & a_2 & & \\ & & \ddots & \\ & & & a_n \end{pmatrix} \begin{pmatrix} b_1 & & & \\ & b_2 & & \\ & & \ddots & \\ & & & b_n \end{pmatrix} = \begin{pmatrix} a_1b_1 & & & \\ & a_2b_2 & & \\ & & \ddots & \\ & & & a_nb_n \end{pmatrix}.$$

Definition 2.8

An nth-order matrix $\boldsymbol{A} = (a_{ij})_{n\times n}$ is called an **upper triangular matrix** if $a_{ij} = 0 \quad (j = 1, 2, \ldots, n-1)$ for $i > j$. It is called a **lower triangular matrix** if $a_{ij} = 0$ for $(j = 2, 3, \ldots, n)$ for $i < j$.

Example 2.7 Product of two upper triangular matrices

Prove that the product of two upper triangular matrices (of same order) is still an upper triangular matrix (of that order).

Proof: Suppose

$$\boldsymbol{A} = \begin{pmatrix} a_{11} & a_{12} & \cdots & a_{1n} \\ & a_{22} & \cdots & a_{2n} \\ & & \ddots & \vdots \\ & & & a_{nn} \end{pmatrix}, \qquad \boldsymbol{B} = \begin{pmatrix} b_{11} & b_{12} & \cdots & b_{1n} \\ & b_{22} & \cdots & b_{2n} \\ & & \ddots & \vdots \\ & & & b_{nn} \end{pmatrix},$$

then

$$\boldsymbol{AB} = \boldsymbol{C} = (c_{ij})_{n\times n}.$$

When $i > j$, $c_{ij} = \sum_{k=1}^{n} a_{ik}b_{kj} = \sum_{k=1}^{i-1} a_{ik}b_{kj} + \sum_{k=i}^{n} a_{ik}b_{kj}$, in the first summation on the *RHS* of the above formula, $a_{ik} = 0$ $(k = 1, 2, \ldots, i-1)$ because $\boldsymbol{A}$ is an upper triangular matrix. With the same argument, in the second summation on the *RHS*, $b_{kj} = 0 \quad (k = i,\ i + 1, \ldots,\ n$, all greater than $j)$. So, $c_{ij} = 0$ when $i > j$, and consequently, $\boldsymbol{C}$ is an upper triangular matrix. ■

In the same way, we can prove that the product of two lower triangular matrices is again a lower triangular matrix.

Using the multiplication of the matrices, a system of linear equations can be expressed concisely in terms of matrix notation.

Consider a system of linear equations,

$$\begin{cases} a_{11}x_1 & + & a_{12}x_2 & + & \cdots & + & a_{1n}x_n & = & b_1, \\ a_{21}x_1 & + & a_{22}x_2 & + & \cdots & + & a_{2n}x_n & = & b_2, \\ \cdots & & \cdots & & \cdots & & \cdots & & \cdots \\ a_{m1}x_1 & + & a_{m2}x_2 & + & \cdots & + & a_{mn}x_n & = & b_m. \end{cases} \tag{2.15}$$

Note that the ith equation of the system can be expressed as

$$(a_{i1}, a_{i2}, \ldots, a_{in}) \begin{pmatrix} x_1 \\ x_2 \\ \vdots \\ x_n \end{pmatrix} = b_i \quad i = 1, 2, \ldots, m.$$

Hence, the system of equations (2.15) as a whole can be denoted as

$$\begin{pmatrix} a_{11} & a_{12} & \cdots & a_{1n} \\ a_{21} & a_{22} & \cdots & a_{2n} \\ \vdots & \vdots & & \vdots \\ a_{m1} & a_{m2} & \cdots & a_{mn} \end{pmatrix} \begin{pmatrix} x_1 \\ x_2 \\ \vdots \\ x_n \end{pmatrix} = \begin{pmatrix} b_1 \\ b_2 \\ \vdots \\ b_m \end{pmatrix}.$$

Denote

$$\boldsymbol{A} = \begin{pmatrix} a_{11} & a_{12} & \cdots & a_{1n} \\ a_{21} & a_{22} & \cdots & a_{2n} \\ \vdots & \vdots & & \vdots \\ a_{m1} & a_{m2} & \cdots & a_{mn} \end{pmatrix}, \quad \boldsymbol{x} = \begin{pmatrix} x_1 \\ x_2 \\ \vdots \\ x_n \end{pmatrix}, \quad \boldsymbol{b} = \begin{pmatrix} b_1 \\ b_2 \\ \vdots \\ b_m \end{pmatrix}. \tag{2.16}$$

Then, the system of linear equations (2.15) can be briefly expressed in the matrix form

$$\boldsymbol{Ax} = \boldsymbol{b}, \tag{2.17}$$

where $\boldsymbol{A}$ is called the **coefficient matrix** of (2.15).

The above result throws some light on the determinant of the product of two matrices.

THEOREM 2.1

Let $\boldsymbol{A}$ and $\boldsymbol{B}$ be two nth-order matrices, then the determinant of product matrix $\boldsymbol{AB}$ is equal to the product of the determinants of $\boldsymbol{A}$ and $\boldsymbol{B}$, that is,

$$|\boldsymbol{AB}| = |\boldsymbol{A}|\,|\boldsymbol{B}|.$$

Proof: Let $\boldsymbol{A} = (a_{ij})_{n\times n}$ and $\boldsymbol{B} = (b_{ij})_{n\times n}$. Using the result of Example 1.11 in §1.2, we have

$$|\boldsymbol{A}|\,|\boldsymbol{B}| = \begin{vmatrix} a_{11} & a_{12} & \cdots & a_{1n} & 0 & 0 & \cdots & 0 \\ a_{21} & a_{22} & \cdots & a_{2n} & 0 & 0 & \cdots & 0 \\ \vdots & \vdots & & \vdots & \vdots & \vdots & & \vdots \\ a_{n1} & a_{n2} & \cdots & a_{nn} & 0 & 0 & \cdots & 0 \\ -1 & 0 & \cdots & 0 & b_{11} & b_{12} & \cdots & b_{1n} \\ 0 & -1 & \cdots & 0 & b_{21} & b_{22} & \cdots & b_{2n} \\ \vdots & \vdots & \ddots & \vdots & \vdots & \vdots & & \vdots \\ 0 & 0 & \cdots & -1 & b_{n1} & b_{n2} & \cdots & b_{nn} \end{vmatrix}_{2n\times 2n}.$$

Adding the $(n+1)$th row multiplied by a_{11} to row ①, adding the $(n+2)$th row multiplied by a_{12} to row ①, $\cdots$, and adding the $(2n)$th row multiplied by a_{1n} to row ①, we get

$$|\boldsymbol{A}|\,|\boldsymbol{B}| = \begin{vmatrix} 0 & 0 & \cdots & 0 & c_{11} & c_{12} & \cdots & c_{1n} \\ a_{21} & a_{22} & \cdots & a_{2n} & 0 & 0 & \cdots & 0 \\ \vdots & \vdots & & \vdots & \vdots & \vdots & & \vdots \\ a_{n1} & a_{n2} & \cdots & a_{nn} & 0 & 0 & \cdots & 0 \\ -1 & 0 & \cdots & 0 & b_{11} & b_{12} & \cdots & b_{1n} \\ 0 & -1 & \cdots & 0 & b_{21} & b_{22} & \cdots & b_{2n} \\ \vdots & \vdots & \ddots & \vdots & \vdots & \vdots & & \vdots \\ 0 & 0 & \cdots & -1 & b_{n1} & b_{n2} & \cdots & b_{nn} \end{vmatrix},$$

where

$$c_{1j} = a_{11}b_{1j} + a_{12}b_{2j} + \cdots + a_{1n}b_{nj} = \sum_{k=1}^{n} a_{1k}b_{kj} \ (j = 1, 2, \ldots, n).$$

That is, $c_{11}, c_{12}, \ldots, c_{1n}$ is the first row of $\boldsymbol{AB}$. Following the above process, transforming all of $a_{21}, a_{22}, \ldots, a_{2n}, \ldots, a_{n1}, a_{n2}, \ldots, a_{nn}$ into zeros in the determinant, we get

$$\begin{aligned} |\boldsymbol{A}|\,|\boldsymbol{B}| &= \begin{vmatrix} \boldsymbol{A} & \boldsymbol{0} \\ -\boldsymbol{I}_n & \boldsymbol{B} \end{vmatrix} = \begin{vmatrix} \boldsymbol{0} & \boldsymbol{AB} \\ -\boldsymbol{I}_n & \boldsymbol{B} \end{vmatrix} = (-1)^n \begin{vmatrix} \boldsymbol{AB} & \boldsymbol{0} \\ \boldsymbol{B} & -\boldsymbol{I}_n \end{vmatrix} \\ &= (-1)^n\,|\boldsymbol{AB}|\,|-\boldsymbol{I}_n| = (-1)^n\,|\boldsymbol{AB}|\,(-1)^n \\ &= |\boldsymbol{AB}|. \end{aligned}$$

□

The following two examples present an application of the above result.

***Example 2.8 Application of Theorem 2.1 (I)**

Let

$$A = \begin{pmatrix} a & -b & -c & -d \\ b & a & -d & c \\ c & d & a & -b \\ d & -c & b & a \end{pmatrix}.$$

Compute $(\det A)^2$ and $\det A$ (i.e., $|A|^2$ and $|A|$).

Solution: Let the rows and columns interchange, and denote the new matrix as A^{T}, that is,

$$A^{\mathrm{T}} = \begin{pmatrix} a & b & c & d \\ -b & a & d & -c \\ -c & -d & a & b \\ -d & c & -b & a \end{pmatrix}.$$

Since $\left|A^{\mathrm{T}}\right| = |A|$, so we have

$$\begin{aligned} |A|^2 &= |A|\left|A^{\mathrm{T}}\right| = \left|AA^{\mathrm{T}}\right| \\ &= \begin{vmatrix} a^2+b^2+c^2+d^2 & 0 & 0 & 0 \\ 0 & a^2+b^2+c^2+d^2 & 0 & 0 \\ 0 & 0 & a^2+b^2+c^2+d^2 & 0 \\ 0 & 0 & 0 & a^2+b^2+c^2+d^2 \end{vmatrix} \\ &= (a^2+b^2+c^2+d^2)^4. \end{aligned}$$

So, $|A| = \pm(a^2+b^2+c^2+d^2)^2$. But, the leading diagonal elements of A are all a and the sign of a^4 in $|A|$ is "+"; hence,

$$|A| = (a^2+b^2+c^2+d^2)^2.$$

■

Example 2.9 Application of Theorem 2.1 (II)

Suppose

$$A = \begin{pmatrix} a_{11} & a_{12} & \cdots & a_{1n} \\ a_{21} & a_{22} & \cdots & a_{2n} \\ \vdots & \vdots & & \vdots \\ a_{n1} & a_{n2} & \cdots & a_{nn} \end{pmatrix}, \quad A^* = \begin{pmatrix} A_{11} & A_{21} & \cdots & A_{n1} \\ A_{12} & A_{22} & \cdots & A_{n2} \\ \vdots & \vdots & & \vdots \\ A_{1n} & A_{2n} & \cdots & A_{nn} \end{pmatrix}, \tag{2.18}$$

where A_{ij} is the algebraic complement of element a_{ij} in determinant $|\boldsymbol{A}|$. Prove that $|\boldsymbol{A}^*| = |\boldsymbol{A}|^{n-1}$ when $|\boldsymbol{A}| \neq 0$.

Proof: Suppose $\boldsymbol{AA}^* = \boldsymbol{C} = (c_{ij})$, in which

$$c_{ij} = a_{i1}A_{j1} + a_{i2}A_{j2} + \cdots + a_{in}A_{jn} = \begin{cases} |\boldsymbol{A}|, & \text{when } j = i \\ 0, & \text{when } j \neq i \end{cases} \quad i, j = 1, 2, \ldots, n.$$

Then,

$$\boldsymbol{AA}^* = \begin{pmatrix} |\boldsymbol{A}| & & & \\ & |\boldsymbol{A}| & & \\ & & \ddots & \\ & & & |\boldsymbol{A}| \end{pmatrix} = |\boldsymbol{A}|\boldsymbol{I}_n. \tag{2.19}$$

So,

$$|\boldsymbol{A}|\,|\boldsymbol{A}^*| = |\boldsymbol{AA}^*| = |\boldsymbol{A}|^n.$$

Since $|\boldsymbol{A}| \neq 0$, we have $|\boldsymbol{A}^*| = |\boldsymbol{A}|^{n-1}$. ∎

Finally, we define the power of a square matrix and the polynomial of a square matrix.

Definition 2.9

For an nth-order matrix $\boldsymbol{A}$ and a positive integer k, the kth power of $\boldsymbol{A}$, written as $\boldsymbol{A}^k$, is

$$\boldsymbol{A}^k = \underbrace{\boldsymbol{AA}\cdots\boldsymbol{A}}_{k \text{ times}}.$$

When m and k are positive integers, from Definition 2.9, we have

$$\boldsymbol{A}^m\boldsymbol{A}^k = \boldsymbol{A}^{m+k}, \tag{2.20}$$

$$(\boldsymbol{A}^m)^k = \boldsymbol{A}^{mk}. \tag{2.21}$$

When $\boldsymbol{AB}$ is not commutative, in general, $(\boldsymbol{AB})^k \neq \boldsymbol{A}^k\boldsymbol{B}^k$. When $\boldsymbol{AB}$ is commutative, $(\boldsymbol{AB})^k = \boldsymbol{A}^k\boldsymbol{B}^k = \boldsymbol{B}^k\boldsymbol{A}^k$, but the converse proposition is not true.

Definition 2.10

Suppose $f(x) = a_k x^k + a_{k-1} x^{k-1} + \cdots + a_1 x + a_0$ is a kth-degree polynomial in x, and $\boldsymbol{A}$ is an nth-order matrix, then

$$f(\boldsymbol{A}) = a_k \boldsymbol{A}^k + a_{k-1} \boldsymbol{A}^{k-1} + \cdots + a_1 \boldsymbol{A} + a_0 \boldsymbol{I},$$

which is called the kth-degree **polynomial** of $\boldsymbol{A}$ (Hint: Notice that the constant term should be changed to $a_0 \boldsymbol{I}$, where $\boldsymbol{I}$ is the nth-order identity matrix).

If $f(x)$ and $g(x)$ are polynomials, $\boldsymbol{A}$ and $\boldsymbol{B}$ are both nth-order matrices, then

$$f(\boldsymbol{A}) g(\boldsymbol{A}) = g(\boldsymbol{A}) f(\boldsymbol{A}).$$

For example,

$$\begin{aligned}(\boldsymbol{A}^2 + \boldsymbol{A} - 2\boldsymbol{I})(\boldsymbol{A} - \boldsymbol{I}) &= (\boldsymbol{A} - \boldsymbol{I})(\boldsymbol{A}^2 + \boldsymbol{A} - 2\boldsymbol{I}) \\ &= \boldsymbol{A}^3 - 3\boldsymbol{A} + 2\boldsymbol{I}.\end{aligned}$$

In general, $f(\boldsymbol{A}) g(\boldsymbol{B}) \neq g(\boldsymbol{B}) f(\boldsymbol{A})$ when $\boldsymbol{AB}$ is not commutative. For example,

$$\begin{aligned}(\boldsymbol{A} + \boldsymbol{B})(\boldsymbol{A} - \boldsymbol{B}) &= \boldsymbol{A}^2 - \boldsymbol{AB} + \boldsymbol{BA} - \boldsymbol{B}^2 (\neq \boldsymbol{A}^2 - \boldsymbol{B}^2) \\ &\neq \boldsymbol{A}^2 - \boldsymbol{BA} + \boldsymbol{AB} - \boldsymbol{B}^2 = (\boldsymbol{A} - \boldsymbol{B})(\boldsymbol{A} + \boldsymbol{B}), \\ (\boldsymbol{A} + \boldsymbol{B})^2 &= (\boldsymbol{A} + \boldsymbol{B})(\boldsymbol{A} + \boldsymbol{B}) \\ &= \boldsymbol{A}^2 + \boldsymbol{AB} + \boldsymbol{BA} + \boldsymbol{B}^2 \neq \boldsymbol{A}^2 + 2\boldsymbol{AB} + \boldsymbol{B}^2.\end{aligned}$$

Since the scalar matrix $\lambda \boldsymbol{I}$ is commutative with any arbitrary square matrix, we can expand the following power matrix by binomial theorem

$$(\boldsymbol{A} + \lambda \boldsymbol{I})^n = \boldsymbol{A}^n + \mathrm{C}_n^1 \lambda \boldsymbol{A}^{n-1} + \mathrm{C}_n^2 \lambda \boldsymbol{A}^{n-2} + \cdots + \mathrm{C}_n^{n-1} \lambda^{n-1} \boldsymbol{A} + \lambda^n \boldsymbol{I}.$$

Notice that $\boldsymbol{A}^2$ has no meaning when $m \neq n$ for an $m \times n$ matrix $\boldsymbol{A}$.

2.3 Transpose of a Matrix and Symmetric Matrix

Definition 2.11

The matrix obtained by interchanging the rows and columns of

$$\boldsymbol{A} = \begin{pmatrix} a_{11} & a_{12} & \cdots & a_{1n} \\ a_{21} & a_{22} & \cdots & a_{2n} \\ \vdots & \vdots & & \vdots \\ a_{m1} & a_{m2} & \cdots & a_{mn} \end{pmatrix},$$

is called the **transpose of matrix *A***, written as $\boldsymbol{A}^{\mathrm{T}}$ or $\boldsymbol{A}'$, that is,

$$\boldsymbol{A}^{\mathrm{T}} = \begin{pmatrix} a_{11} & a_{21} & \cdots & a_{m1} \\ a_{12} & a_{22} & \cdots & a_{m2} \\ \vdots & \vdots & & \vdots \\ a_{1n} & a_{2n} & \cdots & a_{mn} \end{pmatrix}.$$

From Definition 2.11,

$$a_{ji}{}^{\mathrm{T}} = a_{ij} \quad (i = 1, 2, \ldots, m;\ j = 1, 2 \cdots, n)$$

if we denote $\boldsymbol{A} = (a_{ij})_{m\times n}$, $\boldsymbol{A}^{\mathrm{T}} = (a_{ji}{}^{\mathrm{T}})_{n\times m}$.

The transpose operation of a matrix satisfies the following rules:

(i) $(\boldsymbol{A}^{\mathrm{T}})^{\mathrm{T}} = \boldsymbol{A}$;
(ii) $(\boldsymbol{A}+\boldsymbol{B})^{\mathrm{T}} = \boldsymbol{A}^{\mathrm{T}} + \boldsymbol{B}^{\mathrm{T}}$;
(iii) $(k\boldsymbol{A})^{\mathrm{T}} = k\boldsymbol{A}^{\mathrm{T}}$($k$ is a number);
(iv) $(\boldsymbol{AB})^{\mathrm{T}} = \boldsymbol{B}^{\mathrm{T}}\boldsymbol{A}^{\mathrm{T}}$.

As the proofs of (i), (ii), and (iii) are obvious, we will only present the proof of rule (iv).

Proof: Suppose

$$\boldsymbol{A} = (a_{ij})_{m\times n}, \quad \boldsymbol{B} = (b_{ij})_{n\times s}, \quad \boldsymbol{A}^{\mathrm{T}} = (a_{ji}{}^{\mathrm{T}})_{n\times m}, \quad \boldsymbol{B}^{\mathrm{T}} = (b_{ji}{}^{\mathrm{T}})_{s\times n}.$$

Then, $(\boldsymbol{AB})^{\mathrm{T}}$ and $\boldsymbol{B}^{\mathrm{T}}\boldsymbol{A}^{\mathrm{T}}$ are both $s \times m$ matrices, and according to

$$a_{ji}{}^{\mathrm{T}} = a_{ij}, \qquad b_{ji}{}^{\mathrm{T}} = b_{ij},$$

we see that

$$(\boldsymbol{B}^{\mathrm{T}}\boldsymbol{A}^{\mathrm{T}})_{ji} = \sum_{k=1}^{n} b_{jk}{}^{\mathrm{T}} a_{ki}{}^{\mathrm{T}} = \sum_{k=1}^{n} a_{ik}b_{kj} = (\boldsymbol{AB})_{ij} = (\boldsymbol{AB})_{ji}{}^{\mathrm{T}}$$

$$j = 1, 2, \ldots, s;\ i = 1, 2, \ldots, m.$$

So,

$$(\boldsymbol{AB})^{\mathrm{T}} = \boldsymbol{B}^{\mathrm{T}}\boldsymbol{A}^{\mathrm{T}}. \qquad \square$$

From rule (iv), we can prove $(\boldsymbol{A}_1\boldsymbol{A}_2\cdots\boldsymbol{A}_k)^{\mathrm{T}} = \boldsymbol{A}_k{}^{\mathrm{T}}\cdots\boldsymbol{A}_2{}^{\mathrm{T}}\boldsymbol{A}_1{}^{\mathrm{T}}$ by mathematical induction.

Definition 2.12

Let

$$A = \begin{pmatrix} a_{11} & a_{12} & \cdots & a_{1n} \\ a_{21} & a_{22} & \cdots & a_{2n} \\ \vdots & \vdots & & \vdots \\ a_{n1} & a_{n2} & \cdots & a_{nn} \end{pmatrix}$$

be an nth-order matrix. $\boldsymbol{A}$ is called a **symmetric matrix** if $a_{ij} = a_{ji}$ $(i, j = 1, 2, \cdots, n)$, and an **antisymmetric matrix** if $a_{ij} = -a_{ji}$ $(i, j = 1, 2, \ldots, n)$.

The leading diagonal elements a_{ii} of an antisymmetric matrix $\boldsymbol{A}$ are all zeros, since $a_{ii} = -a_{ii}$ $(i = 1, 2, \ldots, n)$. According to Definitions 2.11 and 2.12, we can easily prove that $\boldsymbol{A}$ is a symmetric matrix, if and only if, $\boldsymbol{A}^{\mathrm{T}} = \boldsymbol{A}$; and $\boldsymbol{A}$ is an antisymmetric matrix, if and only if, $\boldsymbol{A}^{\mathrm{T}} = -\boldsymbol{A}$.

Example 2.10 Transpose of symmetric matrices

Let $\boldsymbol{B}$ be an $m \times n$ matrix, then $\boldsymbol{B}^{\mathrm{T}}\boldsymbol{B}$ and $\boldsymbol{B}\boldsymbol{B}^{\mathrm{T}}$ are both symmetric matrices.

Since $\boldsymbol{B}^{\mathrm{T}}\boldsymbol{B}$ is an nth-order matrix, and

$$(\boldsymbol{B}^{\mathrm{T}}\boldsymbol{B})^{\mathrm{T}} = \boldsymbol{B}^{\mathrm{T}}(\boldsymbol{B}^{\mathrm{T}})^{\mathrm{T}} = \boldsymbol{B}^{\mathrm{T}}\boldsymbol{B}.$$

A similar argument shows that $\boldsymbol{B}\boldsymbol{B}^{\mathrm{T}}$ is an mth-order symmetric matrix. ■

Example 2.11 Transpose of antisymmetric matrices

Let $\boldsymbol{A}$ be an nth-order antisymmetric matrix, $\boldsymbol{B}$ an nth-order symmetric matrix, then $\boldsymbol{AB} + \boldsymbol{BA}$ is an nth-order antisymmetric matrix. This is because

$$\begin{aligned}(\boldsymbol{AB} + \boldsymbol{BA})^{\mathrm{T}} &= (\boldsymbol{AB})^{\mathrm{T}} + (\boldsymbol{BA})^{\mathrm{T}} = \boldsymbol{B}^{\mathrm{T}}\boldsymbol{A}^{\mathrm{T}} + \boldsymbol{A}^{\mathrm{T}}\boldsymbol{B}^{\mathrm{T}} \\ &= \boldsymbol{B}(-\boldsymbol{A}) + (-\boldsymbol{A})\boldsymbol{B} = -(\boldsymbol{AB} + \boldsymbol{BA}).\end{aligned}$$ ■

Notice that the product of symmetric matrices is not necessarily a symmetric matrix. We can easily prove that for symmetric matrices $\boldsymbol{A}$ and $\boldsymbol{B}$, the product $\boldsymbol{AB}$ is symmetric, if and only if, $\boldsymbol{AB}$ is commutative.

2.4 Inverse of a Matrix and Invertible Matrix

In matrix operations, we can define subtraction of two matrices after having defined the addition of two matrices and the negative of a matrix. Can we define a "division" of a matrix after having defined the multiplication of two matrices? Since matrix multiplication is not commutative, we cannot define "division" of matrices as we usually do. However, we do know that $aa^{-1}=a^{-1}a=1$, when $a\neq 0$ in the case of operations among numbers, where $a^{-1}=1/a$. We call a^{-1} the reciprocal of a (or the inverse of a).

In the operation of matrix multiplication, the identity matrix $\boldsymbol{I}$ plays the role corresponding to 1 in the case of multiplication operation among numbers. Then, it is quite natural to find whether there exists a matrix $\boldsymbol{A}^{-1}$ corresponding to $\boldsymbol{A}$, such that $\boldsymbol{A}\boldsymbol{A}^{-1}=\boldsymbol{A}^{-1}\boldsymbol{A}=\boldsymbol{I}$? If there exists such a matrix $\boldsymbol{A}^{-1}$, we call $\boldsymbol{A}$ an invertible matrix, and we call $\boldsymbol{A}^{-1}$ the inverse of $\boldsymbol{A}$. Next, we will define the invertible matrix and its inverse matrix, discuss the condition such that a matrix is invertible, and describe methods of finding an inverse matrix.

Definition 2.13

For a matrix $\boldsymbol{A}\in F^{n\times n}$, if there exists a matrix $\boldsymbol{B}\in F^{n\times n}$ such that

$$\boldsymbol{A}\boldsymbol{B}=\boldsymbol{B}\boldsymbol{A}=\boldsymbol{I}_n, \tag{2.22}$$

then, we call $\boldsymbol{A}$ an **invertible matrix** ($\boldsymbol{A}$ is invertible, for short), and $\boldsymbol{B}$ an inverse matrix of $\boldsymbol{A}$. Henceforth, when there is no possibility of confusion, we shall write $\boldsymbol{I}$ for $\boldsymbol{I}_n$.

From Definition 2.13, an invertible matrix and its inverse matrix are square matrices with the same order. In (2.22), the position of $\boldsymbol{A}$ and $\boldsymbol{B}$ is symmetric, so we can also call $\boldsymbol{A}$ an inverse matrix of $\boldsymbol{B}$. Using Definition 2.13 directly, an inverse matrix of identity matrix $\boldsymbol{I}$ is also $\boldsymbol{I}$ itself.

THEOREM 2.2

If $\boldsymbol{A}$ is an invertible matrix, then its inverse is unique.

Proof: Let both $\boldsymbol{B}$ and $\boldsymbol{C}$ be inverse matrices of $\boldsymbol{A}$, then from

$$\boldsymbol{A}\boldsymbol{B}=\boldsymbol{B}\boldsymbol{A}=\boldsymbol{I},$$

and

$$\boldsymbol{A}\boldsymbol{C}=\boldsymbol{C}\boldsymbol{A}=\boldsymbol{I},$$

we have

$$B = IB = (CA)B = C(AB) = CI = C.$$

So, the inverse matrix of A is unique. □

This unique inverse of A is written as A^{-1}, that is, if B is the inverse of A, we write, $A^{-1} = B$.

Next, we will give a necessary and sufficient condition under which a matrix A is invertible.

If A is invertible, from expression (2.22), we have $|A|\,|B| = |I| = 1$ and hence, $|A| \neq 0$. Thus, $|A| \neq 0$ is a necessary condition for the invertibility of A. $|A| \neq 0$ is also a sufficient condition for it. In order to prove this conclusion, we introduce the definition of the adjoint matrix of A.

Definition 2.14

For an nth-order matrix $A = (a_{ij})_{n\times n}$, let A_{ij} be the algebraic complement of a_{ij} in the determinant $\det A$.
Denote

$$\operatorname{cof} A = (A_{ij})_{n\times n}$$

as the matrix of algebraic complements of A. The transpose matrix of $\operatorname{cof} A$ is called the **adjoint matrix of** A, and denoted by $\operatorname{adj} A$ or A^*, that is,

$$A^* = (\operatorname{cof} A)^{\mathrm{T}} = \begin{pmatrix} A_{11} & A_{21} & \cdots & A_{n1} \\ A_{12} & A_{22} & \cdots & A_{n2} \\ \vdots & \vdots & & \vdots \\ A_{1n} & A_{2n} & \cdots & A_{nn} \end{pmatrix}.$$

In Example 2.9 of §2.2, we have proved that $AA^* = |A|\,I$ [see expression (2.19)], and we can prove in the same way, $A^*A = |A|\,I$, where

$$AA^* = A^*A = |A|\,I. \tag{2.23}$$

When $|A| \neq 0$, using (2.23), we can get

$$A\left(\frac{1}{|A|}A^*\right) = \left(\frac{1}{|A|}A^*\right)A = I, \tag{2.24}$$

which shows that A is invertible when $|A| \neq 0$ and

$$A^{-1} = \frac{1}{|A|}A^*. \tag{2.25}$$

In conclusion, we obtain the following theorem.

THEOREM 2.3

$\boldsymbol{A}$ is invertible, if and only if, $|\boldsymbol{A}| \neq 0$ and in that case,

$$\boldsymbol{A}^{-1} = \frac{1}{|\boldsymbol{A}|}\boldsymbol{A}^*.$$

COROLLARY 2.1

If $\boldsymbol{A}$ and $\boldsymbol{B}$ are both nth-order matrices and $\boldsymbol{AB} = \boldsymbol{I}$, then $\boldsymbol{BA} = \boldsymbol{I}$, which means $\boldsymbol{A}$ and $\boldsymbol{B}$ are both invertible, and $\boldsymbol{A}$ and $\boldsymbol{B}$ are inverse matrices of each other.

Proof: Since $\boldsymbol{AB} = \boldsymbol{I}$, we have $|\boldsymbol{A}|\,|\boldsymbol{B}| = 1$, $|\boldsymbol{A}| \neq 0$, $|\boldsymbol{B}| \neq 0$ according to Theorem 2.3. Then, $\boldsymbol{A}$ and $\boldsymbol{B}$ are both invertible and

$$\begin{aligned} &\boldsymbol{AB} = \boldsymbol{I} \\ \Rightarrow\ &\boldsymbol{A}^{-1}\boldsymbol{ABA} = \boldsymbol{A}^{-1}\boldsymbol{IA} \\ \Rightarrow\ &\boldsymbol{BA} = \boldsymbol{I}. \end{aligned}$$

□

We can immediately conclude from Theorem 2.3 that a necessary and sufficient condition for invertibility of diagonal and upper (lower) triangular matrices is that their **leading diagonal elements** $a_{11}, a_{22}, \ldots, a_{nn}$ **are all nonzeros**.

Theorem 2.3 not only gives us the necessary and sufficient condition for an invertible matrix $\boldsymbol{A}$, but also provides a method for finding $\boldsymbol{A}^{-1}$. The corollary of Theorem 2.3 tells us that we only need to check one of the equalities $\boldsymbol{AB} = \boldsymbol{I}$ and $\boldsymbol{BA} = \boldsymbol{I}$ to judge an invertible matrix.

The inverse of a matrix satisfies the following operational rules (let both $\boldsymbol{A}$ and $\boldsymbol{B}$ be invertible, $k \neq 0$ be a number):

(i) $(\boldsymbol{A}^{-1})^{-1} = \boldsymbol{A}$;
(ii) $(k\boldsymbol{A})^{-1} = k^{-1}\boldsymbol{A}^{-1}$;
(iii) $(\boldsymbol{AB})^{-1} = \boldsymbol{B}^{-1}\boldsymbol{A}^{-1}$;
(iv) $(\boldsymbol{A}^{\mathrm{T}})^{-1} = (\boldsymbol{A}^{-1})^{\mathrm{T}}$;
(v) $\det(\boldsymbol{A}^{-1}) = 1/\det\boldsymbol{A}$, that is, $|\boldsymbol{A}^{-1}| = |\boldsymbol{A}|^{-1}$.

Only the proofs of the operational rules (iii), (iv), and (v) are shown here, while the first two are left to readers as exercises.

Since $|\boldsymbol{AB}| = |\boldsymbol{A}|\,|\boldsymbol{B}| \neq 0$, then $\boldsymbol{AB}$ is invertible and

$$(\boldsymbol{AB})(\boldsymbol{B}^{-1}\boldsymbol{A}^{-1}) = \boldsymbol{A}(\boldsymbol{BB}^{-1})\boldsymbol{A}^{-1} = \boldsymbol{AIA}^{-1} = \boldsymbol{AA}^{-1} = \boldsymbol{I}.$$

Thus, $(\boldsymbol{AB})^{-1} = \boldsymbol{B}^{-1}\boldsymbol{A}^{-1}$. Rule (iii) can be extended to a serial product of several invertible matrices,

$$(\boldsymbol{A}_1\boldsymbol{A}_2\cdots\boldsymbol{A}_m)^{-1} = \boldsymbol{A}_m^{-1}\cdots\boldsymbol{A}_2{}^{-1}\boldsymbol{A}_1{}^{-1},$$

where $\boldsymbol{A}_1, \boldsymbol{A}_2, \ldots, \boldsymbol{A}_m$ are all invertible.

As $|\boldsymbol{A}^{\mathrm{T}}| = |\boldsymbol{A}| \neq 0$, so $\boldsymbol{A}^{\mathrm{T}}$ is invertible and

$$(\boldsymbol{AA}^{-1})^{\mathrm{T}} = \boldsymbol{I}, \text{ that is, } (\boldsymbol{A}^{-1})^{\mathrm{T}}\boldsymbol{A}^{\mathrm{T}} = \boldsymbol{I},$$

which shows $(\boldsymbol{A}^{\mathrm{T}})^{-1} = (\boldsymbol{A}^{-1})^{\mathrm{T}}$.

Since $\boldsymbol{AA}^{-1} = \boldsymbol{I}$ implies that $|\boldsymbol{A}|\,|\boldsymbol{A}^{-1}| = 1$, then $|\boldsymbol{A}| \neq 0$ and

$$|\boldsymbol{A}^{-1}| = \frac{1}{|\boldsymbol{A}|} = |\boldsymbol{A}|^{-1}.$$

Note that $\boldsymbol{A}+\boldsymbol{B}$ may not be invertible even though $\boldsymbol{A}$ and $\boldsymbol{B}$ are both invertible. Generally speaking, $(\boldsymbol{A}+\boldsymbol{B})^{-1} \neq \boldsymbol{A}^{-1} + \boldsymbol{B}^{-1}$. For example, the diagonal matrices

$$\boldsymbol{A} = \operatorname{diag}(2, -1), \quad \boldsymbol{B} = \boldsymbol{I}_2, \quad \boldsymbol{C} = \operatorname{diag}(1, 2)$$

are all invertible, but $\boldsymbol{A}+\boldsymbol{B} = \operatorname{diag}(3, 0)$ is not invertible, while $\boldsymbol{A}+\boldsymbol{C} = \operatorname{diag}(3, 1)$ is invertible, whereas

$$(\boldsymbol{A}+\boldsymbol{C})^{-1} = \operatorname{diag}\left(\frac{1}{3}, 1\right) \neq \boldsymbol{A}^{-1} + \boldsymbol{C}^{-1} = \operatorname{diag}\left(\frac{3}{2}, -\frac{1}{2}\right).$$

Example 2.12 Computing an inverse matrix

Are the following matrices invertible? Find the inverse matrix if it is invertible, where

$$\boldsymbol{A} = \begin{pmatrix} 3 & 2 & 1 \\ 1 & 1 & 1 \\ 1 & 0 & 1 \end{pmatrix}, \quad \boldsymbol{B} = \begin{pmatrix} b_1 & & \\ & b_2 & \\ & & b_3 \end{pmatrix}.$$

Solution: Since $|\boldsymbol{A}| = 2$, this shows that $\boldsymbol{A}$ is invertible. The algebraic complements of $\boldsymbol{A}$ are

$$A_{11} = \begin{vmatrix} 1 & 1 \\ 0 & 1 \end{vmatrix} = 1, \quad A_{12} = -\begin{vmatrix} 1 & 1 \\ 1 & 1 \end{vmatrix} = 0, \quad A_{13} = \begin{vmatrix} 1 & 1 \\ 1 & 0 \end{vmatrix} = -1,$$

$$A_{21} = -\begin{vmatrix} 2 & 1 \\ 0 & 1 \end{vmatrix} = -2, \quad A_{22} = \begin{vmatrix} 3 & 1 \\ 1 & 1 \end{vmatrix} = 2, \quad A_{23} = -\begin{vmatrix} 3 & 2 \\ 1 & 0 \end{vmatrix} = 2,$$

$$A_{31} = \begin{vmatrix} 2 & 1 \\ 1 & 1 \end{vmatrix} = 1, \quad A_{32} = -\begin{vmatrix} 3 & 1 \\ 1 & 1 \end{vmatrix} = -2, \quad A_{33} = \begin{vmatrix} 3 & 2 \\ 1 & 1 \end{vmatrix} = 1.$$

So,

$$A^{-1} = \frac{1}{|A|}A^* = \frac{1}{2}\begin{pmatrix} 1 & -2 & 1 \\ 0 & 2 & -2 \\ -1 & 2 & 1 \end{pmatrix}.$$

If $|B| = b_1 b_2 b_3 \neq 0$, then B is invertible, the inverse matrix is still a diagonal matrix, and

$$B^{-1} = \begin{pmatrix} \frac{1}{b_1} & & \\ & \frac{1}{b_2} & \\ & & \frac{1}{b_3} \end{pmatrix}.$$

(Hint: To avoid making mistakes from calculation, readers should verify whether $AA^{-1} = I$ in order to confirm that the result is correct.) ■

Example 2.13 Inverse of a 2 × 2 matrix

Suppose

$$A = \begin{pmatrix} a_{11} & a_{12} \\ a_{21} & a_{22} \end{pmatrix}$$

and the determinant $\det A = a_{11}a_{22} - a_{12}a_{21} = d \neq 0$, then the inverse matrix

$$A^{-1} = \frac{1}{d}A^* = \frac{1}{d}\begin{pmatrix} a_{22} & -a_{12} \\ -a_{21} & a_{11} \end{pmatrix}.$$ ■

Example 2.14 Computing the inverse from a matric polynomial

Suppose $A^2 - 3A - 10I = 0$ for a square matrix A. Prove that A and $A - 4I$ are both invertible and find their inverse matrices.

Proof: We have $A(A - 3I) = 10I$ since $A^2 - 3A - 10I = 0$, that is,

$$A\left[\frac{1}{10}(A - 3I)\right] = I.$$

Then, A is invertible and $A^{-1} = \frac{1}{10}(A - 3I)$.

As $\boldsymbol{A}^2 - 3\boldsymbol{A} - 10\boldsymbol{I} = \boldsymbol{0}$, we have

$$(\boldsymbol{A}+\boldsymbol{I})(\boldsymbol{A}-4\boldsymbol{I}) = 6\boldsymbol{I},$$

where

$$\frac{1}{6}(\boldsymbol{A}+\boldsymbol{I})(\boldsymbol{A}-4\boldsymbol{I}) = \boldsymbol{I}.$$

So, $\boldsymbol{A}-4\boldsymbol{I}$ is invertible and $(\boldsymbol{A}-4\boldsymbol{I})^{-1} = \frac{1}{6}(\boldsymbol{A}+\boldsymbol{I})$. ■

Example 2.15 Solving a nonhomogenous system of linear equations

Consider the nonhomogeneous system of linear equations $\boldsymbol{Ax} = \boldsymbol{b}$. The coefficient matrix $\boldsymbol{A}$ is the one given in Example 2.12 and $\boldsymbol{b} = (5, 1, 1)^{\mathrm{T}}$. Solve the system and provide a solution if it has one.

Solution: $\boldsymbol{A}$ is an invertible matrix and the inverse matrix is unique. Hence, multiply both sides of the equality $\boldsymbol{Ax} = \boldsymbol{b}$ by $\boldsymbol{A}^{-1}$ on the left, that is,

$$\boldsymbol{A}^{-1}(\boldsymbol{Ax}) = \boldsymbol{A}^{-1}\boldsymbol{b}.$$

Then, we obtain a unique solution to the system of equations.

$$\begin{aligned} \boldsymbol{x} = \begin{pmatrix} x_1 \\ x_2 \\ x_3 \end{pmatrix} &= \boldsymbol{A}^{-1}\boldsymbol{b} \\ &= \begin{pmatrix} \frac{1}{2} & -1 & \frac{1}{2} \\ 0 & 1 & -1 \\ \frac{1}{2} & 1 & \frac{1}{2} \end{pmatrix} \begin{pmatrix} 5 \\ 1 \\ 1 \end{pmatrix} \\ &= \begin{pmatrix} 2 \\ 0 \\ -1 \end{pmatrix}. \end{aligned}$$

■

Example 2.16 Inverting an antisymmetric matrix

Prove that $\boldsymbol{A}^{-1}$ is an antisymmetric matrix if $\boldsymbol{A}$ is an invertible antisymmetric matrix.

Proof: Since

$$(A^{-1})^{\mathrm{T}} = (A^{\mathrm{T}})^{-1} = (-A)^{-1} = -A^{-1}$$

therefore, A^{-1} is an antisymmetric matrix. ■

Using the same argument, the inverse of an invertible symmetric matrix is again a symmetric matrix. Next, we give several examples from a synthesis of the above-mentioned ideas.

Example 2.17 Proof of invertibility of a nonzero real matrix

Let $A = (a_{ij})_{n\times n}$ be a nonzero real matrix. Prove that A is an invertible matrix if $A^* = A^{\mathrm{T}}$.

Proof: In order to prove that A is invertible, we only need to prove $|A| \neq 0$. From $A^* = A^{\mathrm{T}}$ and the definition of A^*, it follows that $a_{ij} = A_{ij}$. Now, from the expansion of the determinant $|A|$ along row i, we have

$$|A| = \sum_{j=1}^{n} a_{ij}A_{ij} = \sum_{j=1}^{n} a_{ij}^2.$$

Since A is a nonzero real matrix, the elements $a_{ij}(i, j = 1, 2, \ldots, n)$ of A are real numbers and are not all zeros, so $|A| \neq 0$, which shows that A is invertible. ■

Example 2.18 Identifying invertible matrices with Corollary 2.4

Let A, B, and C be nth-order matrices and $ABC = I$. Which one of ACB, BAC, BCA, CAB, and CBA equals the identity matrix I?

Solution: Based on the associative law and Corollary 2.4 (of Theorem 2.3), we have

$$BCA = (BC)A = I$$

and

$$CAB = C(AB) = I.$$ ■

Example 2.19 Problem involving the inverse of a matrix (I)

Let $\boldsymbol{A}$ be invertible and $\boldsymbol{A}^*\boldsymbol{B}=\boldsymbol{A}^{-1}+\boldsymbol{B}$. Prove that $\boldsymbol{B}$ is invertible and calculate $\boldsymbol{B}$ when

$$\boldsymbol{A}=\begin{pmatrix}2 & 6 & 0\\ 0 & 2 & 6\\ 0 & 0 & 2\end{pmatrix}.$$

Solution: Since $\boldsymbol{A}^*\boldsymbol{B}=\boldsymbol{A}^{-1}+\boldsymbol{B}=\boldsymbol{A}^{-1}+\boldsymbol{I}\boldsymbol{B}$, we have

$$(\boldsymbol{A}^*-\boldsymbol{I})\boldsymbol{B}=\boldsymbol{A}^{-1}. \qquad ①$$

Then, $|\boldsymbol{A}^*-\boldsymbol{I}|\,|\boldsymbol{B}|=|\boldsymbol{A}^{-1}|\neq 0$ implies $\boldsymbol{B}$ and $(\boldsymbol{A}^*-\boldsymbol{I})$ are invertible. From ①, we obtain

$$\begin{aligned}\boldsymbol{B}&=(\boldsymbol{A}^*-\boldsymbol{I})^{-1}\boldsymbol{A}^{-1}\\ &=\left[\boldsymbol{A}(\boldsymbol{A}^*-\boldsymbol{I})\right]^{-1}\\ &=[|\boldsymbol{A}|\boldsymbol{I}-\boldsymbol{A}]^{-1},\end{aligned}$$

where

$$\begin{aligned}|\boldsymbol{A}|\boldsymbol{I}-\boldsymbol{A}&=\begin{pmatrix}8 & & \\ & 8 & \\ & & 8\end{pmatrix}-\begin{pmatrix}2 & 6 & 0\\ 0 & 2 & 6\\ 0 & 0 & 2\end{pmatrix}\\ &=6\begin{pmatrix}1 & -1 & 0\\ 0 & 1 & -1\\ 0 & 0 & 1\end{pmatrix}.\end{aligned}$$

Following the operation rule (ii) of inverse matrix and inversion formula (2.25), we easily obtain

$$\boldsymbol{B}=\frac{1}{6}\begin{pmatrix}1 & 1 & 1\\ 0 & 1 & 1\\ 0 & 0 & 1\end{pmatrix}.$$

■

Example 2.20 Problem involving the inverse of a matrix (II)

If $\boldsymbol{A}$ and $\boldsymbol{B}$ are both nth-order invertible matrices, prove:

(1) $(\boldsymbol{A}\boldsymbol{B})^*=\boldsymbol{B}^*\boldsymbol{A}^*$;
(2) $(\boldsymbol{A}^*)^*=|\boldsymbol{A}|^{n-2}\boldsymbol{A}$.

Proof:

(1) From $|\boldsymbol{AB}| = |\boldsymbol{A}|\,|\boldsymbol{B}| \neq 0$, we know that $\boldsymbol{AB}$ is invertible. According to (2.23), we have $(\boldsymbol{AB})(\boldsymbol{AB})^* = |\boldsymbol{AB}|\,\boldsymbol{I}$.
So,

$$\begin{aligned}(\boldsymbol{AB})^* &= (\boldsymbol{AB})^{-1}\,|\boldsymbol{AB}|\,\boldsymbol{I}\\ &= |\boldsymbol{AB}|\,(\boldsymbol{AB})^{-1}\\ &= |\boldsymbol{A}|\,|\boldsymbol{B}|\,\boldsymbol{B}^{-1}\boldsymbol{A}^{-1}\\ &= |\boldsymbol{B}|\,\boldsymbol{B}^{-1}\,|\boldsymbol{A}|\,\boldsymbol{A}^{-1}\\ &= |\boldsymbol{B}|\,\frac{\boldsymbol{B}^*}{|\boldsymbol{B}|}\,|\boldsymbol{A}|\,\frac{\boldsymbol{A}^*}{|\boldsymbol{A}|}\\ &= \boldsymbol{B}^*\boldsymbol{A}^*.\end{aligned}$$

(2) From $(\boldsymbol{A}^*)^*\boldsymbol{A}^* = |\boldsymbol{A}^*|\,\boldsymbol{I}$, we have $(\boldsymbol{A}^*)^*\,|\boldsymbol{A}|\,\boldsymbol{A}^{-1} = |\boldsymbol{A}|^{n-1}\,\boldsymbol{I}$, thus

$$(\boldsymbol{A}^*)^* = |\boldsymbol{A}|^{n-2}\,\boldsymbol{A}.$$

■

2.5 Elementary Matrix Operations and Elementary Matrices

Gaussian elimination can be used to solve a system of linear equations by performing three types of operations on the augmented matrix:

(i) multiplying any row of the matrix by a nonzero scalar c,
(ii) adding any scalar multiple of a row of the matrix to another row,
(iii) switching any two rows of the matrix.

These three row operations are commonly called the **elementary row operations of a matrix**. Row operation (i) is called the **row multiplication operation**, rule (ii) is called the **row addition operation**, and rule (iii) is called the **row switching operation**.

For other matrix-related problems (such as expanding the determinant of a square matrix), we need to apply these operations on columns. In those cases, the three operations carried out on columns are called **elementary column operations**. Both elementary row and column operations are uniformly called **elementary operations**.

The elementary operations play an important role in matrix theory. Elementary operations can not only be described literally but also be expressed by multiplication operations of matrices. To appreciate this, we need to introduce the concept of the elementary matrices.

Definition 2.15

The matrix obtained from an identity matrix by an elementary operation is called an **elementary matrix**.

Corresponding to the three types of elementary row (or column) operations, we have three types of elementary matrices:

(i) **Elementary multiplication matrix**

$$\boldsymbol{E}_i(c) = \mathrm{diag}(1, \cdots, 1, c, 1, \ldots, 1).$$

$\boldsymbol{E}_i(c)$ is obtained from the identity matrix whose row (or column) i is multiplied by c $(c \neq 0)$;

(ii) **Elementary addition matrix**

$$\boldsymbol{E}_{ij}(c) = \begin{pmatrix} 1 & & & & & \\ & \ddots & & & & \\ & & 1 & & & \\ & & & \ddots & & \\ & & c & & 1 & \\ & & & & & \ddots & \\ & & & & & & 1 \end{pmatrix} \begin{matrix} \\ \\ \text{row } i \\ \\ \text{row } j \\ \\ \\ \end{matrix}.$$

$\boldsymbol{E}_{ij}(c)$ is obtained from the identity matrix by adding the ith row multiplied by c to the jth row, or by adding the jth column multiplied by c to the ith column.

(iii) **Elementary switching matrix**

$$\boldsymbol{E}_{ij} = \begin{pmatrix} 1 & & & & & & \\ & \ddots & & & & & \\ & & 0 & & & 1 & \\ & & & 1 & & & \\ & & & & \ddots & & \\ & & & & & 1 & \\ & & 1 & & & & 0 \\ & & & & & & & \ddots \\ & & & & & & & & 1 \end{pmatrix} \begin{matrix} \\ \\ \text{row } i \\ \\ \\ \\ \text{row } j \\ \\ \\ \end{matrix}.$$

$\boldsymbol{E}_{ij}$ is obtained from the identity matrix by switching row i and row j (or column i and column j).

Example 2.21 Product of elementary matrices

Compute the products of the following elementary matrices and matrices, $\boldsymbol{A} = (a_{ij})_{3\times n}$, $\boldsymbol{C} = (c_{ij})_{3\times 2}$, and $\boldsymbol{B} = (b_{ij})_{3\times 3}$.

Solution:

$$\begin{pmatrix} 1 & 0 & 0 \\ 0 & c & 0 \\ 0 & 0 & 1 \end{pmatrix} \begin{pmatrix} a_{11} & a_{12} & \cdots & a_{1n} \\ a_{21} & a_{22} & \cdots & a_{2n} \\ a_{31} & a_{32} & \cdots & a_{3n} \end{pmatrix} = \begin{pmatrix} a_{11} & a_{12} & \cdots & a_{1n} \\ ca_{21} & ca_{22} & \cdots & ca_{2n} \\ a_{31} & a_{32} & \cdots & a_{3n} \end{pmatrix}.$$

$$\begin{pmatrix} 1 & 0 & d \\ 0 & 1 & 0 \\ 0 & 0 & 1 \end{pmatrix} \begin{pmatrix} c_{11} & c_{12} \\ c_{21} & c_{22} \\ c_{31} & c_{32} \end{pmatrix} = \begin{pmatrix} c_{11} + dc_{31} & c_{12} + dc_{32} \\ c_{21} & c_{22} \\ c_{31} & c_{32} \end{pmatrix}.$$

$$\begin{pmatrix} b_{11} & b_{12} & b_{13} \\ b_{21} & b_{22} & b_{23} \\ b_{31} & b_{32} & b_{33} \end{pmatrix} \begin{pmatrix} 1 & 0 & 0 \\ 0 & 0 & 1 \\ 0 & 1 & 0 \end{pmatrix} = \begin{pmatrix} b_{11} & b_{13} & b_{12} \\ b_{21} & b_{23} & b_{22} \\ b_{31} & b_{33} & b_{32} \end{pmatrix}.$$ ■

From Example 2.21, the result of $\boldsymbol{A}$ and $\boldsymbol{C}$ left multiplied by an elementary matrix corresponds to executing an elementary row operation to $\boldsymbol{A}$ and $\boldsymbol{C}$. The result of $\boldsymbol{B}$ right multiplied by an elementary matrix corresponds to executing an elementary column operation to $\boldsymbol{B}$. Furthermore, if the elementary matrix is obtained from the identity matrix by a particular kind of elementary row (column) operation, $\boldsymbol{A}$ and $\boldsymbol{C}$ left multiplied ($\boldsymbol{B}$ right multiplied) by that elementary matrix corresponds to the same kind of row (column) operation to $\boldsymbol{A}$ and $\boldsymbol{C}(\boldsymbol{B})$.

Readers can easily prove the following general results:

- $\boldsymbol{E}_i(c)\boldsymbol{A}$ represents that the row i of $\boldsymbol{A}$ is being multiplied by c;
- $\boldsymbol{E}_{ij}(c)\boldsymbol{A}$ represents addition of the ith row multiplied by c to the jth row of $\boldsymbol{A}$;
- $\boldsymbol{E}_{ij}\boldsymbol{A}$ represents the switching of the positions of row i and row j;
- $\boldsymbol{B}\boldsymbol{E}_i(c)$ represents that the ith column of $\boldsymbol{B}$ is being multiplied by c;
- $\boldsymbol{B}\boldsymbol{E}_{ij}(c)$ represents the addition of the jth column multiplied by c to the ith column of $\boldsymbol{B}$;
- $\boldsymbol{B}\boldsymbol{E}_{ij}$ represents the switching of the positions of column i and column j of $\boldsymbol{B}$.

None of the determinants of elementary matrices are equal to zero, so the elementary matrices are all invertible. Indeed,

$$\boldsymbol{E}_i\left(\frac{1}{c}\right)\boldsymbol{E}_i(c) = \boldsymbol{I}, \quad \boldsymbol{E}_{ij}(-c)\boldsymbol{E}_{ij}(c) = \boldsymbol{I}, \quad \boldsymbol{E}_{ij}\boldsymbol{E}_{ij} = \boldsymbol{I},$$

show that the inverse matrix of an elementary matrix is a similar type of elementary matrix, that is,

$$E_i^{-1}(c) = E_i\left(\frac{1}{c}\right), \quad E_{ij}^{-1}(c) = E_{ij}(-c), \quad E_{ij}^{-1} = E_{ij}.$$

Example 2.22 Operation on elementary matrices

Given the following elementary matrices

$$P_1 = \begin{pmatrix} 0 & 0 & 1 & 0 \\ 0 & 1 & 0 & 0 \\ 1 & 0 & 0 & 0 \\ 0 & 0 & 0 & 1 \end{pmatrix}, \quad P_2 = \begin{pmatrix} 1 & & & \\ 0 & 1 & & \\ 0 & 0 & 1 & \\ c & 0 & 0 & 1 \end{pmatrix} \quad \text{and} \quad P_3 = \begin{pmatrix} 1 & & & \\ & k & & \\ & & 1 & \\ & & & 1 \end{pmatrix},$$

calculate $P_1P_2P_3$ and $(P_1P_2P_3)^{-1}$.

Solution: P_3 left multiplied by P_2 represents a row addition operation on P_3, P_2P_3 left multiplied by P_1 represents a row switching operation on P_2P_3, where we obtain

$$P_1P_2P_3 = \begin{pmatrix} 0 & 0 & 1 & 0 \\ 0 & 1 & 0 & 0 \\ 1 & 0 & 0 & 0 \\ 0 & 0 & 0 & 1 \end{pmatrix} \begin{pmatrix} 1 & & & \\ 0 & 1 & & \\ 0 & 0 & 1 & \\ c & 0 & 0 & 1 \end{pmatrix} \begin{pmatrix} 1 & & & \\ & k & & \\ & & 1 & \\ & & & 1 \end{pmatrix}$$

$$= \begin{pmatrix} 0 & 0 & 1 & 0 \\ 0 & 1 & 0 & 0 \\ 1 & 0 & 0 & 0 \\ 0 & 0 & 0 & 1 \end{pmatrix} \begin{pmatrix} 1 & & & \\ 0 & k & & \\ 0 & 0 & 1 & \\ c & 0 & 0 & 1 \end{pmatrix}$$

$$= \begin{pmatrix} 0 & 0 & 1 & 0 \\ 0 & k & 0 & 0 \\ 1 & 0 & 0 & 0 \\ c & 0 & 0 & 1 \end{pmatrix}.$$

$$(P_1P_2P_3)^{-1} = P_3^{-1}P_2^{-1}P_1^{-1}$$

$$= \begin{pmatrix} 1 & 0 & 0 & 0 \\ 0 & \frac{1}{k} & 0 & 0 \\ 0 & 0 & 1 & 0 \\ 0 & 0 & 0 & 1 \end{pmatrix} \begin{pmatrix} 1 & & & \\ 0 & 1 & & \\ 0 & 0 & 1 & \\ -c & 0 & 0 & 1 \end{pmatrix} \begin{pmatrix} 0 & 0 & 1 & 0 \\ 0 & 1 & 0 & 0 \\ 1 & 0 & 0 & 0 \\ 0 & 0 & 0 & 1 \end{pmatrix}$$

$$= \begin{pmatrix} 1 & & & \\ & \frac{1}{k} & & \\ & & 1 & \\ & & & 1 \end{pmatrix} \begin{pmatrix} 0 & 0 & 1 & 0 \\ 0 & 1 & 0 & 0 \\ 1 & 0 & 0 & 0 \\ 0 & 0 & -c & 1 \end{pmatrix}$$

$$= \begin{pmatrix} 0 & 0 & 1 & 0 \\ 0 & \frac{1}{k} & 0 & 0 \\ 1 & 0 & 0 & 0 \\ 0 & 0 & -c & 1 \end{pmatrix}.$$

■

In $P_1P_2P_3$, P_1 right multiplied by P_2 is equivalent to a column addition operation on P_1, while P_1P_2 right multiplied by P_3 is equivalent to a column multiplication operation on P_1P_2.

*Example 2.23 The *LU* decomposition of a matrix

Decompose the tri-diagonal matrix

$$A = \begin{pmatrix} 2 & 1 & 0 & 0 \\ 1 & 2 & 1 & 0 \\ 0 & 1 & 2 & 1 \\ 0 & 0 & 1 & 2 \end{pmatrix}$$

to the product of a lower triangular matrix whose leading diagonal elements are all 1 and an upper triangular matrix U, that is $A = LU$ (we call this the LU decomposition of a matrix).

Solution: The elementary addition matrix and its inverse are both the same type of triangular matrices whose leading diagonal elements are all 1. If we transform the elements below the leading diagonal of A to zeros by row addition operation (note that each row addition operation in this case corresponds to some elementary matrix which is a lower triangular matrix with the leading diagonal elements being 1, while A will be changed to an upper triangular matrix), then we will end up decomposing A to LU form. It is as follows:

$$\begin{pmatrix} 1 & & & \\ -\frac{1}{2} & 1 & & \\ & & 1 & \\ & & & 1 \end{pmatrix} A = \begin{pmatrix} 2 & 1 & 0 & 0 \\ 0 & \frac{3}{2} & 1 & 0 \\ 0 & 1 & 2 & 1 \\ 0 & 0 & 1 & 2 \end{pmatrix} \triangleq A_1,$$

$$\begin{pmatrix} 1 & & & \\ & 1 & & \\ & -\frac{2}{3} & 1 & \\ & & & 1 \end{pmatrix} A_1 = \begin{pmatrix} 2 & 1 & 0 & 0 \\ 0 & \frac{3}{2} & 1 & 0 \\ 0 & 0 & \frac{4}{3} & 1 \\ 0 & 0 & 1 & 2 \end{pmatrix} \xlongequal{\text{written as}} A_2,$$

$$\begin{pmatrix} 1 & & & \\ & 1 & & \\ & & 1 & \\ & & -\frac{3}{4} & 1 \end{pmatrix} A_2 = \begin{pmatrix} 2 & 1 & 0 & 0 \\ & \frac{3}{2} & 1 & 0 \\ & & \frac{4}{3} & 1 \\ & & & \frac{5}{4} \end{pmatrix} \underline{\underline{\text{written as}}}\ \boldsymbol{U}.$$

If we denote the matrices in the *LHS* of the above three expressions as L_1, L_2, L_3, then

$$L_3 L_2 L_1 A = U,$$

Thus,

$$A = (L_3 L_2 L_1)^{-1} U = LU,$$

where

$$\begin{aligned} L &= (L_3 L_2 L_1)^{-1} = L_1^{-1} L_2^{-1} L_3^{-1} \\ &= \begin{pmatrix} 1 & & & \\ \frac{1}{2} & 1 & & \\ & & 1 & \\ & & & 1 \end{pmatrix} \begin{pmatrix} 1 & & & \\ & 1 & & \\ & \frac{2}{3} & 1 & \\ & & & 1 \end{pmatrix} \begin{pmatrix} 1 & & & \\ & 1 & & \\ & & 1 & \\ & & \frac{3}{4} & 1 \end{pmatrix} \\ &= \begin{pmatrix} 1 & & & \\ \frac{1}{2} & 1 & & \\ & \frac{2}{3} & 1 & \\ & & \frac{3}{4} & 1 \end{pmatrix}. \end{aligned}$$

■

Now we provide a method for finding the inverse matrix by elementary operations.

THEOREM 2.4

An invertible matrix can be changed into the identity matrix through elementary row operations.

Proof: Gaussian elimination was mentioned in §2.1. The elimination process involves the application of three elementary row operations to the augmented matrix, which finally provides us with a row-reduced echelon matrix. Therefore, there exist elementary matrices $\boldsymbol{P}_1, \boldsymbol{P}_2, \cdots, \boldsymbol{P}_s$, such that

$$\boldsymbol{P}_s \cdots \boldsymbol{P}_2 \boldsymbol{P}_1 \boldsymbol{A} = \boldsymbol{U}.$$

When $\boldsymbol{A}$ is an nth-order invertible matrix, its row-reduced echelon matrix is also an invertible matrix (because elementary matrices are all invertible), thus $\boldsymbol{U}$ is the identity matrix $\boldsymbol{I}$. □

COROLLARY 2.2

An invertible matrix $\boldsymbol{A}$ *can be expressed as the product of elementary matrices.*

Proof: According to Theorem 2.4, there exist elementary matrices $\boldsymbol{P}_1, \boldsymbol{P}_2, \cdots, \boldsymbol{P}_s$, such that

$$\boldsymbol{P}_s \cdots \boldsymbol{P}_2 \boldsymbol{P}_1 \boldsymbol{A} = \boldsymbol{I}, \tag{2.26}$$

thus

$$\boldsymbol{A} = (\boldsymbol{P}_s \cdots \boldsymbol{P}_2 \boldsymbol{P}_1)^{-1} = \boldsymbol{P}_1^{-1} \boldsymbol{P}_2^{-1} \cdots \boldsymbol{P}_s^{-1}, \tag{2.27}$$

where $\boldsymbol{P}_1^{-1}, \boldsymbol{P}_2^{-1}, \cdots, \boldsymbol{P}_s^{-1}$ are still elementary matrices.
Hence, the corollary follows. □

From (2.26), we see that

$$\boldsymbol{A}^{-1} = \boldsymbol{P}_s \cdots \boldsymbol{P}_2 \boldsymbol{P}_1 = \boldsymbol{P}_s \cdots \boldsymbol{P}_2 \boldsymbol{P}_1 \boldsymbol{I}. \tag{2.28}$$

Thus, taking (2.26) and (2.28) into consideration, we obtain the following corollary.

COROLLARY 2.3

If we do the same elementary row operations to an invertible matrix $\boldsymbol{A}$ *and to the identity matrix* $\boldsymbol{I}$ *of the same order, then* $\boldsymbol{I}$ *is changed into* $\boldsymbol{A}^{-1}$ *when* $\boldsymbol{A}$ *is changed into identity matrix, that is,*

$$(\boldsymbol{A}, \boldsymbol{I}) \xrightarrow{\text{elementary row operations}} (\boldsymbol{I}, \boldsymbol{A}^{-1}).$$

Again, from (2.27), we have

$$\boldsymbol{A} \boldsymbol{P}_s \cdots \boldsymbol{P}_2 \boldsymbol{P}_1 = \boldsymbol{I};$$

$$\boldsymbol{I} \boldsymbol{P}_s \cdots \boldsymbol{P}_2 \boldsymbol{P}_1 = \boldsymbol{A}^{-1}.$$

Consequently, the inverse of the matrix $\boldsymbol{A}$ obtained through elementary column operations in a similar way is

$$\begin{pmatrix} \boldsymbol{A} \\ \boldsymbol{I} \end{pmatrix} \xrightarrow{\text{elementary column operations}} \begin{pmatrix} \boldsymbol{I} \\ \boldsymbol{A}^{-1} \end{pmatrix}.$$

Example 2.24 Computing an inverse matrix through elementary row operations

Find the inverse matrix of $\boldsymbol{A}$ through elementary row operations

$$\boldsymbol{A} = \begin{pmatrix} 0 & 2 & -1 \\ 1 & 1 & 2 \\ -1 & -1 & -1 \end{pmatrix}.$$

Solution:

$$(\boldsymbol{A}, \boldsymbol{I}) = \left(\begin{array}{ccc|ccc} 0 & 2 & -1 & 1 & 0 & 0 \\ 1 & 1 & 2 & 0 & 1 & 0 \\ -1 & -1 & -1 & 0 & 0 & 1 \end{array}\right)$$

$$\xrightarrow{①\rightleftharpoons②} \left(\begin{array}{ccc|ccc} 1 & 1 & 2 & 0 & 1 & 0 \\ 0 & 2 & -1 & 1 & 0 & 0 \\ -1 & -1 & -1 & 0 & 0 & 1 \end{array}\right)$$

$$\xrightarrow{③+①} \left(\begin{array}{ccc|ccc} 1 & 1 & 2 & 0 & 1 & 0 \\ 0 & 2 & -1 & 1 & 0 & 0 \\ 0 & 0 & 1 & 0 & 1 & 1 \end{array}\right) \xrightarrow[②+③]{①+③\times(-2)} \left(\begin{array}{ccc|ccc} 1 & 1 & 0 & 0 & -1 & -2 \\ 0 & 2 & 0 & 1 & 1 & 1 \\ 0 & 0 & 1 & 0 & 1 & 1 \end{array}\right)$$

$$\xrightarrow[②\times\left(\frac{1}{2}\right)]{①+②\times\left(-\frac{1}{2}\right)} \left(\begin{array}{ccc|ccc} 1 & 0 & 0 & -\frac{1}{2} & -\frac{3}{2} & -\frac{5}{2} \\ 0 & 1 & 0 & \frac{1}{2} & \frac{1}{2} & \frac{1}{2} \\ 0 & 0 & 1 & 0 & 1 & 1 \end{array}\right).$$

So,

$$\boldsymbol{A}^{-1} = \begin{pmatrix} -\frac{1}{2} & -\frac{3}{2} & -\frac{5}{2} \\ \frac{1}{2} & \frac{1}{2} & \frac{1}{2} \\ 0 & 1 & 1 \end{pmatrix}.$$

■

Example 2.25 Computing a matrix from given conditions

Given $\boldsymbol{ABA}^{\mathrm{T}} = 2\boldsymbol{BA}^{\mathrm{T}} + \boldsymbol{I}$, find $\boldsymbol{B}$, where

$$\boldsymbol{A} = \begin{pmatrix} 1 & 0 & 0 \\ 0 & 1 & 2 \\ 0 & 0 & 1 \end{pmatrix}.$$

Solution: Notice that $2\boldsymbol{BA}^{\mathrm{T}} = 2\boldsymbol{IBA}^{\mathrm{T}}$, then

$$(\boldsymbol{A} - 2\boldsymbol{I})\boldsymbol{BA}^{\mathrm{T}} = \boldsymbol{I}, \text{ that is, } \boldsymbol{BA}^{\mathrm{T}} = (\boldsymbol{A} - 2\boldsymbol{I})^{-1}.$$

So,

$$\begin{aligned} \boldsymbol{B} &= (\boldsymbol{A} - 2\boldsymbol{I})^{-1}(\boldsymbol{A}^{\mathrm{T}})^{-1} \\ &= \left[\boldsymbol{A}^{\mathrm{T}}(\boldsymbol{A} - 2\boldsymbol{I})\right]^{-1} \\ &= (\boldsymbol{A}^{\mathrm{T}}\boldsymbol{A} - 2\boldsymbol{A}^{\mathrm{T}})^{-1}. \end{aligned}$$

Now,

$$\begin{aligned} \boldsymbol{A}^{\mathrm{T}}\boldsymbol{A} - 2\boldsymbol{A}^{\mathrm{T}} &= \begin{pmatrix} 1 & 0 & 0 \\ 0 & 1 & 2 \\ 0 & 2 & 5 \end{pmatrix} - \begin{pmatrix} 2 & 0 & 0 \\ 0 & 2 & 0 \\ 0 & 4 & 2 \end{pmatrix} \\ &= \begin{pmatrix} -1 & 0 & 0 \\ 0 & -1 & 2 \\ 0 & -2 & 3 \end{pmatrix}. \end{aligned}$$

Hence,

$$\boldsymbol{B} = \begin{pmatrix} -1 & 0 & 0 \\ 0 & -1 & 2 \\ 0 & -2 & 3 \end{pmatrix}^{-1} = \begin{pmatrix} -1 & 0 & 0 \\ 0 & 3 & -2 \\ 0 & 2 & -1 \end{pmatrix}.$$ ■

Notice that we only carry out the row operations throughout to get the inverse matrix of an invertible matrix. In the elementary row operations for any square matrix, if a row with all zeros occurs, the corresponding determinant is equal to zero, which means that the matrix is not invertible.

Readers should also practice the method of finding an inverse matrix through elementary column operations.

Example 2.26 Making a matrix not invertible through its elements

Determine a and b so that $\boldsymbol{A}$ is not invertible, where

$$\boldsymbol{A}=\begin{pmatrix}0&1&2&3\\1&4&7&10\\-1&0&1&b\\a&2&3&4\end{pmatrix}.$$

Solution: Applying elementary row and column operations on $\boldsymbol{A}$, we first simplify it to an echelon matrix, and then determine a and b so as to satisfy $|\boldsymbol{A}|=0$. For the sake of simplicity in calculation, we try to keep the elements of a and b at the right lower corner of $\boldsymbol{A}$. Toward this, we switch the columns ①, ②, and ③ twice (where $|\boldsymbol{A}|$ does not change). Then, we do the row addition operations, that is,

$$\boldsymbol{A}\to\begin{pmatrix}1&2&0&3\\4&7&1&10\\0&1&-1&b\\2&3&a&4\end{pmatrix}\xrightarrow[④+①\times(-2)]{②+①\times(-4)}\begin{pmatrix}1&2&0&3\\0&-1&1&-2\\0&1&-1&b\\0&-1&a&-2\end{pmatrix}$$

$$\xrightarrow[④+②\times(-1)]{③+②}\begin{pmatrix}1&2&0&3\\0&-1&1&-2\\0&0&0&b-2\\0&0&a-1&0\end{pmatrix}.$$

For $\boldsymbol{A}$ to be not invertible, a necessary and sufficient condition is

$$|\boldsymbol{A}|=\begin{vmatrix}1&2\\0&-1\end{vmatrix}\begin{vmatrix}0&b-2\\a-1&0\end{vmatrix}$$
$$=(a-1)(b-2)=0.$$

That is, $a=1$ or $b=2$. ■

2.6 Partitioned Matrix

Decomposing a large-scale matrix into several smaller partitioned matrices is an important skill in matrix operations. Smaller partitioned matrices require less computational resources and such small-scale operations can be carried out in a fast and easy manner. In this section, we show readers how to partition a matrix and also describe the operations of partitioned matrices through examples.

We decompose a matrix of order 5

$$A=\left(\begin{array}{cc|ccc} 2 & 1 & 1 & 0 & -1 \\ 1 & 2 & 2 & -3 & 0 \\ \hline 0 & 0 & 1 & 0 & 0 \\ 0 & 0 & 0 & 1 & 0 \\ 0 & 0 & 0 & 0 & 1 \end{array}\right),$$

into four blocks with horizontal and vertical lines. We then denote it as $\boldsymbol{A}$;

$$\begin{pmatrix} 2 & 1 \\ 1 & 2 \end{pmatrix}=\boldsymbol{A}_1, \qquad \begin{pmatrix} 1 & 0 & -1 \\ 2 & -3 & 0 \end{pmatrix}=\boldsymbol{A}_2,$$

$$\begin{pmatrix} 0 & 0 \\ 0 & 0 \\ 0 & 0 \end{pmatrix}=\mathbf{0}, \qquad \begin{pmatrix} 1 & & \\ & 1 & \\ & & 1 \end{pmatrix}=\boldsymbol{I}_3,$$

with the above four blocks written as

$$\boldsymbol{A}=\begin{pmatrix} \boldsymbol{A}_1 & \boldsymbol{A}_2 \\ \mathbf{0} & \boldsymbol{I}_3 \end{pmatrix}.$$

It is called a 2×2 partition of the matrix $\boldsymbol{A}$, in which every smaller matrix is called a submatrix of $\boldsymbol{A}$.

$\boldsymbol{A}$ is called an $s \times t$ **partitioned matrix** if an $m \times n$ matrix $\boldsymbol{A}$ is split into s blocks in the row direction and t blocks in the column direction, which is written as $\boldsymbol{A}=(\boldsymbol{A}_{kl})_{s\times t}$, where $\boldsymbol{A}_{kl}$ $(k=1,2,\ldots,s;\ l=1,2,\ldots,t)$ is called a **submatrix** of $\boldsymbol{A}$. The submatrices can be matrices of varying sizes.

Besides the 2×2 partitioned matrix, some frequently used partitioned matrices have the following forms:

(i) Decomposed along a row.

$$\boldsymbol{A}=\begin{pmatrix} a_{11} & a_{12} & \cdots & a_{1n} \\ a_{21} & a_{22} & \cdots & a_{2n} \\ \vdots & \vdots & & \vdots \\ a_{m1} & a_{m2} & \cdots & a_{mn} \end{pmatrix}=\begin{pmatrix} \boldsymbol{a}_1 \\ \boldsymbol{a}_2 \\ \vdots \\ \boldsymbol{a}_m \end{pmatrix},$$

where $\boldsymbol{a}_i=(a_{i1},a_{i2},\ldots,a_{in}),\ i=1,2,\ldots,m$.

(ii) Decomposed along a column.

$$\boldsymbol{B}=\begin{pmatrix} b_{11} & b_{12} & \cdots & b_{1s} \\ b_{21} & b_{22} & \cdots & b_{2s} \\ \vdots & \vdots & & \vdots \\ b_{n1} & b_{n2} & \cdots & b_{ns} \end{pmatrix}=(\boldsymbol{b}_1, \ \boldsymbol{b}_2, \ \cdots, \ \boldsymbol{b}_s),$$

where $\boldsymbol{b}_j=(b_{1j},b_{2j},\ldots,b_{nj})^{\mathrm{T}},\ j=1,2,\ldots,s$.

When the nonzero elements in the nth-order matrix $\boldsymbol{C}$ are all centralized near the leading diagonal, it can sometimes be decomposed into the following **diagonal block matrix** (also called the almost diagonal matrix)

$$\boldsymbol{C}=\begin{pmatrix}\boldsymbol{C}_1 & & & \\ & \boldsymbol{C}_2 & & \\ & & \ddots & \\ & & & \boldsymbol{C}_m\end{pmatrix},$$

where $\boldsymbol{C}_i$ is r_ith-order square matrix ($i=1,2,\ldots,m$; $\sum_{i=1}^{m} r_i = n$). For instance,

$$\boldsymbol{C}=\left(\begin{array}{cc|cccc|c} 0 & -1 & 0 & 0 & 0 & 0 \\ 1 & 2 & 0 & 0 & 0 & 0 \\ \hline 0 & 0 & 1 & -1 & 0 & 0 \\ 0 & 0 & -1 & 1 & 2 & 0 \\ 0 & 0 & 0 & 2 & -2 & 0 \\ \hline 0 & 0 & 0 & 0 & 0 & 3 \end{array}\right)=\begin{pmatrix}\boldsymbol{C}_1 & & \\ & \boldsymbol{C}_2 & \\ & & \boldsymbol{C}_3\end{pmatrix},$$

where $\boldsymbol{C}_1=\begin{pmatrix}0 & -1\\ 1 & 2\end{pmatrix}$, $\boldsymbol{C}_2=\begin{pmatrix}1 & -1 & 0\\ -1 & 1 & 2\\ 0 & 2 & -2\end{pmatrix}$, $\boldsymbol{C}_3=(3)$.

Next, we discuss the operations of partitioned matrices.

2.6.1 Addition of partitioned matrix

Suppose $\boldsymbol{A}=(\boldsymbol{A}_{kl})_{s\times t}$ and $\boldsymbol{B}=(\boldsymbol{B}_{kl})_{s\times t}$ are partitioned matrices; if the submatrices $\boldsymbol{A}_{kl}$ and $\boldsymbol{B}_{kl}$ corresponding to $\boldsymbol{A}$ and $\boldsymbol{B}$ are matrices of the same size, then

$$\boldsymbol{A}+\boldsymbol{B}=(\boldsymbol{A}_{kl}+\boldsymbol{B}_{kl})_{s\times t}.$$

For example,

$$\begin{pmatrix}\boldsymbol{A}_{11} & \boldsymbol{A}_{12}\\ \boldsymbol{A}_{21} & \boldsymbol{A}_{22}\end{pmatrix}+\begin{pmatrix}\boldsymbol{B}_{11} & \boldsymbol{B}_{12}\\ \boldsymbol{B}_{21} & \boldsymbol{B}_{22}\end{pmatrix}=\begin{pmatrix}\boldsymbol{A}_{11}+\boldsymbol{B}_{11} & \boldsymbol{A}_{12}+\boldsymbol{B}_{12}\\ \boldsymbol{A}_{21}+\boldsymbol{B}_{21} & \boldsymbol{A}_{22}+\boldsymbol{B}_{22}\end{pmatrix},$$

where $\boldsymbol{A}_{11}$ and $\boldsymbol{B}_{11}$; $\boldsymbol{A}_{12}$ and $\boldsymbol{B}_{12}$; $\boldsymbol{A}_{21}$ and $\boldsymbol{B}_{21}$; and $\boldsymbol{A}_{22}$ and $\boldsymbol{B}_{22}$ respectively, are smaller matrices (submatrices) of the same size.

2.6.2 Scalar multiplication of partitioned matrix

Let the partitioned matrix be $\boldsymbol{A}=(\boldsymbol{A}_{kl})_{s\times t}$, and λ is a number, then

$$\lambda\boldsymbol{A}=(\lambda\boldsymbol{A}_{kl})_{s\times t}.$$

2.6.3 Multiplication of partitioned matrix

Suppose $\boldsymbol{A} \in F^{m\times n}$ and $\boldsymbol{B} \in F^{n\times p}$; if $\boldsymbol{A}$ is decomposed into $r \times s$ partitioned matrices $(\boldsymbol{A}_{kl})_{r\times s}$, $\boldsymbol{B}$ is decomposed into $s \times t$ partitioned matrices $(\boldsymbol{B}_{kl})_{s\times t}$, and the partitioning of columns in $\boldsymbol{A}$ with respect to the rows in $\boldsymbol{B}$ are done in a compatible way, then

$$\boldsymbol{AB} = \begin{array}{c} \begin{array}{cccc} \text{col}\, j_1 & \text{col}\, j_2 & \quad & \text{col}\, j_s \end{array} \\ \left(\begin{array}{c|c|c|c} \boldsymbol{A}_{11} & \boldsymbol{A}_{12} & \cdots & \boldsymbol{A}_{1s} \\ \boldsymbol{A}_{21} & \boldsymbol{A}_{22} & \cdots & \boldsymbol{A}_{2s} \\ \vdots & \vdots & & \vdots \\ \boldsymbol{A}_{r1} & \boldsymbol{A}_{r2} & \cdots & \boldsymbol{A}_{rs} \end{array}\right) \end{array} \left(\begin{array}{cccc} \boldsymbol{B}_{11} & \boldsymbol{B}_{12} & \cdots & \boldsymbol{B}_{1t} \\ \hline \boldsymbol{B}_{21} & \boldsymbol{B}_{22} & \cdots & \boldsymbol{B}_{2t} \\ \hline \vdots & \vdots & & \vdots \\ \hline \boldsymbol{B}_{s1} & \boldsymbol{B}_{s2} & \cdots & \boldsymbol{B}_{st} \end{array}\right) \begin{array}{l} \text{row } j_1 \\ \text{row } j_2 \\ \\ \text{row } j_s \end{array}$$

$$= \boldsymbol{C} \overset{\text{written as}}{=\!=\!=} (\boldsymbol{C}_{kl})_{r\times t},$$

where $\boldsymbol{C}$ is an $r \times t$ partitioned matrix, and

$$\boldsymbol{C}_{kl} = \sum_{i=1}^{s} \boldsymbol{A}_{ki}\boldsymbol{B}_{il} \quad (k=1, 2, \cdots, r;\ l=1, 2, \ldots, t).$$

It can be easily proved (not shown) that the two ways of matrix multiplication (multiplication through block multiplication or without partitioning) result in the same $\boldsymbol{AB}$.

Example 2.27 Partitioning and block multiplication

Decompose the following fifth-order matrices $\boldsymbol{A}$ and $\boldsymbol{B}$ into 2×2 partitioned matrices, and compute $\boldsymbol{AB}$ through block multiplication.

$$\boldsymbol{A} = \begin{pmatrix} 1 & 0 & 0 & 0 & 0 \\ 0 & 1 & 0 & 0 & 0 \\ -1 & 2 & 1 & 0 & 0 \\ 1 & 1 & 0 & 1 & 0 \\ -2 & 0 & 0 & 0 & 1 \end{pmatrix}, \qquad \boldsymbol{B} = \begin{pmatrix} 3 & 2 & 0 & 1 & 0 \\ 1 & 3 & 0 & 0 & 1 \\ -1 & 0 & 0 & 0 & 0 \\ 0 & -1 & 0 & 0 & 0 \\ 0 & 0 & -1 & 0 & 0 \end{pmatrix}.$$

Solution: By observation, we can decompose $\boldsymbol{A}$ into the following four submatrices

$$\boldsymbol{A} = \left(\begin{array}{cc|ccc} 1 & 0 & 0 & 0 & 0 \\ 0 & 1 & 0 & 0 & 0 \\ \hline -1 & 2 & 1 & 0 & 0 \\ 1 & 1 & 0 & 1 & 0 \\ -2 & 0 & 0 & 0 & 1 \end{array}\right) = \begin{pmatrix} \boldsymbol{I}_2 & \boldsymbol{0}_{2\times 3} \\ \boldsymbol{A}_1 & \boldsymbol{I}_3 \end{pmatrix},$$

where

$$A_1 = \begin{pmatrix} -1 & 2 \\ 1 & 1 \\ -2 & 0 \end{pmatrix}.$$

Based on the assumption of block multiplication, the partitioned blocks have to be decomposed into a compatible number of columns and rows for $\boldsymbol{A}$ and $\boldsymbol{B}$, but the columns of $\boldsymbol{B}$ can be decomposed arbitrarily. In order to compute conveniently, we can decompose matrix $\boldsymbol{B}$ as follows:

$$\boldsymbol{B} = \left(\begin{array}{ccc|cc} 3 & 2 & 0 & 1 & 0 \\ 1 & 3 & 0 & 0 & 1 \\ \hline -1 & 0 & 0 & 0 & 0 \\ 0 & -1 & 0 & 0 & 0 \\ 0 & 0 & -1 & 0 & 0 \end{array}\right) = \begin{pmatrix} \boldsymbol{B}_1 & \boldsymbol{I}_2 \\ -\boldsymbol{I}_3 & \boldsymbol{0}_{3\times 2} \end{pmatrix},$$

where

$$\boldsymbol{B}_1 = \begin{pmatrix} 3 & 2 & 0 \\ 1 & 3 & 0 \end{pmatrix}.$$

Then,

$$\boldsymbol{AB} = \begin{pmatrix} \boldsymbol{I}_2 & \boldsymbol{0} \\ \boldsymbol{A}_1 & \boldsymbol{I}_3 \end{pmatrix} \begin{pmatrix} \boldsymbol{B}_1 & \boldsymbol{I}_2 \\ -\boldsymbol{I}_3 & \boldsymbol{0} \end{pmatrix} = \begin{pmatrix} \boldsymbol{B}_1 & \boldsymbol{I}_2 \\ \boldsymbol{A}_1\boldsymbol{B}_1 - \boldsymbol{I}_3 & \boldsymbol{A}_1 \end{pmatrix},$$

where

$$\boldsymbol{A}_1\boldsymbol{B}_1 - \boldsymbol{I}_3 = \begin{pmatrix} -2 & 4 & 0 \\ 4 & 4 & 0 \\ -6 & -4 & -1 \end{pmatrix}.$$

So,

$$\boldsymbol{AB} = \begin{pmatrix} 3 & 2 & 0 & 1 & 0 \\ 1 & 3 & 0 & 0 & 1 \\ -2 & 4 & 0 & -1 & 2 \\ 4 & 4 & 0 & 1 & 1 \\ -6 & -4 & -1 & -2 & 0 \end{pmatrix}.$$

We can easily verify that the result of $\boldsymbol{AB}$ by direct multiplication is the same as that by block multiplication. ■

Example 2.28 Partitioning of a matrix

Let $\boldsymbol{A}$ be an $m \times n$ matrix and $\boldsymbol{B}$ be an $n \times s$ matrix; $\boldsymbol{B}$ is then decomposed into an $1 \times s$ partitioned matrix along its columns. Let $\boldsymbol{A}$ be a 1×1 partitioned matrix. Then,

$$\boldsymbol{AB} = \boldsymbol{A}(\boldsymbol{b}_1, \boldsymbol{b}_2, \ldots, \boldsymbol{b}_s) = (\boldsymbol{Ab}_1, \boldsymbol{Ab}_2, \ldots, \boldsymbol{Ab}_s).$$

If $\boldsymbol{AB} = \boldsymbol{0}$ ($m \times n$ zero matrix), then $\boldsymbol{Ab}_j = \boldsymbol{0}$ ($n \times 1$ zero matrix), $j = 1, 2, \ldots, s$. Hence, every column $\boldsymbol{b}_j$ of $\boldsymbol{B}$ is a solution of linear equations $\boldsymbol{Ax} = \boldsymbol{0}$. ■

Example 2.29 Decomposition of matrices into diagonal block matrices

If nth-order matrices $\boldsymbol{C}$ and $\boldsymbol{D}$ can be decomposed into diagonal block matrices of the same type, that is,

$$\boldsymbol{C} = \begin{pmatrix} \boldsymbol{C}_1 & & & \\ & \boldsymbol{C}_2 & & \\ & & \ddots & \\ & & & \boldsymbol{C}_m \end{pmatrix}, \quad \boldsymbol{D} = \begin{pmatrix} \boldsymbol{D}_1 & & & \\ & \boldsymbol{D}_2 & & \\ & & \ddots & \\ & & & \boldsymbol{D}_m \end{pmatrix},$$

where $\boldsymbol{C}_i$ and $\boldsymbol{D}_i$ are square matrices ($i = 1, 2, \ldots, m$) of the same order, then

$$\boldsymbol{CD} = \begin{pmatrix} \boldsymbol{C}_1\boldsymbol{D}_1 & & & \\ & \boldsymbol{C}_2\boldsymbol{D}_2 & & \\ & & \ddots & \\ & & & \boldsymbol{C}_m\boldsymbol{D}_m \end{pmatrix}.$$

■

The block multiplication plays an important role in proving some important propositions. Observe the following example carefully.

Example 2.30 Upper triangular matrix and its inverse

Prove that if an upper triangular matrix A of order n is invertible, then the inverse matrix A^{-1} is also an upper triangular matrix.

Proof: The proof is by mathematical induction on n. If $n = 1$, $(a)^{-1} = \left(\frac{1}{a}\right)$, then the conclusion is true. A first-order matrix is seen as an upper (lower) triangular, diagonal and symmetric matrix. Assume the proposition is true for an $(n-1)$th-order invertible upper triangular matrix. Then, we consider the case of nth-order.
Suppose

$$A = \left(\begin{array}{c|ccc} a_{11} & a_{12} & \cdots & a_{1n} \\ \hline 0 & a_{22} & \cdots & a_{2n} \\ \vdots & \vdots & \ddots & \vdots \\ 0 & 0 & \cdots & a_{nn} \end{array}\right) \underline{\underline{\text{partitioning}}} \begin{pmatrix} a_{11} & \alpha \\ \mathbf{0} & A_1 \end{pmatrix} (a_{ii} \neq 0, i = 1, 2, \dots, n),$$

where A_1 is an $(n-1)$th-order invertible upper triangular matrix.

Let the inverse matrix of A be

$$B = \left(\begin{array}{c|ccc} b_{11} & b_{12} & \cdots & b_{1n} \\ \hline b_{21} & b_{22} & \cdots & b_{2n} \\ \vdots & \vdots & \ddots & \vdots \\ b_{n1} & b_{n2} & \cdots & b_{nn} \end{array}\right) \underline{\underline{\text{partitioning}}} \begin{pmatrix} b_{11} & \beta \\ \gamma & B_1 \end{pmatrix},$$

then

$$AB = \begin{pmatrix} a_{11} & \alpha \\ \mathbf{0} & A_1 \end{pmatrix} \begin{pmatrix} b_{11} & \beta \\ \gamma & B_1 \end{pmatrix} = I_n = \begin{pmatrix} 1 & \mathbf{0} \\ \mathbf{0} & I_{n-1} \end{pmatrix}.$$

$$AB = \begin{pmatrix} a_{11}b_{11} + \alpha\gamma & a_{11}\beta + \alpha B_1 \\ A_1\gamma & A_1 B_1 \end{pmatrix} = \begin{pmatrix} 1 & \mathbf{0} \\ \mathbf{0} & I_{n-1} \end{pmatrix},$$

implies

$$A_1\gamma = \mathbf{0} \Rightarrow \gamma = A^{-1}\mathbf{0} = \mathbf{0},$$

$$A_1 B_1 = I_{n-1} \Rightarrow B_1 = A_1^{-1},$$

where B_1 is an $(n-1)$th-order upper triangular matrix according to inductive hypothesis.

Thus,

$$A^{-1} = B = \begin{pmatrix} b_{11} & \beta \\ \mathbf{0} & B_1 \end{pmatrix}$$

is an upper triangular matrix (where $b_{11} = a_{11}^{-1}$; $\beta = -a_{11}^{-1}\alpha A_1^{-1}$). ■

2.6.4 Transposition of partitioned matrix

The transposed matrix of the partitioned matrix $A = (A_{kl})_{s\times t}$ is

$$A^{\mathrm{T}} = (B_{lk})_{t\times s},$$

where $B_{lk} = {A_{kl}}^{\mathrm{T}}, l = 1, 2, \ldots, t; k = 1, 2, \ldots, s$.

For example,

$$A = \begin{pmatrix} A_{11} & A_{12} & A_{13} \\ A_{21} & A_{22} & A_{23} \end{pmatrix},$$

then

$$A^{\mathrm{T}} = \begin{pmatrix} {A_{11}}^{\mathrm{T}} & {A_{21}}^{\mathrm{T}} \\ {A_{12}}^{\mathrm{T}} & {A_{22}}^{\mathrm{T}} \\ {A_{13}}^{\mathrm{T}} & {A_{23}}^{\mathrm{T}} \end{pmatrix},$$

$$B \xlongequal{\text{partitioning on rows}} \begin{pmatrix} b_1 \\ b_2 \\ \vdots \\ b_m \end{pmatrix},$$

then

$$B^{\mathrm{T}} = ({b_1}^{\mathrm{T}}, {b_2}^{\mathrm{T}}, \cdots, {b_m}^{\mathrm{T}}).$$

2.6.5 The inverse of a partitioned matrix

Consider the diagonal block matrix (almost diagonal matrix):

$$A = \begin{pmatrix} A_1 & & & \\ & A_2 & & \\ & & \ddots & \\ & & & A_m \end{pmatrix}.$$

The determinant of A is $|A| = |A_1||A_2|\cdots|A_m|$, therefore, a necessary and sufficient condition of a diagonal block matrix being invertible is

$$|A_i| \neq 0, \qquad i = 1, 2, \cdots, m.$$

According to the multiplication of the diagonal block matrices, we can easily get the inverse matrix of A,

$$A^{-1} = \begin{pmatrix} A_1^{-1} & & & \\ & A_2^{-1} & & \\ & & \ddots & \\ & & & A_m^{-1} \end{pmatrix}.$$

Using the partitioned matrix to compute the inverse of a matrix, readers can convert the problem into finding the inverse of lower order matrices after changing the higher order matrix into lower ones. The inverse of a 2×2 partitioned matrix can be obtained by this method through solving matrix equations in terms of the definition of the inverse matrix.

Example 2.31 Computing the inverse of a partitioned matrix

Suppose

$$A = \begin{pmatrix} B & 0 \\ C & D \end{pmatrix},$$

where B and D are both invertible square matrices. Show that A is invertible and determine A^{-1}.

Solution: As $|A| = |B|\,|D| \neq 0$, so A is invertible.

Suppose

$$A^{-1} = \begin{pmatrix} X & Y \\ Z & T \end{pmatrix},$$

where X and B, T and D are the same order matrices respectively. Then,

$$\begin{pmatrix} B & 0 \\ C & D \end{pmatrix}\begin{pmatrix} X & Y \\ Z & T \end{pmatrix} = \begin{pmatrix} BX & BY \\ CX + DZ & CY + DT \end{pmatrix} = \begin{pmatrix} I_m & 0 \\ 0 & I_n \end{pmatrix},$$

we obtain

$$\begin{aligned} BX = I_m \quad &\text{then} \quad X = B^{-1}; \\ BY = 0 \quad &\text{then} \quad Y = B^{-1}0 = 0; \\ CX + DZ = 0 \text{ and } DZ = -CX = -CB^{-1} \quad &\text{then} \quad Z = -D^{-1}CB^{-1}; \\ CY + DT = I_n \text{ and } DT = I_n \quad &\text{then} \quad T = D^{-1}. \end{aligned}$$

Thus,

$$A^{-1} = \begin{pmatrix} B^{-1} & 0 \\ -D^{-1}CB^{-1} & D^{-1} \end{pmatrix}.$$

■

*2.6.6 Partitioned elementary matrix and elementary operations of the partitioned matrix

Here, we consider only 2×2 partitioned matrix as an example. For a partitioned matrix

$$A = \begin{pmatrix} A_{11} & A_{12} \\ A_{21} & A_{22} \end{pmatrix},$$

we define three types of elementary row operations and column operations in the same way, and define three types of partitioned elementary matrices accordingly.

(i) Partitioned multiplication matrix (here, $\boldsymbol{C}_1$ and $\boldsymbol{C}_2$ are invertible matrices)

$$\begin{pmatrix} \boldsymbol{C}_1 & \boldsymbol{0} \\ \boldsymbol{0} & \boldsymbol{I}_n \end{pmatrix} \quad \text{or} \quad \begin{pmatrix} \boldsymbol{I}_m & \boldsymbol{0} \\ \boldsymbol{0} & \boldsymbol{C}_2 \end{pmatrix}.$$

(ii) Partitioned addition matrix

$$\begin{pmatrix} \boldsymbol{I}_m & \boldsymbol{0} \\ \boldsymbol{C}_3 & \boldsymbol{I}_n \end{pmatrix} \quad \text{or} \quad \begin{pmatrix} \boldsymbol{I}_m & \boldsymbol{C}_4 \\ \boldsymbol{0} & \boldsymbol{I}_n \end{pmatrix}.$$

(iii) Partitioned switching matrix

$$\begin{pmatrix} \boldsymbol{0} & \boldsymbol{I}_m \\ \boldsymbol{I}_n & \boldsymbol{0} \end{pmatrix}.$$

Partitioned elementary matrices are naturally square matrices. Their operational effect is the same as that of the elementary matrices described in §2.5 under the assumption that their left (or right) multiplication is well defined.

The elementary operation of a partitioned matrix is also an important skill of matrix operations, and it is comparatively suitable for some problems.

Example 2.32 Multiplication of a partitioned elementary matrix

Let nth-order matrix $\boldsymbol{A}$ be denoted by blocks

$$\boldsymbol{A} = \begin{pmatrix} \boldsymbol{A}_{11} & \boldsymbol{A}_{12} \\ \boldsymbol{A}_{21} & \boldsymbol{A}_{22} \end{pmatrix},$$

where $\boldsymbol{A}_{11}$ and $\boldsymbol{A}_{22}$ are square matrices, $\boldsymbol{A}$ and $\boldsymbol{A}_{11}$ are invertible. Prove that $\boldsymbol{A}_{22} - \boldsymbol{A}_{21}\boldsymbol{A}_{11}^{-1}\boldsymbol{A}_{12}$ is invertible and find $\boldsymbol{A}^{-1}$.

Solution: In order to transform the partitioned matrix $\boldsymbol{A}$ into an upper triangular partition matrix, perform the elementary row operations by left multiplying

$$\boldsymbol{P}_1 = \begin{pmatrix} \boldsymbol{I}_1 & \boldsymbol{0} \\ -\boldsymbol{A}_{21}\boldsymbol{A}_{11}^{-1} & \boldsymbol{I}_2 \end{pmatrix},$$

on $\boldsymbol{A}$, where $\boldsymbol{I}_1$ and $\boldsymbol{I}_2$ are identity matrices whose orders are equal to those of $\boldsymbol{A}_{11}$ and $\boldsymbol{A}_{22}$ respectively. Thus,

$$\boldsymbol{P}_1\boldsymbol{A} = \begin{pmatrix} \boldsymbol{A}_{11} & \boldsymbol{A}_{12} \\ \boldsymbol{0} & \boldsymbol{A}_{22} - \boldsymbol{A}_{21}\boldsymbol{A}_{11}^{-1}\boldsymbol{A}_{12} \end{pmatrix} \overset{\text{written as}}{=\!=\!=} \boldsymbol{B}$$

$$|\boldsymbol{P}_1\boldsymbol{A}| = |\boldsymbol{A}_{11}|\,\left|\boldsymbol{A}_{22} - \boldsymbol{A}_{21}\boldsymbol{A}_{11}^{-1}\boldsymbol{A}_{12}\right|.$$

Since $|\boldsymbol{P}_1| = 1$, $|\boldsymbol{A}| \neq 0$, $|\boldsymbol{A}_{11}| \neq 0$, hence

$$\left|\boldsymbol{A}_{22} - \boldsymbol{A}_{21}\boldsymbol{A}_{11}{}^{-1}\boldsymbol{A}_{12}\right| = \frac{|\boldsymbol{A}|}{|\boldsymbol{A}_{11}|} \neq 0.$$

Thus, $(\boldsymbol{A}_{22} - \boldsymbol{A}_{21}\boldsymbol{A}_{11}{}^{-1}\boldsymbol{A}_{12})$ is invertible.

Denote $\boldsymbol{Q} = \boldsymbol{A}_{22} - \boldsymbol{A}_{21}\boldsymbol{A}_{11}{}^{-1}\boldsymbol{A}_{12}$. Perform row operations on $\boldsymbol{B}$ to change it into a diagonal block matrix.

Let

$$\boldsymbol{P}_2 = \begin{pmatrix} \boldsymbol{I}_1 & -\boldsymbol{A}_{12}\boldsymbol{Q}^{-1} \\ \boldsymbol{0} & \boldsymbol{I}_2 \end{pmatrix},$$

and then

$$\boldsymbol{P}_2\boldsymbol{B} = \begin{pmatrix} \boldsymbol{A}_{11} & \boldsymbol{0} \\ \boldsymbol{0} & \boldsymbol{Q} \end{pmatrix} \underline{\underline{\text{written as}}}\ \boldsymbol{C},$$

which is $\boldsymbol{P}_2\boldsymbol{P}_1\boldsymbol{A} = \boldsymbol{C}$.

Inverting the two sides we get $\boldsymbol{A}^{-1}(\boldsymbol{P}_2\boldsymbol{P}_1)^{-1} = \boldsymbol{C}^{-1}$, then

$$\begin{aligned} \boldsymbol{A}^{-1} &= \boldsymbol{C}^{-1}(\boldsymbol{P}_2\boldsymbol{P}_1) \\ &= \begin{pmatrix} \boldsymbol{A}_{11}{}^{-1} & \boldsymbol{0} \\ \boldsymbol{0} & \boldsymbol{Q}^{-1} \end{pmatrix} \begin{pmatrix} \boldsymbol{I}_1 & -\boldsymbol{A}_{12}{}^{-1}\boldsymbol{Q}^{-1} \\ \boldsymbol{0} & \boldsymbol{I}_2 \end{pmatrix} \begin{pmatrix} \boldsymbol{I}_1 & \boldsymbol{0} \\ -\boldsymbol{A}_{21}\boldsymbol{A}_{11}{}^{-1} & \boldsymbol{I}_2 \end{pmatrix}. \end{aligned}$$

Denote

$$\boldsymbol{A}^{-1} = \begin{pmatrix} \boldsymbol{D}_{11} & \boldsymbol{D}_{12} \\ \boldsymbol{D}_{21} & \boldsymbol{D}_{22} \end{pmatrix},$$

where

$$\boldsymbol{D}_{11} = \boldsymbol{A}_{11}{}^{-1} + \boldsymbol{A}_{11}{}^{-1}\boldsymbol{A}_{12}\boldsymbol{Q}^{-1}\boldsymbol{A}_{21}\boldsymbol{A}_{11}{}^{-1},$$

$$\boldsymbol{D}_{22} = \boldsymbol{Q}^{-1},$$

$$\boldsymbol{D}_{12} = -\boldsymbol{A}_{11}{}^{-1}\boldsymbol{A}_{12}\boldsymbol{Q}^{-1},$$

$$\boldsymbol{D}_{21} = -\boldsymbol{Q}^{-1}\boldsymbol{A}_{21}\boldsymbol{A}_{11}{}^{-1},$$

and

$$\boldsymbol{Q} = \boldsymbol{A}_{22} - \boldsymbol{A}_{21}\boldsymbol{A}_{11}{}^{-1}\boldsymbol{A}_{12}.$$ ■

Example 2.33 Application of a partitioned elementary matrix

Suppose

$$Q = \begin{pmatrix} A & B \\ C & D \end{pmatrix},$$

and A is invertible. Prove that $\det Q = |A|\,|D - CA^{-1}B|$.

Proof: First, we left multiply Q by a partitioned addition matrix

$$P = \begin{pmatrix} I_n & 0 \\ -CA^{-1} & I_m \end{pmatrix},$$

where I_n and A have the same order, I_m and D have the same order, we have

$$PQ = \begin{pmatrix} A & B \\ 0 & D - CA^{-1}B \end{pmatrix}.$$

Taking determinants of both sides, we get $|P|\,|Q| = |A|\,|D - CA^{-1}B|$. Since $|P| = 1$, the proposition is proved. ■

Example 2.34 Determinants of a partitioned elementary matrix

Let A and B be both nth-order matrices. Prove that

$$\begin{vmatrix} A & B \\ B & A \end{vmatrix} = |A + B|\,|A - B|.$$

Proof: Add the first row of the partitioned matrix $P = \begin{pmatrix} A & B \\ B & A \end{pmatrix}$ to its second row, then add the second column multiplied by $-I$ to the first column. An upper triangular matrix is obtained;

$$\begin{pmatrix} I & 0 \\ I & I \end{pmatrix}\begin{pmatrix} A & B \\ B & A \end{pmatrix}\begin{pmatrix} I & 0 \\ -I & I \end{pmatrix} = \begin{pmatrix} A & B \\ A+B & A+B \end{pmatrix}\begin{pmatrix} I & 0 \\ -I & I \end{pmatrix}$$
$$= \begin{pmatrix} A-B & B \\ 0 & A+B \end{pmatrix}.$$

Taking determinants of both sides

$$\begin{vmatrix} A & B \\ B & A \end{vmatrix} = \begin{vmatrix} A-B & B \\ 0 & A+B \end{vmatrix} = |A - B|\,|A + B|.$$ ■

*2.7 Number Fields, Propositions, and Quantifiers

*2.7.1 Number fields

A set of numbers F containing the numbers 0 and 1 is called a number field if it is closed under the four arithmetic operations, namely, addition, subtraction, multiplication, and division (the divisor is not zero). That is, the sum, difference, product, and quotient of two numbers of F are still in F.

Obviously, each of the sets of rational, real, and complex numbers form number fields. They are called rational, real, and complex number fields respectively, which are written as $\mathbb{Q}$, $\mathbb{R}$, and $\mathbb{C}$. However, the set $\mathbb{Z}$, consisting of integral numbers is not a number field because the quotient of two integers may not be an integer.

There are still some other number sets, like $Q(\sqrt{2}) = \{a + b\sqrt{2} \,|\, a, b \in \mathbb{Q}\}$, which make number fields as well. Here, we point out that the rational number field $\mathbb{Q}$ is the smallest one, that is, any number field F contains $\mathbb{Q}$. In fact, $1 + 1 + \cdots + 1 (n \text{ copies}) = n \in F$ and $0 - n \in F$ as $0, 1 \in F$, where the integer number set $\mathbb{Z} \subseteq F$. Then, for arbitrary $p, q \in F$, $p \neq 0$, we always have $q/p \in F$, while $q/p \in \mathbb{Q}$. So, $\mathbb{Q} \subseteq F$.

*2.7.2 Propositions

A so-called "proposition" is just a Declarative Sentence. Strictly speaking, a proposition is not the statement of the sentence itself, but the meaning the statement expresses. "Sentence" is a linguistic concept, while "proposition" is a logical concept.

Consider the following statements:

① $3 < 5$;
② snow is white;
③ snow is not white;
④ $3 < 5$ and 3 is an aliquot of 5;
⑤ he studies English or he studies French;
⑥ if it is not raining, I will go out for a walk;
⑦ the two triangles are similar, if and only if, their three inner angles are respectively equal.

They are all propositions, while ①–③ are simple propositions, ④–⑦ are composite propositions with two propositions and a logical connective. A negation to a proposition is also a proposition. Proposition ③ is the negation of proposition ②. The logical connectives mentioned here are: $\wedge$ (conjunction), $\vee$ (disjunction), $\rightarrow$ (implication), $\Leftrightarrow$ (double implication), and $\neg$ (negation). We look at the following short examples to understand their meaning (note that p and q are propositions in the following examples).

(i) $p \wedge q$ represents "p **and** q."

(ii) $p \vee q$ represents "p **or** q."
(iii) $p \rightarrow q$ represents "**if** p **then** q" (or to say p **implies** q).
(iv) $p \Leftrightarrow q$ represents "p **if and only if** q" (or to say p and q are **equivalent**).
(v) $\neg p$ represents "**not** p," that is, the **negative proposition** of p.

Notice the implication proposition $p \rightarrow q$ and its inverse negative proposition $(\neg q) \rightarrow (\neg p)$ (written as $\neg q \rightarrow \neg p$ for short), that is, "if not q then not p," are equivalent. Let us understand this by reviewing the following examples.

- "If it is raining, then I shall take the umbrella" is logically equivalent to "if I haven't took the umbrella, then it is not raining."
- "If the three sides of the two triangles are correspondingly equal, then the two triangles are congruent" is equivalent to "if the two triangles are not congruent, then the three sides of the two triangles are not correspondingly equal"; that is, at least one pair of corresponding sides are unequal.

To prove a mathematical theorem "if p then q" by reduction to absurdity is equivalent to proving the proposition (i.e., the inverse negative proposition) "if not q, then not p."

In a course of linear algebra, we often prove a theorem (implication proposition) by reduction to absurdity. Thus, we should be good at expressing the negative proposition. Here are more examples.

The negative proposition to "there exist numbers $x_1, x_2, \cdots, x_n$, not all equal to zero, such that $x_1\xi_1 + x_2\xi_2 + \cdots + x_n\xi_n = \mathbf{0}$ holds" is "any collection of numbers $x_1, x_2, \ldots, x_n$, not all equal to zero, must satisfy $x_1\xi_1 + x_2\xi_2 + \cdots + x_n\xi_n \neq \mathbf{0}$", that is, "only when $x_1, x_2, \ldots, x_n$ are all equal to zero, $x_1\xi_1 + x_2\xi_2 + \cdots + x_n\xi_n = \mathbf{0}$ is true."

"If p then q" is also called a conditional proposition. In this case, we say that the necessary condition for a true p is a true q (or call q the necessary condition of p, for short). A true p is the sufficient condition for a true q (or call p the sufficient condition of q).

*2.7.3 Quantifiers

Some propositions usually use two assertions: "every element of the set X has the property p," "there exists an element x of the set X which has the property p." For simplicity, we use the logical symbols "$\forall x \in X, p$" [or "$(\forall x \in X)p$"], "$\exists x \in X, p$" [or "$(\exists x \in X)p$"] to express the above two assertions. Here, "$\forall x$" expresses "any element x" and "$\forall$" is called an arbitrary quantifier where its notation is the mirror rotation of the horizontal axis of "A" for "any." "$\exists x$" expresses that "there is an element x (or exists an element x)" and "$\exists$" is called an existential quantifier which is the mirror rotation of the vertical axis of "E" for "exist."

For example, $A \subset B$ means "if $a \in A$, then $a \in B$" for sets A and B, which can be expressed as "$\forall a \in A, a \in B$."

The negative of $A \subset B$ is $A \not\subset B$, its meaning is "$\exists a \in A, a \notin B$."

Generally speaking, the negative proposition which includes quantifiers satisfies the following two basic equivalence rules:

$$\text{Not } (\forall x \in X)p \text{ is equivalent to } (\exists x \in X) \text{ not } p, \tag{2.29}$$

$$\text{Not } (\exists x \in X)p \text{ is equivalent to } (\forall x \in X) \text{ not } p. \tag{2.30}$$

***Example 2.35 Propositions and quantifiers (I)**

Let X and Y be sets of the students of class A and B, respectively. Let $p(x, y)$ express that $x \in X$ and $y \in Y$ have the same surname. Then, proposition 1°

$$(\forall x \in X)(\exists y \in Y)p(x, y)$$

represents "any student of X can find a student of Y with the same surname." According to the above two equivalent rules (2.29) and (2.30), the proposition "not 1°" is equivalent to

$$(\exists x \in X) \text{ not } (\exists y \in Y)p(x, y) \Leftrightarrow (\exists x \in X)(\forall y \in Y) \text{ not } p(x, y)$$

where $\Leftrightarrow$ is the double implication symbol, that is, the meaning of equivalence. We know from the latter operation that the meaning of proposition "not 1°" is: "there is a student in class A with a surname which is not the same as that of any student in class B." ■

***Example 2.36 Propositions and quantifiers (II)**

Suppose a is the limit of sequence $\{u_n\}$. Defining it in terms of $(\varepsilon - N)$, we get:

"for arbitrary $\varepsilon > 0$, there exists $N > 0$, such that $|u_n - a| < \varepsilon$ for any $n > N$," which can be assembled to "proposition 2°" by several quantifiers:

$$(\forall \varepsilon > 0)\{(\exists N > 0)[(\forall n > N)\,|u_n - a| < \varepsilon]\}.$$

Then, we arrive at the proposition: a is not the limit of sequence $\{u_n\}$ and is "not 2°," that is,

$$\begin{aligned}\text{not } 2° &\Leftrightarrow (\exists \varepsilon > 0) \text{ not } \{(\exists N > 0)[(\forall n > N)\,|u_n - a| < \varepsilon]\}\\ &\Leftrightarrow (\exists \varepsilon > 0)\{(\forall N > 0) \text{ not } [(\forall n > N)\,|u_n - a| < \varepsilon]\}\\ &\Leftrightarrow (\exists \varepsilon > 0)\{(\forall N > 0)[(\exists n > N) \text{ not } |u_n - a| < \varepsilon]\}\\ &\Leftrightarrow (\exists \varepsilon > 0)\{(\forall N > 0)[(\exists n > N)\,|u_n - a| \geq \varepsilon]\}.\end{aligned}$$

That is, "there exists $\varepsilon > 0$, exists $n > N$ for any $N > 0$, such that $|u_n - a| \geq \varepsilon$." ■

From the above examples, it is clear that for a mathematical proposition including several quantifiers, one needs to only modify "∀" in the original proposition to "∃" and "∃" to "∀," while changing "property with p" to "property with not p" to know the meaning of the negative proposition according to the equivalent rules (2.29) and (2.30).

Exercises and Supplementary Exercises

Exercises

Q2.1–2.4: Solve the following systems of linear equations using Gaussian elimination.

2.1 $$\begin{cases} 2x_1 - \frac{1}{2}x_2 - \frac{1}{2}x_3 = 0, \\ -\frac{1}{2}x_1 + 2x_2 - \frac{1}{2}x_4 = 3, \\ -\frac{1}{2}x_1 + 2x_3 - \frac{1}{2}x_4 = 3, \\ -\frac{1}{2}x_2 - \frac{1}{2}x_3 + 2x_4 = 0. \end{cases}$$

2.2 $$\begin{cases} x_1 - 2x_2 + 3x_3 - 4x_4 = 4, \\ x_2 - x_3 + x_4 = -3, \\ x_1 + 3x_2 - 3x_4 = 1, \\ -7x_2 + 3x_3 + x_4 = -3. \end{cases}$$

2.3 $$\begin{cases} 2x_1 + 3x_2 + 5x_3 + x_4 = 3, \\ 3x_1 + 4x_2 + 2x_3 + 3x_4 = -2, \\ x_1 + 2x_2 + 8x_3 - x_4 = 8, \\ 7x_1 + 9x_2 + x_3 + 8x_4 = 0. \end{cases}$$

2.4 $$\begin{cases} x_1 - 10x_2 + 11x_3 - 11x_4 = 0, \\ 2x_1 + 4x_2 - 5x_3 + 7x_4 = 0, \\ 3x_1 - 3x_2 + 3x_3 - 2x_4 = 0, \\ 5x_1 + x_2 - 2x_3 + 5x_4 = 0. \end{cases}$$

Q2.5–2.6: In each of the following systems of equations, consider p and q such that the system of equations has a solution or no solution. In the case of having solutions, compute all the solutions.

2.5 $$\begin{cases} px_1 + x_2 + x_3 = 1, \\ x_1 + px_2 + x_3 = p, \\ x_1 + x_2 + px_3 = p^2. \end{cases}$$

2.6 $$\begin{cases} x_1 - 3x_2 - 6x_3 + 2x_4 = -1, \\ x_1 - x_2 - 2x_3 + 3x_4 = 0, \\ x_1 + 5x_2 + 10x_3 - x_4 = q, \\ 3x_1 + x_2 + px_3 + 4x_4 = 1. \end{cases}$$

2.7 A general needs to conduct a roll-call on his troops. If the soldiers are assembled in the formation 3×3, there are two soldiers left in the last row. If he wants the formation to be 5×5, three soldiers remain in the last row. If the formation is again changed to 7×7, two solders will be left. How many soldiers are there in his troop? (Hint: the answer is between 500 and 1000)

2.8 An ancient Chinese riddle on 100 chickens: It costs \$5 per hen, \$3 per cockerel, and \$1 for three chicks. A customer paid \$100 and bought a total of 100 hens, cockerels, and chicks. How many hens, cockerels, and chicks did the customer buy respectively?

2.9 Suppose $\boldsymbol{A} = \begin{pmatrix} 1 & 0 \\ 2 & -1 \end{pmatrix}$, $\boldsymbol{B} = \begin{pmatrix} 3 & 0 \\ 1 & 2 \end{pmatrix}$, compute $2\boldsymbol{A}$, $3\boldsymbol{B}$, $\boldsymbol{A}+\boldsymbol{B}$, $2\boldsymbol{A}-3\boldsymbol{B}$, $\boldsymbol{AB}-\boldsymbol{BA}$.

2.10 Suppose $\boldsymbol{A} = \begin{pmatrix} 3 & 1 & 1 \\ 2 & 1 & 2 \\ 1 & 2 & 3 \end{pmatrix}$,

$\boldsymbol{B} = \begin{pmatrix} 1 & 1 & 1 \\ 2 & -1 & 0 \\ 1 & 0 & 1 \end{pmatrix}$. Find $\boldsymbol{AB}-\boldsymbol{BA}$.

Q2.11–2.20: Compute the multiplication of matrices.

2.11 $$\begin{pmatrix} 3 & -2 \\ 0 & 1 \\ 2 & 4 \\ -1 & 0 \end{pmatrix} \begin{pmatrix} 2 & 1 & -1 \\ 0 & -1 & 0 \end{pmatrix}.$$

2.12 $$\begin{pmatrix} 1 & 2 & -1 \\ -2 & 1 & 0 \\ 1 & 0 & 3 \end{pmatrix} \begin{pmatrix} 2 & 3 \\ 1 & -1 \\ 2 & 4 \end{pmatrix}.$$

2.13 $$(1, -1, 2) \begin{pmatrix} 2 & -1 & 0 \\ 1 & 1 & 3 \\ 4 & 2 & 1 \end{pmatrix}.$$

2.14 $\begin{pmatrix} a & b \\ ma & mb \end{pmatrix} \begin{pmatrix} -1 & 1 \\ \frac{a}{b} & -\frac{a}{b} \end{pmatrix}$.

2.15 $(y_1, y_2, \ldots, y_n) \begin{pmatrix} a_{11} & a_{12} & \cdots & a_{1n} \\ a_{21} & a_{22} & \cdots & a_{2n} \\ \vdots & \vdots & & \vdots \\ a_{n1} & a_{n2} & \cdots & a_{nn} \end{pmatrix}$.

2.16 $(x_1, x_2) \begin{pmatrix} a_{11} & a_{12} \\ a_{21} & a_{22} \end{pmatrix} \begin{pmatrix} x_1 \\ x_2 \end{pmatrix}$.

2.17 $\begin{pmatrix} 1 & & & & \\ & 1 & & & \\ & & 0 & & \\ & & & \ddots & \\ & & & & 0 \end{pmatrix} \begin{pmatrix} a_{11} & a_{12} & \cdots & a_{1n} \\ a_{21} & a_{22} & \cdots & a_{2n} \\ \vdots & \vdots & & \vdots \\ a_{n1} & a_{n2} & \cdots & a_{nn} \end{pmatrix}$.

2.18 $\begin{pmatrix} a_{11} & a_{12} & \cdots & a_{1n} \\ a_{21} & a_{22} & \cdots & a_{2n} \\ \vdots & \vdots & & \vdots \\ a_{n1} & a_{n2} & \cdots & a_{nn} \end{pmatrix} \begin{pmatrix} 1 & & & & \\ & 2 & & & \\ & & 0 & & \\ & & & \ddots & \\ & & & & 0 \end{pmatrix}$.

2.19 $\begin{pmatrix} 1 & 0 & 0 \\ -2 & 1 & 0 \\ 0 & 0 & 1 \end{pmatrix} \begin{pmatrix} a_1 & a_2 & a_3 \\ b_1 & b_2 & b_3 \\ c_1 & c_2 & c_3 \end{pmatrix}$.

2.20 $\begin{pmatrix} a_1 & a_2 & a_3 & a_4 \\ b_1 & b_2 & b_3 & b_4 \\ c_1 & c_2 & c_3 & c_4 \end{pmatrix} \begin{pmatrix} 0 & 0 & 1 & 0 \\ 0 & 1 & 0 & 0 \\ 1 & 0 & 0 & 0 \\ 0 & 0 & 0 & 1 \end{pmatrix}$.

2.21 Given $\boldsymbol{A} = \boldsymbol{P}\boldsymbol{\Lambda}\boldsymbol{Q}$, where

$$\boldsymbol{P} = \begin{pmatrix} 2 & 3 \\ 1 & 2 \end{pmatrix},\ \boldsymbol{\Lambda} = \begin{pmatrix} 1 & 0 \\ 0 & -1 \end{pmatrix},$$

$$\boldsymbol{Q} = \begin{pmatrix} 2 & -3 \\ -1 & 2 \end{pmatrix},\ \boldsymbol{QP} = \boldsymbol{I}_2,$$

compute $\boldsymbol{A}^8$, $\boldsymbol{A}^9$, $\boldsymbol{A}^{2n}$, $\boldsymbol{A}^{2n+1}$ (n is a positive integer).

2.22 Compute $\begin{pmatrix} a & & \\ & -b & \\ & & c \end{pmatrix}^n$ (n is a positive integer).

2.23 Calculate $\begin{pmatrix} \cos\phi & \sin\phi \\ -\sin\phi & \cos\phi \end{pmatrix}^n$ and $\begin{pmatrix} 0 & 1 \\ -1 & 0 \end{pmatrix}^n$.

2.24 $\boldsymbol{A}$ and $\boldsymbol{B}$ are both nth-order matrices. When are the following equations true?

(i) $(\boldsymbol{A}+\boldsymbol{B})^3 = \boldsymbol{A}^3 + 3\boldsymbol{A}^2\boldsymbol{B} + 3\boldsymbol{A}\boldsymbol{B}^2 + \boldsymbol{B}^3$;
(ii) $(\boldsymbol{A}+\boldsymbol{B})(\boldsymbol{A}-\boldsymbol{B}) = \boldsymbol{A}^2 - \boldsymbol{B}^2$.

2.25 If $\boldsymbol{AB} = \boldsymbol{BA}$ and $\boldsymbol{AC} = \boldsymbol{CA}$, prove that $\boldsymbol{A}$, $\boldsymbol{B}$, and $\boldsymbol{C}$ are matrices of the same order, and $\boldsymbol{A}(\boldsymbol{B}+\boldsymbol{C}) = (\boldsymbol{B}+\boldsymbol{C})\boldsymbol{A}$ and $\boldsymbol{A}(\boldsymbol{BC}) = (\boldsymbol{BC})\boldsymbol{A}$.

2.26 Find all second-order matrices such that their squares are zero.

2.27 Find all second-order matrices which commutes with $\boldsymbol{A} = \begin{pmatrix} 1 & 1 \\ 0 & 1 \end{pmatrix}$ under matrix multiplication.

2.28 Find all third-order matrices which commutes with $\boldsymbol{A} = \begin{pmatrix} 1 & 0 & 0 \\ 0 & 1 & 2 \\ 0 & 1 & -2 \end{pmatrix}$ under matrix multiplication.

2.29 Suppose $\boldsymbol{A}$ is an nth-order diagonal matrix with distinct diagonal elements, that is,

$$\boldsymbol{A} = \begin{pmatrix} a_1 & & & \\ & a_2 & & \\ & & \ddots & \\ & & & a_n \end{pmatrix},$$

$$a_i \neq a_j (i, j = 1, 2, \ldots, n) \text{ if } i \neq j.$$

Prove that a matrix that commutes with $\boldsymbol{A}$ under matrix multiplication is diagonal.

2.30 Prove that the product of two lower triangular matrices of order n is still a lower triangular matrix.

2.31 If $\boldsymbol{A}$ is an upper triangular matrix with all the leading diagonal elements zero, prove that $\boldsymbol{A}^2$ is also an upper triangular matrix with all the leading diagonal elements zero.

2.32 Prove that the product of upper triangular matrices with all the leading diagonal elements 1 is also an upper triangular matrix with all the leading diagonal elements 1.

2.33 Suppose

$$A=\begin{pmatrix}5 & -2 & 1\\ 3 & 4 & -1\end{pmatrix},\qquad B=\begin{pmatrix}-3 & 2 & 0\\ -2 & 0 & 1\end{pmatrix},$$

calculate AB^{T}, $B^{\mathrm{T}}A$, $A^{\mathrm{T}}A$, $(BB^{\mathrm{T}}+AB^{\mathrm{T}})$.

2.34 Prove that $(A_1A_2\cdots A_k)^{\mathrm{T}}=A_k{}^{\mathrm{T}}\cdots A_2{}^{\mathrm{T}}A_1{}^{\mathrm{T}}$.

2.35 If A and B are both symmetric matrices of order n, prove $(A+B)$ and $(A-2B)$ are also symmetric matrices.

2.36 For an arbitrary nth-order matrix A, prove that

(i) $(A+A^{\mathrm{T}})$ is a symmetric matrix, $(A-A^{\mathrm{T}})$ is an antisymmetric matrix;

(ii) A can be expressed as the sum of a symmetric matrix and an antisymmetric matrix.

2.37 Prove that a necessary and sufficient condition for AB being a symmetric matrix is that A and B are commutative, where A and B are both nth-order symmetric matrices.

2.38 Let A be a real symmetric matrix and $A^2=\mathbf{0}$. Prove that $A=\mathbf{0}$.

2.39 Let A be a symmetric matrix of order n and B be an antisymmetric matrix of order n.

(i) Are A^k and B^k symmetric or antisymmetric matrices?

(ii) Prove that $(AB+BA)$ is also an antisymmetric matrix.

2.40 Compute the inverses of the following matrices:

(i) $\begin{pmatrix}8 & -4\\ -5 & 3\end{pmatrix}$;

(ii) $\begin{pmatrix}\cos\theta & \sin\theta\\ -\sin\theta & \cos\theta\end{pmatrix}$;

(iii) $\begin{pmatrix}1 & 2 & 2\\ 2 & 1 & -2\\ 2 & -2 & 1\end{pmatrix}$;

(iv) $\begin{pmatrix}2 & 3 & -1\\ 1 & 2 & 0\\ -1 & 2 & -2\end{pmatrix}$;

(v) $\begin{pmatrix}1 & 0 & 0\\ 1 & 1 & 0\\ 1 & 1 & 1\end{pmatrix}$;

(vi) $\begin{pmatrix}1 & 1 & 0 & 0\\ 0 & 1 & 1 & 0\\ 0 & 0 & 1 & 1\\ 0 & 0 & 0 & 1\end{pmatrix}$.

2.41 Solve the following matrix equations by inverse matrices:

(i) $\begin{pmatrix}2 & 5\\ 1 & 3\end{pmatrix}B=\begin{pmatrix}1 & 1\\ -1 & 0\end{pmatrix}$;

(ii) $\begin{pmatrix}2 & 3 & -1\\ 1 & 2 & 0\\ -1 & 2 & -2\end{pmatrix}X=\begin{pmatrix}2 & 1\\ -1 & 0\\ 3 & 1\end{pmatrix}$;

(iii) $A\begin{pmatrix}1 & 1 & 1\\ 0 & 1 & 1\\ 0 & 0 & 1\end{pmatrix}=\begin{pmatrix}1 & -2 & 1\\ 0 & 1 & -1\end{pmatrix}$.

2.42 Solve the system of linear equations by using inverse matrix.

$$\begin{cases} x_1 + x_2 + x_3 = 1,\\ \phantom{x_1+{}} 2x_2 + 2x_3 = 1,\\ x_1 - x_2 \phantom{{}+2x_3} = 2.\end{cases}$$

2.43 Let A, B, and C be square matrices of the same order.

(i) When is the proposition "if $AB=AC$, then $B=C$ (cancellation law)" true?

(ii) Is $AB\neq AC$ true if $B\neq C$?

2.44 Let both A and B be nth-order matrices. Are the following propositions true? If so, provide the proof. Otherwise, give a counter example.

(i) If both A and B are not invertible, neither is $(A+B)$.

(ii) If AB is invertible, then both A and B are invertible.

(iii) If AB is not invertible, neither A nor B are invertible.

(iv) If A is invertible, then kA is also invertible (k is a number).

2.45 Suppose a square matrix A satisfies $A^2 - A - 2I = 0$. Prove that

(i) both A and $(A - I)$ are invertible and find their inverse matrices,

(ii) both $(A + I)$ and $(A - 2I)$ are not invertible.

2.46 Suppose the square matrix A satisfies $A^2 - 2A + 4I = 0$. Prove that both $(A + I)$ and $(A - 3I)$ are invertible and give their inverse matrices.

2.47 Prove that the inverse matrix of an invertible symmetric matrix is still symmetric.

2.48 Find a necessary and sufficient condition such that an upper (lower) triangular matrix is invertible. Then, prove that the inverse matrix of an invertible upper (lower) triangular matrix is also an upper (lower) triangular matrix.

Q2.49–2.53: Find the inverses of these matrices using elementary operations.

2.49 $\begin{pmatrix} 1 & 2 & 2 \\ 2 & 1 & -2 \\ 2 & -2 & 1 \end{pmatrix}$.

2.50 $\begin{pmatrix} 1 & 2 & 3 & 4 \\ 2 & 3 & 1 & 2 \\ 1 & 1 & 1 & -1 \\ 1 & 0 & -2 & -6 \end{pmatrix}$.

2.51 $\begin{pmatrix} 1 & 0 & 0 & 0 \\ 1 & 1 & 0 & 0 \\ 1 & 1 & 1 & 0 \\ 1 & 1 & 1 & 1 \end{pmatrix}$.

2.52 $\begin{pmatrix} 1 & a & a^2 & a^3 \\ 0 & 1 & a & a^2 \\ 0 & 0 & 1 & a \\ 0 & 0 & 0 & 1 \end{pmatrix}$.

2.53 $\begin{pmatrix} 0 & a_1 & 0 & \cdots & 0 \\ 0 & 0 & a_2 & \cdots & 0 \\ \vdots & \vdots & \vdots & \ddots & \vdots \\ 0 & 0 & 0 & \cdots & a_{n-1} \\ a_n & 0 & 0 & \cdots & 0 \end{pmatrix}$,

where $a_i \neq 0, i = 1, 2, \ldots, n$

Q2.54–2.56: Solve the matrix equations.

2.54 $\begin{pmatrix} 1 & 2 \\ 3 & 4 \end{pmatrix} X = \begin{pmatrix} 3 & 5 \\ 5 & 9 \end{pmatrix}$.

2.55 $X \begin{pmatrix} 1 & 2 & -3 \\ 3 & 2 & -4 \\ 2 & -1 & 0 \end{pmatrix} = \begin{pmatrix} 1 & -3 & 0 \\ 10 & 2 & 7 \\ 10 & 7 & 8 \end{pmatrix}$.

2.56 $\begin{pmatrix} 1 & 1 & 1 & \cdots & 1 \\ 0 & 1 & 1 & \cdots & 1 \\ 0 & 0 & 1 & \cdots & 1 \\ \vdots & \vdots & \vdots & \ddots & \vdots \\ 0 & 0 & 0 & \cdots & 1 \end{pmatrix} X = \begin{pmatrix} 1 & 2 & 3 & \cdots & n \\ 0 & 1 & 2 & \cdots & n-1 \\ 0 & 0 & 1 & \cdots & n-2 \\ \vdots & \vdots & \vdots & \ddots & \vdots \\ 0 & 0 & 0 & \cdots & 1 \end{pmatrix}$.

***2.57** Make a LU decomposition to each of the following matrices, where L is a lower triangular matrix with the main diagonal elements as 1, and U is an upper triangular matrix:

(i) $\begin{pmatrix} 1 & -2 & 3 \\ 2 & 4 & 2 \\ 0 & 1 & 1 \end{pmatrix}$;

(ii) $\begin{pmatrix} 3 & 1 & 0 & 0 \\ 2 & 3 & 1 & 0 \\ 0 & 2 & 3 & 1 \\ 0 & 0 & 2 & 3 \end{pmatrix}$.

2.58 Calculate the products of the following matrices by block multiplication matrices.

(i) $A = \begin{pmatrix} 1 & 3 & 0 & 0 & 0 \\ 2 & 8 & 0 & 0 & 0 \\ 0 & 0 & 1 & 0 & 1 \\ 0 & 0 & 2 & 3 & 2 \\ 0 & 0 & 3 & 1 & 1 \end{pmatrix}$, $B = \begin{pmatrix} 1 & 3 & 0 & 0 & 0 \\ 2 & 8 & 0 & 0 & 0 \\ 1 & 0 & 1 & 0 & 1 \\ 0 & 1 & 2 & 3 & 2 \\ 2 & 3 & 3 & 1 & 1 \end{pmatrix}$.

Compute AB.

(ii) $A = \begin{pmatrix} 1 & 0 & 1 & 0 & 0 \\ 0 & 2 & -1 & 0 & 0 \\ 3 & 1 & 0 & 0 & 0 \\ 0 & 0 & 0 & -2 & 0 \\ 0 & 0 & 0 & 0 & -2 \end{pmatrix}$,

$$B = \begin{pmatrix} 1 & 0 & 1 & 0 & 0 \\ 0 & 2 & 0 & 0 & 0 \\ 0 & 0 & 3 & 0 & 0 \\ 0 & 0 & 0 & -1 & 3 \\ 0 & 0 & 0 & 4 & 2 \end{pmatrix}.$$

Compute AB.

2.59 Let A be an $m \times n$ matrix, B be an $n \times s$ matrix, and x be an $n \times 1$ matrix. Prove that a necessary and sufficient condition for $AB = 0$ is that each column of B is a solution to the homogeneous system of linear equations $Ax = 0$.

2.60 Let C be an invertible matrix of order n, D be a $3 \times n$ matrix, and

$$D = \begin{pmatrix} 1 & 2 & \cdots & n \\ 0 & 0 & \cdots & 0 \\ 0 & 0 & \cdots & 0 \end{pmatrix}.$$

Find an $n \times (n+3)$ matrix A such that $A\begin{pmatrix} C \\ D \end{pmatrix} = I_n$ by block multiplication.

2.61 Suppose $A = \begin{pmatrix} 0 & B \\ C & 0 \end{pmatrix}$, where B is an invertible matrix of order n, and C an invertible matrix with order m. Prove that A is invertible and find A^{-1}.

2.62 Prove that the following matrices are invertible by using the method of partitioning matrix, and give their inverse matrices.

(i) $\begin{pmatrix} 1 & 2 & 0 & 0 & 0 \\ 2 & 5 & 0 & 0 & 0 \\ 0 & 0 & 3 & 0 & 0 \\ 0 & 0 & 0 & 1 & 0 \\ 0 & 0 & 0 & 0 & 1 \end{pmatrix}$;

(ii) $\begin{pmatrix} 0 & 0 & 0 & 4 & 4 \\ 0 & 0 & 0 & 7 & 8 \\ 1 & 1 & 1 & 0 & 0 \\ 0 & 1 & 1 & 0 & 0 \\ 0 & 0 & 1 & 0 & 0 \end{pmatrix}$;

(iii) $\begin{pmatrix} 0 & a_1 & 0 & \cdots & 0 \\ 0 & 0 & a_2 & \cdots & 0 \\ \vdots & \vdots & \vdots & \ddots & \vdots \\ 0 & 0 & 0 & \cdots & a_{n-1} \\ a_n & 0 & 0 & \cdots & 0 \end{pmatrix}$;

(iv) $\begin{pmatrix} 2 & 0 & 1 & 0 & 2 \\ 0 & 2 & 0 & 1 & 3 \\ 0 & 0 & 1 & 0 & 0 \\ 0 & 0 & 0 & 1 & 0 \\ 0 & 0 & 0 & 0 & 1 \end{pmatrix}$.

***2.63** Lct A, B, C, and D be nth-order matrices, $|A| \neq 0$, $AC = CA$. Prove that,

$$\begin{vmatrix} A & B \\ C & D \end{vmatrix} = |AD - CB|.$$

***2.64** Suppose $A = \begin{pmatrix} 0 & B \\ C & D \end{pmatrix}$, where B and C are individually second- and third-order invertible matrices, and B^{-1} and C^{-1} are known. Compute A^{-1}.

***2.65** Partition the nth-order matrix A into

$$A = \begin{pmatrix} A_{n-1} & b \\ c & a_{nn} \end{pmatrix},$$

where A_{n-1} is an invertible matrix of order $(n-1)$, denoted as A_{n-1}^{-1}. If A is invertible, find A^{-1} (this method of using A_{n-1}^{-1} to find A^{-1} is called the bordering method).

***2.66** Using the bordering method in **Q2.65**, find the inverse matrices of

$$A = \begin{pmatrix} 1 & 2 & 2 \\ 2 & 1 & -2 \\ 2 & -2 & 1 \end{pmatrix}, \quad B = \begin{pmatrix} 1 & 2 & 3 & 4 \\ 2 & 3 & 1 & 2 \\ 1 & 1 & 1 & -1 \\ 1 & 0 & -2 & -6 \end{pmatrix}.$$

Supplementary exercises

2.67 Let both $\boldsymbol{A}$ and $\boldsymbol{B}$ be matrices of order 4, $|\boldsymbol{A}|=-2$, $|\boldsymbol{B}|=3$. Calculate

(i) $\left|\frac{1}{2}\boldsymbol{AB}^{-1}\right|$;
(ii) $\left|-\boldsymbol{AB}^{\mathrm{T}}\right|$;
(iii) $\left|(\boldsymbol{AB})^{-1}\right|$;
(iv) $\det\left[(\boldsymbol{AB})^{\mathrm{T}}\right]^{-1}$;
(v) $\left|-3\boldsymbol{A}^*\right|$ ($\boldsymbol{A}^*$ is the adjoint matrix of $\boldsymbol{A}$).

2.68 Suppose $\boldsymbol{\alpha}=(1,-2,3)^{\mathrm{T}}$, $\boldsymbol{\beta}=\left(-1,\frac{1}{2},0\right)^{\mathrm{T}}$, $\boldsymbol{A}=\boldsymbol{\alpha\beta}^{\mathrm{T}}$. Calculate $\left|\boldsymbol{A}^{100}\right|$.

2.69 Suppose $\boldsymbol{A}$ is a fourth-order matrix. When $|\boldsymbol{A}|=a\neq 0$, calculate $\det(\left|\boldsymbol{A}^*\right|\boldsymbol{A})$.

2.70 Suppose both $\boldsymbol{A}$ and $\boldsymbol{B}$ are nth-order matrices, $\boldsymbol{I}$ is an nth-order unit matrix; which of the following propositions is true?

(i) $\boldsymbol{A}=\boldsymbol{I}\Leftrightarrow|\boldsymbol{A}|=1$;
(ii) $\boldsymbol{AB}\neq\boldsymbol{0}\Leftrightarrow\boldsymbol{A}\neq\boldsymbol{0}$ and $\boldsymbol{B}\neq\boldsymbol{0}$;
(iii) $\boldsymbol{A}^2=\boldsymbol{I}$, then $\boldsymbol{A}=\boldsymbol{I}$ or $\boldsymbol{A}=-\boldsymbol{I}$;
(iv) $(\boldsymbol{A}+\boldsymbol{I})(\boldsymbol{A}-\boldsymbol{I})=(\boldsymbol{A}-\boldsymbol{I})(\boldsymbol{A}+\boldsymbol{I})$;
(v) $\boldsymbol{AB}$ is invertible $\Leftrightarrow$ both $\boldsymbol{A}$ and $\boldsymbol{B}$ are invertible;
(vi) $\left|-2\boldsymbol{A}^*\boldsymbol{B}\right|=-2\,|\boldsymbol{A}|\,|\boldsymbol{B}|$.

2.71 Suppose $\boldsymbol{\alpha}=(a,b,c)$, $\boldsymbol{\beta}=(x,y,z)$, and

$$\boldsymbol{\alpha}^{\mathrm{T}}\boldsymbol{\beta}=\begin{pmatrix}-2&4&-6\\1&-2&3\\-1&2&-3\end{pmatrix}.$$

Calculate $\boldsymbol{\alpha\beta}^{\mathrm{T}}$.

2.72 Suppose $\boldsymbol{\alpha}=(x_1,x_2,\ldots,x_n)^{\mathrm{T}}$, $\boldsymbol{\beta}=(y_1,y_2,\ldots,y_n)^{\mathrm{T}}$, $\boldsymbol{\alpha}^{\mathrm{T}}\boldsymbol{\beta}=3$, $\boldsymbol{B}=\boldsymbol{\alpha\beta}^{\mathrm{T}}$ and $\boldsymbol{A}=\boldsymbol{I}-\boldsymbol{B}$. Prove that

(i) $\boldsymbol{B}^k=3^{k-1}\boldsymbol{B}$ ($k\geq 2$ is a positive integer),
(ii) either $(\boldsymbol{A}+2\boldsymbol{I})$ or $(\boldsymbol{A}-\boldsymbol{I})$ is not invertible,
(iii) both $\boldsymbol{A}$ and $(\boldsymbol{A}+\boldsymbol{I})$ are invertible.

2.73 Suppose $\boldsymbol{A}$ is a third-order matrix, $|\boldsymbol{A}|>0$, $\boldsymbol{A}^*=\mathrm{diag}(1,-1,-4)$, and $\boldsymbol{ABA}^{-1}=\boldsymbol{BA}^{-1}+3\boldsymbol{I}$. Calculate $\boldsymbol{B}$.

2.74 Suppose that the nth-order matrix $\boldsymbol{A}$ satisfies $\boldsymbol{A}^{\mathrm{T}}\boldsymbol{A}=\boldsymbol{I}$ and $|\boldsymbol{A}|<0$. Find $|\boldsymbol{A}+\boldsymbol{I}|$.

2.75 Suppose that $\boldsymbol{A}$ is an invertible matrix of odd order, $\boldsymbol{A}^{-1}=\boldsymbol{A}^{\mathrm{T}}$ and $|\boldsymbol{A}|=1$. Find $|\boldsymbol{I}-\boldsymbol{A}|$.

2.76 Suppose both $\boldsymbol{A}$ and $\boldsymbol{B}$ are nth-order matrices and $\boldsymbol{A}=\boldsymbol{B}^{\mathrm{T}}$. Which of the following expressions is a simplified form of $\boldsymbol{A}^{\mathrm{T}}(\boldsymbol{B}^{-1}\boldsymbol{A}^{-1}+\boldsymbol{I})^{\mathrm{T}}$?

(a) $\boldsymbol{A}+\boldsymbol{B}$;
(b) $\boldsymbol{B}+\boldsymbol{A}^{-1}$;
(c) $\boldsymbol{A}^{\mathrm{T}}\boldsymbol{B}$;
(d) $\boldsymbol{A}+\boldsymbol{A}^{-1}$;
(e) $\boldsymbol{A}+\boldsymbol{B}^{-1}$;
(f) $\boldsymbol{AA}^{\mathrm{T}}$.

2.77 Suppose $\boldsymbol{\alpha}=(1,0,-1)^{\mathrm{T}}$, k is a positive integer, and $\boldsymbol{A}=\boldsymbol{\alpha\alpha}^{\mathrm{T}}$. Find $\left|k\boldsymbol{I}-\boldsymbol{A}^n\right|$.

2.78 A fourth-order matrix $\boldsymbol{A}$ satisfies: $(2\boldsymbol{I}-\boldsymbol{C}^{-1}\boldsymbol{B})\boldsymbol{A}^{\mathrm{T}}=\boldsymbol{C}^{-1}$. Calculate $\boldsymbol{A}$ where

$$\boldsymbol{B}=\begin{pmatrix}1&2&-3&-2\\0&1&2&-3\\0&0&1&2\\0&0&0&1\end{pmatrix},\quad\boldsymbol{C}=\begin{pmatrix}1&2&0&1\\0&1&2&0\\0&0&1&2\\0&0&0&1\end{pmatrix}.$$

2.79 Suppose

$$\boldsymbol{B}=\begin{pmatrix}-1&1&0\\0&0&2\\0&0&2\end{pmatrix},\qquad\boldsymbol{C}=\begin{pmatrix}1&1&0\\0&2&2\\0&0&3\end{pmatrix},$$

and $\boldsymbol{A}$, $\boldsymbol{B}$, and $\boldsymbol{C}$ satisfy $(\boldsymbol{I}-\boldsymbol{C}^{-1}\boldsymbol{B})^{\mathrm{T}}\boldsymbol{C}^{\mathrm{T}}\boldsymbol{A}=\boldsymbol{I}$. Find $\boldsymbol{A}$ and $\boldsymbol{A}^{-1}$.

2.80 Suppose $\boldsymbol{B}$ is an nth-order ($n\geq 2$) matrix with elements 1. Prove that

(i) $\boldsymbol{B}^k=n^{k-1}\boldsymbol{B}$ ($k\geq 2$, positive integer),
(ii) $(\boldsymbol{I}-\boldsymbol{B})^{-1}=\boldsymbol{I}-\frac{1}{n-1}\boldsymbol{B}$.

2.81 Suppose $\boldsymbol{A}$ is a real symmetric matrix of order 3, its main diagonal elements are all 0, and $\boldsymbol{B}=\mathrm{diag}(0,1,2)$. When is $(\boldsymbol{AB}+\boldsymbol{I})$ an invertible matrix?

2.82 Let both $\boldsymbol{P}$ and $\boldsymbol{A}$ be nth-order matrices, and $\boldsymbol{P}^{-1}\boldsymbol{A}\boldsymbol{P}=\text{diag}(1, 1, \ldots, 1, 0, \cdots, 0)$ (having r number of 1's). Calculate $|\boldsymbol{A}+2\boldsymbol{I}|$.

2.83 Suppose $\boldsymbol{A}$ is an invertible matrix with order n ($n\geq 2$). Prove that

(i) $(\boldsymbol{A}^*)^{-1}=(\boldsymbol{A}^{-1})^*$;

(ii) $(\boldsymbol{A}^{\mathrm{T}})^*=(\boldsymbol{A}^*)^{\mathrm{T}}$;

(iii) $(k\boldsymbol{A})^*=k^{n-1}\boldsymbol{A}^*$ (k is a nonzero constant number).

2.84 Calculate the following matrices

(i) $\begin{pmatrix}1 & 1 & 0\\ 0 & 1 & 1\\ 0 & 0 & 1\end{pmatrix}^n$;

(ii) $\begin{pmatrix}a & 1 & 0 & 0\\ 0 & a & 1 & 0\\ 0 & 0 & a & 1\\ 0 & 0 & 0 & a\end{pmatrix}^n$.

2.85 Prove that a matrix which is commutative with any matrix of order n is an nth-order scalar matrix.

2.86 The sum of main diagonal elements of an nth-order matrix $\boldsymbol{A}=(a_{ij})$ is called the trace of the matrix $\boldsymbol{A}$, written as $\text{tr}(\boldsymbol{A})$, that is,

$$\text{tr}(\boldsymbol{A})=\sum_{i=1}^{n} a_{ii}.$$

Prove that if $\boldsymbol{A}$ is an $m\times n$ matrix and $\boldsymbol{B}$ an $n\times m$ matrix, then

$$\text{tr}(\boldsymbol{AB})=\text{tr}(\boldsymbol{BA}).$$

2.87 Prove $\boldsymbol{AB}-\boldsymbol{BA}\neq \boldsymbol{I}_n$ always holds for two arbitrary matrices $\boldsymbol{A}$ and $\boldsymbol{B}$ of order n.

2.88 An nth-order matrix $\boldsymbol{A}$ is called a nilpotent matrix if there exists a positive integer k such that $\boldsymbol{A}^k=\boldsymbol{0}$. Suppose a nilpotent matrix $\boldsymbol{A}$ satisfies $\boldsymbol{A}^k=\boldsymbol{0}$ (k is a positive integer), prove that $(\boldsymbol{I}-\boldsymbol{A})$ is invertible and find its inverse matrix.

2.89 Suppose $\boldsymbol{A}=\begin{pmatrix}a & 1 & 0 & 0\\ 0 & a & 1 & 0\\ 0 & 0 & a & 1\\ 0 & 0 & 0 & a\end{pmatrix}$, $f(x)=(x-b)^n$. Calculate $f(\boldsymbol{A})$ and find its inverse when $f(\boldsymbol{A})$ is invertible.

2.90 Suppose $\boldsymbol{A}=\begin{pmatrix}0 & 1\\ 3 & -2\end{pmatrix}$,

$$f(x)=\begin{vmatrix}x-1 & x & 0\\ 0 & x-1 & -3\\ 1 & 1 & 1\end{vmatrix},$$

$$g(x)=\begin{vmatrix}x & -1\\ -3 & x+2\end{vmatrix}.$$

Calculate $f(\boldsymbol{A})$ and $g(\boldsymbol{A})$.

2.91 Prove that the inverse of an upper (lower) triangular matrix with all main diagonal elements as 1, is still an upper (lower) triangular matrix.

2.92 Prove that a necessary condition for an nth-order antisymmetric matrix to be invertible is that n is an even number. Give an example to illustrate that n being an even number is not a sufficient condition for an antisymmetric matrix to be invertible.

2.93 Suppose $\boldsymbol{P}=\begin{pmatrix}\boldsymbol{A} & \boldsymbol{B}\\ \boldsymbol{0} & \boldsymbol{C}\end{pmatrix}$, both $\boldsymbol{A}$ and $\boldsymbol{C}$ are invertible matrices. Prove that $\boldsymbol{P}$ is invertible, and give $\boldsymbol{P}^{-1}$.

2.94 Prove that the inverse matrix of a lower triangular invertible matrix is also a lower triangular matrix.

2.95 Prove any pair of polynomials $f(\boldsymbol{A})$ and $g(\boldsymbol{A})$ of nth-order matrix $\boldsymbol{A}$ are commutative.

2.96 Prove if nth-order matrices $\boldsymbol{A}$ and $\boldsymbol{B}$ are commutative, then any pair of polynomials $f(\boldsymbol{A})$ and $g(\boldsymbol{B})$ are commutative.

3 Systems of Linear Equations

CHAPTER OVERVIEW

In this chapter, we will learn about the fundamental theory of systems of linear equations and their solutions. Necessary and sufficient conditions are provided for:

- a nonhomogeneous system of linear equations to have a solution or
- a homogeneous system to have a nonzero solution.

Specific forms of their respective solutions and their structures are considered as well.

Previously, we learnt that Gaussian elimination can be used as a basic tool to solve linear equations. It does not, however, tell us under what conditions the row-reduced echelon matrix of $\boldsymbol{Ax} = \boldsymbol{b}$ has $d_{r+1} = 0$ (i.e., the system of linear equations has a solution), where d_{r+1} is a leading element of echelon matrix $(\boldsymbol{C}, \boldsymbol{d})$ obtained from reducing $(\boldsymbol{A}, \boldsymbol{b})$ by row elimination (see §2.1 for procedure).

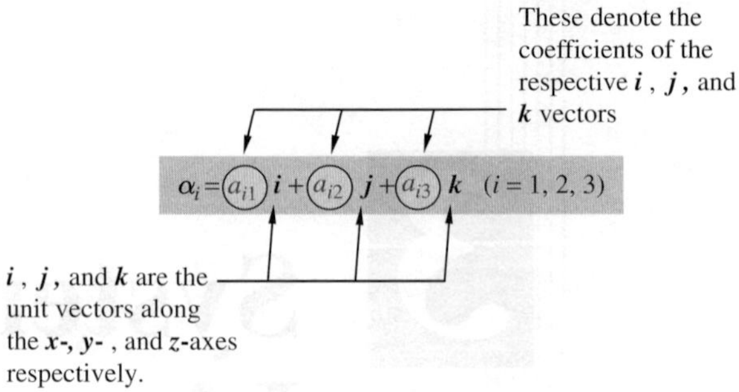

Figure 3.1 Notations used for systems of linear equations

Furthermore, we are also keen to learn if the numbers of nonzero rows of various echelon matrices, obtained from different types of elimination processes, could be uniquely determined. Are we certain that the number of free variables can be uniquely determined? Free variables can be chosen selectively and there is a solution set with respect to such choices. However, are the sets the same? To get a clear answer to all these questions, we shall now introduce n-dimensional vectors, define linear operations among them, study their linear dependence, and then introduce the concept of rank for these vectors.

Readers are encouraged to devote more time for the study of this chapter as we will be introducing many novel and tough concepts of linear algebra. Mastering Chapters 2 and 3 will build up readers' foundation on linear algebra.

3.1 n-dimensional Vectors and Linear Dependence

In the analytic geometry of space, consider the following system of homogeneous linear equations in three unknowns,

$$\begin{cases} a_{11}x_1 + a_{12}x_2 + a_{13}x_3 = 0, \\ a_{21}x_1 + a_{22}x_2 + a_{23}x_3 = 0, \\ a_{31}x_1 + a_{32}x_2 + a_{33}x_3 = 0. \end{cases} \tag{3.1}$$

The geometric interpretation of the solution(s) to this system may be discussed in terms of vectors as a tool. Denote coefficient vectors (i.e., vectors along the normal to the respective planes) of the three equations by

$$\boldsymbol{\alpha}_i = a_{i1}\boldsymbol{i} + a_{i2}\boldsymbol{j} + a_{i3}\boldsymbol{k} \ (i = 1, 2, 3),$$

or in brief, by

$$\boldsymbol{\alpha}_i = (a_{i1}, a_{i2}, a_{i3}) \ (i = 1, 2, 3).$$

The solution to the system of linear equations (3.1), written in vector form, is generally represented as

$$\boldsymbol{x} = x_1\boldsymbol{i} + x_2\boldsymbol{j} + x_3\boldsymbol{k} \quad \text{or} \quad \boldsymbol{x} = (x_1, x_2, x_3).$$

Clearly, the solution vector $\boldsymbol{x}$ of the system (3.1) must be perpendicular to each of $\boldsymbol{\alpha}_1$, $\boldsymbol{\alpha}_2$, and $\boldsymbol{\alpha}_3$ (because the corresponding dot product is zero). Hence,

(i) if $\boldsymbol{\alpha}_1$, $\boldsymbol{\alpha}_2$, and $\boldsymbol{\alpha}_3$ are *not coplanar*, the zero vector being the only vector perpendicular to each one of them, then the system of linear equations (3.1) has only the zero solution.

(ii) if $\boldsymbol{\alpha}_1$, $\boldsymbol{\alpha}_2$, and $\boldsymbol{\alpha}_3$ are *coplanar but not collinear*, then all the vectors perpendicular to this plane are solution vectors of system (3.1), where (3.1) has infinitely many solution vectors that are parallel to each other.

(iii) if $\boldsymbol{\alpha}_1$, $\boldsymbol{\alpha}_2$, and $\boldsymbol{\alpha}_3$ are *collinear*, then all the vectors on the plane which are perpendicular to this line (and through the origin) are solution vectors of system (3.1). In this case, any solution vector can be expressed as

$$\boldsymbol{x} = k_1\boldsymbol{x}^{(1)} + k_2\boldsymbol{x}^{(2)},$$

where $\boldsymbol{x}^{(1)}$ and $\boldsymbol{x}^{(2)}$ are two nonzero solution vectors of the system of linear equations (3.1), which are not collinear and k_1 and k_2 are arbitrary constants.

For a system of linear equations in n unknowns $\boldsymbol{Ax} = \boldsymbol{b}$,

(i) if we consider each row in the coefficient matrix or each row in the augmented matrix as a vector respectively, and

(ii) if we define the operation of addition and scalar multiplication among each row as 3-dimensional geometric vectors,

then the behavior of solution(s) of the system of linear equations depend on the *mutual relation* among all the row vectors of $\boldsymbol{A}$ and those of $(\boldsymbol{A}, \boldsymbol{b})$.

We will now introduce the concept of an n-dimensional vector. Further, we will define linear operations among them and discuss the concept of linear dependence of vectors.

Definition 3.1

An ordered n-tuple $(a_1, a_2, \ldots, a_n)$ composed of n numbers $a_1, a_2, \ldots, a_n$ from a number field F is called an ***n*-vector** (more commonly called an ***n*-dimensional vector**) over F, denoted as

$$\boldsymbol{\alpha} = (a_1, a_2, \ldots, a_n), \tag{3.2}$$

where a_i is called the ith component of $\boldsymbol{\alpha}$.

A vector written as (3.2) is called a **row vector**, whereas a vector written as a column (also expressed by the transpose notation of matrix),

$$\boldsymbol{\alpha} = (a_1, a_2, \ldots, a_n)^{\mathrm{T}}, \tag{3.3}$$

is called a **column vector**.

The set of all n-vectors over F is denoted by F^n.

Definition 3.2

Let $\boldsymbol{\alpha} = (a_1, a_2, \ldots, a_n)$, $\boldsymbol{\beta} = (b_1, b_2, \ldots, b_n) \in F^n$, $k \in F$.
Define

(i) $\boldsymbol{\alpha} = \boldsymbol{\beta}$, if and only if, $a_i = b_i$ $(i = 1, 2, \ldots, n)$,
(ii) addition of vectors (or the sum of $\boldsymbol{\alpha}$ and $\boldsymbol{\beta}$) is

$$\boldsymbol{\alpha} + \boldsymbol{\beta} = (a_1 + b_1, a_2 + b_2, \ldots, a_n + b_n),$$

(iii) multiplication of a vector by a number (or scalar multiplication) is

$$k\boldsymbol{\alpha} = (ka_1, ka_2, \ldots, ka_n).$$

When $k = -1$ in Definition (3.2) (iii), then

$$(-1)\,\boldsymbol{\alpha} = (-a_1, -a_2, \ldots, -a_n). \tag{3.4}$$

The vector in the *RHS* of (3.4) is called a **negative vector** of $\boldsymbol{\alpha}$, denoted by $-\boldsymbol{\alpha}$. The subtraction of vectors is defined as

$$\boldsymbol{\beta} - \boldsymbol{\alpha} = \boldsymbol{\beta} + (-\boldsymbol{\alpha}).$$

The n-dimensional vector $(0, 0, \ldots, 0)$ with all components as zero is called the n-dimensional **zero vector**, denoted by $\mathbf{0}_n$ or $\mathbf{0}$, for short.

We name the above-mentioned operations of addition and scalar multiplication defined in F^n as linear operations of vectors. Suppose $\boldsymbol{\alpha}, \boldsymbol{\beta}, \boldsymbol{\gamma} \in F^n$, and $1, k, l \in F$; we can easily verify by definition that they satisfy the following rules of linear operations.

(i) $\boldsymbol{\alpha} + \boldsymbol{\beta} = \boldsymbol{\beta} + \boldsymbol{\alpha}$ (commutative law of addition);
(ii) $(\boldsymbol{\alpha} + \boldsymbol{\beta}) + \boldsymbol{\gamma} = \boldsymbol{\alpha} + (\boldsymbol{\beta} + \boldsymbol{\gamma})$ (associative law of addition);
(iii) $\boldsymbol{\alpha} + \mathbf{0}_n = \boldsymbol{\alpha}$, for any vector $\boldsymbol{\alpha}$;
(iv) There is a negative vector $-\boldsymbol{\alpha}$, such that $\boldsymbol{\alpha} + (-\boldsymbol{\alpha}) = \mathbf{0}_n$ for any vector $\boldsymbol{\alpha}$;
(v) $1\boldsymbol{\alpha} = \boldsymbol{\alpha}$;
(vi) $k\,(l\boldsymbol{\alpha}) = (kl)\,\boldsymbol{\alpha}$ (associative law of scalar multiplication);
(vii) $k\,(\boldsymbol{\alpha} + \boldsymbol{\beta}) = k\boldsymbol{\alpha} + k\boldsymbol{\beta}$ (distributive law of scalar multiplication);
(viii) $(k + l)\,\boldsymbol{\alpha} = k\boldsymbol{\alpha} + l\boldsymbol{\alpha}$ (distributive law of scalar multiplication).

Besides the above eight rules, the following rules on linear operation can be verified easily.

(ix) $0\boldsymbol{\alpha} = \mathbf{0}_n$, $k\mathbf{0}_n = \mathbf{0}_n$ (where 0 is the number zero, k is an arbitrary number);
(x) If $k\boldsymbol{\alpha} = \mathbf{0}_n$, then either $k = 0$ or $\boldsymbol{\alpha} = \mathbf{0}_n$;
(xi) The vector equation $\boldsymbol{\alpha} + x = \boldsymbol{\beta}$ has a unique solution $x = \boldsymbol{\beta} - \boldsymbol{\alpha}$.

Definition 3.3

The set of all n-dimensional vectors over a number field F endowed with the well-defined addition and scalar multiplication is called an ***n*-dimensional vector space** over the number field F, which is still denoted by F^n. It is called an ***n* dimension real vector space** when $F = \mathbb{R}$ (the field of real numbers), which is denoted as $\mathbb{R}^n$.

Definition 3.4

Suppose $\boldsymbol{\alpha}_i \in F^n$, $k_i \in F$ $(i = 1, 2, \ldots, m)$. Then, the vector

$$\sum_{i=1}^{m} k_i \boldsymbol{\alpha}_i = k_1\boldsymbol{\alpha}_1 + k_2\boldsymbol{\alpha}_2 + \cdots + k_m\boldsymbol{\alpha}_m$$

is called a **linear combination** of vectors $\boldsymbol{\alpha}_1, \boldsymbol{\alpha}_2, \ldots, \boldsymbol{\alpha}_m$ over F. A vector $\boldsymbol{\beta}$ is said to be **linearly expressed** by $\boldsymbol{\alpha}_1, \boldsymbol{\alpha}_2, \ldots, \boldsymbol{\alpha}_m$ if $\boldsymbol{\beta} = \sum_{i=1}^{m} k_i\boldsymbol{\alpha}_i$.

The linear dependence of vectors, a key concept in linear algebra, is a property of vectors under linear operations. To better explain this concept, we will introduce a few concepts in geometry with regard to 3-dimensional real vector space. Then, we will then provide the formal definition of linear dependence of vectors.

For any three coplanar nonzero vectors $\boldsymbol{\alpha}_1, \boldsymbol{\alpha}_2, \boldsymbol{\alpha}_3$, there is at least one vector among them, which can be linearly expressed by the other two. For example, in Figure 3.2, observe that $\boldsymbol{\alpha}_3 = l_1\boldsymbol{\alpha}_1 + l_2\boldsymbol{\alpha}_2$ and in Figure 3.3, $\boldsymbol{\alpha}_1 = 0\boldsymbol{\alpha}_2 + l_3\boldsymbol{\alpha}_3$. These two different statements can be represented by the following equivalent expression: There exist numbers k_1, k_2, k_3, not all 0, such that $k_1\boldsymbol{\alpha}_1 + k_2\boldsymbol{\alpha}_2 + k_3\boldsymbol{\alpha}_3 = \mathbf{0}_3$. If $\boldsymbol{\alpha}_1, \boldsymbol{\alpha}_2, \boldsymbol{\alpha}_3$ are not coplanar (as shown in Figure 3.4), none of the three can be linearly expressed by the other two, that is, $k_1\boldsymbol{\alpha}_1 + k_2\boldsymbol{\alpha}_2 + k_3\boldsymbol{\alpha}_3 = \mathbf{0}_3$ only if $k_1 = k_2 = k_3 = 0$.

The previously discussed property of 3-dimensional vectors under linear operations is precisely what determines the linear dependence of vectors (i.e.,

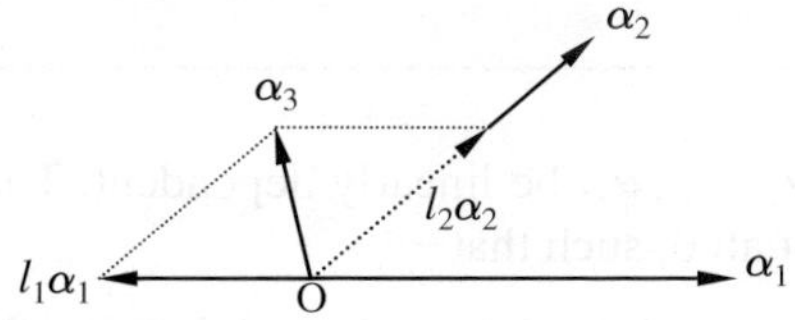

Figure 3.2 $\alpha_3 = l_1\alpha_1 + l_2\alpha_2$

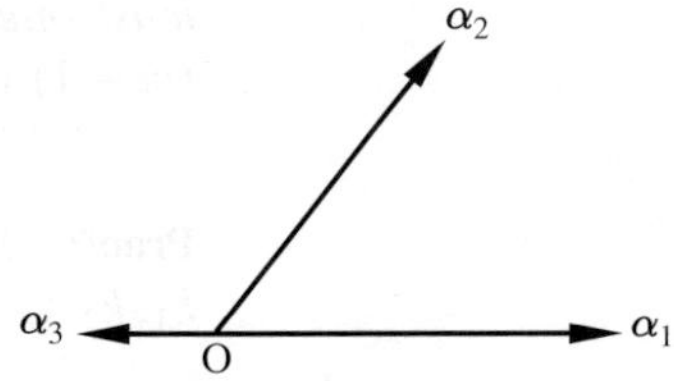

Figure 3.3 $\alpha_1 = 0\alpha_2 + l_3\alpha_3$

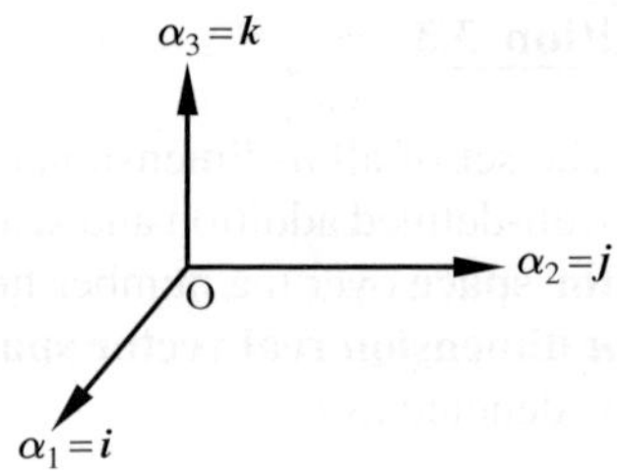

Figure 3.4 $k_1\alpha_1 + k_2\alpha_2 + k_3\alpha_3 = 0_3$ only if $k_1 = k_2 = k_3 = 0$

in a set of vectors, whether there exists a vector which can be linearly expressed by the others, or in other words, whether there exist coefficients, not all 0, such that a linear combination of all the vectors belonging to that set is equal to the zero vector).

Definition 3.5

For m vectors $\boldsymbol{\alpha}_1, \boldsymbol{\alpha}_2, \ldots, \boldsymbol{\alpha}_m \in F^n$, if there exist m numbers $k_1, k_2, \ldots, k_m \in F$, not all 0, such that

$$k_1\boldsymbol{\alpha}_1 + k_2\boldsymbol{\alpha}_2 + \cdots + k_m\boldsymbol{\alpha}_m = \mathbf{0} \tag{3.5}$$

then $\boldsymbol{\alpha}_1, \boldsymbol{\alpha}_2, \ldots, \boldsymbol{\alpha}_m$ are said to be **linearly dependent** over F. Otherwise, $\boldsymbol{\alpha}_1, \boldsymbol{\alpha}_2, \ldots, \boldsymbol{\alpha}_m$ are said to be **linearly independent** over F.

Readers should pay attention to the mathematical essence of the word "otherwise" in the definition of linear independence, which means "there do not exist numbers $k_1, k_2, \ldots, k_m$, not all 0, such that (3.5) holds." The logically equivalent statements are "(3.5) holds, only if $k_1, k_2, \ldots, k_m$ are all 0" or "$k_1, k_2, \ldots, k_m$ must be all 0 if (3.5) holds."

We will now express $\mathbf{0}_n$ as $\mathbf{0}$ (Hint: Do not confuse it with 0, they have different meanings).

THEOREM 3.1

$\boldsymbol{\alpha}_1, \boldsymbol{\alpha}_2, \ldots, \boldsymbol{\alpha}_m$ ($m \geq 2$) are linearly dependent, if and only if, there exists at least one vector among them which can be linearly expressed by the other ($m-1$) vectors.

Proof: Let $\boldsymbol{\alpha}_1, \boldsymbol{\alpha}_2, \ldots, \boldsymbol{\alpha}_m$ be linearly dependent. Thus, there are m numbers $k_1, k_2, \ldots, k_m$, not all 0, such that

$$k_1\boldsymbol{\alpha}_1 + k_2\boldsymbol{\alpha}_2 + \cdots + k_m\boldsymbol{\alpha}_m = \mathbf{0}.$$

Without any loss of generality, letting $k_1 \neq 0$, we get

$$\boldsymbol{\alpha}_1 = -\frac{k_2}{k_1}\boldsymbol{\alpha}_2 - \frac{k_3}{k_1}\boldsymbol{\alpha}_3 - \cdots - \frac{k_m}{k_1}\boldsymbol{\alpha}_m$$

by the rules of linear operations of vectors. Thus, the condition of necessity is proved.

To prove its sufficiency, suppose, without any loss of generality, that $\boldsymbol{\alpha}_1$ be linearly expressed by $\boldsymbol{\alpha}_2, \boldsymbol{\alpha}_3, \ldots, \boldsymbol{\alpha}_m$. Then,

$$\boldsymbol{\alpha}_1 = l_2\boldsymbol{\alpha}_2 + l_3\boldsymbol{\alpha}_3 + \cdots + l_m\boldsymbol{\alpha}_m,$$

where

$$1\boldsymbol{\alpha}_1 - l_2\boldsymbol{\alpha}_2 - l_3\boldsymbol{\alpha}_3 - \cdots - l_m\boldsymbol{\alpha}_m = \mathbf{0}.$$

Obviously, $1, -l_2, -l_3, \ldots, -l_m$ are not all zeros. So, $\boldsymbol{\alpha}_1, \boldsymbol{\alpha}_2, \ldots, \boldsymbol{\alpha}_m$ are linearly dependent. □

An equivalent proposition (contrapositive proposition) to Theorem 3.1 is: The vectors $\boldsymbol{\alpha}_1, \boldsymbol{\alpha}_2, \ldots, \boldsymbol{\alpha}_m$ $(m \geq 2)$ are linearly independent, if and only if, none of the vectors can be linearly expressed by the others.

Example 3.1 Proof of linearly independent vectors

$\boldsymbol{\varepsilon}_i = (0, 0, \ldots, 0, 1, 0, \ldots, 0)$, where the ith component is 1 and the others are 0, is an n-dimensional vector. Prove that $\boldsymbol{\varepsilon}_1, \boldsymbol{\varepsilon}_2, \ldots, \boldsymbol{\varepsilon}_n$ are linearly independent.

Proof: Suppose there exist n numbers $k_1, k_2, \ldots, k_n$ such that

$$k_1\boldsymbol{\varepsilon}_1 + k_2\boldsymbol{\varepsilon}_2 + \cdots + k_n\boldsymbol{\varepsilon}_n = \mathbf{0}.$$

Then,

$$(k_1, k_2, \ldots, k_n) = \mathbf{0},$$

where $k_1 = k_2 = \cdots = k_n = 0$. So, $\boldsymbol{\varepsilon}_1, \boldsymbol{\varepsilon}_2, \ldots, \boldsymbol{\varepsilon}_n$ are linearly independent. ■

Later on, we shall call $\boldsymbol{\varepsilon}_1, \boldsymbol{\varepsilon}_2, \ldots, \boldsymbol{\varepsilon}_n$ basic vectors, since any vector $\boldsymbol{\alpha} = (a_1, a_2, \ldots, a_n)$ in F^n can be linearly expressed by $\boldsymbol{\varepsilon}_1, \boldsymbol{\varepsilon}_2, \ldots, \boldsymbol{\varepsilon}_n$ as follows:

$$\boldsymbol{\alpha} = a_1\boldsymbol{\varepsilon}_1 + a_2\boldsymbol{\varepsilon}_2 + \cdots + a_n\boldsymbol{\varepsilon}_n.$$

Example 3.2 Zero vector is linearly dependent

A set of vectors containing the zero vector is linearly dependent.

Proof: Let the vectors be $\boldsymbol{\alpha}_1, \boldsymbol{\alpha}_2, \ldots, \boldsymbol{\alpha}_m$ (where $\boldsymbol{\alpha}_1 = \mathbf{0}$). Then, there exist numbers 1, 0, $\cdots$, 0, clearly not all equal to 0, such that

$$1\boldsymbol{\alpha}_1 + 0\boldsymbol{\alpha}_2 + \cdots + 0\boldsymbol{\alpha}_m = \mathbf{0},$$

where $\boldsymbol{\alpha}_1, \boldsymbol{\alpha}_2, \ldots, \boldsymbol{\alpha}_m$ are linearly dependent. ■

By virtue of Definition 3.5, readers can easily prove that, a single vector $\boldsymbol{\alpha}$ is linearly dependent (independent), if and only if, $\boldsymbol{\alpha}$ is a zero (nonzero) vector.

Example 3.3 Linear dependence of a collection of vectors

If some of the vectors among $\boldsymbol{\alpha}_1, \boldsymbol{\alpha}_2, \ldots, \boldsymbol{\alpha}_m$, are linearly dependent, then the whole collection is also linearly dependent.

Proof: Without loss of generality, let $\boldsymbol{\alpha}_1, \boldsymbol{\alpha}_2, \ldots, \boldsymbol{\alpha}_j$ $(j < m)$ be linearly dependent. So, there exist numbers $k_1, k_2, \ldots, k_j$, not all 0, such that

$$k_1\boldsymbol{\alpha}_1 + k_2\boldsymbol{\alpha}_2 + \cdots + k_j\boldsymbol{\alpha}_j = \mathbf{0}.$$

Thus, there are numbers $k_1, k_2, \ldots, k_j, 0, 0, \ldots, 0$, not all 0, such that

$$k_1\boldsymbol{\alpha}_1 + k_2\boldsymbol{\alpha}_2 + \cdots + k_j\boldsymbol{\alpha}_j + 0\boldsymbol{\alpha}_{j+1} + 0\boldsymbol{\alpha}_{j+2} + \cdots + 0\boldsymbol{\alpha}_m = \mathbf{0},$$

where $\boldsymbol{\alpha}_1, \boldsymbol{\alpha}_2, \ldots, \boldsymbol{\alpha}_m$ are also linearly dependent. ■

An equivalent proposition (i.e., contrapositive proposition) to Example 3.3 is: If $\boldsymbol{\alpha}_1, \boldsymbol{\alpha}_2, \ldots, \boldsymbol{\alpha}_m$ are linearly independent, then any collection of vectors made out of them is also linearly independent.

In brief, vectors in a set are linearly dependent, if vectors in even one of its subsets are linearly dependent, while vectors in any subset of a linearly independent set of vectors are linearly independent.

One word of warning here is do not interpret Theorem 3.1 as: *Any vector among some linearly dependent vectors can be linearly expressed by the other vectors*. To illustrate, observe that $\boldsymbol{\alpha}_1 = (0, 1)$, $\boldsymbol{\alpha}_2 = (0, 2)$, $\boldsymbol{\alpha}_3 = (1, 1)$ are linearly dependent (since $\boldsymbol{\alpha}_1$ and $\boldsymbol{\alpha}_2$ are linearly dependent), but $\boldsymbol{\alpha}_3$ cannot be linearly expressed by $\boldsymbol{\alpha}_1$ and $\boldsymbol{\alpha}_2$, that is, $\boldsymbol{\alpha}_3 \neq k_1\boldsymbol{\alpha}_1 + k_2\boldsymbol{\alpha}_2$ for any choice of k_1, k_2 whatsoever.

If two nonzero vectors $\boldsymbol{\alpha} = (\alpha_1, \alpha_2, \ldots, \alpha_n)$ and $\boldsymbol{\beta} = (b_1, b_2, \ldots, b_n)$ are linearly dependent, there must be nonzero numbers k_1 and k_2 such that

$$k_1\boldsymbol{\alpha} + k_2\boldsymbol{\beta} = \mathbf{0}.$$

Thus, $\boldsymbol{\alpha} = k\boldsymbol{\beta}$ (or $\boldsymbol{\beta} = k\boldsymbol{\alpha}$), that is, $a_i = kb_i$ $(i = 1, 2, \ldots, n)$. The n components of $\boldsymbol{\alpha}$ and $\boldsymbol{\beta}$ are proportional.

THEOREM 3.2

Let $\boldsymbol{\alpha}_1, \boldsymbol{\alpha}_2, \ldots, \boldsymbol{\alpha}_r \in F^n$, where

$$\boldsymbol{\alpha}_1 = (a_{11}, a_{21}, \ldots, a_{n1})^{\mathrm{T}},$$
$$\boldsymbol{\alpha}_2 = (a_{12}, a_{22}, \ldots, a_{n2})^{\mathrm{T}},$$
$$\ldots$$
$$\boldsymbol{\alpha}_r = (a_{1r}, a_{2r}, \ldots, a_{nr})^{\mathrm{T}}.$$

$\boldsymbol{\alpha}_1, \boldsymbol{\alpha}_2, \ldots, \boldsymbol{\alpha}_r$ are linearly dependent, if and only if, the homogeneous system of linear equations,

$$\boldsymbol{A}\boldsymbol{x} = \mathbf{0}, \tag{3.6}$$

has a nonzero solution, where

$$\boldsymbol{A} = (\boldsymbol{\alpha}_1, \boldsymbol{\alpha}_2, \ldots, \boldsymbol{\alpha}_r) = \begin{pmatrix} a_{11} & a_{12} & \cdots & a_{1r} \\ a_{21} & a_{22} & \cdots & a_{2r} \\ \vdots & \vdots & & \vdots \\ a_{n1} & a_{n2} & \cdots & a_{nr} \end{pmatrix},$$

$$\boldsymbol{x} = \begin{pmatrix} x_1 \\ x_2 \\ \vdots \\ x_r \end{pmatrix}.$$

Proof: Suppose that $\boldsymbol{\alpha}_1, \boldsymbol{\alpha}_2, \ldots, \boldsymbol{\alpha}_r$ are linearly dependent. Then, there certainly exist numbers $x_1, x_2, \ldots, x_r$, not all 0, such that,

$$x_1\boldsymbol{\alpha}_1 + x_2\boldsymbol{\alpha}_2 + \cdots + x_r\boldsymbol{\alpha}_r = \mathbf{0}_n, \tag{3.7}$$

that is,

$$x_1 \begin{pmatrix} a_{11} \\ a_{21} \\ \vdots \\ a_{n1} \end{pmatrix} + x_2 \begin{pmatrix} a_{12} \\ a_{22} \\ \vdots \\ a_{n2} \end{pmatrix} + \cdots + x_r \begin{pmatrix} a_{1r} \\ a_{2r} \\ \vdots \\ a_{nr} \end{pmatrix} = \begin{pmatrix} 0 \\ 0 \\ \vdots \\ 0 \end{pmatrix}. \tag{3.8}$$

Clearly, (3.8) and (3.6) are equivalent, using different matrix forms. This means, the homogeneous system of linear equations (3.6) has a nonzero solution. Conversely, if the linear equations system (3.6) has a nonzero solution, that is, there are numbers $x_1, x_2, \ldots, x_r$, not all 0, such that (3.8) holds, then $\boldsymbol{\alpha}_1, \boldsymbol{\alpha}_2, \ldots, \boldsymbol{\alpha}_r$ are clearly linearly dependent. Hence, the theorem follows. □

An equivalent proposition to Theorem 3.2 is: $\boldsymbol{\alpha}_1, \boldsymbol{\alpha}_2, \ldots, \boldsymbol{\alpha}_r$ are linearly independent, if and only if, the homogeneous system of linear equations (3.6) has only the zero solution.

If $n < r$ in Theorem 3.2, then (3.6) must have a free variable in the solution process via Gaussian elimination, that is, it must have a nonzero solution. So any $(n+1)$ vectors of dimension n are linearly dependent. Therefore, any set of linearly independent vectors in $\mathbb{R}^n$ contains at most n vectors.

THEOREM 3.3

Suppose $\boldsymbol{\alpha}_1, \boldsymbol{\alpha}_2, \ldots, \boldsymbol{\alpha}_r$ be linearly independent, but $\boldsymbol{\beta}, \boldsymbol{\alpha}_1, \boldsymbol{\alpha}_2, \ldots, \boldsymbol{\alpha}_r$ are linearly dependent. Then, $\boldsymbol{\beta}$ can be linearly expressed by $\boldsymbol{\alpha}_1, \boldsymbol{\alpha}_2, \ldots, \boldsymbol{\alpha}_r$ and the expression is unique.

Proof: Since $\boldsymbol{\beta}, \boldsymbol{\alpha}_1, \boldsymbol{\alpha}_2, \ldots, \boldsymbol{\alpha}_r$ are linearly dependent, there are numbers $k, k_1, k_2, \ldots, k_r$, not all zero, such that

$$k\boldsymbol{\beta} + k_1\boldsymbol{\alpha}_1 + k_2\boldsymbol{\alpha}_2 + \cdots + k_r\boldsymbol{\alpha}_r = \mathbf{0}, \tag{3.9}$$

where $k \neq 0$. As otherwise, $k = 0$ would imply that all of $k_1, k_2, \ldots, k_r$ are 0, since $\boldsymbol{\alpha}_1, \boldsymbol{\alpha}_2, \ldots, \boldsymbol{\alpha}_r$ are linearly independent, contradicting the choice of $k, k_1, k_2, \ldots, k_r$ such that not all are 0. Hence, $\boldsymbol{\beta}$ can be linearly expressed as

$$\boldsymbol{\beta} = -\frac{k_1}{k}\boldsymbol{\alpha}_1 - \frac{k_2}{k}\boldsymbol{\alpha}_2 - \cdots - \frac{k_r}{k}\boldsymbol{\alpha}_r.$$

Next, we shall prove the uniqueness of this expression. For two expressions

$$\begin{aligned}\boldsymbol{\beta} &= l_1\boldsymbol{\alpha}_1 + l_2\boldsymbol{\alpha}_2 + \cdots + l_r\boldsymbol{\alpha}_r \\ &= h_1\boldsymbol{\alpha}_1 + h_2\boldsymbol{\alpha}_2 + \cdots + h_r\boldsymbol{\alpha}_r.\end{aligned}$$

we have

$$(l_1 - h_1)\boldsymbol{\alpha}_1 + (l_2 - h_2)\boldsymbol{\alpha}_2 + \cdots + (l_r - h_r)\boldsymbol{\alpha}_r = \mathbf{0}.$$

But, as $\boldsymbol{\alpha}_1, \boldsymbol{\alpha}_2, \ldots, \boldsymbol{\alpha}_r$ are linearly independent, so

$$l_i - h_i = 0,$$

that is, $l_i = h_i$ for all $i = 1, 2, \ldots, r$.

Hence, the uniqueness follows. □

From Theorems 3.2 and 3.5, we immediately get the following corollary.

COROLLARY 3.1

For n linearly independent vectors $\boldsymbol{\alpha}_1, \boldsymbol{\alpha}_2, \ldots, \boldsymbol{\alpha}_n$ in F^n, any vector $\boldsymbol{\alpha}$ in F^n can be linearly expressed by $\boldsymbol{\alpha}_1, \boldsymbol{\alpha}_2, \ldots, \boldsymbol{\alpha}_n$ uniquely.

Example 3.4 Verification of Theorem 3.2

Let $\boldsymbol{\alpha}_1 = (1, -1, 1)$, $\boldsymbol{\alpha}_2 = (1, 2, 0)$, $\boldsymbol{\alpha}_3 = (1, 0, 3)$, and $\boldsymbol{\alpha}_4 = (2, -3, 7)$.

(i) Are $\boldsymbol{\alpha}_1, \boldsymbol{\alpha}_2$, and $\boldsymbol{\alpha}_3$ linearly dependent?
(ii) Can $\boldsymbol{\alpha}_4$ be linearly expressed by $\boldsymbol{\alpha}_1, \boldsymbol{\alpha}_2$, and $\boldsymbol{\alpha}_3$? If yes, find the expression.

Solution:

(i) Transpose $\boldsymbol{\alpha}_1, \boldsymbol{\alpha}_2$, and $\boldsymbol{\alpha}_3$ to column vectors to make up the matrix

$$\boldsymbol{A} = (\boldsymbol{\alpha}_1^{\mathrm{T}}, \boldsymbol{\alpha}_2^{\mathrm{T}}, \boldsymbol{\alpha}_3^{\mathrm{T}}) = \begin{pmatrix} 1 & 1 & 1 \\ -1 & 2 & 0 \\ 1 & 0 & 3 \end{pmatrix}.$$

It can be seen that $\boldsymbol{A}$ is invertible from $|\boldsymbol{A}| = 7$, so the system of equations $\boldsymbol{Ax} = \boldsymbol{0}$ has only the zero solution, where $\boldsymbol{\alpha}_1, \boldsymbol{\alpha}_2$, and $\boldsymbol{\alpha}_3$ are linearly independent from Theorem 3.2.

(ii) From Corollary 3.1, $\boldsymbol{\alpha}_4$ can be linearly expressed by $\boldsymbol{\alpha}_1, \boldsymbol{\alpha}_2$, and $\boldsymbol{\alpha}_3$ uniquely. Let

$$x_1\boldsymbol{\alpha}_1 + x_2\boldsymbol{\alpha}_2 + x_3\boldsymbol{\alpha}_3 = \boldsymbol{\alpha}_4.$$

That is,

$$x_1\,(1, -1, 1,) + x_2\,(1, 2, 0) + x_3\,(1, 0, 3) = (2, -3, 7)\,.$$

Thus, we get

$$(\boldsymbol{\alpha}_1^{\mathrm{T}}, \boldsymbol{\alpha}_2^{\mathrm{T}}, \boldsymbol{\alpha}_3^{\mathrm{T}}) \begin{pmatrix} x_1 \\ x_2 \\ x_3 \end{pmatrix} = \begin{pmatrix} 1 & 1 & 1 \\ -1 & 2 & 0 \\ 1 & 0 & 3 \end{pmatrix} \begin{pmatrix} x_1 \\ x_2 \\ x_3 \end{pmatrix} = \begin{pmatrix} 2 \\ -3 \\ 7 \end{pmatrix},$$

that is, $\boldsymbol{Ax} = \boldsymbol{\alpha}_4^{\mathrm{T}}$. Solving this system of equations yields the unique solution $x_1 = 1, x_2 = -1, x_3 = 2$, so, $\boldsymbol{\alpha}_4 = \boldsymbol{\alpha}_1 - \boldsymbol{\alpha}_2 + 2\boldsymbol{\alpha}_3$. ■

Example 3.5 Proof of linear dependence of a set of vectors

Suppose the vectors $\boldsymbol{\alpha}_1, \boldsymbol{\alpha}_2$, and $\boldsymbol{\alpha}_3$ are linearly independent, and $\boldsymbol{\beta}_1 = \boldsymbol{\alpha}_1 + \boldsymbol{\alpha}_2 + 2\boldsymbol{\alpha}_3$, $\boldsymbol{\beta}_2 = \boldsymbol{\alpha}_1 - \boldsymbol{\alpha}_2$, $\boldsymbol{\beta}_3 = \boldsymbol{\alpha}_1 + \boldsymbol{\alpha}_3$. Prove that $\boldsymbol{\beta}_1, \boldsymbol{\beta}_2$, and $\boldsymbol{\beta}_3$ are linearly dependent.

Proof: We are done if there exist x_1, x_2, and x_3, not all 0, such that $x_1\boldsymbol{\beta}_1 + x_2\boldsymbol{\beta}_2 + x_3\boldsymbol{\beta}_3 = \mathbf{0}$.

Suppose

$$x_1\boldsymbol{\beta}_1 + x_2\boldsymbol{\beta}_2 + x_3\boldsymbol{\beta}_3 = \mathbf{0}. \tag{3.10}$$

Then,

$$x_1(\boldsymbol{\alpha}_1 + \boldsymbol{\alpha}_2 + 2\boldsymbol{\alpha}_3) + x_2(\boldsymbol{\alpha}_1 - \boldsymbol{\alpha}_2) + x_3(\boldsymbol{\alpha}_1 + \boldsymbol{\alpha}_3) = \mathbf{0},$$

$$(x_1 + x_2 + x_3)\boldsymbol{\alpha}_1 + (x_1 - x_2)\boldsymbol{\alpha}_2 + (2x_1 + x_3)\boldsymbol{\alpha}_3 = \mathbf{0}.$$

Because $\boldsymbol{\alpha}_1, \boldsymbol{\alpha}_2$, and $\boldsymbol{\alpha}_3$ are linearly independent, all the coefficients in the last equation must be 0, so we have

$$\begin{cases} x_1 \quad\quad x_2 + x_3 = 0, \\ x_1 - x_2 \quad\quad = 0, \\ 2x_1 \quad\quad + x_3 = 0. \end{cases}$$

Now, $(-1, -1, 2)$ is a nonzero solution of this system and hence, $\boldsymbol{\beta}_1, \boldsymbol{\beta}_2$, and $\boldsymbol{\beta}_3$ are linearly dependent. ∎

Denote

$$\boldsymbol{A} = (\boldsymbol{\alpha}_1, \boldsymbol{\alpha}_2, \ldots, \boldsymbol{\alpha}_s), \quad \boldsymbol{B} = (\boldsymbol{\alpha}_1^*, \boldsymbol{\alpha}_2^*, \ldots, \boldsymbol{\alpha}_s^*),$$

where $\boldsymbol{\alpha}_1^*, \boldsymbol{\alpha}_2^*, \ldots, \boldsymbol{\alpha}_s^*$ are obtained by arbitrarily adding m components to n-dimensional $\boldsymbol{\alpha}_1, \boldsymbol{\alpha}_2, \ldots, \boldsymbol{\alpha}_s$ respectively. If $\boldsymbol{\alpha}_1, \boldsymbol{\alpha}_2, \ldots, \boldsymbol{\alpha}_s$ are linearly independent, then the $(n+m)$-dimensional vectors $\boldsymbol{\alpha}_1^*, \boldsymbol{\alpha}_2^*, \ldots, \boldsymbol{\alpha}_s^*$ are also linearly independent by virtue of Theorem 3.2. In fact, if $\boldsymbol{\alpha}_1, \boldsymbol{\alpha}_2, \ldots, \boldsymbol{\alpha}_s$ are linearly independent, that is, if the homogeneous system of linear equations $\boldsymbol{Ax} = \mathbf{0}$ has only the zero solution, then $\boldsymbol{Bx} = \mathbf{0}$ obviously has only the zero solution (i.e., $\boldsymbol{\alpha}_1^*, \boldsymbol{\alpha}_2^*, \ldots, \boldsymbol{\alpha}_s^*$ are also linearly independent) as well.

Conversely, if $\boldsymbol{\alpha}_1^*, \boldsymbol{\alpha}_2^*, \ldots, \boldsymbol{\alpha}_s^*$ are linearly dependent, that is, $\boldsymbol{Bx} = \mathbf{0}$ has a nonzero solution, then $\boldsymbol{Ax} = \mathbf{0}$ also has a nonzero solution (i.e., $\boldsymbol{\alpha}_1, \boldsymbol{\alpha}_2, \ldots, \boldsymbol{\alpha}_s$ are also linearly dependent). A consequent result is: If $\boldsymbol{\alpha}_1, \boldsymbol{\alpha}_2, \ldots, \boldsymbol{\alpha}_s$ are linearly dependent, then vectors obtained by deleting their components from the same positions are also linearly dependent.

3.2 Rank of Vectors and Its Maximal Linearly Independent Subsets

In this section, we shall use linearly independent vectors to define the "rank of a set of vectors" and consider the maximal number of linearly independent vectors among a set of vectors.

Firstly, let us take a look at an example.

In $\mathbb{R}^3$, four coplanar vectors $\boldsymbol{\alpha}_1, \boldsymbol{\alpha}_2, \boldsymbol{\alpha}_3$, and $\boldsymbol{\alpha}_4$ are given as shown in Figure 3.5. Obviously, they are linearly dependent. Out of these vectors, two are linearly independent and any vector can be linearly expressed by those two. For example, $\boldsymbol{\alpha}_1$ and $\boldsymbol{\alpha}_2$ are linearly independent, $\boldsymbol{\alpha}_3$ and $\boldsymbol{\alpha}_4$ can be linearly expressed by $\boldsymbol{\alpha}_1$ and $\boldsymbol{\alpha}_2$. Furthermore, any three of them are linearly dependent, that is, any linearly independent subset of the given set contains at most "2" vectors. The number "2" here is called the **rank** of the set of vectors. The formal definition of the rank of a set of vectors is given below.

Definition 3.6

If there are r linearly independent vectors in a set of vectors $\{\boldsymbol{\alpha}_1, \boldsymbol{\alpha}_2, \ldots, \boldsymbol{\alpha}_s\}$ and any vector in the set can be linearly expressed by these r vectors, then the number r is called the **rank of the set of vectors** and is denoted by $\text{rank}\,\{\boldsymbol{\alpha}_1, \boldsymbol{\alpha}_2, \ldots, \boldsymbol{\alpha}_s\} = r$.

Obviously, $\text{rank}\,\{\boldsymbol{\alpha}_1, \boldsymbol{\alpha}_2, \ldots, \boldsymbol{\alpha}_s\} = s$ if $\boldsymbol{\alpha}_1, \boldsymbol{\alpha}_2, \ldots, \boldsymbol{\alpha}_s$ are linearly independent, while the rank of a set of vectors is zero, if the set has only the zero vector.

In order to prove and prescribe that any $(r+1)$ vectors are linearly dependent if chosen from a set of vectors of rank r, we give the following definition and a theorem.

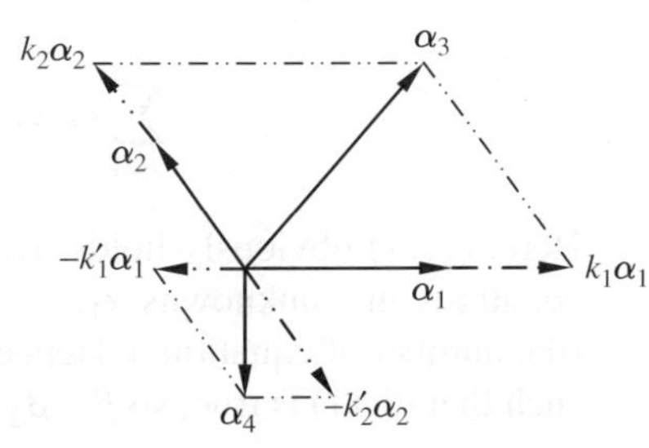

Figure 3.5 Four coplanar vectors in $\mathbb{R}^3$

Definition 3.7

If each of $\boldsymbol{\beta}_1, \boldsymbol{\beta}_2, \ldots, \boldsymbol{\beta}_t$ can be linearly expressed by $\boldsymbol{\alpha}_1, \boldsymbol{\alpha}_2, \ldots, \boldsymbol{\alpha}_s$, then it is said that the set of vectors $\{\boldsymbol{\beta}_1, \boldsymbol{\beta}_2, \ldots, \boldsymbol{\beta}_t\}$ can be linearly expressed by the set of vectors $\{\boldsymbol{\alpha}_1, \boldsymbol{\alpha}_2, \ldots, \boldsymbol{\alpha}_s\}$. If the two sets of vectors can be linearly expressed by each other, then they are said to be **equivalent vector sets**.

THEOREM 3.4

If a set of vectors $\{\boldsymbol{\beta}_1, \boldsymbol{\beta}_2, \ldots, \boldsymbol{\beta}_t\}$ can be linearly expressed by a set of vectors $\{\boldsymbol{\alpha}_1, \boldsymbol{\alpha}_2, \ldots, \boldsymbol{\alpha}_s\}$ and $t > s$, then $\boldsymbol{\beta}_1, \boldsymbol{\beta}_2, \ldots, \boldsymbol{\beta}_t$ are linearly dependent.

Proof: Let $\boldsymbol{\beta}_j = \sum_{i=1}^{s} k_{ij}\boldsymbol{\alpha}_i$ $(j = 1, 2, \ldots, t)$. In order to prove $\boldsymbol{\beta}_1, \boldsymbol{\beta}_2, \ldots, \boldsymbol{\beta}_t$ are linearly dependent, we only need to prove that there exist numbers $x_1, x_2, \ldots, x_t$, not all 0, such that

$$x_1\boldsymbol{\beta}_1 + x_2\boldsymbol{\beta}_2 + \cdots + x_t\boldsymbol{\beta}_t = \mathbf{0}, \tag{3.11}$$

which is

$$\begin{aligned}\sum_{j=1}^{t} x_j\boldsymbol{\beta}_j &= \sum_{j=1}^{t} x_j \sum_{i=1}^{s} k_{ij}\boldsymbol{\alpha}_i \\ &= \sum_{j=1}^{t}\sum_{i=1}^{s} x_j k_{ij}\boldsymbol{\alpha}_i \\ &= \sum_{i=1}^{s}\left(\sum_{j=1}^{t} k_{ij}x_j\right)\boldsymbol{\alpha}_i = \mathbf{0},\end{aligned}$$

when the coefficients satisfy

$$\sum_{j=1}^{t} k_{ij}x_j = 0 \ (i = 1, 2, \ldots, s). \tag{3.12}$$

Here, (3.11) obviously holds. But, (3.12) is a homogeneous system of linear equations in t unknowns $x_1, x_2, \ldots, x_t$. It has a nonzero solution since $t > s$ (the number of equations). Hence, there exist numbers $x_1, x_2, \ldots, x_t$, not all 0, such that (3.11) is true, so $\boldsymbol{\beta}_1, \boldsymbol{\beta}_2, \ldots, \boldsymbol{\beta}_t$ are linearly dependent. □

An equivalent proposition to Theorem 3.4 is Corollary 3.2.

COROLLARY 3.2

If vectors $\boldsymbol{\beta}_1, \boldsymbol{\beta}_2, \ldots, \boldsymbol{\beta}_t$ *can be linearly expressed by vectors* $\boldsymbol{\alpha}_1, \boldsymbol{\alpha}_2, \ldots, \boldsymbol{\alpha}_s$ *and* $\boldsymbol{\beta}_1, \boldsymbol{\beta}_2, \ldots, \boldsymbol{\beta}_t$ *are linearly independent, then* $t \leq s$.

COROLLARY 3.3

If $\operatorname{rank}\{\boldsymbol{\alpha}_1, \boldsymbol{\alpha}_2, \ldots, \boldsymbol{\alpha}_s\} = r$, *then any* $(r+1)$ *vectors of* $\boldsymbol{\alpha}_1, \boldsymbol{\alpha}_2, \ldots, \boldsymbol{\alpha}_s$ *are linearly dependent.*

Proof: Without any loss of generality, let $\boldsymbol{\alpha}_1, \boldsymbol{\alpha}_2, \ldots, \boldsymbol{\alpha}_r$ be r linearly independent vectors among $\{\boldsymbol{\alpha}_1, \boldsymbol{\alpha}_2, \ldots, \boldsymbol{\alpha}_s\}$. Any vector of $\boldsymbol{\alpha}_1, \boldsymbol{\alpha}_2, \ldots, \boldsymbol{\alpha}_s$ can be linearly expressed by $\boldsymbol{\alpha}_1, \boldsymbol{\alpha}_2, \ldots, \boldsymbol{\alpha}_r$ from Theorem 3.4. Then, any $(r+1)$ vectors in the whole set are linearly dependent. □

Now, the rank of a set of vectors can be equivalently defined as: If there are r linearly independent vectors in a set of vectors, such that any $(r+1)$ vectors of that set are linearly dependent, then the number r is called the **rank of the set of vectors**.

For a set of vectors of rank r, any linearly independent subset has at most r vectors. So, a linearly independent subset comprising r vectors in a set of vectors of rank r is called a **maximal linearly independent subset** of that set of vectors. In general, the maximal linearly independent subset of a set of vectors is not unique, but the number of vectors in such a subset is the same. In Figure 3.5, $\{\boldsymbol{\alpha}_1, \boldsymbol{\alpha}_2\}$ and $\{\boldsymbol{\alpha}_1, \boldsymbol{\alpha}_3\}$ are both maximal linearly independent subsets.

COROLLARY 3.4

Let $\operatorname{rank}\{\boldsymbol{\alpha}_1, \boldsymbol{\alpha}_2, \ldots, \boldsymbol{\alpha}_s\} = p$ *and* $\operatorname{rank}\left\{\boldsymbol{\beta}_1, \boldsymbol{\beta}_2, \ldots, \boldsymbol{\beta}_t\right\} = r$. *If* $\boldsymbol{\beta}_1, \boldsymbol{\beta}_2, \ldots, \boldsymbol{\beta}_t$ *can be linearly expressed by* $\boldsymbol{\alpha}_1, \boldsymbol{\alpha}_2, \ldots, \boldsymbol{\alpha}_s$, *then* $r \leq p$.

Proof: Without any loss of generality, let $\boldsymbol{\alpha}_1, \boldsymbol{\alpha}_2, \ldots, \boldsymbol{\alpha}_p$ and $\boldsymbol{\beta}_1, \boldsymbol{\beta}_2, \ldots, \boldsymbol{\beta}_r$ be maximal linearly independent subsets of the two given sets of vectors. So, we have

$$\boldsymbol{\alpha}_i = \sum_{j=1}^{p} c_{ij}\boldsymbol{\alpha}_j (i = 1, 2, \ldots, s),$$

and

$$\boldsymbol{\beta}_k = \sum_{i=1}^{s} b_{ki}\boldsymbol{\alpha}_i \ (k = 1, 2, \ldots, r, \ldots, t).$$

So,

$$\boldsymbol{\beta}_k = \sum_{i=1}^{s} b_{ki} \left(\sum_{j=1}^{p} c_{ij} \boldsymbol{\alpha}_j \right) = \sum_{j=1}^{p} \left(\sum_{i=1}^{s} b_{ki} c_{ij} \right) \boldsymbol{\alpha}_j,$$

where $\boldsymbol{\beta}_1, \boldsymbol{\beta}_2, \ldots, \boldsymbol{\beta}_r$ can be linearly expressed by $\boldsymbol{\alpha}_1, \boldsymbol{\alpha}_2, \ldots, \boldsymbol{\alpha}_p$. Hence, $r \leq p$ from Corollary 3.2. □

From Corollary 3.2, we can immediately deduce that the ranks of equivalent vector sets are equal.

In the next section, we will discuss how to find the rank and maximal linearly independent subset of a given set of vectors.

3.3 Rank of a Matrix, *Equivalent Normal-form

For a matrix $\boldsymbol{A}$, a row (column) is called a row (column) vector of $\boldsymbol{A}$. The rank of row (column) vectors is called the row (column) rank of $\boldsymbol{A}$. Obviously, for an $m \times n$ matrix $\boldsymbol{A}$, the row rank $\leq m$, the column rank $\leq n$.

Consider an echelon matrix

$$\boldsymbol{A} = \begin{pmatrix} a_{11} & a_{12} & a_{13} & a_{14} & a_{15} \\ 0 & 0 & a_{23} & a_{24} & a_{25} \\ 0 & 0 & 0 & a_{34} & a_{35} \\ 0 & 0 & 0 & 0 & 0 \end{pmatrix}$$

(where $a_{11} \neq 0, a_{23} \neq 0, a_{34} \neq 0$). Its row rank $= 3$ and column rank $= 3$. This is because $\boldsymbol{A}$ can be partitioned along its row and column respectively as

$$\boldsymbol{A} = \begin{pmatrix} \boldsymbol{\alpha}_1 \\ \boldsymbol{\alpha}_2 \\ \boldsymbol{\alpha}_3 \\ \boldsymbol{\alpha}_4 \end{pmatrix}, \boldsymbol{A} = \begin{pmatrix} \boldsymbol{\beta}_1, & \boldsymbol{\beta}_2, & \boldsymbol{\beta}_3, & \boldsymbol{\beta}_4, & \boldsymbol{\beta}_5 \end{pmatrix}.$$

Thus,

(i) from $x_1\boldsymbol{\alpha}_1 + x_2\boldsymbol{\alpha}_2 + x_3\boldsymbol{\alpha}_3 = \mathbf{0}$, we deduce that numbers x_1, x_2, and x_3 must be all zeros, so that $\boldsymbol{\alpha}_1, \boldsymbol{\alpha}_2$, and $\boldsymbol{\alpha}_3$ are linearly independent. Together with $\boldsymbol{\alpha}_4 = \mathbf{0}$, the row rank of $\boldsymbol{A}$ is 3.

(ii) from $y_1\boldsymbol{\beta}_1 + y_3\boldsymbol{\beta}_3 + y_4\boldsymbol{\beta}_4 = \mathbf{0}$, we see that y_1, y_3, and y_4 must all be zeros, where $\boldsymbol{\beta}_1, \boldsymbol{\beta}_3$, and $\boldsymbol{\beta}_4$ are linearly independent. It is also easy to see that any four column vectors of $\boldsymbol{A}$ are linearly dependent [since any 4 vectors with dimension 3 (the last elements are all zeros) are linearly dependent], thus the column rank of $\boldsymbol{A}$ is also 3.

This example gives readers an indication of the result typically derived. The row rank of an echelon matrix is equal to its column rank, which is the number of nonzero rows of the echelon matrix.

Gaussian elimination can be used to solve a system of linear equations by applying elementary row operations on the augmented matrix $(\boldsymbol{A}, \boldsymbol{b})$ so as to transform it into an echelon matrix. Now, let us examine whether an elementary operation changes the row rank and column rank of a matrix.

THEOREM 3.5

*If a matrix **A** is reduced to **B** by elementary row operations, then their row ranks are the same.*

Proof: We only consider three elementary row operations.
Let $\boldsymbol{A}$ be an $m \times n$ matrix. The m row vectors of $\boldsymbol{A}$ are denoted by $\boldsymbol{\alpha}_1, \boldsymbol{\alpha}_2, \ldots, \boldsymbol{\alpha}_m$.

(i) Switching the positions of any two rows of $\boldsymbol{A}$. The m rows of the matrix $\boldsymbol{B}$ are still the m row vectors of $\boldsymbol{A}$. The row rank of $\boldsymbol{B}$ is equal to that of $\boldsymbol{A}$.

(ii) Multiplying the ith row of $\boldsymbol{A}$ by a nonzero constant c. The m row vectors of $\boldsymbol{B}$ are $\boldsymbol{\alpha}_1, \boldsymbol{\alpha}_2, \ldots, c\boldsymbol{\alpha}_i, \ldots, \boldsymbol{\alpha}_m$. Obviously, the set of row vectors of $\boldsymbol{B}$ is equivalent to the set of row vectors of $\boldsymbol{A}$. Their row ranks are the same, with reference to Corollary 3.2 of Theorem 3.4.

(iii) $$\boldsymbol{A} = \begin{pmatrix} \boldsymbol{\alpha}_1 \\ \vdots \\ \boldsymbol{\alpha}_i \\ \vdots \\ \boldsymbol{\alpha}_j \\ \vdots \\ \boldsymbol{\alpha}_m \end{pmatrix} \xrightarrow[\text{to row } j]{\text{add } c \text{ times row } i} \begin{pmatrix} \boldsymbol{\alpha}_1 \\ \vdots \\ \boldsymbol{\alpha}_i \\ \vdots \\ c\boldsymbol{\alpha}_i + \boldsymbol{\alpha}_j \\ \vdots \\ \boldsymbol{\alpha}_m \end{pmatrix} \text{ written as } \begin{pmatrix} \boldsymbol{\beta}_1 \\ \vdots \\ \boldsymbol{\beta}_i \\ \vdots \\ \boldsymbol{\beta}_j \\ \vdots \\ \boldsymbol{\beta}_m \end{pmatrix} = \boldsymbol{B}.$$

Obviously, the row vectors of $\boldsymbol{B}$ can be linearly expressed by the row vectors of $\boldsymbol{A}$. On the other hand, $\boldsymbol{\alpha}_k = \boldsymbol{\beta}_k$ $(k \neq j)$, $\boldsymbol{\alpha}_j = -c\boldsymbol{\beta}_i + \boldsymbol{\beta}_j$, the row vectors of $\boldsymbol{A}$ can also be linearly expressed by row vectors of $\boldsymbol{B}$. So, row ranks of $\boldsymbol{A}$ and $\boldsymbol{B}$ are equal. □

Applying Theorem 3.5 for the augmented matrix $(\boldsymbol{A}, \boldsymbol{b})$ of the system of linear equations $\boldsymbol{A}\boldsymbol{x} = \boldsymbol{b}$, the number of nonzero rows of the echelon matrix is equal to the row rank of $(\boldsymbol{A}, \boldsymbol{b})$, irrespective of the manner in which row operations are performed.

Elementary row operations do not change column rank of a matrix as well, and this is because:

THEOREM 3.6

If a matrix $\boldsymbol{A}$ is changed to $\boldsymbol{B}$ by elementary row operations, then any corresponding set of column vectors in $\boldsymbol{A}$ and $\boldsymbol{B}$ have the same nature with regard to linear dependence. That is, if

$$\boldsymbol{A}=(\boldsymbol{\alpha}_1,\boldsymbol{\alpha}_2,\ldots,\boldsymbol{\alpha}_n)\xrightarrow{\text{elementary row operations}}(\boldsymbol{\zeta}_1,\boldsymbol{\zeta}_2,\ldots,\boldsymbol{\zeta}_n)=\boldsymbol{B},$$

the sets of column vectors $\boldsymbol{\alpha}_{i_1},\boldsymbol{\alpha}_{i_2},\ldots,\boldsymbol{\alpha}_{i_r}$ and $\boldsymbol{\zeta}_{i_1},\boldsymbol{\zeta}_{i_2},\ldots,\boldsymbol{\zeta}_{i_r}$ $(1\le i_1<i_2<\cdots<i_r\le n)$ have the same nature with regard to linear dependence.

Proof: To reduce $\boldsymbol{A}$ to $\boldsymbol{B}$ by elementary row operations is to multiply $\boldsymbol{A}$ on the left by some elementary matrices $\boldsymbol{P}_1,\boldsymbol{P}_2,\ldots,\boldsymbol{P}_s$ successively. Hence, $\boldsymbol{P}=\boldsymbol{P}_s\cdots\boldsymbol{P}_2\boldsymbol{P}_1$ finally leads to

$$\boldsymbol{PA}=\boldsymbol{B}.$$

So,

$$\boldsymbol{P\alpha}_i=\boldsymbol{\zeta}_i,\qquad i=1,2,\ldots,n.$$

Let

$$\begin{aligned}\boldsymbol{A}_1&=\left(\boldsymbol{\alpha}_{i_1},\boldsymbol{\alpha}_{i_2},\ldots,\boldsymbol{\alpha}_{i_r}\right),\\ \boldsymbol{B}_1&=(\boldsymbol{\zeta}_{i_1},\boldsymbol{\zeta}_{i_2},\ldots,\boldsymbol{\zeta}_{i_r}),\\ \boldsymbol{x}_1&=\left(x_{i_1},x_{i_1},\ldots,x_{i_r}\right)^{\mathrm{T}},\end{aligned}$$

then the homogeneous systems of linear equations $\boldsymbol{A}_1\boldsymbol{x}_1=\mathbf{0}$ and $\boldsymbol{B}_1\boldsymbol{x}_1=\mathbf{0}$ (i.e., $\boldsymbol{PA}_1\boldsymbol{x}_1=\mathbf{0}$) are explicitly equivalent systems. So, the sets of column vectors in $\boldsymbol{A}_1$ and $\boldsymbol{B}_1$ have the same nature regarding linear dependence, according to Theorem 3.2. □

Theorem 3.6 provides a concise and effective method toward finding the rank and a maximal linearly independent subset of a set of vectors as well.

Example 3.6 Rank and maximal linearly independent subset

Given the vectors

$$\boldsymbol{\alpha}_1=(-1,-1,0,0)^{\mathrm{T}},\ \boldsymbol{\alpha}_2=(1,2,1,-1)^{\mathrm{T}},$$
$$\boldsymbol{\alpha}_3=(0,1,1,-1)^{\mathrm{T}},\ \boldsymbol{\alpha}_4=(1,3,2,1)^{\mathrm{T}},$$
$$\boldsymbol{\alpha}_5=(2,6,4,-1)^{\mathrm{T}}.$$

Find the rank of the set of vectors made out of them and also a maximal linearly independent subset of that set. Then, express the remaining vectors by that maximal linearly independent subset.

Solution: Construct a matrix $\boldsymbol{A}=(\boldsymbol{\alpha}_1,\boldsymbol{\alpha}_2,\boldsymbol{\alpha}_3,\boldsymbol{\alpha}_4,\boldsymbol{\alpha}_5)$ (if the vectors are given in a row, write them also in column form $\boldsymbol{A}$). Apply elementary row operations and reduce $\boldsymbol{A}$ to an echelon matrix as follows:

$$\boldsymbol{A}=\begin{pmatrix}-1&1&0&1&2\\-1&2&1&3&6\\0&1&1&2&4\\0&-1&-1&1&-1\end{pmatrix}\xrightarrow[②+①]{①\times(-1)}\begin{pmatrix}1&-1&0&-1&-2\\0&1&1&2&4\\0&1&1&2&4\\0&-1&-1&1&-1\end{pmatrix}$$

$$\xrightarrow[④+②]{③+②\times(-1)}\begin{pmatrix}1&-1&0&-1&-2\\0&1&1&2&4\\0&0&0&0&0\\0&0&0&3&3\end{pmatrix}$$

$$\xrightarrow[③\rightleftharpoons④]{④\times(\frac{1}{3})}\begin{pmatrix}1&-1&0&-1&-2\\0&1&1&2&4\\0&0&0&1&1\\0&0&0&0&0\end{pmatrix}$$

$$\xrightarrow[②+③\times(-2)]{①+③}\begin{pmatrix}1&-1&0&0&1\\0&1&1&0&2\\0&0&0&1&1\\0&0&0&0&0\end{pmatrix}$$

$$\xrightarrow{①+②}\begin{pmatrix}1&0&1&0&1\\0&1&1&0&2\\0&0&0&1&1\\0&0&0&0&0\end{pmatrix}=\boldsymbol{U}.$$

Denote the last echelon matrix $\boldsymbol{U}$ as $(\boldsymbol{\zeta}_1,\boldsymbol{\zeta}_2,\boldsymbol{\zeta}_3,\boldsymbol{\zeta}_4,\boldsymbol{\zeta}_5)$.

It is easy to see that vectors $\boldsymbol{\zeta}_1,\boldsymbol{\zeta}_2$, and $\boldsymbol{\zeta}_4$ form a maximal linearly independent subset of column vectors of $\boldsymbol{U}$. So, $\boldsymbol{\alpha}_1,\boldsymbol{\alpha}_2$, and $\boldsymbol{\alpha}_4$ also form a maximal linearly independent subset of column vectors of $\boldsymbol{A}$. Thus, $\operatorname{rank}\{\boldsymbol{\alpha}_1,\boldsymbol{\alpha}_2,\boldsymbol{\alpha}_3,\boldsymbol{\alpha}_4,\boldsymbol{\alpha}_5\}=3$.

Let

$$x_1\boldsymbol{\alpha}_1 + x_2\boldsymbol{\alpha}_2 + x_4\boldsymbol{\alpha}_4 = \boldsymbol{\alpha}_3,$$

$$y_1\boldsymbol{\alpha}_1 + y_2\boldsymbol{\alpha}_2 + y_4\boldsymbol{\alpha}_4 = \boldsymbol{\alpha}_5.$$

Solving the two systems of linear equations by Gaussian elimination provides the following solutions. To obtain the solutions, we use the echelon matrix $\boldsymbol{U}$ (e.g., for the second equation system, the first, second, fourth, and fifth columns may be used) to get the solutions. Thus,

$$\boldsymbol{\alpha}_3 = \boldsymbol{\alpha}_1 + \boldsymbol{\alpha}_2,$$

$$\boldsymbol{\alpha}_5 = \boldsymbol{\alpha}_1 + 2\boldsymbol{\alpha}_2 + \boldsymbol{\alpha}_4.$$

■

To determine the rank of the vectors and a maximal linearly independent subset, we only need to reduce $\boldsymbol{A}$ to a general echelon matrix other than the row-reduced echelon matrix.

One can deduce from Theorems 3.5 and 3.6 that elementary column operations also do not change the row rank and column rank of a matrix because to apply column operations on $\boldsymbol{A}$ is equivalent to applying row operations on $\boldsymbol{A}^{\mathrm{T}}$. Summarizing the above statements leads to the following theorem.

THEOREM 3.7

Elementary operations do not change the row rank and column rank of a matrix.

From Theorems 3.5 and 3.6, we can also deduce the following theorem.

THEOREM 3.8

The row rank of a matrix $\boldsymbol{A}$ *is equal to its column rank.*

Proof: $\boldsymbol{A}$ is changed to an echelon matrix $\boldsymbol{U}$ by elementary row operations, then,

$$\text{row rank of } \boldsymbol{A} = \text{row rank of } \boldsymbol{U} = \text{column rank of } \boldsymbol{U} = \text{column rank of } \boldsymbol{A}. \qquad \square$$

Since the row and column ranks of $\boldsymbol{A}$ are equal, the following definitions hold.

Definition 3.8

The row rank of a matrix $\boldsymbol{A}$ is called the **rank of matrix** $\boldsymbol{A}$, which is denoted as $rank(\boldsymbol{A})$ or $\mathrm{r}(\boldsymbol{A})$. An nth-order matrix of $rank(\boldsymbol{A}) = n$ is also called a **full rank matrix**.

THEOREM 3.9

The rank of an nth-order matrix $\boldsymbol{A}$ *is equal to n, if and only if,* $\boldsymbol{A}$ *is a nonsingular matrix (i.e.,* $|\boldsymbol{A}| \neq 0$*).*

Proof: If $\mathrm{r}(\boldsymbol{A}) = n$, then $\boldsymbol{A}$ can be reduced to a row-reduced echelon matrix with n nonzero rows (i.e., the identity matrix $\boldsymbol{I}$) by elementary row operations. Indeed, there exists an invertible matrix $\boldsymbol{P}$ such that $\boldsymbol{PA} = \boldsymbol{I}$, that is, $|\boldsymbol{P}|\,|\boldsymbol{A}| = 1$, and therefore $|\boldsymbol{A}| \neq 0$. Conversely, if $|\boldsymbol{A}| \neq 0$, then the homogeneous system of linear equations $\boldsymbol{Ax} = \boldsymbol{0}$ has only the zero solution, that is, $\boldsymbol{x} = \boldsymbol{A}^{-1}\boldsymbol{0} = \boldsymbol{0}$. So, n column vectors of $\boldsymbol{A}$ are linearly independent, that is, $\mathrm{r}(\boldsymbol{A}) = n$. □

For a general $m \times n$ matrix $\boldsymbol{A}$, its rank is closely related to its subdeterminants.

Definition 3.9

A kth-order determinant

$$\begin{vmatrix} a_{i_1 j_1} & a_{i_1 j_2} & \cdots & a_{i_1 j_k} \\ a_{i_2 j_1} & a_{i_2 j_2} & \cdots & a_{i_2 j_k} \\ \vdots & \vdots & & \vdots \\ a_{i_k j_1} & a_{i_k j_2} & \cdots & a_{i_k j_k} \end{vmatrix}, \tag{3.13}$$

is called a kth-order **minor determinant** of $\boldsymbol{A}$, or a kth-order **subdeterminant** of $\boldsymbol{A}$, where the k^2 elements are sequentially selected from the intersection points of k rows (rows $i_1, i_2, \ldots, i_k$) and k columns (columns $j_1, j_2, \ldots, j_k$) of matrix $\boldsymbol{A}$. When the determinant (3.13) is (not) zero, it is called a kth-order **zero (nonzero) subdeterminant**. The determinant (3.13) is called a kth-order **principal subdeterminant** when $j_1 = i_1, j_2 = i_2, \ldots, j_k = i_k$.

r is called the highest nonzero subdeterminant order of matrix $\boldsymbol{A}$ if there exists a rth-order nonzero subdeterminant and all $(r+1)$th-order subdeterminants (if exist) of matrix $\boldsymbol{A}$ are 0.

THEOREM 3.10

rank$(\boldsymbol{A}) = r$, *if and only if, the highest nonzero subdeterminant order of* $\boldsymbol{A}$ *is* r.

Proof: First, we prove the necessary condition. Let rank$(\boldsymbol{A}) = r$, that is, row rank of $\boldsymbol{A}$ is r. Without any loss of generality, suppose the rank of the first r rows of $\boldsymbol{A}$, denoted as $\boldsymbol{A}_1$, be r. Then, its column rank is also r. Further, without any loss of generality, suppose the first r columns of $\boldsymbol{A}_1$ are linearly independent. From Theorem 3.9, we know that the subdeterminant at the upper-left corner of $\boldsymbol{A}$ is a nonzero subdeterminant. Also, any $(r+1)$ row vectors of $\boldsymbol{A}$ are linearly dependent, so any $(r+1)$th-order subdeterminant is a zero (since there exists a row, which can be linearly expressed by other rows). So, the highest order of nonzero subdeterminants is r.

Next, we prove the sufficiency condition. Suppose r is the highest order of nonzero subdeterminants of $\boldsymbol{A}$ and without any loss of generality, further suppose that $|\boldsymbol{A}_r| \neq 0$, which is the rth-order subdeterminant at the upper-left corner of $\boldsymbol{A}$. Thus, $\boldsymbol{A}_r$ is invertible and its r row vectors are linearly independent. The r rows of $\boldsymbol{A}$ containing $\boldsymbol{A}_r$ must also be linearly independent. While any $(r+1)$ row vectors of $\boldsymbol{A}$ are linearly dependent [otherwise, from the proof of the necessity condition, we can see that there exists an $(r+1)$th-order nonzero subdeterminant in $\boldsymbol{A}$, which is a contradiction to the initial assumption]. Therefore, the row rank of $\boldsymbol{A} = \mathrm{r}(\boldsymbol{A}) = r$. □

Conclusions for properties about the rank of a matrix are:

(i) rank of the matrix = row rank of the matrix = column rank of the matrix = the highest order of nonzero subdeterminants of the matrix,
(ii) elementary operations preserve the rank of a matrix.

The rank of a matrix is very important for describing solutions of a system of linear equations. We shall consider it in the next two sections.

We will now list the properties of rank of a matrix.

Property 3.1

$$\mathrm{r}(\boldsymbol{A}+\boldsymbol{B}) \leq \mathrm{r}(\boldsymbol{A}) + \mathrm{r}(\boldsymbol{B}).$$

Proof: Let both $\boldsymbol{A}$ and $\boldsymbol{B}$ be $m \times n$ matrices, $\mathrm{r}(\boldsymbol{A}) = p$, $\mathrm{r}(\boldsymbol{B}) = q$. $\boldsymbol{A}$ and $\boldsymbol{B}$ are partitioned along a column as $\boldsymbol{A} = (\boldsymbol{\alpha}_1, \boldsymbol{\alpha}_2, \ldots, \boldsymbol{\alpha}_n)$ and $\boldsymbol{B} = (\boldsymbol{\beta}_1, \boldsymbol{\beta}_2, \ldots, \boldsymbol{\beta}_n)$. So,

$$\boldsymbol{A} + \boldsymbol{B} = (\boldsymbol{\alpha}_1 + \boldsymbol{\beta}_1, \boldsymbol{\alpha}_2 + \boldsymbol{\beta}_2, \ldots, \boldsymbol{\alpha}_n + \boldsymbol{\beta}_n).$$

Without any loss of generality, suppose that the maximal linearly independent subsets of column vectors of $\boldsymbol{A}$ and $\boldsymbol{B}$ are $\boldsymbol{\alpha}_1, \boldsymbol{\alpha}_2, \ldots, \boldsymbol{\alpha}_p$ and $\boldsymbol{\beta}_1, \boldsymbol{\beta}_2, \ldots, \boldsymbol{\beta}_q$ respectively. Then, the column vectors of $\boldsymbol{A} + \boldsymbol{B}$ can be linearly expressed by vectors $\boldsymbol{\alpha}_1, \boldsymbol{\alpha}_2, \ldots, \boldsymbol{\alpha}_p$ and $\boldsymbol{\beta}_1, \boldsymbol{\beta}_2, \ldots, \boldsymbol{\beta}_q$.

So,

$$\begin{aligned} \mathrm{r}(\boldsymbol{A} + \boldsymbol{B}) &= \text{column rank of } \boldsymbol{A} + \boldsymbol{B} \\ &\leq \operatorname{rank}(\boldsymbol{\alpha}_1, \boldsymbol{\alpha}_2, \ldots, \boldsymbol{\alpha}_p, \boldsymbol{\beta}_1, \boldsymbol{\beta}_2, \ldots, \boldsymbol{\beta}_1) \leq p + q. \end{aligned}$$

□

Property 3.2

$$\mathrm{r}(\boldsymbol{AB}) \leq \min[\mathrm{r}(\boldsymbol{A}), \mathrm{r}(\boldsymbol{B})].$$

Proof: Let $\boldsymbol{A}, \boldsymbol{B}$ be $m \times n, n \times s$ matrices, respectively. $\boldsymbol{A}$ is partitioned along a column, then

$$\boldsymbol{AB} = (\boldsymbol{\alpha}_1, \boldsymbol{\alpha}_2, \ldots, \boldsymbol{\alpha}_n) \begin{pmatrix} b_{11} & b_{12} & \cdots & b_{1s} \\ b_{21} & b_{22} & \cdots & b_{2s} \\ \vdots & \vdots & & \vdots \\ b_{n1} & b_{n2} & \cdots & b_{ns} \end{pmatrix}.$$

The column vectors $\boldsymbol{\gamma}_1, \boldsymbol{\gamma}_2, \ldots, \boldsymbol{\gamma}_s$ of $\boldsymbol{AB}$ can be linearly expressed by column vectors $\boldsymbol{\alpha}_1, \boldsymbol{\alpha}_2, \ldots, \boldsymbol{\alpha}_n$ of $\boldsymbol{A}$, and $\mathrm{r}(\boldsymbol{AB}) =$ column rank of $\boldsymbol{AB} \leq$ column rank of $\boldsymbol{A} = \mathrm{r}(\boldsymbol{A})$. Similarly, partitioning $\boldsymbol{B}$ along a row, we have $\mathrm{r}(\boldsymbol{AB}) \leq \mathrm{r}(\boldsymbol{B})$. □

Property 3.3

Let $\boldsymbol{A}$ be an $m \times n$ matrix and $\boldsymbol{P}, \boldsymbol{Q}$ be mth-order, nth-order invertible matrices, respectively, then

$$\mathrm{r}(\boldsymbol{A}) = \mathrm{r}(\boldsymbol{PA}) = \mathrm{r}(\boldsymbol{AQ}) = \mathrm{r}(\boldsymbol{PAQ}).$$

Proof: Since the invertible matrices $\boldsymbol{P}$ and $\boldsymbol{Q}$ can be expressed as products of some elementary matrices, and elementary operations do not change the rank of a matrix, so the result is true. Or $\boldsymbol{A} = \boldsymbol{P}^{-1}(\boldsymbol{PA})$, from Property 3.2, we have

$$\mathrm{r}(\boldsymbol{A}) = \mathrm{r}[\boldsymbol{P}^{-1}(\boldsymbol{PA})] \leq \mathrm{r}(\boldsymbol{PA}) \leq \mathrm{r}(\boldsymbol{A}).$$

So, $\mathrm{r}(\boldsymbol{PA}) = \mathrm{r}(\boldsymbol{A})$. The other equalities can be proved similarly. □

Example 3.7 Proof of $|A^{\mathrm{T}}A| = 0$

Let A be an $m \times n$ matrix, $m < n$, prove that $|\boldsymbol{A}^{\mathrm{T}}\boldsymbol{A}| = 0$.

Proof: Since $\mathrm{r}(A) = \mathrm{r}(A^{\mathrm{T}}) \leq \min(m, n) < n$, according to Property 3.2, we have

$$\mathrm{r}\left(\boldsymbol{A}^{\mathrm{T}}\boldsymbol{A}\right) \leq \min[\mathrm{r}(\boldsymbol{A}^{\mathrm{T}}), \mathrm{r}(\boldsymbol{A})] < n,$$

and $\boldsymbol{A}^{\mathrm{T}}\boldsymbol{A}$ is an nth-order matrix. From Theorem 3.9 or Theorem 3.10, we have $|\boldsymbol{A}^{\mathrm{T}}\boldsymbol{A}| = 0$. ■

Lastly, let us consider the simplest form attainable by a matrix with rank r, after elementary operations are applied to it.

Definition 3.10

If a matrix $\boldsymbol{A}$ is changed to $\boldsymbol{B}$ by elementary operations (or if there exist invertible matrices $\boldsymbol{P}$ and $\boldsymbol{Q}$ such that $\boldsymbol{PAQ} = \boldsymbol{B}$), then $\boldsymbol{A}$ is said to be **equivalent** to $\boldsymbol{B}$, and is written as $\boldsymbol{A} \cong \boldsymbol{B}$.

From the definition, we have following equivalence relations:

(i) Reflexivity, $\boldsymbol{A} \cong \boldsymbol{A}$.
(ii) Symmetry, if $\boldsymbol{A} \cong \boldsymbol{B}$, then $\boldsymbol{B} \cong \boldsymbol{A}$ ($\boldsymbol{A}$ and $\boldsymbol{B}$ are said to be equivalent).
(iii) Transitivity, if $\boldsymbol{A} \cong \boldsymbol{B}$, $\boldsymbol{B} \cong \boldsymbol{C}$, then $\boldsymbol{A} \cong \boldsymbol{C}$.

THEOREM 3.11

If $\boldsymbol{A}$ is an $m \times n$ matrix and $\mathrm{r}(\boldsymbol{A}) = r$, then there exist invertible matrices $\boldsymbol{P}$ (mth-order) and $\boldsymbol{Q}$ (nth-order) such that

$$\boldsymbol{PAQ} = \begin{pmatrix} \boldsymbol{I}_r & \boldsymbol{0} \\ \boldsymbol{0} & \boldsymbol{0} \end{pmatrix}_{m \times n} = \boldsymbol{U}, \tag{3.14}$$

where $\boldsymbol{I}_r$ is the rth-order identity matrix.

Proof: After applying elementary row operations to $\boldsymbol{A}$, a row-reduced echelon matrix $\boldsymbol{U}_1$ with r nonzero rows is obtained. In other words, there are elementary matrices $\boldsymbol{P}_1, \boldsymbol{P}_2, \ldots, \boldsymbol{P}_s$, such that

$$\boldsymbol{P}_s \cdots \boldsymbol{P}_2\boldsymbol{P}_1\boldsymbol{A} = \boldsymbol{U}_1.$$

Further, U_1 is reduced to the matrix U of (3.14) by elementary operations on columns, which indicates that there are elementary matrices $Q_1, Q_2, \ldots, Q_t$, such that

$$U_1 Q_1 Q_2 \cdots Q_t = U.$$

Denote $P_s \cdots P_2 P_1 = P$, $Q_1 Q_2 \cdots Q_t = Q$ (P, Q are both invertible), then

$$PAQ = U.$$ □

The matrix in the *RHS* of (3.14) is called the **equivalent normal-form** of A. Given Theorem 3.11, matrices with the same rank and the same size must be equivalent to a similar equivalent normal-form. Therefore, any two matrices with the same rank and same size are equivalent to each other.

*Example 3.8 Left inverse of a rectangular matrix

Let A be an $m \times n$ matrix ($m > n$) and $\text{rank}(A) = n$. Prove that there exists an $n \times m$ matrix B such that $BA = I_n$.

Proof: According to Theorem 3.11, there exists mth-order invertible matrix P and nth-order invertible matrix Q, such that

$$PAQ = \begin{pmatrix} I_n \\ 0_1 \end{pmatrix},$$

and

$$PA = \begin{pmatrix} I_n \\ 0_1 \end{pmatrix} Q^{-1} = \begin{pmatrix} Q^{-1} \\ 0_1 \end{pmatrix},$$

where 0_1 is the $(m-n) \times n$ zero matrix. Let $C = (Q, 0_2)$ and 0_2 be an $n \times (m-n)$ zero matrix. Then,

$$CPA = (Q, 0_2) \begin{pmatrix} Q^{-1} \\ 0_1 \end{pmatrix} = QQ^{-1} + 0_2 0_1 = I_n.$$

Denote $B = CP$. Then, $BA = I_n$. ■

By the way, A can be reduced to the normal-form by row elementary operations only for A in Example 3.8, that is, there exists an mth-order invertible matrix P such that

$$PA = \begin{pmatrix} I_n \\ 0_1 \end{pmatrix}.$$

Let us take a look at the following examples about the rank of a matrix or vectors again.

Example 3.9 Calculating the rank of a set of vectors

Let $\boldsymbol{\alpha}_1 = (1,3,1,2), \boldsymbol{\alpha}_2 = (2,5,3,3), \boldsymbol{\alpha}_3 = (0,1,-1,a)$, $\boldsymbol{\alpha}_4 = (3,10,k,4)$. Find the rank of the set of vectors $\{\boldsymbol{\alpha}_1, \boldsymbol{\alpha}_2, \boldsymbol{\alpha}_3, \boldsymbol{\alpha}_4\}$ and the linear expression of $\boldsymbol{\alpha}_4$ by $\boldsymbol{\alpha}_1, \boldsymbol{\alpha}_2$, and $\boldsymbol{\alpha}_3$.

Solution: We first arrange the four vectors as columns to a matrix $\boldsymbol{A}$, and reduce $\boldsymbol{A}$ to an echelon matrix $\boldsymbol{U}$ by elementary row operations.

$$\boldsymbol{A} = \begin{pmatrix} 1 & 2 & 0 & 3 \\ 3 & 5 & 1 & 10 \\ 1 & 3 & -1 & k \\ 2 & 3 & a & 4 \end{pmatrix} \xrightarrow[\text{row operations}]{\text{elementary}} \begin{pmatrix} 1 & 2 & 0 & 3 \\ 0 & -1 & 1 & 1 \\ 0 & 0 & a-1 & -3 \\ 0 & 0 & 0 & k-2 \end{pmatrix} = \boldsymbol{U}.$$

When $a = 1$ or $k = 2$, the echelon matrix $\boldsymbol{U}$ has only three nonzero rows, so

$$\text{rank}\,(\boldsymbol{A}) = \text{rank}\,\{\boldsymbol{\alpha}_1, \boldsymbol{\alpha}_2, \boldsymbol{\alpha}_3, \boldsymbol{\alpha}_4\} = 3.$$

When $a \neq 1$ and $k \neq 2$, $\text{rank}\,(\boldsymbol{U}) = \text{rank}\,(\boldsymbol{A}) = \text{rank}\,\{\boldsymbol{\alpha}_1, \boldsymbol{\alpha}_2, \boldsymbol{\alpha}_3, \boldsymbol{\alpha}_4\} = 4$.

Let $x_1\boldsymbol{\alpha}_1 + x_2\boldsymbol{\alpha}_2 + x_3\boldsymbol{\alpha}_3 = \boldsymbol{\alpha}_4$. It is a nonhomogeneous system of linear equations with augmented matrix $\boldsymbol{A}$. $\boldsymbol{A}$ is reduced to $\boldsymbol{U}$ by Gaussian elimination (i.e., elementary row operations are applied to $\boldsymbol{A}$). When $k = 2$ and $a \neq 1$, $\boldsymbol{\alpha}_4$ can be linearly expressed by $\boldsymbol{\alpha}_1, \boldsymbol{\alpha}_2$, and $\boldsymbol{\alpha}_3$ and

$$x_3 = \frac{3}{1-a}, \quad x_2 = -1 + x_3 = \frac{2+a}{1-a}, \quad x_1 = 3 - 2x_2 = -\frac{1+5a}{1-a}.$$

When $k \neq 2$ or $a = 1$, $\boldsymbol{\alpha}_4$ cannot be linearly expressed by $\boldsymbol{\alpha}_1, \boldsymbol{\alpha}_2$, and $\boldsymbol{\alpha}_3$. ■

Example 3.10 Using elementary row operations and concept of rank

Let $\boldsymbol{A} = \begin{pmatrix} 1 & 2 & 1 \\ 2 & 2 & -2 \\ -1 & t & 5 \\ 1 & 0 & -3 \end{pmatrix}$, $\quad \text{r}\,(\boldsymbol{A}) = 2$. Determine t.

Solution: A is reduced to B by elementary row operations:

$$A \to \begin{pmatrix} 1 & 2 & 1 \\ 0 & -2 & -4 \\ 0 & 2+t & 6 \\ 0 & 0 & 0 \end{pmatrix} = B.$$

The second and third rows in B are definitely proportional since $\text{rank}(A) = \text{rank}(B) = 2$. So, we get $\frac{-2}{2+t} = -\frac{4}{6}$, where, $t = 1$. ■

Example 3.11 Deriving conditions for the rank of product matrix

Given $r(B) = 2$, $A = \begin{pmatrix} 1 & 2 & 0 \\ 0 & a & 1 \\ 1 & 3 & b \end{pmatrix}$,

(i) determine the suitable condition(s) on a and b for $r(AB) = 2$, and
(ii) determine the suitable condition(s) on A and B such that $r(AB) = 1$.

Solution:

(i) If $|A| = ab - 1 \neq 0$, then A is invertible and $r(AB) = r(B) = 2$.
(ii) If $|A| = ab - 1 = 0$, then A is not invertible and $r(A) = 2$ (because there exists a second-order nonzero subdeterminant $\begin{vmatrix} 1 & 2 \\ 1 & 3 \end{vmatrix} = 1$, or there exist only two columns that are not proportional and hence, linearly independent). In this case, the column vectors of A are linearly dependent and the homogeneous system of linear equations $Ax = \mathbf{0}$ has a nonzero solution from Theorem 3.2. If the three column vectors of B (which is assumed to be a 3×3 matrix as well, without loss of generality) are x_1, x_2, and x_3 [i.e., $B = (x_1, x_2, x_3)$], where x_1 and x_2 are proportional to each other and are solutions of $Ax = \mathbf{0}$, while x_3 is not a solution of $Ax = \mathbf{0}$ (i.e., $Ax_3 = \beta \neq \mathbf{0}$), then

$$AB = A(x_1, x_2, x_3) = (\mathbf{0}, \mathbf{0}, \beta).$$

Therefore, $r(AB) = 1$. In this case, $B = (x_1, x_2, x_3)$ still satisfies $r(B) = 2$ since x_1 or x_2 is not proportional to x_3 (otherwise, $x_3 = k_1 x_1$, or $x_3 = k_2 x_2$ is a solution of $Ax = \mathbf{0}$ indeed). Furthermore, $r(B) \neq 3$ [otherwise, $r(AB) = 2$], so there are only two column vectors linearly independent in B. Hence, $r(AB) = 1$ if A and B satisfy the above conditions. ■

3.4 Solutions of a Homogeneous System of Linear Equations

Consider a homogeneous system of linear equations with an $m \times n$ coefficient matrix $\boldsymbol{A}$

$$\boldsymbol{A}\boldsymbol{x} = \boldsymbol{0}. \tag{3.15}$$

If $\boldsymbol{A} = (\boldsymbol{\alpha}_1, \boldsymbol{\alpha}_2, \ldots, \boldsymbol{\alpha}_n)$ is partitioned in columns, then (3.15) can be expressed as a vector equation

$$x_1\boldsymbol{\alpha}_1 + x_2\boldsymbol{\alpha}_2 + \cdots + x_n\boldsymbol{\alpha}_n = \boldsymbol{0}. \tag{3.16}$$

So, (3.15) has a nonzero solution, if and only if, $\boldsymbol{\alpha}_1, \boldsymbol{\alpha}_2, \ldots, \boldsymbol{\alpha}_n$ are linearly dependent, that is

$$\text{rank}\,(\boldsymbol{A}) = \text{rank}\,(\boldsymbol{\alpha}_1, \boldsymbol{\alpha}_2, \ldots, \boldsymbol{\alpha}_n) < n.$$

Thus, we have the following theorem.

THEOREM 3.12

Let $\boldsymbol{A}$ be an $m \times n$ matrix. The homogeneous system of linear equations $\boldsymbol{A}\boldsymbol{x} = \boldsymbol{0}$ has a nonzero solution, if and only if, $\text{r}(\boldsymbol{A}) < n$.

Proof: Suppose rank $(\boldsymbol{A}) = r$. Then, $\boldsymbol{A}$ has r linearly independent row vectors, and the remaining $(m - r)$ row vectors can be linearly expressed by these r vectors. $\boldsymbol{A}$ can therefore be reduced to the following echelon matrix with r nonzero rows by elementary row operations.

$$\boldsymbol{U} = \begin{pmatrix} c_{11} & \cdots & c_{1i_2} & \cdots & c_{1i_r} & \cdots & c_{1n} \\ 0 & \cdots & c_{2i_2} & \cdots & c_{2i_r} & \cdots & c_{2n} \\ \vdots & & \vdots & & \vdots & & \vdots \\ 0 & \cdots & 0 & \cdots & c_{ri_r} & \cdots & c_{rn} \\ 0 & \cdots & 0 & \cdots & 0 & \cdots & 0 \\ \vdots & & \vdots & & \vdots & & \vdots \\ 0 & \cdots & 0 & \cdots & 0 & \cdots & 0 \end{pmatrix}.$$

Then, $\boldsymbol{U}\boldsymbol{x} = \boldsymbol{0}$ and $\boldsymbol{A}\boldsymbol{x} = \boldsymbol{0}$ are equivalent systems of linear equations. Thus, the homogeneous system of linear equations $\boldsymbol{U}\boldsymbol{x} = \boldsymbol{0}$ has a nonzero solution, if and only if, $r < n$. □

An equivalent proposition to Theorem 3.12 is: A homogeneous system of linear equations $\boldsymbol{A}\boldsymbol{x} = \boldsymbol{0}$ has only the zero solution, if and only if, rank$(\boldsymbol{A}) =$ number of columns of $\boldsymbol{A}$.

For an nth-order matrix $\boldsymbol{A}$, $\boldsymbol{Ax} = \boldsymbol{0}$ has a nonzero solution (has only the zero solution), if and only if, $|\boldsymbol{A}| = 0$ ($|\boldsymbol{A}| \neq 0$) as narrated before.

Example 3.12 Conditional existence of matrix $B \neq 0$ such that, $AB = 0$

Let $\boldsymbol{A}$ be an nth-order matrix. Prove that there exists an $n \times s$ matrix $\boldsymbol{B} \neq \boldsymbol{0}$ such that, $\boldsymbol{AB} = \boldsymbol{0}$, if and only if, $|\boldsymbol{A}| = 0$.

Proof: $\boldsymbol{B}$ is partitioned along its column as $\boldsymbol{B} = (\boldsymbol{b}_1, \boldsymbol{b}_2, \dots, \boldsymbol{b}_s)$, then $\boldsymbol{AB} = \boldsymbol{0}$ is equivalent to

$$\boldsymbol{Ab}_j = \boldsymbol{0}, \quad j = 1, 2, \dots, s,$$

that is, every column of $\boldsymbol{B}$ is a solution of the homogeneous system of linear equations $\boldsymbol{Ax} = \boldsymbol{0}$.

If $\boldsymbol{AB} = \boldsymbol{0}$ and $\boldsymbol{B} \neq \boldsymbol{0}$, then $\boldsymbol{Ax} = \boldsymbol{0}$ has a nonzero solution. So $|\boldsymbol{A}| = 0$. Conversely, if $|\boldsymbol{A}| = 0$, take s nonzero solutions of $\boldsymbol{Ax} = \boldsymbol{0}$ as s columns of $\boldsymbol{B}$. Then $\boldsymbol{B} \neq \boldsymbol{0}$ but $\boldsymbol{AB} = \boldsymbol{0}$. ■

To study the solution structure for a homogeneous system of linear equations, we will first discuss the properties of solutions and then give a concept of system of fundamental solutions.

THEOREM 3.13

If $\boldsymbol{x}_1$ and $\boldsymbol{x}_2$ are two solutions of homogeneous system of linear equations $\boldsymbol{Ax} = \boldsymbol{0}$, then $k_1\boldsymbol{x}_1 + k_2\boldsymbol{x}_2$ (k_1 and k_2 are arbitrary constants) is also a solution.

Proof: Because $\boldsymbol{A}(k_1\boldsymbol{x}_1 + k_2\boldsymbol{x}_2) = k_1\boldsymbol{Ax}_1 + k_2\boldsymbol{Ax}_2 = k_1\boldsymbol{0} + k_2\boldsymbol{0} = \boldsymbol{0}$, so $k_1\boldsymbol{x}_1 + k_2\boldsymbol{x}_2$ is a solution of $\boldsymbol{Ax} = \boldsymbol{0}$. □

Theorem 3.13 can be generalized to the case of finitely many solutions.

Definition 3.11

Let $\boldsymbol{x}_1, \boldsymbol{x}_2, \dots, \boldsymbol{x}_p$ be the solution vectors of $\boldsymbol{Ax} = \boldsymbol{0}$. We call $\boldsymbol{x}_1, \boldsymbol{x}_2, \dots, \boldsymbol{x}_p$ a **system of fundamental solutions (basis)** of $\boldsymbol{Ax} = \boldsymbol{0}$ if

(i) $\boldsymbol{x}_1, \boldsymbol{x}_2, \dots, \boldsymbol{x}_p$ are linearly independent,
(ii) any solution of $\boldsymbol{Ax} = \boldsymbol{0}$ can be linearly expressed by $\boldsymbol{x}_1, \boldsymbol{x}_2, \dots, \boldsymbol{x}_p$.

For a system of fundamental solutions $\boldsymbol{x}_1, \boldsymbol{x}_2, \ldots, \boldsymbol{x}_p$ of $\boldsymbol{A}\boldsymbol{x}=\boldsymbol{0}$, any $k_1\boldsymbol{x}_1 + k_2\boldsymbol{x}_2 + \cdots + k_p\boldsymbol{x}_p$ with arbitrary constants $k_1, k_2, \ldots, k_p$ is a solution of $\boldsymbol{A}\boldsymbol{x}=\boldsymbol{0}$. We are now going to prove the existence of a system of fundamental solutions for a homogeneous system of linear equations which has a nonzero solution.

THEOREM 3.14

Let $\boldsymbol{A}$ be an $m \times n$ matrix. If $\mathrm{r}(\boldsymbol{A}) = r < n$, then the homogeneous system of linear equations $\boldsymbol{A}\boldsymbol{x} = \boldsymbol{0}$ has a system of fundamental solutions with $(n-r)$ solutions.

Proof: Let us first prove the existence of the $(n-r)$ linearly independent solutions. Without any loss of generality, we may assume that a row-reduced echelon matrix $\boldsymbol{U}$ is obtained by elementary row operations from $\boldsymbol{A}$ and let it be denoted by

$$\boldsymbol{U} = \begin{pmatrix} 1 & 0 & \cdots & 0 & c_{1,r+1} & \cdots & c_{1n} \\ 0 & 1 & \cdots & 0 & c_{2,r+1} & \cdots & c_{2n} \\ \vdots & \vdots & & \vdots & \vdots & & \vdots \\ 0 & 0 & \cdots & 1 & c_{r,r+1} & \cdots & c_{rn} \\ 0 & 0 & \cdots & 0 & 0 & \cdots & 0 \\ \vdots & \vdots & & \vdots & \vdots & & \vdots \\ 0 & 0 & \cdots & 0 & 0 & \cdots & 0 \end{pmatrix}.$$

An equivalent form of $\boldsymbol{U}\boldsymbol{x}=\boldsymbol{0}$ is

$$\begin{cases} x_1 \quad\quad\quad + c_{1,r+1}x_{r+1} + \cdots + c_{1n}x_n = 0, \\ \quad x_2 \quad\quad + c_{2,r+1}x_{r+1} + \cdots + c_{2n}x_n = 0, \\ \quad\quad \ddots \quad\quad \vdots \quad\quad\quad \vdots \quad\quad\quad \vdots \quad\quad \vdots \\ \quad\quad\quad x_r + c_{r,r+1}x_{r+1} + \cdots + c_{rn}x_n = 0, \end{cases} \tag{3.17}$$

which is equivalent to $\boldsymbol{A}\boldsymbol{x} = \boldsymbol{0}$. Let $\boldsymbol{x}_{r+1}, \boldsymbol{x}_{r+2}, \ldots, \boldsymbol{x}_n$ be free variables and substitute these last $(n-r)$ variables with

$$(1, 0, \ldots, 0), (0, 1, \ldots, 0), \ldots, (0, 0, \ldots, 1)$$

into (3.17), respectively. Then, we get the values of the first r variables, which are denoted by $d_{1i}, d_{2i}, \ldots, d_{ri}$ for $i = 1, 2, \ldots, n-r$. Denote

$$\begin{aligned} \boldsymbol{x}_1 &= (d_{11}, d_{21}, \ldots, d_{r1}, 1, 0, \ldots, 0)^{\mathrm{T}}, \\ \boldsymbol{x}_2 &= (d_{12}, d_{22}, \ldots, d_{r2}, 0, 1, \ldots, 0)^{\mathrm{T}}, \\ &\cdots \\ \boldsymbol{x}_{n-r} &= \left(d_{1,n-r}, d_{2,n-r}, \ldots, d_{r,n-r}, 0, 0, \ldots, 1\right)^{\mathrm{T}}. \end{aligned}$$

Clearly, the solutions $\boldsymbol{x}_1, \boldsymbol{x}_2, \ldots, \boldsymbol{x}_{n-r}$ are linearly independent (since from $\lambda_1\boldsymbol{x}_1 + \lambda_2\boldsymbol{x}_2 + \cdots + \lambda_{n-r}\boldsymbol{x}_{n-r} = \boldsymbol{0}$, we can obtain $\lambda_1 = \lambda_2 = \cdots = \lambda_{n-r} = 0$).

Next, we prove that any solution $\boldsymbol{x}$ of $\boldsymbol{Ax} = \boldsymbol{0}$ can be linearly expressed by $\boldsymbol{x}_1, \boldsymbol{x}_2, \ldots, \boldsymbol{x}_{n-r}$. Any numbers $k_1, k_2, \ldots, k_{n-r}$ substituted respectively for $\boldsymbol{x}_{r+1}, \boldsymbol{x}_{r+2}, \ldots, \boldsymbol{x}_n$ into (3.17) yields a solution

$$\boldsymbol{x} = (d_1, d_2, \ldots, d_r, k_1, k_2, \ldots, k_{n-r})^{\mathrm{T}},$$

and

$$\boldsymbol{x}^* = k_1\boldsymbol{x}_1 + k_2\boldsymbol{x}_2 + \cdots + k_{n-r}\boldsymbol{x}_{n-r}$$

is also a solution. So,

$$\boldsymbol{x} - \boldsymbol{x}^* = \begin{pmatrix} d_1 \\ d_2 \\ \vdots \\ d_r \\ k_1 \\ k_2 \\ \vdots \\ k_{n-r} \end{pmatrix} - k_1 \begin{pmatrix} d_{11} \\ d_{21} \\ \vdots \\ d_{r1} \\ 1 \\ 0 \\ \vdots \\ 0 \end{pmatrix} - k_2 \begin{pmatrix} d_{12} \\ d_{22} \\ \vdots \\ d_{r2} \\ 0 \\ 1 \\ \vdots \\ 0 \end{pmatrix} - \cdots - k_{n-r} \begin{pmatrix} d_{1,n-r} \\ d_{2,n-r} \\ \vdots \\ d_{r,n-r} \\ 0 \\ 0 \\ \vdots \\ 1 \end{pmatrix} = \begin{pmatrix} d_1' \\ d_2' \\ \vdots \\ d_r' \\ 0 \\ 0 \\ \vdots \\ 0 \end{pmatrix}$$

is a solution of $\boldsymbol{Ax} = \boldsymbol{0}$ corresponding to the free unknowns $\boldsymbol{x}_{r+1}, \boldsymbol{x}_{r+2}, \cdots, \boldsymbol{x}_n$ all being zero. In this case, (3.17) provides us with only the zero solution.

So,

$$\boldsymbol{x} - \boldsymbol{x}^* = \boldsymbol{0},$$

$$\boldsymbol{x} = \boldsymbol{x}^* = k_1\boldsymbol{x}_1 + k_2\boldsymbol{x}_2 + \cdots + k_{n-r}\boldsymbol{x}_{n-r}.$$

Thus, $\boldsymbol{x}_1, \boldsymbol{x}_2, \ldots, \boldsymbol{x}_{n-r}$ is a solution basis of $\boldsymbol{Ax} = \boldsymbol{0}$, with $(n-r)$ solutions. □

The above proof provides us a method of finding a system of fundamental solutions of $\boldsymbol{Ax} = \boldsymbol{0}$ by choosing the $(n-r)$ free variables to obtain $(n-r)$ linearly independent solutions. But the choice of free variables is not unique. Indeed, any r variables with their coefficient determinant not equal to zero from $\boldsymbol{Ux} = \boldsymbol{0}$ can be chosen as basic variables and the remaining $(n-r)$ variables as free variables. So, the system of fundamental solutions is not unique, the precise proof is easy and is left to readers as exercise.

If $\boldsymbol{x}_1, \boldsymbol{x}_2, \ldots, \boldsymbol{x}_p$ and $\boldsymbol{x}_1^*, \boldsymbol{x}_2^*, \ldots, \boldsymbol{x}_p^*$ are two solution bases of $\boldsymbol{Ax} = \boldsymbol{0}$, then the two sets

$$S = \{k_1\boldsymbol{x}_1 + k_2\boldsymbol{x}_2 + \cdots + k_p\boldsymbol{x}_p \,|\, k_1, k_2, \ldots, k_p \text{ are arbitrary constants}\}$$

and

$$S^* = \{l_1\boldsymbol{x}_1^* + l_2\boldsymbol{x}_2^* + \cdots + l_p\boldsymbol{x}_p^* \,|\, l_1, l_2, \ldots, l_p \text{ are arbitrary constants}\}$$

are equal (indeed, we can prove: $S \subseteq S^*$ and $S^* \subseteq S$).

Here, S and S^* are obviously solution sets of $\boldsymbol{Ax}=\mathbf{0}$. We can express the solution set in the form of a general expression such as

$$\boldsymbol{x}=k_1\boldsymbol{x}_1+k_2\boldsymbol{x}_2+\cdots+k_p\boldsymbol{x}_p$$

also called the **general solution** of $\boldsymbol{Ax}=\mathbf{0}$, which clearly describes the structure of the solutions. To solve a homogeneous system of linear equations having a nonzero solution, we usually find a system of fundamental solutions first, and then present its general solution.

Example 3.13 General solution to a homogeneous system of linear equations

Find the general solution of the homogeneous system of linear equations $\boldsymbol{Ax}=\mathbf{0}$, whose coefficient matrix is

$$\boldsymbol{A}=\begin{pmatrix} 1 & 2 & 1 & 1 & 1 \\ 2 & 4 & 3 & 1 & 1 \\ -1 & -2 & 1 & 3 & -3 \\ 0 & 0 & 2 & 4 & -2 \end{pmatrix}.$$

Solution: Matrix $\boldsymbol{A}$ is reduced by elementary row operations to a row-reduced echelon matrix

$$\boldsymbol{U}=\begin{pmatrix} 1 & 2 & 0 & 0 & 2 \\ 0 & 0 & 1 & 0 & -1 \\ 0 & 0 & 0 & 1 & 0 \\ 0 & 0 & 0 & 0 & 0 \end{pmatrix}.$$

Take x_2 and x_5 as free variables. Let $x_2=1$, $x_5=0$ and $x_2=0$, $x_5=1$; then, we have a system of fundamental solutions

$$\boldsymbol{x}_1=(-2,1,0,0,0)^{\mathrm{T}}, \quad \boldsymbol{x}_2=(-2,0,1,0,1)^{\mathrm{T}}.$$

Thus, the general solution of $\boldsymbol{Ax}=\mathbf{0}$ is

$$\boldsymbol{x}=k_1\boldsymbol{x}_1+k_2\boldsymbol{x}_2.$$

That is,

$$\boldsymbol{x}=\begin{pmatrix} x_1 \\ x_2 \\ x_3 \\ x_4 \\ x_5 \end{pmatrix}=k_1\begin{pmatrix} -2 \\ 1 \\ 0 \\ 0 \\ 0 \end{pmatrix}+k_2\begin{pmatrix} -2 \\ 0 \\ 1 \\ 0 \\ 1 \end{pmatrix},$$

where k_1 and k_2 are arbitrary constants. ■

Example 3.14 The rank of A and B when $AB = 0$

Let A and B be $m \times n$ and $n \times s$ matrices, respectively, and $AB = 0$. Prove that $r(A) + r(B) \leq n$.

Proof: B is partitioned along the column as $B = (b_1, b_2, \ldots, b_s)$. Since $AB = 0$, then

$$Ab_j = 0, \quad j = 1, 2, \ldots, s,$$

that is, each column of B is a solution of $Ax = 0$. A system of fundamental solutions of $Ax = 0$ consists of $[n - r(A)]$ solutions, that is to say, any set of solutions for $Ax = 0$ contains at most $[n - r(A)]$ linearly independent solutions. So,

$$\text{rank}\,(B) = \text{rank}\,(b_1, b_2, \ldots, b_s) \leq n - r\,(A)$$

$$r\,(A) + r\,(B) \leq n. \qquad \blacksquare$$

***Example 3.15 $r(A^TA) = r(A)$**

Let A be a real $m \times n$ matrix. Prove that $r(A^TA) = r(A)$.

Proof: $r(A^TA) \leq r(A)$ is proved by invoking Property 3.2 in §3.3. We have to prove $r(A) \leq r(A^TA)$. An equivalent statement is: The solution set of $(A^TA)x = 0$ is contained in $Ax = 0$.

$(A^TA)x = 0$ $(x \in \mathbb{R}^n)$ implies $x^T(A^TA)x = 0$, that is, $(Ax)^TAx = 0$. Let

$$Ax = (b_1, b_2, \ldots, b_m)^T.$$

Thus, $b_1^2 + b_2^2 + \cdots + b_m^2 = 0$. Then, $b_1 = b_2 = \cdots = b_m = 0$, that is, $Ax = 0$. In brief, a solution of the system of equations $(A^TA)x = 0$ is certainly a solution of $Ax = 0$. Therefore,

$$n - r(A^TA) \leq n - r(A),$$

where

$$r(A) \leq r(A^TA). \qquad \blacksquare$$

Theorem 3.14 describes a relationship between the rank of matrix A and solutions of $Ax = 0$. This theorem is not only important for finding solutions of $\boldsymbol{Ax}=\boldsymbol{0}$, but also can be used to determine the rank of a matrix based on the solutions found, as shown in Examples 3.14 and 3.15.

3.5 Solutions of a Nonhomogeneous System of Linear Equations

The nonhomogeneous system of linear equations with an $m \times n$ matrix $\boldsymbol{A}$ as coefficient matrix is

$$\boldsymbol{Ax}=\boldsymbol{b}, \tag{3.18}$$

which can be expressed by the vector equation

$$x_1\boldsymbol{\alpha}_1+x_2\boldsymbol{\alpha}_2+\cdots+x_n\boldsymbol{\alpha}_n=\boldsymbol{b}, \tag{3.19}$$

where $\boldsymbol{\alpha}_1, \boldsymbol{\alpha}_2, \ldots, \boldsymbol{\alpha}_n$ are n column vectors of $\boldsymbol{A}$. So, (3.18) has a solution, if and only if, $\boldsymbol{b}$ can be linearly expressed by the column vectors of $\boldsymbol{A}$. Thus,

$$\text{rank}\,\{\boldsymbol{\alpha}_1, \boldsymbol{\alpha}_2, \ldots, \boldsymbol{\alpha}_n, \boldsymbol{b}\}=\text{rank}\,\{\boldsymbol{\alpha}_1, \boldsymbol{\alpha}_2, \ldots, \boldsymbol{\alpha}_n\}$$

that is,

$$\text{r}(\boldsymbol{A}, \boldsymbol{b})=\text{r}(\boldsymbol{A}).$$

Consequently, we have the following theorem.

THEOREM 3.15

For a nonhomogeneous system of linear equations $\boldsymbol{Ax}=\boldsymbol{b}$, the following propositions are equivalent:

(i) $\boldsymbol{Ax}=\boldsymbol{b}$ *has a solution (or is consistent),*
(ii) $\boldsymbol{b}$ *can be linearly expressed by the column vectors of* $\boldsymbol{A}$*,*
(iii) *the rank of augmented matrix* $(\boldsymbol{A}, \boldsymbol{b})$ *is equal to the rank of coefficient matrix* $\boldsymbol{A}$*.*

This theorem can be proved with the following argument.

Proof: The augmented matrix $(\boldsymbol{A}, \boldsymbol{b})$ is reduced by elementary row operations to an echelon matrix

$$(\boldsymbol{C}, \boldsymbol{d}) = \begin{pmatrix} 1 & \cdots & c_{1r} & c_{1,r+1} & \cdots & c_{1r} & d_1 \\ \vdots & & \vdots & \vdots & & \vdots & \vdots \\ 0 & \cdots & 1 & c_{r,r+1} & \cdots & c_{rn} & d_n \\ 0 & \cdots & 0 & 0 & \cdots & 0 & d_{r+1} \\ 0 & \cdots & 0 & 0 & \cdots & 0 & 0 \\ \vdots & & \vdots & \vdots & & \vdots & \vdots \\ 0 & \cdots & 0 & 0 & \cdots & 0 & 0 \end{pmatrix}, \tag{3.20}$$

whereby $\boldsymbol{Cx} = \boldsymbol{d}$ and $\boldsymbol{Ax} = \boldsymbol{b}$ are equivalent. So, $\boldsymbol{Ax} = \boldsymbol{b}$ has a solution (i.e., $\boldsymbol{Cx} = \boldsymbol{d}$ has a solution) $\Leftrightarrow d_{r+1} = 0$. And

$$d_{r+1} = 0 \Leftrightarrow \mathrm{r}(\boldsymbol{C}, \boldsymbol{d}) = \mathrm{r}(\boldsymbol{C}).$$

Together with the fact that $\mathrm{r}(\boldsymbol{C}, \boldsymbol{d}) = \mathrm{r}(\boldsymbol{A}, \boldsymbol{b})$, and $\mathrm{r}(\boldsymbol{C}) = \mathrm{r}(\boldsymbol{A})$, we have $\boldsymbol{Ax} = \boldsymbol{b}$ has a solution $\Leftrightarrow \mathrm{r}(\boldsymbol{A}, \boldsymbol{b}) = \mathrm{r}(\boldsymbol{A})$. That is,

$$\mathrm{rank}\,\{\boldsymbol{\alpha}_1, \boldsymbol{\alpha}_2, \ldots, \boldsymbol{\alpha}_n, \boldsymbol{b}\} = \mathrm{rank}\,\{\boldsymbol{\alpha}_1, \boldsymbol{\alpha}_2, \ldots, \boldsymbol{\alpha}_n\}, \tag{3.21}$$

where $\{\boldsymbol{\alpha}_1, \boldsymbol{\alpha}_2, \ldots, \boldsymbol{\alpha}_n\}$ is the set of column vectors of $\boldsymbol{A}$. Obviously, (3.21) holds, if and only if, $\boldsymbol{b}$ can be linearly expressed by $\{\boldsymbol{\alpha}_1, \boldsymbol{\alpha}_2, \ldots, \boldsymbol{\alpha}_n\}$. Otherwise, a contradiction is

$$\mathrm{rank}\,\{\boldsymbol{\alpha}_1, \boldsymbol{\alpha}_2, \ldots, \boldsymbol{\alpha}_n, \boldsymbol{b}\} = \mathrm{rank}\,\{\boldsymbol{\alpha}_1, \boldsymbol{\alpha}_2, \ldots, \boldsymbol{\alpha}_n\} + 1. \qquad \square$$

COROLLARY 3.5

$\boldsymbol{Ax} = \boldsymbol{b}$ has a unique solution, if and only if,

$$\mathrm{r}(\boldsymbol{A}, \boldsymbol{b}) = \mathrm{r}(\boldsymbol{A}) = \textit{the number of columns of } \boldsymbol{A}. \tag{3.22}$$

Because $\boldsymbol{b}$ can be linearly expressed by column vectors $\boldsymbol{\alpha}_1, \boldsymbol{\alpha}_2, \ldots, \boldsymbol{\alpha}_n$ of $\boldsymbol{A}$, and the expression is unique, if and only if, $\boldsymbol{\alpha}_1, \boldsymbol{\alpha}_2, \ldots, \boldsymbol{\alpha}_n$ are linearly independent. Or from (3.20), we have

$$\boldsymbol{Ax} = \boldsymbol{b} \text{ has a unique solution } \Leftrightarrow d_{r+1} = 0 \text{ and } r = n \Leftrightarrow \text{ (3.22) is true.}$$

Next, let us discuss the structure of solutions of $\boldsymbol{Ax} = \boldsymbol{b}$. For this purpose, we first discuss the properties of the solutions of $\boldsymbol{Ax} = \boldsymbol{b}$.

Note that if $\boldsymbol{x}_1$ and $\boldsymbol{x}_2$ are two solutions of $\boldsymbol{Ax} = \boldsymbol{b}$, then $k_1\boldsymbol{x}_1 + k_2\boldsymbol{x}_2$ is not a solution of $\boldsymbol{Ax} = \boldsymbol{b}$ in general, because

$$\begin{aligned} \boldsymbol{A}\,(k_1\boldsymbol{x}_1 + k_2\boldsymbol{x}_2) &= k_1\boldsymbol{Ax}_1 + k_2\boldsymbol{Ax}_2 \\ &= k_1\boldsymbol{b} + k_2\boldsymbol{b} = (k_1 + k_2)\,\boldsymbol{b} \neq \boldsymbol{b}. \end{aligned}$$

However, solutions of a nonhomogeneous system of linear equations have the following properties.

THEOREM 3.16

If $\boldsymbol{x}_1$ and $\boldsymbol{x}_2$ are solutions of $\mathbf{A}\boldsymbol{x} = \boldsymbol{b}$, then $\boldsymbol{x}_1 - \boldsymbol{x}_2$ is a solution of the corresponding homogeneous system of linear equations $\mathbf{A}\boldsymbol{x} = \mathbf{0}$.

Proof: Since $\boldsymbol{A}(\boldsymbol{x}_1 - \boldsymbol{x}_2) = \boldsymbol{A}\boldsymbol{x}_1 - \boldsymbol{A}\boldsymbol{x}_2 = \boldsymbol{b} - \boldsymbol{b} = \mathbf{0}$, hence, $(\boldsymbol{x}_1 - \boldsymbol{x}_2)$ is a solution of $\boldsymbol{A}\boldsymbol{x} = \mathbf{0}$. □

THEOREM 3.17

If $\mathbf{A}\boldsymbol{x} = \boldsymbol{b}$ has a solution, then its general solution is

$$\boldsymbol{x} = \boldsymbol{x}_0 + \bar{\boldsymbol{x}},$$

where $\boldsymbol{x}_0$ is a particular solution (a single solution) of $\mathbf{A}\boldsymbol{x} = \boldsymbol{b}$, while

$$\bar{\boldsymbol{x}} = k_1\boldsymbol{x}_1 + k_2\boldsymbol{x}_2 + \cdots + k_p\boldsymbol{x}_p$$

is the general solution of $\mathbf{A}\boldsymbol{x} = \mathbf{0}$ (called the derived system of $\mathbf{A}\boldsymbol{x} = \boldsymbol{b}$).

Proof: Since $\boldsymbol{A}(\boldsymbol{x}_0 + \bar{\boldsymbol{x}}) = \boldsymbol{A}\boldsymbol{x}_0 + \boldsymbol{A}\bar{\boldsymbol{x}} = \boldsymbol{b}$, so $\boldsymbol{x}_0 + \bar{\boldsymbol{x}}$ is a solution of $\boldsymbol{A}\boldsymbol{x} = \boldsymbol{b}$. Let $\boldsymbol{x}^*$ be any solution of $\boldsymbol{A}\boldsymbol{x} = \boldsymbol{b}$. Then, $(\boldsymbol{x}^* - \boldsymbol{x}_0)$ is a solution of $\boldsymbol{A}\boldsymbol{x} = \mathbf{0}$ and

$$\boldsymbol{x}^* = \boldsymbol{x}_0 + \left(\boldsymbol{x}^* - \boldsymbol{x}_0\right).$$

So, $\boldsymbol{x}^*$ can be expressed as $(\boldsymbol{x}_0 + \bar{\boldsymbol{x}})$, which is therefore the general solution of $\boldsymbol{A}\boldsymbol{x} = \boldsymbol{b}$. □

Example 3.16 General solution of $Ax = b$

For a nonhomogeneous system of linear equations $\boldsymbol{A}\boldsymbol{x} = \boldsymbol{b}$ with augmented matrix

$$(\boldsymbol{A}, \boldsymbol{b}) = \left(\begin{array}{cccccc|c} 1 & 1 & 1 & 0 & 0 & 0 \\ 1 & 1 & -1 & -1 & -2 & 1 \\ 2 & 2 & 0 & -1 & -2 & 1 \\ 5 & 5 & -3 & -4 & -8 & 4 \end{array}\right),$$

find the general solution of $\boldsymbol{A}\boldsymbol{x} = \boldsymbol{b}$.

Solution:

$$(\boldsymbol{A},\boldsymbol{b}) \xrightarrow[\text{row operations}]{\text{elementary}} \left(\begin{array}{ccccc|c} 1 & 1 & 0 & -\frac{1}{2} & -1 & \frac{1}{2} \\ 0 & 0 & 1 & \frac{1}{2} & 1 & -\frac{1}{2} \\ 0 & 0 & 0 & 0 & 0 & 0 \\ 0 & 0 & 0 & 0 & 0 & 0 \end{array}\right) = (\boldsymbol{U},\boldsymbol{d}).$$

Let $x_2 = x_4 = x_5 = 0$. Substituting them into $\boldsymbol{Ux} = \boldsymbol{d}$ yields a particular solution of $\boldsymbol{Ax} = \boldsymbol{b}$,

$$\boldsymbol{x}_0 = \left(\frac{1}{2}, 0, -\frac{1}{2}, 0, 0\right)^{\mathrm{T}}.$$

Let (x_2, x_4, x_5) be chosen as (1, 0, 0), (0, 1, 0), (0, 0, 1), respectively, and substitute them in turn into $\boldsymbol{Ux} = \boldsymbol{0}$ (Hint: Do not substitute into $\boldsymbol{Ux} = \boldsymbol{d}$!). Then we have a basis of $\boldsymbol{Ax} = \boldsymbol{0}$,

$$\begin{aligned} \boldsymbol{x}_1 &= (-1, 1, 0, 0, 0)^{\mathrm{T}}, \\ \boldsymbol{x}_2 &= \left(\frac{1}{2}, 0, -\frac{1}{2}, 1, 0\right)^{\mathrm{T}}, \\ \boldsymbol{x}_3 &= (1, 0, -1, 0, 1)^{\mathrm{T}}. \end{aligned}$$

So, the general solution of $\boldsymbol{Ax} = \boldsymbol{b}$ is

$$\begin{aligned} \boldsymbol{x} &= \boldsymbol{x}_0 + k_1\boldsymbol{x}_1 + k_2\boldsymbol{x}_2 + k_3\boldsymbol{x}_3 \\ &= \begin{pmatrix} \frac{1}{2} \\ 0 \\ -\frac{1}{2} \\ 0 \\ 0 \end{pmatrix} + k_1 \begin{pmatrix} -1 \\ 1 \\ 0 \\ 0 \\ 0 \end{pmatrix} + k_2 \begin{pmatrix} \frac{1}{2} \\ 0 \\ -\frac{1}{2} \\ 1 \\ 0 \end{pmatrix} + k_3 \begin{pmatrix} 1 \\ 0 \\ -1 \\ 0 \\ 1 \end{pmatrix}, \end{aligned}$$

where k_1, k_2, and k_3 are arbitrary constants. (Hint: In solving a system of linear equations, one good method of checking is to substitute $\boldsymbol{x}_0$, $\boldsymbol{x}_1$, $\boldsymbol{x}_2$, and $\boldsymbol{x}_3$ into the original equations to avoid computational mistakes.) The general solution obtained here is exactly the same as that of letting $x_2 = k_1$, $x_4 = k_2$, $x_5 = k_3$ by Gaussian elimination. ■

Obviously,

$$x_1^* = (-1, 1, 0, 0, 0)^{\mathrm{T}},$$
$$x_2^* = (1, 0, -1, 2, 0)^{\mathrm{T}},$$
$$x_3^* = (1, 0, -1, 0, 1)^{\mathrm{T}},$$

is a solution basis of $\boldsymbol{Ax} = \boldsymbol{0}$.

Example 3.17 Consistency of a system of linear equations

Consider a system of linear equations

$$\begin{cases} px_1 + x_2 + x_3 = 4, \\ x_1 + tx_2 + x_3 = 3, \\ x_1 + 2tx_2 + x_3 = 4. \end{cases}$$

Discuss its consistency with respect to p and t, and give solutions, whenever that may exist.

Solution: The augmented matrix $(\boldsymbol{A}, \boldsymbol{b})$ is reduced by elementary row operations to

$$(\boldsymbol{A}, \boldsymbol{b}) \xrightarrow[\text{rows}]{\text{interchange}} \begin{pmatrix} 1 & t & 1 & \vdots & 3 \\ 1 & 2t & 1 & \vdots & 4 \\ p & 1 & 1 & \vdots & 4 \end{pmatrix}$$

$$\xrightarrow[③+①\times(-p)]{②+①\times(-1)} \begin{pmatrix} 1 & t & 1 & \vdots & 3 \\ 0 & t & 0 & \vdots & 1 \\ 0 & 1-pt & 1-p & \vdots & 4-3p \end{pmatrix}$$

$$\xrightarrow{③+②\times p} \begin{pmatrix} 1 & t & 1 & \vdots & 3 \\ 0 & t & 0 & \vdots & 1 \\ 0 & 1 & 1-p & \vdots & 4-2p \end{pmatrix}$$

$$\xrightarrow[②\leftrightarrow③]{②+③\times(-t)} \begin{pmatrix} 1 & t & 1 & \vdots & 3 \\ 0 & 1 & 1-p & \vdots & 4-2p \\ 0 & 0 & (p-1)t & \vdots & 1-4t+2pt \end{pmatrix}.$$

(i) If $(p-1)t \neq 0$ (i.e., $p \neq 1$, $t \neq 0$), it has the unique solution

$$x_1 = \frac{2t-1}{(p-1)t},$$
$$x_2 = \frac{1}{t},$$
$$x_3 = \frac{1-4t+2pt}{(p-1)t}.$$

(ii) If $p=1$ and $1-4t+2pt=1-2t=0$, that is, $t=\frac{1}{2}$, it has infinitely many solutions. And

$$(\boldsymbol{A},\boldsymbol{b}) \Rightarrow \left(\begin{array}{ccc|c} 1 & \frac{1}{2} & 1 & 3 \\ 0 & 1 & 0 & 2 \\ 0 & 0 & 0 & 0 \end{array}\right) \Rightarrow \left(\begin{array}{ccc|c} 1 & 0 & 1 & 2 \\ 0 & 1 & 0 & 2 \\ 0 & 0 & 0 & 0 \end{array}\right).$$

So, the general solution is

$$\boldsymbol{x}=\begin{pmatrix} 2 \\ 2 \\ 0 \end{pmatrix}+k\begin{pmatrix} -1 \\ 0 \\ 1 \end{pmatrix} \quad (k \text{ is an arbitrary constant}).$$

(iii) When $p=1$ but $1-4t+2pt=1-2t\neq 0$,that is, $t\neq\frac{1}{2}$, it has no solution.

(iv) When $t=0$, $1-4t+2pt=1\neq 0$, it also has no solution. In this case, the second equation of the system contradicts the third one. ■

***Example 3.18 Solution of a nonhomogeneous system in terms of a solution basis of corresponding homogeneous system**

If $\boldsymbol{x}_0$ is a particular solution of $\boldsymbol{Ax}=\boldsymbol{b}$ $(\boldsymbol{b}\neq\boldsymbol{0})$ and $\{\boldsymbol{x}_1,\boldsymbol{x}_2,\ldots,\boldsymbol{x}_p\}$ is a solution basis of $\boldsymbol{Ax}=\boldsymbol{0}$, prove that $\boldsymbol{x}_0,\boldsymbol{x}_0+\boldsymbol{x}_1,\ldots,\boldsymbol{x}_0+\boldsymbol{x}_p$ are linearly independent and any solution of $\boldsymbol{Ax}=\boldsymbol{b}$ can be expressed as

$$\boldsymbol{x}=k_0\boldsymbol{x}_0+\sum_{i=1}^{p}k_i(\boldsymbol{x}_0+\boldsymbol{x}_i),$$

where $k_0+k_1+\cdots+k_p=1$.

Proof: Suppose

$$c_0\boldsymbol{x}_0+c_1(\boldsymbol{x}_0+\boldsymbol{x}_1)+\cdots+c_p(\boldsymbol{x}_0+\boldsymbol{x}_p)=\boldsymbol{0},$$

that is,

$$(c_0+c_1+\cdots+c_p)\boldsymbol{x}_0+c_1\boldsymbol{x}_1+\cdots+c_p\boldsymbol{x}_p=\boldsymbol{0}.$$

Then, $c_0+c_1+\cdots+c_p=0$ (otherwise, $\boldsymbol{x}_0$ is a solution of $\boldsymbol{Ax}=\boldsymbol{0}$, which is a contradiction to $\boldsymbol{b}\neq\boldsymbol{0}$.) Furthermore, $c_1,c_2,\ldots,c_p$ must be all zeros as $\boldsymbol{x}_1,\boldsymbol{x}_2,\ldots,\boldsymbol{x}_p$ are linearly independent, and so $c_0=0$. Hence, $\boldsymbol{x}_0,\boldsymbol{x}_0+\boldsymbol{x}_1,\ldots,\boldsymbol{x}_0+\boldsymbol{x}_p$ are linearly independent.

From Theorem 3.17, any solution $\boldsymbol{x}$ of $\boldsymbol{Ax}=\boldsymbol{b}$ can be expressed as

$$\begin{aligned}\boldsymbol{x} &= \boldsymbol{x}_0 + k_1\boldsymbol{x}_1 + \cdots + k_p\boldsymbol{x}_p \\ &= (1-k_1-\cdots-k_p)\boldsymbol{x}_0 + k_1(\boldsymbol{x}_0+\boldsymbol{x}_1) + \cdots + k_p(\boldsymbol{x}_0+\boldsymbol{x}_p).\end{aligned}$$

Denote $1-k_1-\cdots-k_p = k_0$, then $k_0+k_1+\cdots+k_p = 1$. Hence, the second part is proved. ■

In Example 3.18, $\boldsymbol{x}_1+\boldsymbol{x}_0, \ldots, \boldsymbol{x}_p+\boldsymbol{x}_0$ are all solutions of $\boldsymbol{Ax}=\boldsymbol{b}$. Generally speaking, if (number of columns of $\boldsymbol{A}$)$-\mathrm{r}(\boldsymbol{A}) = p$, then the nonhomogeneous system $\boldsymbol{Ax}=\boldsymbol{b}$ has $(p+1)$ linearly independent solutions $\boldsymbol{x}_0, \boldsymbol{x}_1^*, \ldots, \boldsymbol{x}_p^*$, and any solution of $\boldsymbol{Ax}=\boldsymbol{b}$ can be expressed as

$$\boldsymbol{x} = k_0\boldsymbol{x}_0 + k_1\boldsymbol{x}_1^* + \cdots + k_p\boldsymbol{x}_p^*\boldsymbol{x}, \tag{3.23}$$

where the sum of the coefficients $k_0+k_1+\cdots+k_p = 1$. Notice that the coefficients here are not all arbitrary and (3.23) is not called a general solution for $\boldsymbol{Ax}=\boldsymbol{b}$.

Example 3.19 General solution of $Ax=b$ from a solution base of $Ax=0$

Let $\boldsymbol{A}$ be a 5×4 matrix with $\mathrm{r}(\boldsymbol{A})=2$. Suppose $\boldsymbol{x}_1=(1,2,0,1)^{\mathrm{T}}$ and $\boldsymbol{x}_2=(2,1,1,3)^{\mathrm{T}}$ are two solutions of $\boldsymbol{Ax}=\boldsymbol{b}$, and $\boldsymbol{x}_3=(1,0,1,0)^{\mathrm{T}}$ is a solution of the corresponding homogeneous system $\boldsymbol{Ax}=\boldsymbol{0}$. Find a general solution of $\boldsymbol{Ax}=\boldsymbol{b}$.

Solution: A solution basis of $\boldsymbol{Ax}=\boldsymbol{0}$ consists of $4-\mathrm{r}(\boldsymbol{A})=2$ vectors. $\boldsymbol{x}_4=\boldsymbol{x}_2-\boldsymbol{x}_1=(1,-1,1,2)^{\mathrm{T}}$ and $\boldsymbol{x}_3$ are two linearly independent solutions (i.e., a solution base) of $\boldsymbol{Ax}=\boldsymbol{0}$. So, a general solution of $\boldsymbol{Ax}=\boldsymbol{b}$ is

$$\begin{aligned}\boldsymbol{x} &= k_1\boldsymbol{x}_3 + k_2(\boldsymbol{x}_2-\boldsymbol{x}_1) + \boldsymbol{x}_1 \\ &= k_1(1,0,1,0)^{\mathrm{T}} + k_2(1,-1,1,2)^{\mathrm{T}} + (1,2,0,1)^{\mathrm{T}},\end{aligned}$$

where k_1 and k_2 are arbitrary constants. ■

Example 3.20 Intersection of two solution sets

Let a homogeneous system of linear equations (I) in four variables be

$$\begin{cases} x_1 + x_3 = 0, \\ x_3 - x_4 = 0. \end{cases}$$

A solution basis of a homogeneous system of linear equations (II) in four variables is $\boldsymbol{x}_3 = (0, 1, 2, 0)^{\mathrm{T}}$ and $\boldsymbol{x}_4 = (-1, -3, -3, 1)^{\mathrm{T}}$.

(i) Give a general solution of the system of linear equations (I).

*(ii) Does (I) and (II) have a common nonzero solution? If yes, find all nonzero common solutions. If not, why? [Hint: By answering these questions, we can find the intersection of the two solution sets.]

Solution:

(i) Let us apply elementary row operations on the coefficient matrix $\boldsymbol{A}$ of (I),

$$\boldsymbol{A} = \begin{pmatrix} 1 & 0 & 1 & 0 \\ 0 & 0 & 1 & -1 \end{pmatrix} \xrightarrow[\text{row operations}]{\text{elementary}} \begin{pmatrix} 1 & 0 & 0 & 1 \\ 0 & 0 & 1 & -1 \end{pmatrix} = \boldsymbol{U}.$$

An equivalent equation system to (I) is $\boldsymbol{Ux} = \boldsymbol{0}$, that is,

$$\begin{cases} x_1 + x_4 = 0, \\ x_3 - x_4 = 0. \end{cases}$$

Let x_2 and x_4 be free variables and we take $x_2 = 1, x_4 = 0$ and $x_2 = 0$, $x_4 = 1$, respectively. We obtain a solution basis

$$\boldsymbol{x}_1 = (0, 1, 0, 0)^{\mathrm{T}},$$

$$\boldsymbol{x}_2 = (-1, 0, 1, 1)^{\mathrm{T}}.$$

Therefore, a general solution is $\boldsymbol{x} = k_1\boldsymbol{x}_1 + k_2\boldsymbol{x}_2$, where $\boldsymbol{x}_1$ and $\boldsymbol{x}_2$ are as shown above and k_1 and k_2 are arbitrary constants.

*(ii) The general solution for the linear equations system (II) is $\boldsymbol{x} = k_3\boldsymbol{x}_3 + k_4\boldsymbol{x}_4$. If (I) and (II) have a common solution $\boldsymbol{x}_0$, then we must have

$$\boldsymbol{x}_0 = k_1\boldsymbol{x}_1 + k_2\boldsymbol{x}_2 = k_3\boldsymbol{x}_3 + k_4\boldsymbol{x}_4, \qquad ①$$

thus

$$k_1\boldsymbol{x}_1 + k_2\boldsymbol{x}_2 - k_3\boldsymbol{x}_3 - k_4\boldsymbol{x}_4 = \boldsymbol{0}.$$

It is a homogeneous system of linear equations $\boldsymbol{Bk} = \boldsymbol{0}$ in four unknowns k_1, k_2, k_3, and k_4, where $\boldsymbol{k} = (k_1, k_2, k_3, k_4)^{\mathrm{T}}$, and the coefficient matrix is

$$\boldsymbol{B} = (\boldsymbol{x}_1, \boldsymbol{x}_2, -\boldsymbol{x}_3, -\boldsymbol{x}_4) = \begin{pmatrix} 0 & -1 & 0 & 1 \\ 1 & 0 & -1 & 3 \\ 0 & 1 & -2 & 3 \\ 0 & 1 & 0 & -1 \end{pmatrix}$$

$$\xrightarrow[\text{elementary row operation}]{\text{interchange 1st and 2nd rows}} \begin{pmatrix} 1 & 0 & 0 & 1 \\ 0 & 1 & 0 & -1 \\ 0 & 0 & 1 & -2 \\ 0 & 0 & 0 & 0 \end{pmatrix}.$$

Taking k_4 as a free variable and letting $k_4 = 1$ yield a basis of solutions (which consists of only one solution) for $\boldsymbol{Bk} = \boldsymbol{0}$,

$$\boldsymbol{k}_1 = (-1, 1, 2, 1)^{\mathrm{T}}.$$

The general solution is

$$\boldsymbol{k} = (k_1, k_2, k_3, k_4)^{\mathrm{T}} = \lambda(-1, 1, 2, 1)^{\mathrm{T}} (\lambda \text{ is an arbitrary constant}).$$

Thus, letting $k_1 = -\lambda$, $k_2 = \lambda$ and substituting them into expression ① lead to all nonzero common solutions for (I) and (II),

$$\begin{aligned} \boldsymbol{x}_0 &= -\lambda \boldsymbol{x}_1 + \lambda \boldsymbol{x}_2 \\ &= \lambda(-1, -1, 2, 1)^{\mathrm{T}} \ (\lambda \text{ is an arbitrary constant}). \end{aligned}$$ ■

If $\boldsymbol{Bx} = \boldsymbol{0}$ has only the zero solution, then (I) and (II) have no nonzero common solution.

If the system of linear equations (II) is given in the form of a homogeneous system of linear equations in four unknowns, then the intersection of solution sets for (I) and (II) is the solution set for a system consisting of all equations in (I) and (II).

Exercises and Supplementary Exercises

Exercises

Q3.1–3.2: Express $\boldsymbol{\alpha}$ linearly in terms of $\boldsymbol{\alpha}_1, \boldsymbol{\alpha}_2, \boldsymbol{\alpha}_3$, and $\boldsymbol{\alpha}_4$.

3.1 $\boldsymbol{\alpha} = \begin{pmatrix} 1 \\ 2 \\ 1 \\ 1 \end{pmatrix}, \boldsymbol{\alpha}_1 = \begin{pmatrix} 1 \\ 1 \\ 1 \\ 1 \end{pmatrix}, \boldsymbol{\alpha}_2 = \begin{pmatrix} 1 \\ 1 \\ -1 \\ -1 \end{pmatrix},$

$\boldsymbol{\alpha}_3 = \begin{pmatrix} 1 \\ -1 \\ 1 \\ -1 \end{pmatrix}, \boldsymbol{\alpha}_4 = \begin{pmatrix} 1 \\ -1 \\ -1 \\ 1 \end{pmatrix}.$

3.2 $\boldsymbol{\alpha} = (0, 0, 0, 1), \boldsymbol{\alpha}_1 = (1, 1, 0, 1),$
$\boldsymbol{\alpha}_2 = (2, 1, 3, 1), \boldsymbol{\alpha}_3 = (1, 1, 0, 0),$
$\boldsymbol{\alpha}_4 = (0, 1, -1, -1).$

Q3.3–3.4: Determine the linear dependence for each question.

3.3 $\boldsymbol{\alpha}_1 = (1, 1, 1)^{\mathrm{T}}, \boldsymbol{\alpha}_2 = (0, 2, 5)^{\mathrm{T}},$
$\boldsymbol{\alpha}_3 = (1, 3, 6)^{\mathrm{T}}.$

3.4 $\boldsymbol{\beta}_1 = (1, -1, 2, 4)^{\mathrm{T}}, \boldsymbol{\beta}_2 = (0, 3, 1, 2)^{\mathrm{T}},$
$\boldsymbol{\beta}_3 = (3, 0, 7, 14)^{\mathrm{T}}.$

3.5 When is a single vector $\boldsymbol{\alpha} = (\alpha_1, \alpha_2, \ldots, \alpha_n)$ linearly dependent or linearly independent?

3.6 Prove that if a set of vectors is linearly independent, then any subset of it is linearly independent.

3.7 Prove that if $\boldsymbol{\alpha}_1$ and $\boldsymbol{\alpha}_2$ are linearly independent, then $\boldsymbol{\alpha}_1 + \boldsymbol{\alpha}_2$ and $\boldsymbol{\alpha}_1 - \boldsymbol{\alpha}_2$ are linearly independent.

3.8 Suppose there are two sets of vectors $\boldsymbol{\alpha}_1, \boldsymbol{\alpha}_2, \ldots, \boldsymbol{\alpha}_s$ and $\boldsymbol{\beta}_1, \boldsymbol{\beta}_2, \ldots, \boldsymbol{\beta}_s$, where

$$\boldsymbol{\alpha}_1 = \begin{pmatrix} a_{11} \\ a_{21} \\ \vdots \\ a_{k1} \end{pmatrix}, \boldsymbol{\alpha}_2 = \begin{pmatrix} a_{12} \\ a_{22} \\ \vdots \\ a_{k2} \end{pmatrix}, \ldots, \boldsymbol{\alpha}_s = \begin{pmatrix} a_{1s} \\ a_{2s} \\ \vdots \\ a_{ks} \end{pmatrix},$$

$\boldsymbol{\beta}_1, \boldsymbol{\beta}_2, \ldots, \boldsymbol{\beta}_s$ are $(k + m)$-dimensional vectors which are formed by arbitrarily increasing m components $b_{1j}, b_{2j}, \ldots, b_{mj}$ $(j = 1, 2, \ldots, s)$ after the components of each vector $\boldsymbol{\alpha}_1, \boldsymbol{\alpha}_2, \ldots, \boldsymbol{\alpha}_s$ respectively. Prove that

(i) if $\boldsymbol{\alpha}_1, \boldsymbol{\alpha}_2, \ldots, \boldsymbol{\alpha}_s$ are linearly independent, then $\boldsymbol{\beta}_1, \boldsymbol{\beta}_2, \ldots, \boldsymbol{\beta}_s$ are linearly independent.

(ii) if $\boldsymbol{\beta}_1, \boldsymbol{\beta}_2, \ldots, \boldsymbol{\beta}_s$ are linearly dependent, then $\boldsymbol{\alpha}_1, \boldsymbol{\alpha}_2, \ldots, \boldsymbol{\alpha}_s$ are also linearly dependent.

In addition, give a geometric interpretation by taking a pair of vectors in $\mathbb{R}^2$ and $\mathbb{R}^3$ respectively (i.e., assign $s = 2, k = 2, m = 1$).

3.9 Prove that $\boldsymbol{\alpha}_1 + \boldsymbol{\alpha}_2$, $\boldsymbol{\alpha}_2 + \boldsymbol{\alpha}_3$, and $\boldsymbol{\alpha}_3 + \boldsymbol{\alpha}_1$ are linearly independent, if and only if, $\boldsymbol{\alpha}_1, \boldsymbol{\alpha}_2$, and $\boldsymbol{\alpha}_3$ are linearly independent.

3.10 Are the following statements true? If yes, prove it. Otherwise give a counterexample.

(i) $\boldsymbol{\alpha}_1, \boldsymbol{\alpha}_2, \ldots, \boldsymbol{\alpha}_m (m > 2)$ are linearly independent, if and only if, any two of the vectors are linearly independent.

(ii) $\boldsymbol{\alpha}_1, \boldsymbol{\alpha}_2, \ldots, \boldsymbol{\alpha}_m (m > 2)$ are linearly dependent, if and only if, there are $(m - 1)$ vectors that are linearly dependent.

(iii) If $\boldsymbol{\alpha}_1$ and $\boldsymbol{\alpha}_2$ are linearly dependent, $\boldsymbol{\beta}_1$ and $\boldsymbol{\beta}_2$ are linearly dependent, then there are numbers k_1 and k_2 not all zero such that, $k_1\boldsymbol{\alpha}_1 + k_2\boldsymbol{\alpha}_2 = \mathbf{0}$, $k_1\boldsymbol{\beta}_1 + k_2\boldsymbol{\beta}_2 = \mathbf{0}$ and thus $k_1(\boldsymbol{\alpha}_1 + \boldsymbol{\beta}_1) + k_2(\boldsymbol{\alpha}_2 + \boldsymbol{\beta}_2) = \mathbf{0}$. So, $\boldsymbol{\alpha}_1 + \boldsymbol{\beta}_1$, $\boldsymbol{\alpha}_2 + \boldsymbol{\beta}_2$ are linearly dependent.

(iv) If $\boldsymbol{\alpha}_1, \boldsymbol{\alpha}_2$, and $\boldsymbol{\alpha}_3$ are linearly independent, then $\boldsymbol{\alpha}_1 - \boldsymbol{\alpha}_2, \boldsymbol{\alpha}_2 - \boldsymbol{\alpha}_3$, and $\boldsymbol{\alpha}_3 - \boldsymbol{\alpha}_1$ are linearly independent.

(v) If $\boldsymbol{\alpha}_1, \boldsymbol{\alpha}_2, \boldsymbol{\alpha}_3$, and $\boldsymbol{\alpha}_4$ are linearly independent, then $\boldsymbol{\alpha}_1 + \boldsymbol{\alpha}_2$, $\boldsymbol{\alpha}_2 + \boldsymbol{\alpha}_3$, $\boldsymbol{\alpha}_3 + \boldsymbol{\alpha}_4$, and $\boldsymbol{\alpha}_4 + \boldsymbol{\alpha}_1$ are linearly independent (Hint: Show the general result for n vectors).

(vi) If $\boldsymbol{\alpha}_1, \boldsymbol{\alpha}_2, \ldots, \boldsymbol{\alpha}_n$ are linearly dependent, then $\boldsymbol{\alpha}_1 + \boldsymbol{\alpha}_2, \boldsymbol{\alpha}_2 + \boldsymbol{\alpha}_3, \ldots, \boldsymbol{\alpha}_{n-1} + \boldsymbol{\alpha}_n, \boldsymbol{\alpha}_n + \boldsymbol{\alpha}_1$ are linearly dependent.

3.11 If $\boldsymbol{\alpha}_1, \boldsymbol{\alpha}_2, \boldsymbol{\alpha}_3$, and $\boldsymbol{\alpha}_4$ are linearly dependent, but any three vectors of them are linearly independent, prove that there are numbers k_1, k_2, k_3, and k_4 which are not all zero such that $k_1\boldsymbol{\alpha}_1 + k_2\boldsymbol{\alpha}_2 + k_3\boldsymbol{\alpha}_3 + k_4\boldsymbol{\alpha}_4 = \mathbf{0}$.

3.12 If $\boldsymbol{\alpha}_1, \boldsymbol{\alpha}_2, \ldots, \boldsymbol{\alpha}_r$ are linearly independent, prove that $\boldsymbol{\beta}, \boldsymbol{\alpha}_1, \boldsymbol{\alpha}_2, \ldots, \boldsymbol{\alpha}_r$ are linearly independent, if and only if, $\boldsymbol{\beta}$ cannot be linearly expressed by $\boldsymbol{\alpha}_1, \boldsymbol{\alpha}_2, \ldots, \boldsymbol{\alpha}_r$.

3.13 Give the rank of each of the following vector sets as well as a maximal linearly independent subset, and linearly express the other vectors by the maximal linearly independent subset.

(i) $\boldsymbol{\alpha}_1 = (6, 4, 1, 9, 2)$, $\boldsymbol{\alpha}_2 = (1, 0, 2, 3, -4)$, $\boldsymbol{\alpha}_3 = (1, 4, -9, -6, 22)$, $\boldsymbol{\alpha}_4 = (7, 1, 0, -1, 3)$.

(ii) $\boldsymbol{\alpha}_1 = (1, -1, 2, 4)$, $\boldsymbol{\alpha}_2 = (0, 3, 1, 2)$, $\boldsymbol{\alpha}_3 = (3, 0, 7, 14)$, $\boldsymbol{\alpha}_4 = (2, 1, 5, 6)$, $\boldsymbol{\alpha}_5 = (1, -1, 2, 0)$.

(iii) $\boldsymbol{\alpha}_1 = (1, 1, 1)$, $\boldsymbol{\alpha}_2 = (1, 1, 0)$, $\boldsymbol{\alpha}_3 = (1, 0, 0)$, $\boldsymbol{\alpha}_4 = (1, 2, -3)$.

3.14 Consider the vector set: $\boldsymbol{\xi}_1 = (1, -1, 2, 4)$, $\boldsymbol{\xi}_2 = (0, 3, 1, 2)$, $\boldsymbol{\xi}_3 = (3, 0, 7, 14)$, $\boldsymbol{\xi}_4 = (1, -1, 2, 0)$, $\boldsymbol{\xi}_5 = (2, 1, 5, 6)$.

(i) Prove that $\boldsymbol{\xi}_1$ and $\boldsymbol{\xi}_2$ are linearly independent.

(ii) Find a maximal linearly independent subset of the vector set including $\boldsymbol{\xi}_1$ and $\boldsymbol{\xi}_2$.

***3.15** Let $\boldsymbol{A}$ and $\boldsymbol{B}$ be both nth-order matrices, $\mathrm{r}(\boldsymbol{A}) \le n$, $\mathrm{r}(\boldsymbol{B}) \le n$. Prove that,

(i) $\mathrm{rank}\begin{pmatrix} \boldsymbol{A} & \mathbf{0} \\ \mathbf{0} & \boldsymbol{B} \end{pmatrix} = \mathrm{r}(\boldsymbol{A}) + \mathrm{r}(\boldsymbol{B})$,

(ii) $\mathrm{rank}\begin{pmatrix} \boldsymbol{A} & \boldsymbol{C} \\ \mathbf{0} & \boldsymbol{B} \end{pmatrix} \ge \mathrm{r}(\boldsymbol{A}) + \mathrm{r}(\boldsymbol{B})$ ($\boldsymbol{C}$ is an arbitrary nth-order matrix).

***3.16** Prove that $\mathrm{r}(\boldsymbol{AB}) \le \min[\mathrm{r}(\boldsymbol{A}), \mathrm{r}(\boldsymbol{B})]$.

3.17 Let $\boldsymbol{A}$ be an $m \times n$ matrix, $\boldsymbol{B}$ an $n \times m$ matrix, $n < m$. Prove that a homogeneous system of linear equations $(\boldsymbol{AB})\boldsymbol{x} = \mathbf{0}$ has a nonzero solution.

***3.18** Let $\boldsymbol{A}$ be an $s \times n$ matrix, $\boldsymbol{B}$ be an $n \times m$ matrix made up by the first m rows of $\boldsymbol{A}$. Prove that $\mathrm{r}(\boldsymbol{B}) \ge r + m - s$ if the row rank of $\boldsymbol{A}$ is r.

Q3.19–3.22: Find the rank of each of the following matrices and show a nonzero subdeterminant with the highest order in the matrix.

3.19 $\begin{pmatrix} 1 & 2 & 3 & 4 & 5 \\ 0 & 0 & -1 & -2 & -3 \\ 0 & 0 & 0 & 0 & 4 \\ 0 & 0 & 1 & 2 & -1 \end{pmatrix}$.

3.20 $\begin{pmatrix} 1 & -1 & 2 & 1 & 0 \\ 2 & -2 & 4 & -2 & 0 \\ 3 & 0 & 6 & -1 & 1 \\ 0 & 3 & 0 & 0 & 1 \end{pmatrix}$.

3.21 $\begin{pmatrix} 3 & 2 & -1 & -3 & -2 \\ 2 & -1 & 3 & 1 & -3 \\ 4 & 5 & -5 & -6 & 1 \end{pmatrix}$.

3.22 $\begin{pmatrix} 1 & 1 & 0 & 0 \\ 2 & 1 & 1 & 0 \\ 0 & 2 & 1 & 1 \\ 0 & 0 & 2 & 1 \end{pmatrix}$.

3.23 Let $\boldsymbol{A}$ be an $m \times n$ matrix. Prove that there is an $n \times s$ nonzero matrix $\boldsymbol{B}$, such that $\boldsymbol{AB} = \mathbf{0}$, if and only if, $\mathrm{r}(\boldsymbol{A}) < n$.

***3.24** Let $\boldsymbol{A}$ and $\boldsymbol{B}$ be matrices of same size. Prove that $\boldsymbol{A}$ and $\boldsymbol{B}$ are equivalent, if and only if, $\mathrm{r}(\boldsymbol{A}) = \mathrm{r}(\boldsymbol{B})$.

⋆3.25 Let A be an $m \times n$ matrix $(m < n)$ and $\mathrm{r}(A) = m$. Prove that there exists an $n \times m$ matrix B such that $AB = I_m$.

⋆3.26 Prove that there exists an nth-order square matrix B with rank $(n-r)$ such that $BA = 0$, if the rank of the square matrix A with order n is equal to r.

⋆3.27 Prove that any matrix with rank r can be expressed as a sum of r matrices with rank 1, but cannot be expressed as a sum of less than r matrices with rank 1.

3.28 Find a solution basis for each of the following homogeneous systems of linear equations and determine its general solution.

$$\text{(i)} \begin{cases} x_1 - x_2 + 5x_3 - x_4 = 0, \\ x_1 + x_2 - 2x_3 + 3x_4 = 0, \\ 3x_1 - x_2 + 8x_3 + x_4 = 0, \\ x_1 + 3x_2 - 9x_3 + 7x_4 = 0. \end{cases}$$

$$\text{(ii)} \begin{cases} 3x_1 + x_2 - 8x_3 + 2x_4 + x_5 = 0, \\ 2x_1 - 2x_2 - 3x_3 - 7x_4 + 2x_5 = 0, \\ x_1 + 11x_2 - 12x_3 + 34x_4 - 5x_5 = 0, \\ x_1 - 5x_2 + 2x_3 - 16x_4 + 3x_5 = 0. \end{cases}$$

3.29 Give the general solutions for the following nonhomogeneous systems of linear equations.

$$\text{(i)} \begin{cases} 2x_1 + 7x_2 + 3x_3 + x_4 = 6, \\ 3x_1 + 5x_2 + 2x_3 + 2x_4 = 4, \\ 9x_1 + 4x_2 + x_3 + 7x_4 = 2. \end{cases}$$

$$\text{(ii)} \begin{cases} x_1 + x_2 + x_3 + x_4 + x_5 = 7, \\ 3x_1 + 2x_2 + x_3 + x_4 - 3x_5 = -2, \\ x_2 + 2x_3 + 2x_4 + 6x_5 = 23, \\ 5x_1 + 4x_2 + 3x_3 + 3x_4 - x_5 = 12. \end{cases}$$

3.30 Discuss the possible values of p, and/or q regarding the solvability of the following systems of linear equations. Provide their solutions when such solutions exist.

$$\text{(i)} \begin{cases} (p+3)x_1 + x_2 + 2x_3 = p, \\ px_1 + (p-1)x_2 + x_3 = 2p, \\ 3(p+1)x_1 + px_2 + (p+3)x_3 = 3. \end{cases}$$

$$\text{(ii)} \begin{cases} x_1 + x_2 + x_3 + x_4 + x_5 = 1, \\ 3x_1 + 2x_2 + x_3 + x_4 - 3x_5 = p, \\ x_2 + 2x_3 + 2x_4 + 6x_5 = 3, \\ 5x_1 + 4x_2 + 3x_3 + 3x_4 - x_5 = q. \end{cases}$$

$$\text{(iii)} \begin{cases} x_1 + x_2 + 2x_3 - x_4 = 1, \\ x_1 - x_2 - 2x_3 - 7x_4 = 3, \\ x_2 + px_3 + qx_4 = q-3, \\ x_1 + x_2 + 2x_3 + (q-2)x_4 = q+3. \end{cases}$$

3.31 Let A be an $m \times n$ matrix. Prove that if $Ax = 0$ for any n-dimensional vector x, then $A = 0$.

3.32 Let A be an $m \times s$ matrix, B be an $s \times n$ matrix, x be an n-dimensional column vector. Prove that $\mathrm{r}(AB) = \mathrm{r}(B)$ if $(AB)x = 0$ and $Bx = 0$ are equivalent systems of linear equations.

3.33 Let A be an $m \times n$ matrix, B an $n \times s$ matrix. Prove that $\mathrm{r}(A) + \mathrm{r}(B) \le n$ if $AB = 0$ (Hint: Column vectors of B are solutions of $Ax = 0$).

⋆3.34 Let A^* be the adjoint matrix of an nth-order matrix A. Prove that,

$$\text{(i) } \mathrm{r}(A^*) = \begin{cases} n, & \text{if } \mathrm{r}(A) = n, \\ 1, & \text{if } \mathrm{r}(A) = n-1, \\ 0, & \text{if } \mathrm{r}(A) < n-1. \end{cases}$$

(ii) $|A^*| = |A|^{n-1}$.

3.35 Let A be an nth-order invertible matrix $(n > 2)$. Prove that $(A^*)^* = |A|^{n-2}A$.

3.36 Let A be an nth-order matrix. Prove that a nonhomogeneous system of linear equations $Ax = b$ has a solution for any b, if $|A| \neq 0$.

3.37 Let $x_1 - x_2 = a_1, x_2 - x_3 = a_2, x_3 - x_4 = a_3, x_4 - x_5 = a_4$, and $x_5 - x_1 = a_5$. Prove that this system of linear equation has a solution, if and only if, $\sum_{i=1}^{5} a_i = 0$. If there are solutions, find its general solution.

Supplementary exercises

Q3.38–3.44: For these multiple-choice questions, select the correct answer.

3.38 $\boldsymbol{\beta}_1$ and $\boldsymbol{\beta}_2$ are two different solutions of the system of equations $\boldsymbol{Ax} = \boldsymbol{b}$, $\{\boldsymbol{\alpha}_1, \boldsymbol{\alpha}_2\}$ is a solution basis of the corresponding homogeneous system of equations $\boldsymbol{Ax} = \boldsymbol{0}$, then a general solution of $\boldsymbol{Ax} = \boldsymbol{b}$ is:

(a) $k_1\boldsymbol{\alpha}_1 + k_2(\boldsymbol{\alpha}_1 + \boldsymbol{\alpha}_2) + \dfrac{\boldsymbol{\beta}_1 - \boldsymbol{\beta}_2}{2}$,

(b) $k_1\boldsymbol{\alpha}_1 + k_2(\boldsymbol{\alpha}_2 - \boldsymbol{\alpha}_1) + \dfrac{\boldsymbol{\beta}_1 + \boldsymbol{\beta}_2}{2}$,

(c) $k_1\boldsymbol{\alpha}_1 + k_2(\boldsymbol{\beta}_1 + \boldsymbol{\beta}_2) + \dfrac{\boldsymbol{\beta}_1 - \boldsymbol{\beta}_2}{2}$,

(d) $k_1\boldsymbol{\alpha}_1 + k_2(\boldsymbol{\beta}_1 - \boldsymbol{\beta}_2) + \dfrac{\boldsymbol{\beta}_1 + \boldsymbol{\beta}_2}{2}$.

3.39 Given that $\boldsymbol{Q} = \begin{pmatrix} 1 & 2 & 3 \\ 2 & 4 & t \\ 3 & 6 & 9 \end{pmatrix}$, $\boldsymbol{P}$ is a nonzero matrix of order 3, $\boldsymbol{PQ} = \boldsymbol{0}$, then

(a) $\mathrm{r}(\boldsymbol{P}) = 1$ when $t = 6$,

(b) $\mathrm{r}(\boldsymbol{P}) = 2$ when $t = 6$,

(c) $\mathrm{r}(\boldsymbol{P}) = 1$ when $t \neq 6$,

(d) $\mathrm{r}(\boldsymbol{P}) = 2$ when $t \neq 6$.

3.40 Suppose $\boldsymbol{\alpha}_1 = (a_1, a_2, a_3)^{\mathrm{T}}$, $\boldsymbol{\alpha}_2 = (b_1, b_2, b_3)^{\mathrm{T}}$, $\boldsymbol{\alpha}_3 = (c_1, c_2, c_3)^{\mathrm{T}}$, then the three straight lines $a_i x + b_i y + c_i = 0$ $(a_i^2 + b_i^2 \neq 0)$ $(i = 1, 2, 3)$ intersect at a point, if and only if,

(a) $\boldsymbol{\alpha}_1, \boldsymbol{\alpha}_2, \boldsymbol{\alpha}_3$ are linearly dependent,

(b) $\boldsymbol{\alpha}_1, \boldsymbol{\alpha}_2, \boldsymbol{\alpha}_3$ are linearly independent,

(c) $\mathrm{r}(\boldsymbol{\alpha}_1, \boldsymbol{\alpha}_2, \boldsymbol{\alpha}_3) = \mathrm{r}(\boldsymbol{\alpha}_1, \boldsymbol{\alpha}_2)$,

(d) $\boldsymbol{\alpha}_1, \boldsymbol{\alpha}_2, \boldsymbol{\alpha}_3$ are linearly dependent, $\boldsymbol{\alpha}_1$ and $\boldsymbol{\alpha}_2$ are linearly independent.

3.41 Let $\boldsymbol{A}$ be an $m \times n$ matrix, $\mathrm{r}(\boldsymbol{A}) = m (m < n)$, $\boldsymbol{B}$ be an nth-order matrix. Which one is true?

(a) Any one of mth-order subdeterminants in $\boldsymbol{A}$ is not 0.

(b) Any m columns in $\boldsymbol{A}$ are linearly independent.

(c) $|\boldsymbol{A}^{\mathrm{T}}\boldsymbol{A}| \neq 0$.

(d) $\boldsymbol{B} = \boldsymbol{0}$ if $\boldsymbol{AB} = \boldsymbol{0}$.

(e) $\mathrm{r}(\boldsymbol{AB}) = m$ if $\mathrm{r}(\boldsymbol{B}) = n$.

3.42 Let $\boldsymbol{\alpha}_1, \boldsymbol{\alpha}_2, \ldots, \boldsymbol{\alpha}_m (\boldsymbol{\alpha}_i \in \mathbb{R}^n; i = 1, 2, \ldots, m; m > 2)$ be linearly independent. Which one of the following holds?

(a) $k_1\boldsymbol{\alpha}_1 + k_2\boldsymbol{\alpha}_2 + \cdots + k_m\boldsymbol{\alpha}_m = \boldsymbol{0}$ for arbitrary constants $k_1, k_2, \ldots, k_m$.

(b) Any $k (k < m)$ vectors $\boldsymbol{\alpha}_{i_1}, \boldsymbol{\alpha}_{i_2}, \ldots, \boldsymbol{\alpha}_{i_k}$ are linearly dependent.

(c) $\boldsymbol{\alpha}_1, \boldsymbol{\alpha}_2, \ldots, \boldsymbol{\alpha}_m, \boldsymbol{\beta}$ are linearly dependent for any $\boldsymbol{\beta} \in \mathbb{R}^n$.

(d) Any $k (k < m)$ vectors $\boldsymbol{\alpha}_{i_1}, \boldsymbol{\alpha}_{i_2}, \ldots, \boldsymbol{\alpha}_{i_k}$ are linearly independent.

3.43 Let vectors $\boldsymbol{\alpha}, \boldsymbol{\beta}$, and $\boldsymbol{\gamma}$ be linearly independent, $\boldsymbol{\alpha}, \boldsymbol{\beta}$, and $\boldsymbol{\delta}$ be linearly dependent. Which one of the following is true?

(a) $\boldsymbol{\alpha}$ can be linearly expressed by $\boldsymbol{\beta}, \boldsymbol{\gamma}$, and $\boldsymbol{\delta}$.

(b) $\boldsymbol{\beta}$ cannot be linearly expressed by $\boldsymbol{\alpha}, \boldsymbol{\gamma}$, and $\boldsymbol{\delta}$.

(c) $\boldsymbol{\delta}$ can be linearly expressed by $\boldsymbol{\alpha}, \boldsymbol{\beta}$, and $\boldsymbol{\gamma}$.

(d) $\boldsymbol{\delta}$ cannot be linearly expressed by $\boldsymbol{\alpha}, \boldsymbol{\beta}$, and $\boldsymbol{\gamma}$.

3.44 Let $\boldsymbol{A}$ be a 4×3 matrix, $\mathrm{r}(\boldsymbol{A}) = 1$, $\boldsymbol{\xi}_1, \boldsymbol{\xi}_2$, and $\boldsymbol{\xi}_3$ be three linearly independent solutions for the nonhomogeneous system of linear equations $\boldsymbol{Ax} = \boldsymbol{b}$. Which one of the following is a solution basis of $\boldsymbol{Ax} = \boldsymbol{0}$?

(a) $\boldsymbol{\xi}_1 + \boldsymbol{\xi}_2 + \boldsymbol{\xi}_3$,

(b) $\boldsymbol{\xi}_1 + \boldsymbol{\xi}_2 - 2\boldsymbol{\xi}_3$,

(c) $\boldsymbol{\xi}_2 - \boldsymbol{\xi}_1, \boldsymbol{\xi}_3 - \boldsymbol{\xi}_2$,

(d) $\boldsymbol{\xi}_1 + \boldsymbol{\xi}_2, \boldsymbol{\xi}_2 + \boldsymbol{\xi}_3$.

3.45 Let the vector set $\{\boldsymbol{\alpha}_1, \boldsymbol{\alpha}_2, \boldsymbol{\alpha}_3\}$ be linearly dependent, whereas $\{\boldsymbol{\alpha}_2, \boldsymbol{\alpha}_3, \boldsymbol{\alpha}_4\}$ is linearly independent. Answer the following questions with proper justification.

(i) Can $\boldsymbol{\alpha}_1$ be linearly expressed by $\{\boldsymbol{\alpha}_2, \boldsymbol{\alpha}_3\}$?

(ii) Can $\boldsymbol{\alpha}_4$ be linearly expressed by $\{\boldsymbol{\alpha}_1, \boldsymbol{\alpha}_2, \boldsymbol{\alpha}_3\}$?

3.46 Let $\boldsymbol{A}$ be an nth-order matrix. If there exists a positive integer $k (k \geq 2)$ such that $\boldsymbol{A}^k\boldsymbol{\alpha} = \boldsymbol{0}$ but $\boldsymbol{A}^{k-1}\boldsymbol{\alpha} \neq \boldsymbol{0}$ (where $\boldsymbol{\alpha}$ is an n-dimensional nonzero column vector), prove that $\boldsymbol{\alpha}, \boldsymbol{A}\boldsymbol{\alpha}, \ldots, \boldsymbol{A}^{k-1}\boldsymbol{\alpha}$ are linearly independent.

3.47 Let $\boldsymbol{A}, \boldsymbol{B}$ be $n \times m$, $m \times n$ matrices $(n < m)$, respectively, and $\boldsymbol{AB} = \boldsymbol{I}$ (unit matrix of order n). Prove that the set of column vectors of $\boldsymbol{B}$ is linearly independent.

3.48 Given that, rank$\{\boldsymbol{\alpha}_1, \boldsymbol{\alpha}_2, \boldsymbol{\alpha}_3\} =$ rank$\{\boldsymbol{\beta}_1, \boldsymbol{\beta}_2, \boldsymbol{\beta}_3\}$, where

$$\boldsymbol{\alpha}_1 = (1, 2, -3)^{\mathrm{T}}, \ \boldsymbol{\alpha}_2 = (3, 0, 1)^{\mathrm{T}},$$
$$\boldsymbol{\alpha}_3 = (9, 6, -7)^{\mathrm{T}};$$
$$\boldsymbol{\beta}_1 = (0, 1, -1)^{\mathrm{T}}, \ \boldsymbol{\beta}_2 = (a, 2, 1)^{\mathrm{T}},$$
$$\boldsymbol{\beta}_3 = (b, 1, 0)^{\mathrm{T}};$$

and $\boldsymbol{\beta}_3$ can be linearly expressed by $\boldsymbol{\alpha}_1, \boldsymbol{\alpha}_2$, and $\boldsymbol{\alpha}_3$, determine a and b.

3.49 Let $\boldsymbol{A} = \begin{pmatrix} 1 & a & \cdots & a \\ a & 1 & \cdots & a \\ \vdots & \vdots & \ddots & \vdots \\ a & a & \cdots & 1 \end{pmatrix}$ be an nth-order matrix $(n \geq 3)$, $a \in \mathbb{R}$, and $\mathrm{r}(\boldsymbol{A}) = n - 1$, determine a.

3.50 Suppose the sums of elements in every row in $\boldsymbol{A}$ are all 0, and $\mathrm{r}(\boldsymbol{A}) = n - 1$. Give a general solution of the homogeneous system of linear equations $\boldsymbol{Ax} = \boldsymbol{0}$.

3.51 The following systems (I) and (II) of linear equations are equivalent systems. Determine parameters m, n, and t.

$$\textbf{(I)}: \begin{cases} x_1 + x_2 - 2x_4 = -6, \\ 4x_1 - x_2 - x_3 - x_4 = 1, \\ 3x_1 - x_2 - x_3 = 3. \end{cases}$$

$$\textbf{(II)}: \begin{cases} x_1 + mx_2 - x_3 - x_4 = -5, \\ nx_2 - x_3 - 2x_4 = -11, \\ x_3 - 2x_4 = -t + 1. \end{cases}$$

3.52 Suppose $\boldsymbol{\alpha} = (1, 2, 1)^{\mathrm{T}}, \boldsymbol{\beta} = (1, \frac{1}{2}, 0)^{\mathrm{T}}$, $\boldsymbol{\gamma} = (0, 0, 8)^{\mathrm{T}}, \boldsymbol{A} = \boldsymbol{\alpha}\boldsymbol{\beta}^{\mathrm{T}}, \ \boldsymbol{B} = \boldsymbol{\beta}^{\mathrm{T}}\boldsymbol{\alpha}$. Solvc $2\boldsymbol{B}^2\boldsymbol{A}^2\boldsymbol{x} = \boldsymbol{A}^4\boldsymbol{x} + \boldsymbol{B}^4\boldsymbol{x} + \boldsymbol{\gamma}$.

3.53 Suppose $\boldsymbol{A} = (\boldsymbol{\alpha}_1, \boldsymbol{\alpha}_2, \ldots, \boldsymbol{\alpha}_n)$ is an nth-order matrix, of which the determinant $|\boldsymbol{A}| \neq 0$. The $n \times (n-1)$ matrix formed by the first $(n-1)$ columns in $\boldsymbol{A}$ is denoted as $\boldsymbol{A}_1 = (\boldsymbol{\alpha}_1, \boldsymbol{\alpha}_2, \ldots, \boldsymbol{\alpha}_{n-1})$. Does the system of linear equations $\boldsymbol{A}_1\boldsymbol{x} = \boldsymbol{\alpha}_n$ have a solution? Justify your answer.

3.54 Suppose $\boldsymbol{\alpha}$ and $\boldsymbol{\beta}$ are both n-dimensional nonzero column vectors, $\boldsymbol{A} = \boldsymbol{\alpha}\boldsymbol{\beta}^{\mathrm{T}}$. Prove that any two rows (columns) in $\boldsymbol{A}$ are proportional.

3.55 Let the nth-order matrix $\boldsymbol{A}$ be partitioned to

$$\boldsymbol{A} = \begin{pmatrix} \boldsymbol{A}_{11} & \boldsymbol{A}_{12} \\ \boldsymbol{A}_{21} & \boldsymbol{A}_{22} \end{pmatrix},$$

where $\boldsymbol{A}_{11}$ is a kth-order invertible matrix $(k < n)$. Prove that there exists an upper triangular matrix $\boldsymbol{U}$ with main diagonal elements 1 and a lower triangular matrix $\boldsymbol{L}$, such that

$$\boldsymbol{LAU} = \begin{pmatrix} \boldsymbol{A}_{11} & \boldsymbol{0} \\ \boldsymbol{0} & \boldsymbol{B} \end{pmatrix},$$

where $\boldsymbol{B}$ is an $(n-k)$th-order matrix.

3.56 Suppose $\boldsymbol{A}$ and $\boldsymbol{B}$ are both nth-order matrices. Prove that,

(i) $\begin{vmatrix} \boldsymbol{I} & \boldsymbol{B} \\ \boldsymbol{A} & \boldsymbol{I} \end{vmatrix} = |\boldsymbol{I} - \boldsymbol{AB}|$;

(ii) $|\boldsymbol{I} - \boldsymbol{AB}| = |\boldsymbol{I} - \boldsymbol{BA}|$;

(iii) $\det(\lambda\boldsymbol{I} - \boldsymbol{AB}) = \det(\lambda\boldsymbol{I} - \boldsymbol{BA})$ (λ is an arbitrary constant).

3.57 Prove that there exist an $m \times n$ matrix $\boldsymbol{B}$, an $r \times n$ matrix $\boldsymbol{C}$, and $\mathrm{r}(\boldsymbol{B}) = \mathrm{r}(\boldsymbol{C}) = r$ such that $\boldsymbol{A} = \boldsymbol{BC}$ if $\boldsymbol{A}$ is an $m \times n$ matrix and $\mathrm{r}(\boldsymbol{A}) = r$. (Hint: Use equivalent normal-forms).

3.58 Let $\boldsymbol{A}$ and $\boldsymbol{B}$ be nth-order matrices, $\mathrm{r}(\boldsymbol{A}) + \mathrm{r}(\boldsymbol{B}) < n$. Prove that there exists an invertible matrix $\boldsymbol{Q}$ such that $\boldsymbol{AQB} = \boldsymbol{0}$. [Hint: Use equivalent normal-forms, $\boldsymbol{A} \cong \mathrm{diag}(1, \dots, 1, 0, \dots, 0)$, $\boldsymbol{B} \cong \mathrm{diag}(0, \dots, 0, 1, \dots, 1)$]

3.59 Prove that $\boldsymbol{\alpha}_1, \boldsymbol{\alpha}_2, \dots, \boldsymbol{\alpha}_r$ (where $\boldsymbol{\alpha}_1 \neq 0$) are linearly dependent, if and only if, there exists an $\boldsymbol{\alpha}_i (1 \le i \le r)$ such that $\boldsymbol{\alpha}_i$ can be linearly expressed by $\boldsymbol{\alpha}_1, \boldsymbol{\alpha}_2, \dots, \boldsymbol{\alpha}_{i-1}$ and the expression is unique.

3.60 Prove that the set of vectors $\boldsymbol{\alpha}_1, \boldsymbol{\alpha}_2, \dots, \boldsymbol{\alpha}_s$ are linearly independent, if and only if,

$$\boldsymbol{\alpha}_i \neq \sum_{j=1}^{i-1} k_j \boldsymbol{\alpha}_j \quad (i = 2, 3, \dots, s).$$

3.61 Let the set of vectors $\boldsymbol{\alpha}_1, \boldsymbol{\alpha}_2, \dots, \boldsymbol{\alpha}_r$ be linearly independent and add a vector $\boldsymbol{\beta}$ at the beginning of the vector set. Prove that there is at most one vector $\boldsymbol{\alpha}_i (1 \le i \le r)$ that can be linearly expressed by the first i vectors $\boldsymbol{\beta}, \boldsymbol{\alpha}_1, \boldsymbol{\alpha}_2, \dots, \boldsymbol{\alpha}_{i-1}$. Give a geometric interpretation in $\mathbb{R}^3$.

3.62 Prove that if a vector $\boldsymbol{\alpha}$ can be linearly expressed by $\boldsymbol{\alpha}_1, \boldsymbol{\alpha}_2, \dots, \boldsymbol{\alpha}_s$ in the n-dimensional vector space $\mathbb{R}^n$, then the expression is unique, if and only if, vectors $\boldsymbol{\alpha}_1, \boldsymbol{\alpha}_2, \dots, \boldsymbol{\alpha}_s$ are linearly independent.

3.63 Let $\boldsymbol{A}$ be an nth-order matrix, $\mathrm{r}(\boldsymbol{A}) = 1$. Prove that

(i) $\boldsymbol{A} = \begin{pmatrix} a_1 \\ a_2 \\ \vdots \\ a_n \end{pmatrix} \begin{pmatrix} b_1, & b_2, & \cdots, & b_n \end{pmatrix}$, where a_i and b_i are numbers

(ii) $\boldsymbol{A}^2 = k\boldsymbol{A}$.

3.64 Suppose

$$\boldsymbol{A} = \begin{pmatrix} a_{11} & a_{12} & \cdots & a_{1n} \\ a_{21} & a_{22} & \cdots & a_{2n} \\ \vdots & \vdots & & \vdots \\ a_{m1} & a_{m2} & \cdots & a_{mn} \end{pmatrix}, \quad \boldsymbol{y} = \begin{pmatrix} y_1 \\ y_2 \\ \vdots \\ y_n \end{pmatrix},$$

$$\boldsymbol{b} = (b_1, b_2, \dots, b_m)^{\mathrm{T}}, \quad x = (x_1, x_2, \dots, x_m)^{\mathrm{T}}.$$

(i) Prove that any solution $\boldsymbol{x}_1, \boldsymbol{x}_2, \dots, \boldsymbol{x}_m$ of $\boldsymbol{A}^{\mathrm{T}}\boldsymbol{x} = \boldsymbol{0}$ certainly satisfies the equation $b_1\boldsymbol{x}_1 + b_2\boldsymbol{x}_2 + \cdots + b_m\boldsymbol{x}_m = \boldsymbol{0}$ if $\boldsymbol{Ay} = \boldsymbol{b}$ has a solution.

(ii) The system of linear equations $\boldsymbol{Ay} = \boldsymbol{b}$ has a solution, if and only if, the system of linear equations

$$\begin{pmatrix} \boldsymbol{A}^{\mathrm{T}} \\ \boldsymbol{b}^{\mathrm{T}} \end{pmatrix} \boldsymbol{x} = \begin{pmatrix} \boldsymbol{0} \\ \boldsymbol{1} \end{pmatrix}$$

has no solution (where $\boldsymbol{0}$ is the $n \times 1$ zero matrix).

3.65 Suppose $\boldsymbol{A}$ is an $m \times n$ matrix, $m < n$, $\mathrm{r}(\boldsymbol{A}) = m$, a solution basis for the homogeneous system of linear equations $\boldsymbol{Ax} = \boldsymbol{0}$ is

$$\boldsymbol{b}_i = (\boldsymbol{b}_{i1}, \boldsymbol{b}_{i2}, \dots, \boldsymbol{b}_{in})^{\mathrm{T}}, i = 1, 2, \dots, n - m.$$

Determine the number of solutions in a solution basis for the homogeneous system of linear equations

$$\sum_{j=1}^{n} b_{ij} y_j = 0, \quad i = 1, 2, \dots, n - m.$$

and give a basis.

3.66 Suppose m row vectors of an $m \times n$ matrix $\boldsymbol{A}$ is a solution basis for the homogeneous system of linear equations $\boldsymbol{Cx} = \boldsymbol{0}$, and $\boldsymbol{B}$ be an mth-order invertible matrix. Prove that the row vectors of $\boldsymbol{BA}$ is also a solution basis for $\boldsymbol{Cx} = \boldsymbol{0}$.

3.67 Prove that if $\boldsymbol{A}$ is an nth-order matrix ($n > 1$), and $|\boldsymbol{A}| = 0$, then the algebraic complements (i.e., subdeterminants with proper sign) of corresponding elements of any two rows (columns) in $|\boldsymbol{A}|$ are proportional.

3.68 Let $\boldsymbol{A}$ be an $(n-1) \times n$ matrix, $|\boldsymbol{A}_j|$ denote the determinant obtained by deleting the jth column from $\boldsymbol{A}$. Prove:

(i) $(-|\boldsymbol{A}_1|, |\boldsymbol{A}_2|, \dots, (-1)^n |\boldsymbol{A}_n|)^{\mathrm{T}}$ is a solution of $\boldsymbol{Ax} = \boldsymbol{0}$,

(ii) If $|\boldsymbol{A}_j|$ $(j = 1, 2, \dots, n)$ are not all 0, then the solution in (i) is a solution basis for $\boldsymbol{Ax} = \boldsymbol{0}$.

3.69 If $\boldsymbol{A}$ is an nth-order matrix, and $\boldsymbol{A}^2 = \boldsymbol{A}$, prove that $\mathrm{r}(\boldsymbol{A}) + \mathrm{r}(\boldsymbol{A} - \boldsymbol{I}) = n$.

3.70 If $\boldsymbol{A}$ is an nth-order matrix, and $\boldsymbol{A}^2 = \boldsymbol{A}$, prove that $\mathrm{r}(\boldsymbol{A} + \boldsymbol{I}) + \mathrm{r}(\boldsymbol{A} - \boldsymbol{I}) = n$.

3.71 Suppose $\boldsymbol{A}$ and $\boldsymbol{B}$ are both nth-order square matrices. Prove that $\mathrm{r}(\boldsymbol{AB}) \geq \mathrm{r}(\boldsymbol{A}) + \mathrm{r}(\boldsymbol{B}) - n$. Is the above result true if $\boldsymbol{A} = (a_{ij})_{s\times n}$, $\boldsymbol{B} = (b_{ij})_{n\times m}$?
(Hint: Use Theorem 3.11, Property 3.4 in §3.3 as well as the result of **Q3.18**. Or use the result of **Q3.15**.)

3.72 Consider the vectors $\boldsymbol{\alpha}_j = (a_{1j}, a_{2j}, \ldots, a_{nj})^{\mathrm{T}}$ $(j = 1, 2, \ldots, n)$. Prove that if

$$|a_{ii}| > \sum_{\substack{j=1 \\ j\neq i}}^{n} \left|a_{ij}\right|, \ i = 1, 2, \ldots, n,$$

then $\boldsymbol{\alpha}_1, \boldsymbol{\alpha}_2, \ldots, \boldsymbol{\alpha}_n$ are linearly independent.
(Hint: Use the reduction to absurdity. Let $\boldsymbol{A} = (\boldsymbol{\alpha}_1, \boldsymbol{\alpha}_2, \ldots, \boldsymbol{\alpha}_n)$ and suppose the component i such that $|x_i| \geq \left|x_j\right|, j \neq i$, of the nonzero solution $\boldsymbol{x}$ of $\boldsymbol{Ax} = \boldsymbol{0}$.)

4 Vector Spaces and Linear Transformations

CHAPTER OVERVIEW

Vector spaces and linear transformations are considered as part of the core concepts of linear algebra. We can only introduce some basic concepts due to limited hours assigned for this subject.

First, based on the definition of n-dimensional real vector space $\mathbb{R}^n$ and the concept of linear dependence introduced in the previous chapter, we will discuss the basis of $\mathbb{R}^n$ and the coordinates of a vector with respect to such a basis, as well as the basis transformation and the coordinate transformation. Furthermore, we will introduce the operation of an inner product in $\mathbb{R}^n$, while establishing the notion of n-dimensional Euclidean space whose structure will be discussed as well. Moreover, the generalized notion of linear space (vector space) will be abstracted from $\mathbb{R}^n$. We will then discuss the structure of finite dimensional linear space in greater detail. However, there will only be a brief introduction to the notion of inner product space and generalized Euclidean space in Appendix A.

As for linear transformations, we will mainly discuss the linear transformation of $\mathbb{R}^n$ and its matrix representation.

4.1 Basis of $\mathbb{R}^n$ and Coordinates of Vectors under a Basis

We know that unit vectors $\boldsymbol{\varepsilon}_i = (0, \ldots, 0, 1, 0, \ldots, 0)(i = 1, 2, \ldots, n)$ in $\mathbb{R}^n$ are linearly independent, and n row vectors and n column vectors of an nth-order real nonsingular matrix $\boldsymbol{A} = (a_{ij})_{n\times n}$ (i.e., $|\boldsymbol{A}| \neq 0$) are linearly independent. Furthermore, as we have seen in the last chapter, any $(n+1)$ vectors in $\mathbb{R}^n$ are linearly dependent. So, according to Theorem 3.3, every vector $\boldsymbol{\alpha}$ in $\mathbb{R}^n$ can be uniquely written as a linear combination of n independent vectors. Such a connection between vectors in $\mathbb{R}^n$ is characterized by the notion of "basis" and "coordinate" that will be discussed later in this chapter.

Definition 4.1

Let $B = \{\boldsymbol{\beta}_1, \boldsymbol{\beta}_2, \ldots, \boldsymbol{\beta}_n\} \subset \mathbb{R}^n$ be a set of vectors. If B is linearly independent, then every vector $\boldsymbol{\alpha}$ in $\mathbb{R}^n$ can be written as a combination of vectors in B, that is,

$$\boldsymbol{\alpha} = a_1\boldsymbol{\beta}_1 + a_2\boldsymbol{\beta}_2 + \cdots + a_n\boldsymbol{\beta}_n. \tag{4.1}$$

Such a set B is called a basis of $\mathbb{R}^n$. The list $(a_1, a_2, \ldots, a_n)$ is called the **coordinates** of the vector $\boldsymbol{\alpha}$ relative to basis B (or under basis B), which can be denoted as a row vector $\boldsymbol{\alpha}_B = (a_1, a_2, \ldots, a_n)$ or as a column vector $\boldsymbol{\alpha}_B = (a_1, a_2, \ldots, a_n)^{\mathrm{T}}$, called the **coordinate vector** of $\boldsymbol{\alpha}$ as well.

Obviously, the basis of $\mathbb{R}^n$ is not unique, but the **coordinates** are uniquely determined for any vector $\boldsymbol{\alpha}$ with respect to a given basis B. From now on, we shall call the basis composed of n unit vectors as the **natural basis** or the **standard basis**.

The standard basis of 3-dimensional geometric space $\mathbb{R}^3$ is denoted by $\{\boldsymbol{i}, \boldsymbol{j}, \boldsymbol{k}\}$ and any vector $\boldsymbol{\alpha}$ in $\mathbb{R}^3$ can be written uniquely in the form

$$\boldsymbol{\alpha} = a_1\boldsymbol{i} + a_2\boldsymbol{j} + a_3\boldsymbol{k}.$$

The ordered triple (a_1, a_2, a_3) is called the coordinates of $\boldsymbol{\alpha}$ relative to the basis $\{\boldsymbol{i}, \boldsymbol{j}, \boldsymbol{k}\}$. If the initial point of $\boldsymbol{\alpha}$ is the origin, then (a_1, a_2, a_3) will be the Cartesian coordinate of the terminal point P (hereafter, we shall usually give geometrical interpretation in $\mathbb{R}^3$ to some problems with their general results in

$\mathbb{R}^n$ by using the one-to-one correspondence between vector $\boldsymbol{\alpha}$ in $\mathbb{R}^3$ and space point P). In this book, we generally use the column vector $(a_1, a_2, \ldots, a_n)^{\mathrm{T}}$ to depict a vector and its coordinates, expression (4.1) can be expressed in the form

$$\boldsymbol{\alpha} = (\boldsymbol{\beta}_1, \boldsymbol{\beta}_2, \ldots, \boldsymbol{\beta}_n)\begin{pmatrix} a_1 \\ a_2 \\ \vdots \\ a_n \end{pmatrix}. \tag{4.2}$$

Example 4.1 Computing the coordinates of a basis

Let two bases of $\mathbb{R}^n$ be the natural basis B_1 and $B_2 = \{\boldsymbol{\beta}_1, \boldsymbol{\beta}_2, \ldots, \boldsymbol{\beta}_n\}$, where

$$\begin{aligned}
\boldsymbol{\beta}_1 &= (1, -1, 0, \ldots, 0)^{\mathrm{T}}, \\
\boldsymbol{\beta}_2 &= (0, 1, -1, \ldots, 0)^{\mathrm{T}}, \\
&\cdots \\
\boldsymbol{\beta}_{n-1} &= (0, 0, \ldots, 1, -1)^{\mathrm{T}}, \\
\boldsymbol{\beta}_n &= (0, 0, \ldots, 0, 1)^{\mathrm{T}}.
\end{aligned}$$

Find the coordinates of $(a_1, a_2, \ldots, a_n)^{\mathrm{T}}$ under the two bases respectively.

Solution: Under the natural basis $B_1 = \{\boldsymbol{\varepsilon}_1, \boldsymbol{\varepsilon}_2, \ldots, \boldsymbol{\varepsilon}_n\}$, we have

$$\boldsymbol{\alpha} = a_1\boldsymbol{\varepsilon}_1 + a_2\boldsymbol{\varepsilon}_2 + \cdots + a_n\boldsymbol{\varepsilon}_n,$$

therefore, $\boldsymbol{\alpha}_{B_1} = (a_1, a_2, \ldots, a_n)^{\mathrm{T}}$.

Express $\boldsymbol{\alpha}$ as a linear combination of vectors in B_2,

$$\boldsymbol{\alpha} = x_1\boldsymbol{\beta}_1 + x_2\boldsymbol{\beta}_2 + \cdots + x_n\boldsymbol{\beta}_n = (\boldsymbol{\beta}_1, \boldsymbol{\beta}_2, \ldots, \boldsymbol{\beta}_n)\begin{pmatrix} x_1 \\ x_2 \\ \vdots \\ x_n \end{pmatrix}. \quad ①$$

Substituting columns $\boldsymbol{\alpha}, \boldsymbol{\beta}_1, \boldsymbol{\beta}_2, \cdots, \boldsymbol{\beta}_{n-1}$ into expression ① yields

$$\begin{pmatrix} 1 & 0 & \cdots & 0 & 0 \\ -1 & 1 & \cdots & 0 & 0 \\ 0 & -1 & \cdots & 0 & 0 \\ \vdots & \vdots & & \vdots & \vdots \\ 0 & 0 & \cdots & 1 & 0 \\ 0 & 0 & \cdots & -1 & 1 \end{pmatrix}\begin{pmatrix} x_1 \\ x_2 \\ \vdots \\ x_{n-1} \\ x_n \end{pmatrix} = \begin{pmatrix} a_1 \\ a_2 \\ \vdots \\ a_{n-1} \\ a_n \end{pmatrix}. \quad ②$$

Solve the nonhomogeneous system of linear equations ②, and this yields

$$\boldsymbol{\alpha}_{B_2} = \begin{pmatrix} x_1 \\ x_2 \\ \vdots \\ x_{n-1} \\ x_n \end{pmatrix} = \begin{pmatrix} a_1 \\ a_1 + a_2 \\ \vdots \\ a_1 + a_2 + \cdots + a_{n-1} \\ a_1 + a_2 + \cdots + a_{n-1} + a_n \end{pmatrix}.$$

■

It can be seen from Example 4.1 that, for the same vector, there will usually be different coordinates in terms of different bases. So the problems about basis transformation and coordinate transformation need to be discussed as a whole rather than separately. In order to obtain the relation between coordinates of the same vector in $\mathbb{R}^n$ under two bases, we first prove the following theorem.

THEOREM 4.1

Let $B = \{\boldsymbol{\alpha}_1, \boldsymbol{\alpha}_2, \ldots, \boldsymbol{\alpha}_n\}$ *be a basis of* $\mathbb{R}^n$, *and*

$$\begin{cases} \boldsymbol{\eta}_1 = a_{11}\boldsymbol{\alpha}_1 + a_{21}\boldsymbol{\alpha}_2 + \cdots + a_{n1}\boldsymbol{\alpha}_n, \\ \boldsymbol{\eta}_2 = a_{12}\boldsymbol{\alpha}_1 + a_{22}\boldsymbol{\alpha}_2 + \cdots + a_{n2}\boldsymbol{\alpha}_n, \\ \quad\cdots \\ \boldsymbol{\eta}_n = a_{1n}\boldsymbol{\alpha}_1 + a_{2n}\boldsymbol{\alpha}_2 + \cdots + a_{nn}\boldsymbol{\alpha}_n. \end{cases} \tag{4.3}$$

$\boldsymbol{\eta}_1, \boldsymbol{\eta}_2, \ldots, \boldsymbol{\eta}_n$ *are linearly independent, if and only if,*

$$\det \boldsymbol{A} = \begin{vmatrix} a_{11} & a_{12} & \cdots & a_{1n} \\ a_{21} & a_{22} & \cdots & a_{2n} \\ \vdots & \vdots & & \vdots \\ a_{n1} & a_{n2} & \cdots & a_{nn} \end{vmatrix} \neq 0. \tag{4.4}$$

Proof: From (4.3) we get $\boldsymbol{\eta}_j = \sum_{i=1}^{n} a_{ij}\boldsymbol{\alpha}_i (j = 1, 2, \ldots, n)$. A necessary and sufficient condition of $\boldsymbol{\eta}_1, \boldsymbol{\eta}_2, \ldots, \boldsymbol{\eta}_n$ being linearly independent is that the equation

$$\sum_{j=1}^{n} x_j \boldsymbol{\eta}_j = \sum_{j=1}^{n} x_j \sum_{i=1}^{n} a_{ij}\boldsymbol{\alpha}_i = \sum_{i=1}^{n}\left(\sum_{j=1}^{n} a_{ij}x_j\right)\boldsymbol{\alpha}_i = \mathbf{0} \qquad ①$$

has only the zero solution $x_j = 0$ $(j = 1, 2, \ldots, n)$.

Because $\boldsymbol{\alpha}_1, \boldsymbol{\alpha}_2, \ldots, \boldsymbol{\alpha}_n$ are linearly independent, from ① we obtain

$$\sum_{j=1}^{n} a_{ij}x_j = 0 \quad (i = 1, 2, \ldots, n). \qquad ②$$

As a result, a necessary and sufficient condition of equation ① having only the zero solution (namely homogeneous system of linear equations ② has only the zero solution) is that the coefficient determinant of ② is not 0. Thus, (4.4) follows. □

Let $B_1 = \{\boldsymbol{\alpha}_1, \boldsymbol{\alpha}_2, \ldots, \boldsymbol{\alpha}_n\}$ and $B_2 = \{\boldsymbol{\eta}_1, \boldsymbol{\eta}_2, \ldots, \boldsymbol{\eta}_n\}$ be two bases (namely, old basis and new basis) of $\mathbb{R}^n$, their relations are shown as (4.3). A matrix form of (4.3) is

$$(\boldsymbol{\eta}_1, \boldsymbol{\eta}_2, \ldots, \boldsymbol{\eta}_n) = (\boldsymbol{\alpha}_1, \boldsymbol{\alpha}_2, \ldots, \boldsymbol{\alpha}_n) \begin{pmatrix} a_{11} & a_{12} & \cdots & a_{1n} \\ a_{21} & a_{22} & \cdots & a_{2n} \\ \vdots & \vdots & & \vdots \\ a_{n1} & a_{n2} & \cdots & a_{nn} \end{pmatrix}. \tag{4.5}$$

Let $\boldsymbol{A}$ denote the matrix in the *RHS* of (4.5) [Hint: The transpose form of the coefficient matrix of $\boldsymbol{\alpha}_1, \boldsymbol{\alpha}_2, \ldots, \boldsymbol{\alpha}_n$ in (4.3)]. In brief, the expression (4.5) can be expressed as

$$(\boldsymbol{\eta}_1, \boldsymbol{\eta}_2, \ldots, \boldsymbol{\eta}_n) = (\boldsymbol{\alpha}_1, \boldsymbol{\alpha}_2, \ldots, \boldsymbol{\alpha}_n)\boldsymbol{A}. \tag{4.6}$$

Definition 4.2

Let the two bases

$$B_1 = \{\boldsymbol{\alpha}_1, \boldsymbol{\alpha}_2, \ldots, \boldsymbol{\alpha}_n\} \text{ and } B_2 = \{\boldsymbol{\beta}_1, \boldsymbol{\beta}_2, \ldots, \boldsymbol{\beta}_n\}$$

satisfy a condition similar to (4.5), then the matrix in (4.5) is called the **transition matrix** from the old basis B_1 to the new basis B_2 (or $\boldsymbol{A}$ is the transformation matrix transforming from B_1 to B_2).

According to Theorem 4.1, the transition matrix $\boldsymbol{A}$ is invertible, the jth column in $\boldsymbol{A}$ presents the coordinates of basis vector $\boldsymbol{\beta}_j$ under the old basis $\{\boldsymbol{\alpha}_1, \boldsymbol{\alpha}_2, \ldots, \boldsymbol{\alpha}_n\}$.

THEOREM 4.2

Let the coordinates of vector $\boldsymbol{\alpha}$ under two bases $B_1 = \{\boldsymbol{\alpha}_1, \boldsymbol{\alpha}_2, \ldots, \boldsymbol{\alpha}_n\}$ and $B_2 = \{\boldsymbol{\beta}_1, \boldsymbol{\beta}_2, \ldots, \boldsymbol{\beta}_n\}$ be, respectively,

$$\boldsymbol{x} = (x_1, x_2, \ldots, x_n)^{\mathrm{T}}, \ \boldsymbol{y} = (y_1, y_2, \ldots, y_n)^{\mathrm{T}},$$

the transition matrix from B_1 to B_2 be $\boldsymbol{A}$, then

$$\boldsymbol{A}\boldsymbol{y} = \boldsymbol{x} \text{ or } \boldsymbol{y} = \boldsymbol{A}^{-1}\boldsymbol{x}. \tag{4.7}$$

Proof: Using (4.6) and

$$\begin{aligned}\boldsymbol{\alpha} &= x_1\boldsymbol{\alpha}_1 + x_2\boldsymbol{\alpha}_2 + \cdots + x_n\boldsymbol{\alpha}_n \\ &= y_1\boldsymbol{\beta}_1 + y_2\boldsymbol{\beta}_2 + \cdots + y_n\boldsymbol{\beta}_n,\end{aligned}$$

we have

$$\begin{aligned}\boldsymbol{\alpha} = (\boldsymbol{\alpha}_1, \boldsymbol{\alpha}_2, \ldots, \boldsymbol{\alpha}_n)\begin{pmatrix} x_1 \\ x_2 \\ \vdots \\ x_n \end{pmatrix} &= (\boldsymbol{\beta}_1,\ \boldsymbol{\beta}_2, \ldots, \boldsymbol{\beta}_n)\begin{pmatrix} y_1 \\ y_2 \\ \vdots \\ y_n \end{pmatrix} \\ = (\boldsymbol{\alpha}_1, \boldsymbol{\alpha}_2, \ldots, \boldsymbol{\alpha}_n)\boldsymbol{A}\begin{pmatrix} y_1 \\ y_2 \\ \vdots \\ y_n \end{pmatrix} &= (\boldsymbol{\alpha}_1, \boldsymbol{\alpha}_2, \ldots, \boldsymbol{\alpha}_n)\left(\boldsymbol{A}\begin{pmatrix} y_1 \\ y_2 \\ \vdots \\ y_n \end{pmatrix}\right).\end{aligned}$$

Because the coordinate of $\boldsymbol{\alpha}$ under basis $\boldsymbol{\alpha}_1, \boldsymbol{\alpha}_2, \ldots, \boldsymbol{\alpha}_n$ is unique, so

$$\boldsymbol{A}\boldsymbol{y} = \boldsymbol{x} \text{ or } \boldsymbol{y} = \boldsymbol{A}^{-1}\boldsymbol{x}.$$ □

Example 4.2 Computing a transition matrix from a natural basis

For a basis B_2 of $\mathbb{R}^3$ given by $\boldsymbol{\beta}_1 = (1, 2, 1)^{\mathrm{T}}$, $\boldsymbol{\beta}_2 = (1, -1, 0)^{\mathrm{T}}$, $\boldsymbol{\beta}_3 = (1, 0, -1)^{\mathrm{T}}$, find the transition matrix $\boldsymbol{A}$ from the natural basis $B_1 = \{\boldsymbol{\varepsilon}_1, \boldsymbol{\varepsilon}_2, \boldsymbol{\varepsilon}_3\}$ to B_2.

Solution: From $\begin{cases} \boldsymbol{\beta}_1 = \boldsymbol{\varepsilon}_1 + 2\boldsymbol{\varepsilon}_2 + \boldsymbol{\varepsilon}_3, \\ \boldsymbol{\beta}_2 = \boldsymbol{\varepsilon}_1 - \boldsymbol{\varepsilon}_2, \\ \boldsymbol{\beta}_3 = \boldsymbol{\varepsilon}_1 - \boldsymbol{\varepsilon}_3, \end{cases}$

that is, $(\boldsymbol{\beta}_1, \boldsymbol{\beta}_2, \boldsymbol{\beta}_3) = (\boldsymbol{\varepsilon}_1, \boldsymbol{\varepsilon}_2, \boldsymbol{\varepsilon}_3)\begin{pmatrix} 1 & 1 & 1 \\ 2 & -1 & 0 \\ 1 & 0 & -1 \end{pmatrix}$.

Then, $\boldsymbol{A} = \begin{pmatrix} 1 & 1 & 1 \\ 2 & -1 & 0 \\ 1 & 0 & -1 \end{pmatrix}$. ■

It can be seen from Example 4.2 that the transition matrix $\boldsymbol{A}$ from the natural basis $B_1 = \{\boldsymbol{\varepsilon}_1, \boldsymbol{\varepsilon}_2, \ldots, \boldsymbol{\varepsilon}_n\}$ to a basis $B_2 = \{\boldsymbol{\beta}_1, \boldsymbol{\beta}_2, \ldots, \boldsymbol{\beta}_n\}$ is just the matrix formed by arranging $\boldsymbol{\beta}_1, \boldsymbol{\beta}_2, \ldots, \boldsymbol{\beta}_n$ in a column.

Example 4.3 Transition matrix and coordinate vector

Two bases of $\mathbb{R}^3$ are $B_1 = \{\boldsymbol{\alpha}_1, \boldsymbol{\alpha}_2, \boldsymbol{\alpha}_3\}$ and $B_2 = \{\boldsymbol{\beta}_1, \boldsymbol{\beta}_2, \boldsymbol{\beta}_3\}$, where

$$\boldsymbol{\alpha}_1 = (1, 1, 1)^{\mathrm{T}},\ \boldsymbol{\alpha}_2 = (0, 1, 1)^{\mathrm{T}},\quad \boldsymbol{\alpha}_3 = (0, 0, 1)^{\mathrm{T}},$$

$$\boldsymbol{\beta}_1 = (1, 0, 1)^{\mathrm{T}},\ \boldsymbol{\beta}_2 = (0, 1, -1)^{\mathrm{T}},\ \boldsymbol{\beta}_3 = (1, 2, 0)^{\mathrm{T}}.$$

(i) Find the transition matrix $\boldsymbol{A}$ from B_1 to B_2.
(ii) For a given coordinate vector $(1, -2, -1)^{\mathrm{T}}$ of $\boldsymbol{\alpha}$ under basis B_1, find the coordinate vector of $\boldsymbol{\alpha}$ under basis B_2.

Solution:

(i) Let $(\boldsymbol{\beta}_1, \boldsymbol{\beta}_2, \boldsymbol{\beta}_3) = (\boldsymbol{\alpha}_1, \boldsymbol{\alpha}_2, \boldsymbol{\alpha}_3) \begin{pmatrix} a_{11} & a_{12} & a_{13} \\ a_{21} & a_{22} & a_{23} \\ a_{31} & a_{32} & a_{33} \end{pmatrix}$.

Substituting the two basis vectors written in column vector form into the last expression yields

$$\begin{pmatrix} 1 & 0 & 1 \\ 0 & 1 & 2 \\ 1 & -1 & 0 \end{pmatrix} = \begin{pmatrix} 1 & 0 & 0 \\ 1 & 1 & 0 \\ 1 & 1 & 1 \end{pmatrix} \begin{pmatrix} a_{11} & a_{12} & a_{13} \\ a_{21} & a_{22} & a_{23} \\ a_{31} & a_{32} & a_{33} \end{pmatrix}.$$

Then, the transition matrix

$$\boldsymbol{A} = \begin{pmatrix} a_{11} & a_{12} & a_{13} \\ a_{21} & a_{22} & a_{23} \\ a_{31} & a_{32} & a_{33} \end{pmatrix} = \begin{pmatrix} 1 & 0 & 0 \\ 1 & 1 & 0 \\ 1 & 1 & 1 \end{pmatrix}^{-1} \begin{pmatrix} 1 & 0 & 1 \\ 0 & 1 & 2 \\ 1 & -1 & 0 \end{pmatrix}$$

$$= \begin{pmatrix} 1 & 0 & 0 \\ -1 & 1 & 0 \\ 0 & -1 & 1 \end{pmatrix} \begin{pmatrix} 1 & 0 & 1 \\ 0 & 1 & 2 \\ 1 & -1 & 0 \end{pmatrix} = \begin{pmatrix} 1 & 0 & 1 \\ -1 & 1 & 1 \\ 1 & -2 & -2 \end{pmatrix}.$$

(ii) From (4.7) in Theorem 4.2, we get the coordinates of $\boldsymbol{\alpha}$ under basis B_2,

$$\begin{pmatrix} y_1 \\ y_2 \\ y_3 \end{pmatrix} = \boldsymbol{A}^{-1} \begin{pmatrix} 1 \\ -2 \\ -1 \end{pmatrix} = \begin{pmatrix} 0 & -2 & -1 \\ -1 & -3 & -2 \\ 1 & 2 & 1 \end{pmatrix} \begin{pmatrix} 1 \\ -2 \\ -1 \end{pmatrix} = \begin{pmatrix} 5 \\ 7 \\ -4 \end{pmatrix}.$$

Another method is to compute $\boldsymbol{\alpha}$ first, that is, $\boldsymbol{\alpha} = \boldsymbol{\alpha}_1 - 2\boldsymbol{\alpha}_2 - \boldsymbol{\alpha}_3 = (1, -1, -2)^{\mathrm{T}}$, then calculate and get the coordinates (y_1, y_2, y_3) by $\boldsymbol{\alpha} = y_1\boldsymbol{\beta}_1 + y_2\boldsymbol{\beta}_2 + y_3\boldsymbol{\beta}_3$. ■

Using the results of Theorem 4.2, we can now easily derive the coordinate transformation formula for the rotation of axes in the Cartesian coordinate system.

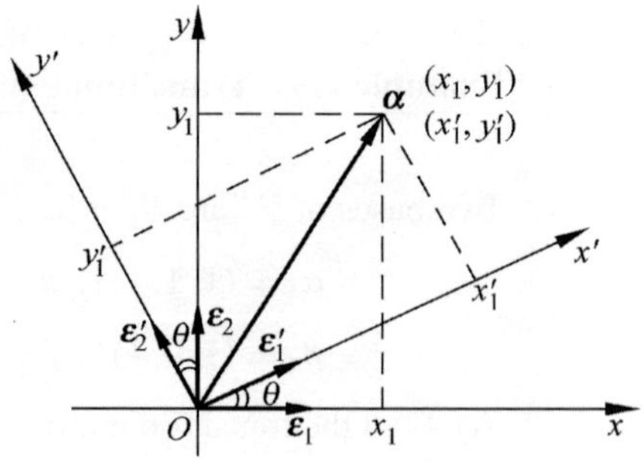

Figure 4.1 Rotation of Cartesian coordinate system

Let the Cartesian coordinate system rotate anticlockwise by an angle θ (as shown in Figure 4.1). Let the basis be $\boldsymbol{\varepsilon}_1 = \boldsymbol{i}, \boldsymbol{\varepsilon}_2 = \boldsymbol{j}$ in the coordinate system Oxy and $\boldsymbol{\varepsilon}_1' = \boldsymbol{i}', \boldsymbol{\varepsilon}_2' = \boldsymbol{j}'$ in the coordinate system O$x'y'$. Then,

$$\begin{cases} \boldsymbol{\varepsilon}_1' = (\cos\theta)\boldsymbol{\varepsilon}_1 + (\sin\theta)\boldsymbol{\varepsilon}_2, \\ \boldsymbol{\varepsilon}_2' = (-\sin\theta)\boldsymbol{\varepsilon}_1 + (\cos\theta)\boldsymbol{\varepsilon}_2. \end{cases}$$

That is,

$$(\boldsymbol{\varepsilon}_1', \boldsymbol{\varepsilon}_2') = (\boldsymbol{\varepsilon}_1, \boldsymbol{\varepsilon}_2)\begin{pmatrix} \cos\theta & -\sin\theta \\ \sin\theta & \cos\theta \end{pmatrix}.$$

So, the coordinates (x_1, y_1) and (x_1', y_1') of vector $\boldsymbol{\alpha}$ under the bases $(\boldsymbol{\varepsilon}_1, \boldsymbol{\varepsilon}_2)$ and $(\boldsymbol{\varepsilon}_1', \boldsymbol{\varepsilon}_2')$ satisfy

$$\begin{pmatrix} x_1' \\ y_1' \end{pmatrix} = \begin{pmatrix} \cos\theta & -\sin\theta \\ \sin\theta & \cos\theta \end{pmatrix}^{-1}\begin{pmatrix} x_1 \\ y_1 \end{pmatrix} = \begin{pmatrix} \cos\theta & \sin\theta \\ -\sin\theta & \cos\theta \end{pmatrix}\begin{pmatrix} x_1 \\ y_1 \end{pmatrix}.$$

4.2 Inner Product in $\mathbb{R}^n$, Orthonormal Basis, and Orthogonal Matrix

In the n-dimensional real vector space discussed above, we have only defined the linear operations for vectors, which cannot characterize metric properties such as length and angle. In 3-dimensional geometrical space, the inner product (namely dot product or scalar product) of vectors provides a relationship between inner product on one hand and the length of a vector as well as angles between vectors, on the other. Starting from the inner product in the 3-dimensional geometrical space, we will provide a general definition of inner product in $\mathbb{R}^n$, while introducing the concepts of orthonormal basis and orthonormal matrix.

4.2.1 Inner product of n-dimensional real vectors, Euclidean space

According to the definition of inner product in the 3-dimensional geometrical space,

$$\boldsymbol{a}\cdot\boldsymbol{b}=\|\boldsymbol{a}\|\ \|\boldsymbol{b}\|\cos\langle\boldsymbol{a},\boldsymbol{b}\rangle,$$

we obtain

$$\cos\langle\boldsymbol{a},\boldsymbol{b}\rangle=\frac{\boldsymbol{a}\cdot\boldsymbol{b}}{\|\boldsymbol{a}\|\ \|\boldsymbol{b}\|},\qquad \|\boldsymbol{a}\|=\sqrt{\boldsymbol{a}\cdot\boldsymbol{a}}.$$

If $\boldsymbol{a}=a_1\boldsymbol{i}+a_2\boldsymbol{j}+a_3\boldsymbol{k}$ is abbreviated as $\boldsymbol{a}=(a_1,a_2,a_3)$ and $\boldsymbol{b}=b_1\boldsymbol{i}+b_2\boldsymbol{j}+b_3\boldsymbol{k}$ is abbreviated as $\boldsymbol{b}=(b_1,b_2,b_3)$, according to the definition and operation properties of the inner product, we get

$$\boldsymbol{a}\cdot\boldsymbol{b}=a_1b_1+a_2b_2+a_3b_3.$$

Now, we generalize the inner product of 3-dimensional geometrical vectors to n-dimensional real vectors. First, we define an inner product operation in n-dimensional real vector space, then introduce the length of a vector and angle between vectors, so that n-dimensional real vectors may become measurable.

Definition 4.3

Let $\boldsymbol{\alpha}=(a_1,a_2,\ldots,a_n)^{\mathrm{T}}$ and $\boldsymbol{\beta}=(b_1,b_2,\ldots,b_n)^{\mathrm{T}}\in\mathbb{R}^n$, an inner product of $\boldsymbol{\alpha}$ and $\boldsymbol{\beta}$ is defined by

$$(\boldsymbol{\alpha},\boldsymbol{\beta})=a_1b_1+a_2b_2+\cdots+a_nb_n,$$

that is, $(\boldsymbol{\alpha},\boldsymbol{\beta})=\boldsymbol{\alpha}^{\mathrm{T}}\boldsymbol{\beta}=\boldsymbol{\beta}^{\mathrm{T}}\boldsymbol{\alpha}$.

Based on Definition 4.3, we can easily prove that the inner product has the following properties:

(i) $(\boldsymbol{\alpha},\boldsymbol{\beta})=(\boldsymbol{\beta},\boldsymbol{\alpha})$,
(ii) $(\boldsymbol{\alpha}+\boldsymbol{\beta},\boldsymbol{\gamma})=(\boldsymbol{\alpha},\boldsymbol{\gamma})+(\boldsymbol{\beta},\boldsymbol{\gamma})$, (4.8)
(iii) $(k\boldsymbol{\alpha},\boldsymbol{\beta})=k(\boldsymbol{\alpha},\boldsymbol{\beta})$,
(iv) $(\boldsymbol{\alpha},\boldsymbol{\alpha})\geq 0$, the equality holds, if and only if, $\boldsymbol{\alpha}=\boldsymbol{0}$, where $\boldsymbol{\alpha},\boldsymbol{\beta},\boldsymbol{\gamma}\in\mathbb{R}^n$, $k\in\mathbb{R}$.

As the inner product of $\boldsymbol{\alpha}$ with itself is a nonnegative number, so the length of an n-dimensional vector $\boldsymbol{\alpha}$ is defined by the inner product as was shown in the case of the 3-dimensional geometrical space.

Definition 4.4

The **length** of a vector $\boldsymbol{\alpha}$ is

$$\|\boldsymbol{\alpha}\| = \sqrt{(\boldsymbol{\alpha}, \boldsymbol{\alpha})}. \tag{4.9}$$

THEOREM 4.3

The inner product of vectors satisfies

$$|(\boldsymbol{\alpha}, \boldsymbol{\beta})| \leq \|\boldsymbol{\alpha}\| \, \|\boldsymbol{\beta}\| . \tag{4.10}$$

We call (4.10) *the Cauchy-Schwarz inequality.*

Proof: If $\boldsymbol{\beta} = \mathbf{0}$, then $(\boldsymbol{\alpha}, \boldsymbol{\beta}) = \mathbf{0}$, $\|\boldsymbol{\beta}\| = 0$. Then, (4.10) is obviously true.
If $\boldsymbol{\beta} \neq \mathbf{0}$,

$$(\boldsymbol{\alpha} + t\boldsymbol{\beta}, \boldsymbol{\alpha} + t\boldsymbol{\beta}) \geq 0.$$

For any vector $\boldsymbol{\alpha} + t\boldsymbol{\beta}$ $(t \in \mathbb{R})$ according to property (iv), which is expanded to, by properties (i), (ii), and (iii):

$$(\boldsymbol{\alpha}, \boldsymbol{\alpha}) + 2\,(\boldsymbol{\alpha}, \boldsymbol{\beta})\, t + (\boldsymbol{\beta}, \boldsymbol{\beta})\, t^2 \geq 0.$$

The *LHS* is a quadratic trinomial with the coefficient of t^2, that is, $(\boldsymbol{\beta}, \boldsymbol{\beta}) > 0$. This inequality also lets the discriminant

$$4(\boldsymbol{\alpha}, \boldsymbol{\beta})^2 - 4\,(\boldsymbol{\alpha}, \boldsymbol{\alpha})\,(\boldsymbol{\beta}, \boldsymbol{\beta}) \leq 0,$$

that is,

$$(\boldsymbol{\alpha}, \boldsymbol{\beta})^2 \leq (\boldsymbol{\alpha}, \boldsymbol{\alpha})\,(\boldsymbol{\beta}, \boldsymbol{\beta}) = \|\boldsymbol{\alpha}\|^2 \|\boldsymbol{\beta}\|^2.$$

So,

$$|\,(\boldsymbol{\alpha}, \boldsymbol{\beta})\,| \leq \|\boldsymbol{\alpha}\| \|\boldsymbol{\beta}\|. \qquad \square$$

It is easy to prove that (4.10) of Theorem 4.3 is an equality, if and only if, $\boldsymbol{\alpha}$ and $\boldsymbol{\beta}$ are linearly dependent.

As $\boldsymbol{\alpha} = (a_1, a_2, \ldots, a_n)^{\mathrm{T}}$, $\boldsymbol{\beta} = (b_1, b_2, \ldots, b_n)^{\mathrm{T}}$, by applying Theorem 4.3, we get

$$\left(\sum_{i=1}^{n} a_i b_i\right)^2 \leq \left(\sum_{i=1}^{n} a_i^2\right)\left(\sum_{i=1}^{n} b_i^2\right). \tag{4.11}$$

Due to the fact that the inner product satisfies Cauchy-Schwarz inequality, we define the angle between vectors by the inner product.

Definition 4.5

Define the angle between vectors $\boldsymbol{\alpha}, \boldsymbol{\beta}$ to be

$$\langle \boldsymbol{\alpha}, \boldsymbol{\beta} \rangle = \arccos \frac{(\boldsymbol{\alpha}, \boldsymbol{\beta})}{\|\boldsymbol{\alpha}\| \, \|\boldsymbol{\beta}\|}. \tag{4.12}$$

$\boldsymbol{\alpha}$ and $\boldsymbol{\beta}$ are called orthogonal (or perpendicular) and denoted as $\boldsymbol{\alpha} \perp \boldsymbol{\beta}$ if $\langle \boldsymbol{\alpha}, \boldsymbol{\beta} \rangle = \frac{\pi}{2}$.

Applying Definition 4.5, we can immediately get:

THEOREM 4.4

Nonzero vectors $\boldsymbol{\alpha}, \boldsymbol{\beta}$ are orthogonal (or perpendicular), if and only if, $(\boldsymbol{\alpha}, \boldsymbol{\beta}) = 0$.

Because the inner product of the zero vector with any vector is zero, we may say that the zero vector is orthogonal to any vector.

In 3-dimensional geometrical space, vectors $\boldsymbol{\alpha}, \boldsymbol{\beta}, (\boldsymbol{\alpha} + \boldsymbol{\beta})$ form a triangle, where the length of the three vectors automatically satisfies the famous triangle inequality, namely,

$$\|\boldsymbol{\alpha} + \boldsymbol{\beta}\| \le \|\boldsymbol{\alpha}\| + \|\boldsymbol{\beta}\|. \tag{4.13}$$

If $\boldsymbol{\alpha} \perp \boldsymbol{\beta}$, their length satisfies the Pythagorean theorem

$$\|\boldsymbol{\alpha} + \boldsymbol{\beta}\|^2 = \|\boldsymbol{\alpha}\|^2 + \|\boldsymbol{\beta}\|^2. \tag{4.14}$$

In the n-dimensional vector space along with the inner product operation, the triangle inequality and Pythagorean theorem are also true. We now give their proofs:

$$\begin{aligned} \|\boldsymbol{\alpha} + \boldsymbol{\beta}\|^2 &= (\boldsymbol{\alpha} + \boldsymbol{\beta}, \boldsymbol{\alpha} + \boldsymbol{\beta}) = (\boldsymbol{\alpha}, \boldsymbol{\alpha}) + 2(\boldsymbol{\alpha}, \boldsymbol{\beta}) + (\boldsymbol{\beta}, \boldsymbol{\beta}) & \text{①} \\ &\le 2\|\boldsymbol{\alpha}\|\|\boldsymbol{\beta}\| + \|\boldsymbol{\alpha}\|^2 + \|\boldsymbol{\beta}\|^2 & \text{②} \\ &= (\|\boldsymbol{\alpha}\| + \|\boldsymbol{\beta}\|)^2. \end{aligned}$$

So,

$$\|\boldsymbol{\alpha} + \boldsymbol{\beta}\| \le \|\boldsymbol{\alpha}\| + \|\boldsymbol{\beta}\|,$$

where the deduction from ① to ② follows from the Cauchy-Schwarz inequality.

If $\boldsymbol{\alpha} \perp \boldsymbol{\beta}$, then $(\boldsymbol{\alpha}, \boldsymbol{\beta}) = 0$ in ① yields

$$\|\boldsymbol{\alpha} + \boldsymbol{\beta}\|^2 = \|\boldsymbol{\alpha}\|^2 + \|\boldsymbol{\beta}\|^2.$$ □

Definition 4.6

An n-dimensional real vector space along with an inner product operation is called an ***n*-dimensional Euclidean space** (or Euclid space for short), and is still denoted by $\mathbb{R}^n$.

4.2.2 Orthonormal basis

In the n-dimensional Euclidean space, vectors of length 1;

$$\boldsymbol{\varepsilon}_1 = (1, 0, 0, \ldots, 0)^{\mathrm{T}},\ \boldsymbol{\varepsilon}_2 = (0, 1, 0, \ldots, 0)^{\mathrm{T}},\ \boldsymbol{\varepsilon}_n = (0, 0, 0, \ldots, 1)^{\mathrm{T}},$$

are obviously pair-wise orthogonal and linearly independent. We call it an orthonormal basis of $\mathbb{R}^n$. However, an orthonormal basis of the n-dimensional Euclidean space is not unique. In order to illustrate this, we first prove the following theorem which leads to a general definition of an orthonormal basis and then we discuss the method of Gram-Schmidt orthogonalization, which turns linearly independent vectors in $\mathbb{R}^n$ into its orthonormal basis.

THEOREM 4.5

If $\boldsymbol{\alpha}_1, \boldsymbol{\alpha}_2, \ldots, \boldsymbol{\alpha}_s$ are pair-wise orthogonal vectors without zero vector in $\mathbb{R}^n$ (called nonzero orthogonal vectors), then they are linearly independent.

Proof:
Let

$$k_1\boldsymbol{\alpha}_1 + k_2\boldsymbol{\alpha}_2 + \cdots + k_n\boldsymbol{\alpha}_n = \mathbf{0}.$$

Then,

$$\left(\boldsymbol{\alpha}_i, \sum_{j=1}^{s} k_j\boldsymbol{\alpha}_j\right) = k_i(\boldsymbol{\alpha}_i, \boldsymbol{\alpha}_i) = 0, \quad i = 1, 2, \ldots, s.$$

Because $(\boldsymbol{\alpha}_i, \boldsymbol{\alpha}_i) > 0$, so $k_i = 0$, $i = 1, 2, \ldots, s$, and $\boldsymbol{\alpha}_1, \boldsymbol{\alpha}_2, \ldots, \boldsymbol{\alpha}_s$ are linearly independent. □

Definition 4.7

Let $\boldsymbol{\alpha}_1, \boldsymbol{\alpha}_2, \ldots, \boldsymbol{\alpha}_n \in \mathbb{R}^n$. If

$$(\boldsymbol{\alpha}_i, \boldsymbol{\alpha}_j) = \begin{cases} 1, & i = j \\ 0, & i \neq j \end{cases} \quad i, j = 1, 2, \ldots, n, \tag{4.15}$$

then $\{\boldsymbol{\alpha}_1, \boldsymbol{\alpha}_2, \ldots, \boldsymbol{\alpha}_n\}$ is called an orthonormal basis in $\mathbb{R}^n$.

Example 4.4 Coordinates of a vector in $\mathbb{R}^n$

Let $B = \{\boldsymbol{\alpha}_1, \boldsymbol{\alpha}_2, \ldots, \boldsymbol{\alpha}_n\}$ be an orthonormal basis of $\mathbb{R}^n$, find the coordinates of a vector $\boldsymbol{\beta}$ in $\mathbb{R}^n$ under B.

Solution: Let $\boldsymbol{\beta} = x_1\boldsymbol{\alpha}_1 + x_2\boldsymbol{\alpha}_2 + \cdots + x_n\boldsymbol{\alpha}_n$, and take the inner product on both sides of this equation with $\boldsymbol{\alpha}_j (j = 1, 2, \ldots, n)$. Then, we get

$$\begin{aligned}(\boldsymbol{\beta}, \boldsymbol{\alpha}_j) &= (x_1\boldsymbol{\alpha}_1 + x_2\boldsymbol{\alpha}_2 + \cdots + x_n\boldsymbol{\alpha}_n, \boldsymbol{\alpha}_j)\\ &= \sum_{i=1}^{n} x_i(\boldsymbol{\alpha}_i, \boldsymbol{\alpha}_j) = x_j(\boldsymbol{\alpha}_j, \boldsymbol{\alpha}_j) = x_j.\end{aligned}$$

So, the jth coordinate of $\boldsymbol{\beta}$ under the orthonormal basis $\boldsymbol{\alpha}_1, \boldsymbol{\alpha}_2, \ldots, \boldsymbol{\alpha}_n$ is

$$x_j = (\boldsymbol{\beta}, \boldsymbol{\alpha}_j), \quad j = 1, 2, \ldots, n. \qquad \blacksquare$$

Take $\boldsymbol{i}, \boldsymbol{j}, \boldsymbol{k}$ as an orthonormal basis in $\mathbb{R}^3$, then x_1, x_2, x_3 in Example 4.4 are the projections of $\boldsymbol{\beta}$ onto $\boldsymbol{i}, \boldsymbol{j}, \boldsymbol{k}$.

4.2.3 Gram-Schmidt orthogonalization procedure

Gram-Schmidt procedure is an algorithm for transforming linearly independent vectors $\boldsymbol{\alpha}_1, \boldsymbol{\alpha}_2, \ldots, \boldsymbol{\alpha}_n$ in $\mathbb{R}^n$ into orthonormal vectors by performing a series of specific linear operations.

Let us first construct an orthonormal basis from a basis $\{\boldsymbol{\alpha}_1, \boldsymbol{\alpha}_2, \boldsymbol{\alpha}_3\}$ in $\mathbb{R}^3$ to illustrate the Gram-Schmidt procedure.

Let $\boldsymbol{\beta}_1 = \boldsymbol{\alpha}_1$. The projection vector of $\boldsymbol{\alpha}_2$ onto $\boldsymbol{\beta}_1$ is

$$(\boldsymbol{\alpha}_2)_{\boldsymbol{\beta}_1} = \boldsymbol{\gamma}_{12} = \|\boldsymbol{\alpha}_2\| \cos\langle\boldsymbol{\alpha}_2, \boldsymbol{\beta}_1\rangle \frac{\boldsymbol{\beta}_1}{\|\boldsymbol{\beta}_1\|} = \frac{(\boldsymbol{\alpha}_2, \boldsymbol{\beta}_1)}{(\boldsymbol{\beta}_1, \boldsymbol{\beta}_1)}\boldsymbol{\beta}_1 \xlongequal{\text{denoted}} k_{12}\boldsymbol{\beta}_1$$

where $k_{12} = \frac{(\boldsymbol{\alpha}_2, \boldsymbol{\beta}_1)}{(\boldsymbol{\beta}_1, \boldsymbol{\beta}_1)}$. Let $\boldsymbol{\beta}_2 = \boldsymbol{\alpha}_2 - \boldsymbol{\gamma}_{12} = \boldsymbol{\alpha}_2 - k_{12}\boldsymbol{\beta}_1$. Then, $\boldsymbol{\beta}_2 \perp \boldsymbol{\beta}_1$(see Figure 4.2). Because $\boldsymbol{\alpha}_3$, $\boldsymbol{\alpha}_1$, and $\boldsymbol{\alpha}_2$ are not coplanar, so $\boldsymbol{\alpha}_3$, $\boldsymbol{\beta}_2$, and $\boldsymbol{\beta}_1$ are not coplanar too. Express the projection vector of $\boldsymbol{\alpha}_3$ onto the plane determined by $\boldsymbol{\beta}_2$ and $\boldsymbol{\beta}_1$ as $\boldsymbol{\gamma}_3$, that is,

$$\boldsymbol{\gamma}_3 = (\boldsymbol{\alpha}_3)_{\boldsymbol{\beta}_1} + (\boldsymbol{\alpha}_3)_{\boldsymbol{\beta}_2} = \boldsymbol{\gamma}_{13} + \boldsymbol{\gamma}_{23} = k_{13}\boldsymbol{\beta}_1 + k_{23}\boldsymbol{\beta}_2,$$

where

$$k_{13} = \frac{(\boldsymbol{\alpha}_3, \boldsymbol{\beta}_1)}{(\boldsymbol{\beta}_1, \boldsymbol{\beta}_1)}, \qquad k_{23} = \frac{(\boldsymbol{\alpha}_3, \boldsymbol{\beta}_2)}{(\boldsymbol{\beta}_2, \boldsymbol{\beta}_2)}.$$

Let

$$\boldsymbol{\beta}_3 = \boldsymbol{\alpha}_3 - \boldsymbol{\gamma}_3 = \boldsymbol{\alpha}_3 - k_{13}\boldsymbol{\beta}_1 - k_{23}\boldsymbol{\beta}_2.$$

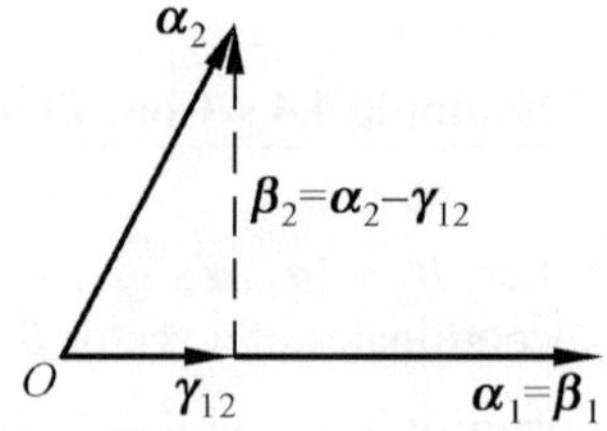

Figure 4.2 $\beta_2 \perp \beta_1$

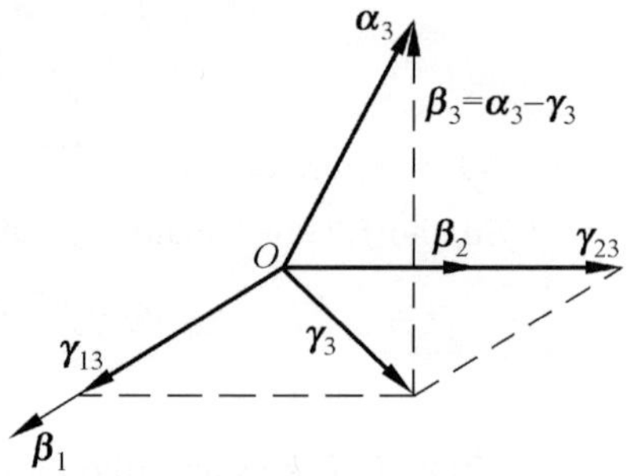

Figure 4.3 $\beta_3 \perp \beta_1$ and $\beta_3 \perp \beta_2$

Then, $\boldsymbol{\beta}_3 \perp \boldsymbol{\beta}_1$ and $\boldsymbol{\beta}_3 \perp \boldsymbol{\beta}_2$ (see Figure 4.3).

$\boldsymbol{\beta}_1, \boldsymbol{\beta}_2$, and $\boldsymbol{\beta}_3$ obtained in this manner are nonzero and pair-wise orthogonal vectors. Finally, $\boldsymbol{\beta}_1, \boldsymbol{\beta}_2$, and $\boldsymbol{\beta}_3$ are normalized to vectors of unit length, that is,

$$\boldsymbol{\eta}_j = \frac{1}{\|\boldsymbol{\beta}_j\|}\boldsymbol{\beta}_j, \quad j = 1, 2, 3.$$

Then, $\{\boldsymbol{\eta}_1, \boldsymbol{\eta}_2, \boldsymbol{\eta}_3\}$ is an orthonormal basis of $\mathbb{R}^3$.

From the intuition derived out of the above orthogonalization process, we may also produce an orthonormal list of vectors $\boldsymbol{\eta}_1, \boldsymbol{\eta}_2, \ldots, \boldsymbol{\eta}_n$ from linearly independent vectors $\boldsymbol{\alpha}_1, \boldsymbol{\alpha}_2, \ldots, \boldsymbol{\alpha}_n$ in $\mathbb{R}^n$ analogously. The process is as follows:

Let

$$\boldsymbol{\beta}_1 = \boldsymbol{\alpha}_1,$$
$$\boldsymbol{\beta}_2 = \boldsymbol{\alpha}_2 + k_{12}\boldsymbol{\beta}_1.$$

Because $\boldsymbol{\beta}_1$ and $\boldsymbol{\alpha}_2$ are linearly independent, so $\boldsymbol{\beta}_2 \neq \mathbf{0}$. In order to make $\boldsymbol{\beta}_1$ and $\boldsymbol{\beta}_2$ orthogonal, that is,

$$(\boldsymbol{\beta}_2, \boldsymbol{\beta}_1) = (\boldsymbol{\alpha}_2 + k_{12}\boldsymbol{\beta}_1, \boldsymbol{\beta}_1) = (\boldsymbol{\alpha}_2, \boldsymbol{\beta}_1) + k_{12}(\boldsymbol{\beta}_1, \boldsymbol{\beta}_1) = 0.$$

We must have

$$k_{12} = -\frac{(\boldsymbol{\alpha}_2, \boldsymbol{\beta}_1)}{(\boldsymbol{\beta}_1, \boldsymbol{\beta}_1)}.$$

Let

$$\boldsymbol{\beta}_3 = \boldsymbol{\alpha}_3 + k_{23}\boldsymbol{\beta}_2 + k_{13}\boldsymbol{\beta}_1,$$

such that $(\boldsymbol{\beta}_3, \boldsymbol{\beta}_1) = (\boldsymbol{\beta}_3, \boldsymbol{\beta}_2) = 0$. We get

$$k_{13} = -\frac{(\boldsymbol{\alpha}_3, \boldsymbol{\beta}_1)}{(\boldsymbol{\beta}_1, \boldsymbol{\beta}_1)}, \quad k_{23} = -\frac{(\boldsymbol{\alpha}_3, \boldsymbol{\beta}_2)}{(\boldsymbol{\beta}_2, \boldsymbol{\beta}_2)}.$$

Continuing with this process, we could presume that nonzero pair-wise orthogonal vectors $\boldsymbol{\beta}_1, \boldsymbol{\beta}_2, \ldots, \boldsymbol{\beta}_{j-1}$ have been found, and take

$$\boldsymbol{\beta}_j = \boldsymbol{\alpha}_j + k_{j-1j}\boldsymbol{\beta}_{j-1} + \cdots + k_{2j}\boldsymbol{\beta}_2 + k_{1j}\boldsymbol{\beta}_1.$$

In order to make $\boldsymbol{\beta}_j$ orthogonal to $\boldsymbol{\beta}_i$ $(i = 1, 2, \ldots, j-1)$, that is,

$$(\boldsymbol{\beta}_j, \boldsymbol{\beta}_i) = (\boldsymbol{\alpha}_j, \boldsymbol{\beta}_i) + k_{ij}(\boldsymbol{\beta}_i, \boldsymbol{\beta}_i) = 0,$$

we must have

$$k_{ij} = -\frac{(\boldsymbol{\alpha}_j, \boldsymbol{\beta}_i)}{(\boldsymbol{\beta}_i, \boldsymbol{\beta}_i)} \quad i = 1, 2, \ldots, j-1.$$

So,

$$\boldsymbol{\beta}_j = \boldsymbol{\alpha}_j - \frac{(\boldsymbol{\alpha}_j, \boldsymbol{\beta}_1)}{(\boldsymbol{\beta}_1, \boldsymbol{\beta}_1)}\boldsymbol{\beta}_1 - \frac{(\boldsymbol{\alpha}_j, \boldsymbol{\beta}_2)}{(\boldsymbol{\beta}_2, \boldsymbol{\beta}_2)}\boldsymbol{\beta}_2 - \cdots - \frac{(\boldsymbol{\alpha}_j, \boldsymbol{\beta}_{j-1})}{(\boldsymbol{\beta}_{j-1}, \boldsymbol{\beta}_{j-1})}\boldsymbol{\beta}_{j-1}. \quad (4.16)$$

Let $\boldsymbol{\beta}_1 = \boldsymbol{\alpha}_1$ and take $j = 2, 3, \ldots, m$ in (4.16). We get pair-wise orthogonal and nonzero vectors $\boldsymbol{\beta}_1, \boldsymbol{\beta}_2, \ldots, \boldsymbol{\beta}_m$ (That all these vectors are nonzero is left for readers to prove), then normalize them into vectors of unit length,

$$\boldsymbol{\eta}_1, \boldsymbol{\eta}_2, \ldots, \boldsymbol{\eta}_m,$$

where $\boldsymbol{\eta}_j = \frac{1}{\|\boldsymbol{\beta}_j\|}\boldsymbol{\beta}_j$, $j = 1, 2, \ldots, m$. Thus, we get orthonormal $\boldsymbol{\eta}_1, \boldsymbol{\eta}_2, \ldots, \boldsymbol{\eta}_m$ from the linearly independent vectors $\boldsymbol{\alpha}_1, \boldsymbol{\alpha}_2, \ldots, \boldsymbol{\alpha}_m$. This process is called **Gram-Schmidt orthogonalization**.

If $\{\boldsymbol{\alpha}_1, \boldsymbol{\alpha}_2, \ldots, \boldsymbol{\alpha}_n\}$ is a basis of $\mathbb{R}^n$, we can certainly produce an orthonormal basis $\{\boldsymbol{\eta}_1, \boldsymbol{\eta}_2, \ldots, \boldsymbol{\eta}_n\}$ of $\mathbb{R}^n$ by applying the Gram-Schmidt orthogonalization procedure. $\mathbb{R}^n$ has not only one orthonormal basis; for example, any three pair-wise orthogonal vectors of unit length build an orthonormal basis of $\mathbb{R}^3$.

Example 4.5 Applying the Gram-Schmidt procedure

Let $B=\{\boldsymbol{\alpha}_1, \boldsymbol{\alpha}_2, \boldsymbol{\alpha}_3\}$ be a basis of $\mathbb{R}^3$, where

$$\boldsymbol{\alpha}_1=(1,-1,0), \boldsymbol{\alpha}_2=(1,\ 0,1), \boldsymbol{\alpha}_3=(1,-1,1).$$

Produce an orthonormal basis of $\mathbb{R}^3$ by applying Gram-Schmidt procedure to B.

Solution:

Let $\boldsymbol{\beta}_1=\boldsymbol{\alpha}_1=(1,-1,0)$,

$$\boldsymbol{\beta}_2=\boldsymbol{\alpha}_2-\frac{(\boldsymbol{\alpha}_2,\boldsymbol{\beta}_1)}{(\boldsymbol{\beta}_1,\boldsymbol{\beta}_1)}\boldsymbol{\beta}_1=(1,0,1)-\frac{1}{2}(1,-1,0)=\left(\frac{1}{2},\frac{1}{2},1\right),$$

$$\boldsymbol{\beta}_3=\boldsymbol{\alpha}_3-\frac{(\boldsymbol{\alpha}_3,\boldsymbol{\beta}_2)}{(\boldsymbol{\beta}_2,\boldsymbol{\beta}_2)}\boldsymbol{\beta}_2-\frac{(\boldsymbol{\alpha}_3,\boldsymbol{\beta}_1)}{(\boldsymbol{\beta}_1,\boldsymbol{\beta}_1)}\boldsymbol{\beta}_1$$

$$=(1,-1,1)-\frac{2}{3}\left(\frac{1}{2},\frac{1}{2},1\right)-\frac{2}{2}(1,-1,0)=\left(-\frac{1}{3},-\frac{1}{3},\frac{1}{3}\right).$$

Normalize $\boldsymbol{\beta}_1$, $\boldsymbol{\beta}_2$, and $\boldsymbol{\beta}_3$ to vectors of unit length. An orthonormal basis of $\mathbb{R}^3$ contains

$$\eta_1=\frac{1}{\|\boldsymbol{\beta}_1\|}\boldsymbol{\beta}_1=\left(\frac{1}{\sqrt{2}},-\frac{1}{\sqrt{2}},0\right),$$

$$\eta_2=\frac{1}{\|\boldsymbol{\beta}_2\|}\boldsymbol{\beta}_2=\left(\frac{1}{\sqrt{6}},\frac{1}{\sqrt{6}},\frac{2}{\sqrt{6}}\right),$$

$$\eta_3=\frac{1}{\|\boldsymbol{\beta}_3\|}\boldsymbol{\beta}_3=\left(-\frac{1}{\sqrt{3}},-\frac{1}{\sqrt{3}},\frac{1}{\sqrt{3}}\right).$$ ■

4.2.4 Orthogonal matrix and its properties

An orthogonal matrix is an important real square matrix such that its column and row vectors are both orthonormal. Its definition and properties are given below.

Definition 4.8

Let $\boldsymbol{A}\in\mathbb{R}^{n\times n}$. $\boldsymbol{A}$ is called an **orthogonal matrix** if $\boldsymbol{A}^{\mathrm{T}}\boldsymbol{A}=\boldsymbol{A}\boldsymbol{A}^{\mathrm{T}}=\boldsymbol{I}$.

THEOREM 4.6

A is an nth-order orthogonal matrix, if and only if, its columns constitute an orthonormal basis of $\mathbb{R}^n$.

Proof:
Let

$$A = \begin{pmatrix} a_{11} & a_{12} & \cdots & a_{1n} \\ a_{21} & a_{22} & \cdots & a_{2n} \\ \vdots & \vdots & & \vdots \\ a_{n1} & a_{n2} & \cdots & a_{nn} \end{pmatrix}.$$

Partitioning it by column as $(\alpha_1, \alpha_2, \ldots, \alpha_n)$, we obtain

$$A^{\mathrm{T}}A = \begin{pmatrix} \alpha_1{}^{\mathrm{T}} \\ \alpha_2{}^{\mathrm{T}} \\ \vdots \\ \alpha_n{}^{\mathrm{T}} \end{pmatrix} (\alpha_1, \alpha_2, \ldots, \alpha_n) = \begin{pmatrix} \alpha_1{}^{\mathrm{T}}\alpha_1 & \alpha_1{}^{\mathrm{T}}\alpha_2 & \cdots & \alpha_1{}^{\mathrm{T}}\alpha_n \\ \alpha_2{}^{\mathrm{T}}\alpha_1 & \alpha_2{}^{\mathrm{T}}\alpha_2 & \cdots & \alpha_2{}^{\mathrm{T}}\alpha_n \\ \vdots & \vdots & & \vdots \\ \alpha_n{}^{\mathrm{T}}\alpha_1 & \alpha_n{}^{\mathrm{T}}\alpha_2 & \cdots & \alpha_n{}^{\mathrm{T}}\alpha_n \end{pmatrix}.$$

A necessary and sufficient condition of $A^{\mathrm{T}}A = I$ is

$$\alpha_i{}^{\mathrm{T}}\alpha_i = (\alpha_i, \alpha_i) = 1, \qquad i = 1, 2, \ldots, n,$$

and

$$\alpha_i{}^{\mathrm{T}}\alpha_j = (\alpha_i, \alpha_j) = 0, \quad j \neq i, \quad i, j = 1, 2, \ldots, n.$$

So, the column set $\{\alpha_1, \alpha_2, \ldots, \alpha_n\}$ is an orthonormal basis of $\mathbb{R}^n$. □

THEOREM 4.7

Let both A and B be nth-order orthogonal matrices, then:

(i) $\det A = 1$ *or* -1,
(ii) $A^{-1} = A^{\mathrm{T}}$,
(iii) A^{T} *(i.e., A^{-1}) is also an orthogonal matrix,*
(iv) AB *is also an orthogonal matrix.*

Proof: The proofs of (i) and (ii) are left for readers as exercises.
(iii) Because $(A^{\mathrm{T}})^{\mathrm{T}}A^{\mathrm{T}} = AA^{\mathrm{T}} = AA^{-1} = I$, then A^{T} (i.e., A^{-1}) is also an orthogonal matrix, thus the row vectors of A build an orthonormal basis of $\mathbb{R}^n$.
(iv) AB is also an orthogonal matrix due to the fact that

$$(AB)^{\mathrm{T}}(AB) = B^{\mathrm{T}}(A^{\mathrm{T}}A)B = B^{\mathrm{T}}B = I.$$

□

THEOREM 4.8

If column vectors $\boldsymbol{x}, \boldsymbol{y} \in \mathbb{R}^n$ *are transformed to* $\boldsymbol{Ax}, \boldsymbol{Ay} \in \mathbb{R}^n$ *by an nth-order orthogonal matrix* $\boldsymbol{A}$*, then the inner product, length, and angle between vectors remain unchanged, that is,*

$$(\boldsymbol{Ax}, \boldsymbol{Ay}) = (\boldsymbol{x}, \boldsymbol{y}), \quad \|\boldsymbol{Ax}\| = \|\boldsymbol{x}\|, \|\boldsymbol{Ay}\| = \|\boldsymbol{y}\|,$$
$$\langle \boldsymbol{Ax}, \boldsymbol{Ay} \rangle = \langle \boldsymbol{x}, \boldsymbol{y} \rangle.$$

Proof:

$$(\boldsymbol{Ax}, \boldsymbol{Ay}) = (\boldsymbol{Ax})^{\mathrm{T}}(\boldsymbol{Ay}) = \boldsymbol{x}^{\mathrm{T}}(\boldsymbol{A}^{\mathrm{T}}\boldsymbol{A})\boldsymbol{y} = \boldsymbol{x}^{\mathrm{T}}\boldsymbol{y} = (\boldsymbol{x}, \boldsymbol{y}).$$

Let $\boldsymbol{x} = \boldsymbol{y}$. We have $(\boldsymbol{Ax}, \boldsymbol{Ax}) = (\boldsymbol{x}, \boldsymbol{x})$, that is, $\|\boldsymbol{Ax}\| = \|\boldsymbol{x}\|$. Similarly, $\|\boldsymbol{Ay}\| = \|\boldsymbol{y}\|$. As

$$\cos\langle \boldsymbol{Ax}, \boldsymbol{Ay} \rangle = \frac{(\boldsymbol{Ax}, \boldsymbol{Ay})}{\|\boldsymbol{Ax}\|\|\boldsymbol{Ay}\|} = \frac{(\boldsymbol{x}, \boldsymbol{y})}{\|\boldsymbol{x}\|\|\boldsymbol{y}\|} = \cos\langle \boldsymbol{x}, \boldsymbol{y} \rangle,$$

the angle between vectors $\boldsymbol{Ax}$ and $\boldsymbol{Ay}$ is equal to that between $\boldsymbol{x}$ and $\boldsymbol{y}$. □

In an Euclidean space, a vector $\boldsymbol{x}$ transformed to $\boldsymbol{Ax}$ by an orthogonal matrix $\boldsymbol{A}$ is usually called an **orthogonal transformation in Euclidean space**. It will play an important role in studying the standard form of quadratic function in Chapter 6.

*4.3 Definition of a Linear Space and Its Simple Properties

The linear space is the first abstract algebraic structure we encounter in this book. It is a concept abstracted and generalized from the 3-dimensional geometrical and n-dimensional vector spaces.

The operations of addition and multiplication between scalars and vectors can characterize certain related properties for some problems of geometry and mechanics as discussed in analytic geometry for 3-dimensional geometrical vectors. In order to study the general theory of solutions for a system of linear equations, we have extended 3-dimensional vectors to n-dimensional ones, and defined the operations of addition and scalar multiplication on them. We now discuss the linear dependence of vectors in n-dimensional vector space in terms of linear operations, which provides us with a description of solutions for a system of linear equations.

In a set formed by all n-dimensional real vectors, if the addition of vectors and the multiplication between vectors and real scalars are defined suitably, the set is called an n-dimensional vector space $\mathbb{R}^n$ over the field of real numbers (denoted by $\mathbb{R}$). $\mathbb{R}^n$ is closed under the two operations that satisfies the eight rules (refer to the formal definition later in this section).

With the same arguments for many different sets, we can also define an addition among the elements of that set and a scalar multiplication between an element of the set and a number from a given number field, such that the set is closed under the two operations and the same eight rules hold. Therefore, besides the particular set and operations already discussed, a conceptual abstract linear space can also be derived from two types of operations, their closure in the relevant set, where certain given rules hold good. This conceptual linear space and its properties have various wide applications.

Definition 4.9

A linear space V over a number field F is a nonempty set along with two operations—called "addition" (denoted by $\boldsymbol{\alpha}+\boldsymbol{\beta}$) and "scalar multiplication" (it is the multiplication between $\boldsymbol{\alpha}$ in V and a number λ in F, denoted by $\lambda\boldsymbol{\alpha}$), such that V is closed under the two operations (the element after the operation is still in V), and satisfies the following eight operation axioms.

(i) $\boldsymbol{\alpha}+\boldsymbol{\beta}=\boldsymbol{\beta}+\boldsymbol{\alpha}$ (Commutative axiom).
(ii) $(\boldsymbol{\alpha}+\boldsymbol{\beta})+\boldsymbol{\gamma}=\boldsymbol{\alpha}+(\boldsymbol{\beta}+\boldsymbol{\gamma})$ (Associative axiom).
(iii) There exists $\boldsymbol{\theta}\in V$ such that $\boldsymbol{\alpha}+\boldsymbol{\theta}=\boldsymbol{\alpha}$, where $\boldsymbol{\theta}$ is called the zero element of V.
(iv) There exists $-\boldsymbol{\alpha}\in V$ such that $\boldsymbol{\alpha}+(-\boldsymbol{\alpha})=\boldsymbol{\theta}$, where $-\boldsymbol{\alpha}$ is called the negative element of $\boldsymbol{\alpha}$.
(v) $1\boldsymbol{\alpha}=\boldsymbol{\alpha}$, $1\in F$.
(vi) $k(l\boldsymbol{\alpha})=(kl)\boldsymbol{\alpha}$.
(vii) $(k+l)\boldsymbol{\alpha}=k\boldsymbol{\alpha}+l\boldsymbol{\alpha}$.
(viii) $k(\boldsymbol{\alpha}+\boldsymbol{\beta})=k\boldsymbol{\alpha}+k\boldsymbol{\beta}$.

$\boldsymbol{\alpha}$, $\boldsymbol{\beta}$, and $\boldsymbol{\gamma}$ are any elements in V, and k, l are any numbers in F in the above eight axioms.

When F is the real (complex) number field, it is called **real (complex) linear space** or **real (complex) space** for short.

The elements in a linear space V are usually called **vectors**, the operations of addition and scalar multiplication are called **linear operations**.

Obviously, 3-dimensional geometrical vector space and $\mathbb{R}^n$ are two particular linear spaces. Let us discuss a few more examples of linear spaces.

Example 4.6 Polynomial space $\mathbb{R}[x]_n$

Over a number field F, a set $F[x]$ of all the polynomials is a linear space with respect to operations of the usual addition of polynomials and multiplication of polynomials by numbers. The zero element of $F[x]$ is the polynomial with all coefficients 0 (called the zero polynomial). The negative element of any polynomial $f(x)$ is $(-1)f(x)$. Polynomials with real coefficient and of degree less than n also form a linear space over the real number field which is denoted by $\mathbb{R}[x]_n$. ∎

Example 4.7 Matrix space $\mathbb{R}^{m\times n}$

The set $\mathbb{R}^{m\times n}$ [or $M_{m\times n}(\mathbb{R})$] of all $m \times n$ real matrices is a linear space under matrix addition and scalar multiplication over the real number field. The zero element in $\mathbb{R}^{m\times n}$ is the zero matrix of size $m \times n$ and the negative element of any element $\boldsymbol{A}$ is $(-\boldsymbol{A})$. ∎

Example 4.8 Linear space $C[a, b]$

The set $C[a, b]$ of all real continuous functions on interval $[a, b]$ is a linear space over the real number field F if the usual definition of addition of two functions and scalar multiplication of a function by a real number are used. The set $C^k[a, b]$ of all the functions whose kth-order derivatives are continuous in (a, b) is also a real linear space with respect to the same addition and scalar multiplication. ∎

For a given nonempty set V and a number field F, if V is not closed under the addition and scalar multiplication defined (namely an operated element does not always belong to V), or one of the eight axioms does not hold, then V is not a linear space over the number field F for the given operations. For instance, all the nth-order real matrices is not a linear space over the complex field with addition and scalar multiplication, all the real nonzero vectors of dimension 3 is not a real linear space along with addition and scalar multiplication of vectors.

From the definition of a linear space, we can get the following simple properties.

Property 4.1

The zero element is unique in a linear space.

Suppose $\boldsymbol{\theta}_1, \boldsymbol{\theta}_2$ are two zero elements in linear space. Then,

$$\boldsymbol{\theta}_1 = \boldsymbol{\theta}_1 + \boldsymbol{\theta}_2 = \boldsymbol{\theta}_2 + \boldsymbol{\theta}_1 = \boldsymbol{\theta}_2.$$

Property 4.2

The negative element of every element $\boldsymbol{\alpha}$ is unique.

Let $\boldsymbol{\beta}_1, \boldsymbol{\beta}_2$ be two negative elements of $\boldsymbol{\alpha}$, that is,

$$\boldsymbol{\alpha} + \boldsymbol{\beta}_1 = \boldsymbol{\alpha} + \boldsymbol{\beta}_2 = \boldsymbol{\theta}.$$

So,

$$\boldsymbol{\beta}_1 = \boldsymbol{\beta}_1 + \boldsymbol{\theta} = \boldsymbol{\beta}_1 + (\boldsymbol{\alpha} + \boldsymbol{\beta}_2) = \boldsymbol{\beta}_1 + \boldsymbol{\alpha} + \boldsymbol{\beta}_2 = \boldsymbol{\theta} + \boldsymbol{\beta}_2 = \boldsymbol{\beta}_2.$$

We define subtraction through the negative element

$$\boldsymbol{\beta} - \boldsymbol{\alpha} = \boldsymbol{\beta} + (-\boldsymbol{\alpha}).$$

Property 4.3

If $\boldsymbol{\alpha}, \boldsymbol{\beta} \in V; k, l \in F$, then

$$k(\boldsymbol{\alpha} - \boldsymbol{\beta}) = k\boldsymbol{\alpha} - k\boldsymbol{\beta},$$
$$(k - l)\boldsymbol{\alpha} = k\boldsymbol{\alpha} - l\boldsymbol{\alpha}.$$

It holds because

$$k(\boldsymbol{\alpha} - \boldsymbol{\beta}) + k\boldsymbol{\beta} = k[(\boldsymbol{\alpha} - \boldsymbol{\beta}) + \boldsymbol{\beta}] = k[\boldsymbol{\alpha} + ((-\boldsymbol{\beta}) + \boldsymbol{\beta})] = k(\boldsymbol{\alpha} + \boldsymbol{\theta}) = k\boldsymbol{\alpha},$$
$$(k - l)\boldsymbol{\alpha} + l\boldsymbol{\alpha} = [(k - l) + l]\boldsymbol{\alpha} = k\boldsymbol{\alpha}.$$

Add $-k\boldsymbol{\beta}$ and $-l\boldsymbol{\alpha}$ to both sides of the above two equations respectively and we will immediately get the desired results.

Let $\boldsymbol{\alpha} = \boldsymbol{\beta}, \boldsymbol{\alpha} = \boldsymbol{\theta}, k = l, k = 0$ and we obtain by applying Property 4.3,

Property 4.4

$k\boldsymbol{\theta} = \boldsymbol{\theta}, k(-\boldsymbol{\beta}) = -(k\boldsymbol{\beta}), 0\boldsymbol{\alpha} = \boldsymbol{\theta}, (-l)\boldsymbol{\alpha} = -(l\boldsymbol{\alpha})$, especially $(-1)\boldsymbol{\alpha} = -\boldsymbol{\alpha}$. Note that $-(l\boldsymbol{\alpha})$ is denoted as $-l\boldsymbol{\alpha}$ for short.

Property 4.5

Let $\boldsymbol{\alpha} \in V, k \in F$. Then, $k = 0$ or $\boldsymbol{\alpha} = \boldsymbol{\theta}$ if $k\boldsymbol{\alpha} = \boldsymbol{\theta}$.
Let $k \neq 0$, then $\boldsymbol{\alpha} = 1\boldsymbol{\alpha} = \frac{1}{k}(k\boldsymbol{\alpha}) = \frac{1}{k}\boldsymbol{\theta} = \boldsymbol{\theta}$.

Because a linear space has the above simple properties, a linear combination of

$$k\boldsymbol{\beta} + k_1\boldsymbol{\alpha}_1 + \cdots + k_r\boldsymbol{\alpha}_r = \boldsymbol{\theta} \quad (k \neq 0),$$

can be solved easily as

$$\boldsymbol{\beta} = -\frac{k_1}{k}\boldsymbol{\alpha}_1 - \cdots - \frac{k_r}{k}\boldsymbol{\alpha}_r.$$

*4.4 Linear Subspace

In a linear space V over a number field F, [$V(F)$ for short], the closure of the two types of operations for a subset W depends on the subset. See the following subsets of $\mathbb{R}^3$,

$$W_1 = \{(x_1, x_2, x_3)\,|x_1 - x_2 + 5x_3 = 0\},$$
$$W_2 = \{(x_1, x_2, x_3)\,|x_1 - x_2 + 5x_3 = 1\}.$$

W_1 is a set of vectors in plane $x_1 - x_2 + 5x_3 = 0$ passing through the origin, W_2 is a set of vectors in plane $x_1 - x_2 + 5x_3 = 1$ not passing through the origin. It is easy to verify that W_1 is closed under vector addition and scalar multiplication, while W_2 is not closed under these two operations.

Obviously, W_2 does not form a linear space with the linear operations in $\mathbb{R}^3$, and so it is not a subspace of $\mathbb{R}^3$, whereas W_1 is a linear subspace of $\mathbb{R}^3$. We now give the definition of a linear subspace.

Definition 4.10

Let $V(F)$ be a linear space and W be a nonempty subset of V. W is called a linear subspace of $V(F)$ (or subspace for short) if W itself is a linear space over the number field F with respect to linear operations defined in $V(F)$.

THEOREM 4.9

The nonempty subset W of a linear space $V(F)$ is a subspace of V, if and only if, W is closed under the two operations in V.

Proof: The "only if" part is obvious. For the "if" part, we only need to prove that vectors in W satisfy the eight axioms. Because W is a nonempty subset of V, rules (i), (ii), (v), (vi), (vii), and (viii) are obviously true. So, we only need to prove $\boldsymbol{\theta} \in W$; then the reciprocal vector $(-\boldsymbol{\alpha})$ of every vector $\boldsymbol{\alpha}$ of W also lies in W due to the fact that W is closed under scalar multiplication. Indeed, $\forall \lambda \in F, \forall \boldsymbol{\alpha} \in W$, we get $\lambda\boldsymbol{\alpha} \in W$. Take $\lambda = 0, -1$, then $0\boldsymbol{\alpha} = \boldsymbol{\theta} \in W$, $(-1)\boldsymbol{\alpha} = -\boldsymbol{\alpha} \in W$. □

Example 4.9 Trivial and nontrivial subspaces

In a linear space V, the subset $\{\boldsymbol{\theta}\}$ which contains only the zero vector is a subspace of V, called the **zero subspace**. V itself is also a subspace of V. These two subspaces are also called **trivial subspaces** of V, while other subspaces of V are called the **nontrivial subspaces**. ■

Example 4.10 Solution space for a system of homogeneous linear equations

For $\boldsymbol{A} \in F^{m\times n}$, the solution set of a system of homogeneous linear equations $\boldsymbol{Ax} = \boldsymbol{0}$,

$$S = \{\boldsymbol{x} \,|\, \boldsymbol{Ax} = \boldsymbol{0}\}$$

is a subspace of F^n, which is called the **solution space** [also called **null space of matrix** $\boldsymbol{A}$, denoted by $N(\boldsymbol{A})$] of the system of homogeneous linear equations. But the solution set of a system of nonhomogeneous linear equations $\boldsymbol{Ax} = \boldsymbol{b}$ is not a subspace of F^n. ■

Example 4.11 Some subspaces of $\mathbb{R}^{n\times n}$

The sets of real scalar matrices, real diagonal matrices, real symmetrical matrices, real upper triangular matrices are subspaces of $\mathbb{R}^{n\times n}$. ■

Example 4.12 Planes or lines passing through the origin are subspaces of $\mathbb{R}^3$

Consider the following subsets of $\mathbb{R}^3$:

$$V_1 = \{(x_1, 0, 0)\,|x_1 \in \mathbb{R}\}, \quad V_2 = \{(1, 0, x_3)\,|x_3 \in \mathbb{R}\}.$$

Then, V_1 is a subspace of $\mathbb{R}^3$, whereas V_2 is not a subspace of $\mathbb{R}^3$. Then, geometrically, we can explain that V_1 consists of all vectors on the x-axis and V_2 consists of all vectors on the line that passes through the point $(1, 0, 0)$ and are parallel to the z-axis. In 3-dimensional geometric vector space, sets of all vectors on a line or a plane passing through the origin are subspaces of $\mathbb{R}^3$, whereas no such set is a subspace of $\mathbb{R}^3$ if the plane or the line does not go through the origin. ■

THEOREM 4.10

Let V be a linear space over number field F, S be a nonempty subset of V. A set consisting of all linear combinations of vectors in S

$$W = \{k_1\boldsymbol{\alpha}_1 + k_2\boldsymbol{\alpha}_2 + \cdots + k_t\boldsymbol{\alpha}_t | \boldsymbol{\alpha}_i \in S, k_i \in F, i = 1, 2, \ldots, t\}$$

is the smallest subspace of V containing S.

Proof: W clearly contains S. For $\boldsymbol{\alpha}, \boldsymbol{\beta} \in W$, there exist $\boldsymbol{\alpha}_1, \boldsymbol{\alpha}_2, \ldots, \boldsymbol{\alpha}_m$, $\boldsymbol{\beta}_1, \boldsymbol{\beta}_2, \ldots, \boldsymbol{\beta}_n \in S$, and $k_1, k_2, \ldots, k_m, l_1, l_2, \ldots, l_n \in F$ such that

$$\boldsymbol{\alpha} = k_1\boldsymbol{\alpha}_1 + k_2\boldsymbol{\alpha}_2 + \cdots + k_m\boldsymbol{\alpha}_m, \quad \boldsymbol{\beta} = l_1\boldsymbol{\beta}_1 + l_2\boldsymbol{\beta}_2 + \cdots + l_n\boldsymbol{\beta}_n,$$

so

$$\boldsymbol{\alpha} + \boldsymbol{\beta} = (k_1\boldsymbol{\alpha}_1 + k_2\boldsymbol{\alpha}_2 + \cdots + k_m\boldsymbol{\alpha}_m) + (l_1\boldsymbol{\beta}_1 + l_2\boldsymbol{\beta}_2 + \cdots + l_n\boldsymbol{\beta}_n) \in W.$$

Also, $\forall k \in F$

$$k\boldsymbol{\alpha} = k(k_1\boldsymbol{\alpha}_1 + k_2\boldsymbol{\alpha}_2 + \cdots + k_m\boldsymbol{\alpha}_m) = kk_1\boldsymbol{\alpha}_1 + kk_2\boldsymbol{\alpha}_2 + \cdots + kk_m\boldsymbol{\alpha}_m \in W.$$

W is therefore a subspace of V.

We then assume W^* to be a subspace containing S; then

$$\forall \boldsymbol{\alpha} = k_1\boldsymbol{\alpha}_1 + \cdots + k_m\boldsymbol{\alpha}_m \in W,$$

as $\boldsymbol{\alpha}_1, \ldots, \boldsymbol{\alpha}_m \in S \subseteq W^*$, one must have $\boldsymbol{\alpha} \in W^*$, where $W \subseteq W^*$. So, W is the smallest subspace of V that contains S. □

We call W of Theorem 4.10 a subspace of V generated by the nonempty subset S of V or say S generates W. As S is a finite subset $\{\boldsymbol{\alpha}_1, \boldsymbol{\alpha}_2, \ldots, \boldsymbol{\alpha}_m\}$, we denote $W = L(\boldsymbol{\alpha}_1, \boldsymbol{\alpha}_2, \ldots, \boldsymbol{\alpha}_m)$, and call W a subspace generated by the vectors $\boldsymbol{\alpha}_1, \boldsymbol{\alpha}_2, \ldots, \boldsymbol{\alpha}_m$.

For example, the solution space of a homogeneous system of linear equations $\boldsymbol{Ax} = \mathbf{0}$ is a subspace generated by a solution basis. The set of vectors on a plane passing through the origin in $\mathbb{R}^3$ is a subspace generated by any two linearly independent vectors in that plane.

THEOREM 4.11

Let W_1, W_2 be two subspaces of a linear space V over the number field F, and $W_1 = L(\boldsymbol{\alpha}_1, \boldsymbol{\alpha}_2, \ldots, \boldsymbol{\alpha}_s)$, $W_2 = L(\boldsymbol{\beta}_1, \boldsymbol{\beta}_2, \ldots, \boldsymbol{\beta}_t)$. Then $W_1 = W_2$, if and only if, the two vector sets $\{\boldsymbol{\alpha}_1, \boldsymbol{\alpha}_2, \ldots, \boldsymbol{\alpha}_s\}$ and $\left\{\boldsymbol{\beta}_1, \boldsymbol{\beta}_2, \ldots, \boldsymbol{\beta}_t\right\}$ are equivalent (that is, any vector in one set can be written as a linear combination of vectors in another set).

Proof: The necessity part is obvious. We give a proof of the sufficiency of the condition.

Let $\boldsymbol{\alpha} = k_1\boldsymbol{\alpha}_1 + k_2\boldsymbol{\alpha}_2 + \cdots + k_s\boldsymbol{\alpha}_s \in W_1$. As $\boldsymbol{\alpha}_i (i = 1, 2, \ldots, s)$ can be written as a linear combination of $\boldsymbol{\beta}_1, \boldsymbol{\beta}_2, \ldots, \boldsymbol{\beta}_t$, so $\boldsymbol{\alpha}$ can be written as a linear combination of $\boldsymbol{\beta}_1, \boldsymbol{\beta}_2, \ldots, \boldsymbol{\beta}_t$, namely, there exist $l_1, l_2, \ldots, l_t \in F$ such that

$$\boldsymbol{\alpha} = l_1\boldsymbol{\beta}_1 + \cdots + l_t\boldsymbol{\beta}_t \in W_2.$$

Hence, $W_1 \subseteq W_2$. In a similar manner we can prove that

$$W_2 \subseteq W_1 \text{ and thus } W_1 = W_2.$$ □

Definition 4.11

Let W_1, W_2 be two subspaces of a linear space V. Then subsets of V,

$$W_1 \cap W_2 = \{\boldsymbol{\alpha} | \boldsymbol{\alpha} \in W_1 \text{ and } \boldsymbol{\alpha} \in W_2\},$$

$$W_1 + W_2 = \{\boldsymbol{\alpha}_1 + \boldsymbol{\alpha}_2 | \boldsymbol{\alpha}_1 \in W_1,\ \boldsymbol{\alpha}_2 \in W_2\},$$

are called the **intersection** and **sum (also called linear sum)** of two subspaces, respectively. If $W_1 \cap W_2 = \{\boldsymbol{\theta}\}$, $W_1 + W_2$ is called a **direct sum** of W_1 and W_2, and is denoted as $W_1 \oplus W_2$.

Notice that $W_1 + W_2$ is a set consisting of sums of vectors in W_1 and W_2, which is usually different from $W_1 \cup W_2$.

THEOREM 4.12

The intersection and sum of two subspaces W_1 and W_2 of a linear space $V(F)$ are also subspaces of V.

Proof: To prove $W_1 + W_2$ is a subspace of V, we only need to prove that $W_1 + W_2$ is closed under linear operations in V.

Let $\boldsymbol{\alpha}, \boldsymbol{\beta} \in W_1 + W_2$, that is, there exist $\boldsymbol{\alpha}_1, \boldsymbol{\beta}_1 \in W_1$; $\boldsymbol{\alpha}_2, \boldsymbol{\beta}_2 \in W_2$ such that

$$\boldsymbol{\alpha} = \boldsymbol{\alpha}_1 + \boldsymbol{\alpha}_2, \quad \boldsymbol{\beta} = \boldsymbol{\beta}_1 + \boldsymbol{\beta}_2.$$

Thus,

$$\boldsymbol{\alpha} + \boldsymbol{\beta} = (\boldsymbol{\alpha}_1 + \boldsymbol{\alpha}_2) + (\boldsymbol{\beta}_1 + \boldsymbol{\beta}_2) = (\boldsymbol{\alpha}_1 + \boldsymbol{\beta}_1) + (\boldsymbol{\alpha}_2 + \boldsymbol{\beta}_2) \in W_1 + W_2.$$

Let $\lambda \in F$. Then, $\lambda\boldsymbol{\alpha} = \lambda(\boldsymbol{\alpha}_1 + \boldsymbol{\alpha}_2) = \lambda\boldsymbol{\alpha}_1 + \lambda\boldsymbol{\alpha}_2 \in W_1 + W_2$. Therefore, $W_1 + W_2$ is also a subspace of V. The proof of the intersection of two subspaces is a subspace is left to readers as an exercise. □

We shall now introduce useful notions of row space and column space of matrix as well as the orthogonal subspace of $\mathbb{R}^n$.

Definition 4.12

The subspace generated by the column (row) vectors of a matrix $\boldsymbol{A}$ is called the column [row] space of $\boldsymbol{A}$, denoted as $R(\boldsymbol{A})[R(\boldsymbol{A}^{\mathrm{T}})]$.

If $\boldsymbol{A} \in M_{m\times n}(\mathbb{R})$, then the column vectors of $\boldsymbol{A}$, $\boldsymbol{\beta}_1, \boldsymbol{\beta}_2, \ldots, \boldsymbol{\beta}_n \in \mathbb{R}^m$, row vectors $\boldsymbol{\alpha}_1, \boldsymbol{\alpha}_2, \ldots, \boldsymbol{\alpha}_m \in \mathbb{R}^n$. So, $R(\boldsymbol{A}) = L(\boldsymbol{\beta}_1, \boldsymbol{\beta}_2, \ldots, \boldsymbol{\beta}_n)$ is a subspace of $\mathbb{R}^m$; $R(\boldsymbol{A}^{\mathrm{T}}) = L(\boldsymbol{\alpha}_1, \boldsymbol{\alpha}_2, \ldots, \boldsymbol{\alpha}_m)$ is a subspace of $\mathbb{R}^n$.

We have learned in Chapter 3, a system of nonhomogeneous linear equations $\boldsymbol{Ax} = \boldsymbol{b}$ has a nonzero solution, if and only if, "$\boldsymbol{b}$ is a linear combination of the column vectors of $\boldsymbol{A}$." According to the definition of column space of matrix, this can also be stated as "$\boldsymbol{b}$ belongs to the column space of $\boldsymbol{A}$, that is, $\boldsymbol{b} \in R(\boldsymbol{A})$."

Definition 4.13

Let a vector $\boldsymbol{\alpha} \in \mathbb{R}^n$ and W be a subspace of $\mathbb{R}^n$. Then, $\boldsymbol{\alpha}$ and subspace W are said to be **orthogonal** and denoted by $\boldsymbol{\alpha} \perp W$ if $(\boldsymbol{\alpha}, \boldsymbol{\gamma}) = 0$ for any $\boldsymbol{\gamma} \in W$.

Definition 4.14

Let V and W be two subspaces of $\mathbb{R}^n$. V and W are said to be orthogonal and denoted by $V \perp W$ if $(\boldsymbol{\alpha}, \boldsymbol{\beta}) = 0$ for all $\boldsymbol{\alpha} \in V$, $\boldsymbol{\beta} \in W$.

For example, the set of all vectors in the plane Oxy in $\mathbb{R}^3$ and the set of all vectors on the z-axis are 2-dimensional and 1-dimensional subspaces of $\mathbb{R}^3$ respectively . They are orthogonal. But, two subspaces consisting of all vectors in two orthogonal planes passing through the origin are not orthogonal since the inner product of a nonzero vector with itself is not zero on the line of intersection of two planes.

A homogeneous system of linear equations $\boldsymbol{Ax} = \boldsymbol{0}$ is

$$\begin{cases} a_{11}x_1 + a_{12}x_2 + \cdots + a_{1n}x_n = 0, \\ a_{21}x_1 + a_{22}x_2 + \cdots + a_{2n}x_n = 0, \\ \qquad \cdots \\ a_{m1}x_1 + a_{m2}x_2 + \cdots + a_{mn}x_n = 0. \end{cases}$$

Any solution vector is orthogonal to each row vector of the coefficient matrix $\boldsymbol{A}$, so the solution space and row space of $\boldsymbol{A}$ is orthogonal. Namely,

$$N(\boldsymbol{A}) \perp R(\boldsymbol{A}^{\mathrm{T}}).$$

THEOREM 4.13

A set consisting of all vectors orthogonal to a subspace V in $\mathbb{R}^n$,

$$W = \{\boldsymbol{\alpha} | \boldsymbol{\alpha} \perp V, \boldsymbol{\alpha} \in \mathbb{R}^n\}$$

is a subspace of $\mathbb{R}^n$.

Proof: Because the zero vector is orthogonal to every subspace, so W is a nonempty set. Let $\boldsymbol{\alpha}_1, \boldsymbol{\alpha}_2 \in W$. If $(\boldsymbol{\alpha}_1, \boldsymbol{\gamma}) = 0$, $(\boldsymbol{\alpha}_2, \boldsymbol{\gamma}) = 0$ for any $\boldsymbol{\gamma} \in V$, then

$$(\boldsymbol{\alpha}_1 + \boldsymbol{\alpha}_2, \boldsymbol{\gamma}) = 0, \ \ (k\boldsymbol{\alpha}_1, \boldsymbol{\gamma}) = 0 \ (k \in \mathbb{R}).$$

So, $(\boldsymbol{\alpha}_1 + \boldsymbol{\alpha}_2) \perp V$, $k\boldsymbol{\alpha}_1 \perp V$, that is, $\boldsymbol{\alpha}_1 + \boldsymbol{\alpha}_2 \in W$, $k\boldsymbol{\alpha}_1 \in W$ and W is, therefore, a subspace of $\mathbb{R}^n$. □

Definition 4.15

A subspace W consisting of all vectors that is orthogonal to a subspace V in $\mathbb{R}^n$ is called an **orthogonal complement** of V, and is denoted as $W = V^{\perp}$.

For instance, the solution space $N(\boldsymbol{A})$ consists of all vectors orthogonal to the row vectors of $\boldsymbol{A}$, so

$$N(\boldsymbol{A}) = (R(\boldsymbol{A}^{\mathrm{T}}))^{\perp}.$$

This is a basic property of the solution space of a homogeneous system of linear equations $\boldsymbol{A}\boldsymbol{x} = \boldsymbol{0}$.

*4.5 Basis and Dimension of a Linear Space, Coordinates of a Vector

In the previous sections of this chapter, we have seen that in the case of F^n, any n linearly independent vectors in F^n is a basis, any vector $\boldsymbol{\alpha}$ can be written as a linear combination of a basis of F^n, and the vector obtained by arranging coefficients in this combination is the coordinates of $\boldsymbol{\alpha}$ under the basis. To explore similar issues in general linear space $V(F)$, the concept of linear dependence of the elements (or vectors) in $V(F)$ must be discussed at first.

The general linear space V with two linear operations, following the eight axioms and certain simple properties, possesses the same definitions and related basic results about linear dependence as in F^n. We will not elaborate further on this point, but readers should be aware that the vectors $\boldsymbol{\alpha}, \boldsymbol{\beta}, \boldsymbol{\gamma}, \ldots,$ in F^n are analogous elements in V, and the zero vector there corresponds the zero element in V.

Example 4.13 Proof of linear independence of some elements in $\mathbb{R}[x]_n$

Prove that the elements

$$f_0 = 1,\ f_1 = x,\ f_2 = x^2, \ldots,\ f_{n-1} = x^{n-1},$$

in the linear space $\mathbb{R}[x]_n$ are linearly independent.

Proof: Let $k_0 f_0 + k_1 f_1 + k_2 f_2 + \cdots + k_{n-1} f_{n-1} = 0(x)$, that is,

$$k_0 + k_1 x_1 + k_2 x^2 + \cdots + k_{n-1} x^{n-1} = 0(x).$$

$0(x)$ in the above equation is the zero element of $\mathbb{R}[x]_n$, that is, the zero polynomial. The equality of the linear combination of $1, x, x^2, \ldots, x^{n-1}$ and the zero polynomial shows that all the coefficients $k_0, k_1, k_2, \ldots, k_{n-1}$ must be zero. So, $1, x, x^2, \ldots, x^{n-1}$ are linearly independent. ∎

Example 4.14 Proof of linear independence of some elements in $\mathbb{R}^{2\times 2}$

Prove that the elements

$$A_1=\begin{pmatrix}1&0\\0&0\end{pmatrix},\ A_2=\begin{pmatrix}1&1\\0&0\end{pmatrix},\ A_3=\begin{pmatrix}1&1\\1&0\end{pmatrix},\ A_4=\begin{pmatrix}1&1\\1&1\end{pmatrix},$$

in $\mathbb{R}^{2\times 2}$ are linearly independent.

Proof:
Let

$$k_1A_1+k_2A_2+k_3A_3+k_4A_4=\mathbf{0}_{2\times 2}, \qquad ①$$

that is,

$$\begin{pmatrix}k_1+k_2+k_3+k_4 & k_2+k_3+k_4\\ k_3+k_4 & k_4\end{pmatrix}=\begin{pmatrix}0&0\\0&0\end{pmatrix},$$

which is

$$\begin{cases}k_1+k_2+k_3+k_4=0,\\ k_2+k_3+k_4=0,\\ k_3+k_4=0,\\ k_4=0.\end{cases}$$

This system of linear equations has only the zero solution. Therefore, ① is true only when $k_1=k_2=k_3=k_4=0$. We conclude that A_1, A_2, A_3, A_4 are linearly independent. ■

Obviously, in $\mathbb{R}^{2\times 2}$, matrices

$$E_{11}=\begin{pmatrix}1&0\\0&0\end{pmatrix},\ E_{12}=\begin{pmatrix}0&1\\0&0\end{pmatrix},\ E_{21}=\begin{pmatrix}0&0\\1&0\end{pmatrix},\ E_{22}=\begin{pmatrix}0&0\\0&1\end{pmatrix},$$

are also linearly independent, and each matrix in $\mathbb{R}^{2\times 2}$,

$$A=\begin{pmatrix}a&b\\c&d\end{pmatrix}=aE_{11}+bE_{12}+cE_{21}+dE_{22}.$$

It is not difficult to prove for readers that, any five elements (second-order matrices) A, B, C, D, Q in $\mathbb{R}^{2\times 2}$ are linearly dependent. If A, B, C, and D are linearly independent, then Q can be uniquely written as a linear combination of A, B, C, and D. These attributes of $\mathbb{R}^{2\times 2}$ are similar to those in $\mathbb{R}^4$. The concepts of basis, dimension, and coordinate of a vector for a linear space are listed below.

Definition 4.16

If there exists a linearly independent set of vectors $B = \{\boldsymbol{\alpha}_1, \boldsymbol{\alpha}_2, \ldots, \boldsymbol{\alpha}_n\}$ in a linear space $V(F)$, and every $\boldsymbol{\alpha} \in V$ can be written as a linear combination of B as

$$\boldsymbol{\alpha} = x_1\boldsymbol{\alpha}_1 + x_2\boldsymbol{\alpha}_2 + \cdots + x_n\boldsymbol{\alpha}_n, \tag{4.17}$$

then V is called an ***n*-dimensional linear space** (or say that the dimension of V is n) and it is denoted as $\dim V = n$, B is called a **basis** of V and the ordered number set $(x_1, x_2, \ldots, x_n)$ is the coordinate of $\boldsymbol{\alpha}$ under basis B, denoted as

$$\boldsymbol{\alpha}_B = (x_1, x_2, \ldots, x_n)^{\mathrm{T}} \in F^n. \tag{4.18}$$

V is called an **infinite-dimensional linear space** if there are arbitrarily many linearly independent vectors in $V(F)$.

It is easy to prove that, $1, x^1, x^2, \ldots, x^n$ (n is a positive integer) in $F[x]$ are linearly independent. Consequently, $F[x]$ is an infinite-dimensional linear space. $C[a,\, b]$ is also an infinite-dimensional linear space. However, in this course, we shall consider only the finite-dimensional linear spaces.

In an n-dimensional linear space V, any $(n+1)$ elements $\boldsymbol{\beta}_1, \boldsymbol{\beta}_2, \ldots, \boldsymbol{\beta}_{n+1}$ can be written as a linear combination of a basis $\boldsymbol{\alpha}_1, \boldsymbol{\alpha}_2, \ldots, \boldsymbol{\alpha}_n$ of V. Therefore, any $(n+1)$ elements in an n-dimensional linear space are linearly dependent from Theorem 3.4 and any n linearly independent vectors construct a basis of n-dimensional linear space V.

For instances, $F[x]_n$ is an n-dimensional linear space with a basis $\{1, x, x^2, \ldots, x^{n-1}\}$, $\mathbb{R}^{2\times 2}$ is a 4-dimensional linear space with two bases $\{\boldsymbol{E}_{11}, \boldsymbol{E}_{12}, \boldsymbol{E}_{21}, \boldsymbol{E}_{22}\}$ and $\{\boldsymbol{A}_1, \boldsymbol{A}_2, \boldsymbol{A}_3, \boldsymbol{A}_4\}$ as shown in Example 4.14. $F^{m\times n}$ is an $(m \times n)$-dimensional linear space with a basis $\boldsymbol{E}_{ij}$ [$\boldsymbol{E}_{ij}$ is a $m \times n$ matrix whose (i, j)th element is 1 and all other elements 0], $i = 1, 2, \ldots, m$; $j = 1, 2, \ldots, n$.

It is left to readers to prove that the $(m \times n)$ elements $\boldsymbol{E}_{ij}$, $i = 1, 2, \ldots, m$; $j = 1, 2, \ldots, n$, are linearly independent and any $(m \times n)$ matrix can be written as a linear combination of these elements.

In a linear space V, the dimension of a subspace $L(\boldsymbol{\alpha}_1, \boldsymbol{\alpha}_2, \ldots, \boldsymbol{\alpha}_s)$ generated by vectors $\boldsymbol{\alpha}_1, \boldsymbol{\alpha}_2, \ldots, \boldsymbol{\alpha}_s$ is equal to the rank of vectors $\boldsymbol{\alpha}_1, \boldsymbol{\alpha}_2, \ldots, \boldsymbol{\alpha}_s$. A maximal linearly independent subset of $\boldsymbol{\alpha}_1, \boldsymbol{\alpha}_2, \ldots, \boldsymbol{\alpha}_s$ is a basis of $L(\boldsymbol{\alpha}_1, \boldsymbol{\alpha}_2, \ldots, \boldsymbol{\alpha}_s)$. The dimensions of the column space $R(\boldsymbol{A})$ and the row space $R(\boldsymbol{A}^{\mathrm{T}})$ of matrix $\boldsymbol{A}$ are both equal to $\operatorname{rank}(\boldsymbol{A})$. The dimension of zero subspace $\{\boldsymbol{\theta}\}$ of V is 0.

A solution basis of a homogeneous system of linear equations $\boldsymbol{Ax} = \boldsymbol{0}$ is a basis of its solution space $N(\boldsymbol{A})$. If $\boldsymbol{A}$ is a $(m \times n)$ matrix and $\operatorname{rank}(\boldsymbol{A}) = r$, then the dimension of solution space $N(\boldsymbol{A})$ is $(n - r)$. So,

$$\dim(R(\boldsymbol{A}^{\mathrm{T}})) + \dim(N(\boldsymbol{A})) = n. \tag{4.19}$$

This is another property of $\boldsymbol{Ax} = \boldsymbol{0}$.

THEOREM 4.14

Let V be a linear space of dimension n, W be an m-dimensional subspace of V, and $B_1 = \{\boldsymbol{\alpha}_1, \boldsymbol{\alpha}_2, \ldots, \boldsymbol{\alpha}_m\}$ be a basis of W. Then, B_1 can be expanded to a basis of V (i.e., to a basis of V by adding $(n-m)$ vectors suitably to basis B_1).

Proof: If $m = n$, then B_1 is a basis of V. If $m < n$, then there exists $\boldsymbol{\alpha}_{m+1} \in V$ such that $\boldsymbol{\alpha}_1, \boldsymbol{\alpha}_2, \ldots, \boldsymbol{\alpha}_m, \boldsymbol{\alpha}_{m+1}$ are linearly independent. Otherwise, $\dim V = m$, which is contrary to $\dim V = n$ as we have taken $m < n$. The theorem is proved for $m+1 = n$. If $m+1 < n$, continue the steps above, and by proceeding in this manner, we can ultimately conclude that, there exist $\boldsymbol{\alpha}_{m+2}, \ldots, \boldsymbol{\alpha}_n \in V$ such that $\{\boldsymbol{\alpha}_1, \boldsymbol{\alpha}_2, \ldots, \boldsymbol{\alpha}_m, \boldsymbol{\alpha}_{m+1}, \ldots, \boldsymbol{\alpha}_n\}$ are linearly independent, which is a basis of V. □

THEOREM 4.15 ***(Formula of subspace dimensions)***

Let W_1 and W_2 be subspaces of $V(F)$, then

$$\dim W_1 + \dim W_2 = \dim(W_1 + W_2) + \dim(W_1 \cap W_2). \tag{4.20}$$

Proof:
Assume

$$\dim W_1 = s, \dim W_2 = t, \dim(W_1 \cap W_2) = r,$$

then without any loss of generality, we may assume that

$$W_1 \cap W_2 = L(\boldsymbol{\alpha}_1, \boldsymbol{\alpha}_2, \ldots, \boldsymbol{\alpha}_r),$$
$$W_1 = L(\boldsymbol{\alpha}_1, \ldots, \boldsymbol{\alpha}_r, \boldsymbol{\beta}_1, \ldots, \boldsymbol{\beta}_{s-r}), \quad W_2 = L(\boldsymbol{\alpha}_1, \ldots, \boldsymbol{\alpha}_r, \boldsymbol{\gamma}_1, \ldots, \boldsymbol{\gamma}_{t-r}).$$

So,

$$W_1 + W_2 = L(\boldsymbol{\alpha}_1, \ldots, \boldsymbol{\alpha}_r, \boldsymbol{\beta}_1, \ldots, \boldsymbol{\beta}_{s-r}, \boldsymbol{\gamma}_1, \ldots, \boldsymbol{\gamma}_{t-r}).$$

We only need to prove that $\dim(W_1 + W_2) = s + t - r$ by showing that $(s + t - r)$ vectors generating $W_1 + W_2$ are linearly independent. Suppose that

$$a_1\boldsymbol{\alpha}_1 + \cdots + a_r\boldsymbol{\alpha}_r + b_1\boldsymbol{\beta}_1 + \cdots + b_{s-r}\boldsymbol{\beta}_{s-r} + c_1\boldsymbol{\gamma}_1 + \cdots + c_{t-r}\boldsymbol{\gamma}_{t-r} = 0. \quad ①$$

Then,

$$a_1\boldsymbol{\alpha}_1 + \cdots + a_r\boldsymbol{\alpha}_r + b_1\boldsymbol{\beta}_1 + \cdots + b_{s-r}\boldsymbol{\beta}_{s-r} = -c_1\boldsymbol{\gamma}_1 - \cdots - c_{t-r}\boldsymbol{\gamma}_{t-r}. \quad ②$$

Because the vectors on the *LHS* and *RHS* of equation ② belong to W_1 and W_2, respectively, they belong to $W_1 \cap W_2$. So,

$$d_1\boldsymbol{\alpha}_1 + \cdots + d_r\boldsymbol{\alpha}_r = -c_1\boldsymbol{\gamma}_1 - \cdots - c_{t-r}\boldsymbol{\gamma}_{t-r},$$

that is,

$$d_1\boldsymbol{\alpha}_1+\cdots+d_r\boldsymbol{\alpha}_r+c_1\boldsymbol{\gamma}_1+\cdots+c_{t-r}\boldsymbol{\gamma}_{t-r}=0,$$

where $\boldsymbol{\alpha}_1,\ldots,\boldsymbol{\alpha}_r,\boldsymbol{\gamma}_1,\ldots,\boldsymbol{\gamma}_{t-r}$ construct a basis of W_2. Hence, their coefficients must all be 0. All the coefficients in the *LHS* of ② become 0 when the above equation is substituted into the *RHS* of it. Vectors in ① are, therefore, linearly independent. □

The properties for the coordinates of a vector under basis B in n-dimensional linear space $V(F)$ are analogous to those of F^n. Main properties in $V(F)$ are listed below.

Property 4.6

The coordinates of a vector under a given basis are uniquely determined.

Property 4.7

The transition matrix defined by Definition 4.2 from basis A to basis B is invertible from Theorem 4.1.

Property 4.8

The formulas of the basis transformation and the coordinate transformation stated in Theorem 4.2 are all applicable in $V(F)$.

Furthermore, for a given basis $B=\{\boldsymbol{\beta}_1,\boldsymbol{\beta}_2,\ldots,\boldsymbol{\beta}_n\}$ of an n-dimensional linear space $V(F)$, vectors in $V(F)$ and their coordinates (vectors in F^n) not only enjoy a one-to-one correspondence but also are invariant under the two linear operations, that is,

$$\boldsymbol{\alpha}+\boldsymbol{\zeta} \text{ in } V(F) \text{ corresponds to } \boldsymbol{\alpha}_B+\boldsymbol{\zeta}_B \text{ in } F^n,$$

$$\lambda\boldsymbol{\alpha} \text{ in } V(F) \text{ corresponds to } \lambda\boldsymbol{\alpha}_B \text{ in } F^n.$$

Actually, if

$$\boldsymbol{\alpha}=x_1\boldsymbol{\beta}_1+x_2\boldsymbol{\beta}_2+\cdots+x_n\boldsymbol{\beta}_n,\ \boldsymbol{\zeta}=y_1\boldsymbol{\beta}_1+y_2\boldsymbol{\beta}_2+\cdots+y_n\boldsymbol{\beta}_n,\ \lambda\in F,$$

then

$$\boldsymbol{\alpha}+\boldsymbol{\zeta}=(x_1+y_1)\boldsymbol{\beta}_1+(x_2+y_2)\boldsymbol{\beta}_2+\cdots+(x_n+y_n)\boldsymbol{\beta}_n,$$
$$\lambda\boldsymbol{\alpha}=(\lambda x_1)\boldsymbol{\beta}_1+(\lambda x_2)\boldsymbol{\beta}_2+\cdots+(\lambda x_n)\boldsymbol{\beta}_n.$$

Hence,

$$(\boldsymbol{\alpha}+\boldsymbol{\zeta})_B=\boldsymbol{\alpha}_B+\boldsymbol{\zeta}_B,\quad (\lambda\boldsymbol{\alpha})_B=\lambda\boldsymbol{\alpha}_B.$$

Two linear spaces $V(F)$ and F^n are called **isomorphic** if they enjoy the above correspondence relations. The above correspondence relations indicate that the study of an n-dimensional linear space can be transferred to that of the n-dimensional vector space F^n under a pre-chosen basis of the linear space and the coordinates of its vectors with respect to that basis. Thus, a unified approach to explore all different n-dimensional linear spaces is to transfer them into the n-dimensional vector space F^n by means of a basis and then use the methodologies in F^n. Therefore, a linear space is usually called a **vector space** and elements of the linear space are also called **vectors**.

Example 4.15 A basis of $\mathbb{R}[x]_n$

Prove that $B=\{f_0,f_1,f_2,\ldots,f_{n-1}\}$ (where $f_0=1$, $f_1=x$, $f_2=x^2$, $\ldots$, $f_{n-1}=x^{n-1}$) is a basis of $\mathbb{R}[x]_n$, and find the coördinates of $p(x)=a_0+a_1x+\cdots+a_{n-1}x^{n-1}$ under the basis B.

Solution: The linear independence of B has been proven in Example 4.13. Now,

$$\forall p(x)=a_0+a_1x+\cdots+a_{n-1}x^{n-1}\in\mathbb{R}[x]_n,$$

we have

$$p(x)=a_0f_0+a_1f_1+a_2f_2+\cdots+a_{n-1}f_{n-1}. \quad ①$$

So, B is a basis of $\mathbb{R}[x]_n$ (sometimes called the natural basis) and $\mathbb{R}[x]_n$ is an n-dimensional real linear space. From ①, we get the coordinates of $p(x)$ under the basis B,

$$[p(x)]_B=(a_0,a_1,a_2,\ldots,a_{n-1})^{\mathrm{T}}.$$

In a matrix multiplication form, ① can be expressed as

$$p(x)=\left(1,x,\ldots,x^{n-1}\right)\begin{pmatrix}a_0\\a_1\\\vdots\\a_{n-1}\end{pmatrix}.$$

■

Example 4.16 Computation of a transition matrix

Let $B_1 = \{g_1, g_2, g_3\}$ and $B_2 = \{h_1, h_2, h_3\}$, where

$$g_1 = 1, \quad g_2 = -1 + x, \quad g_3 = 1 - x + x^2,$$

$$h_1 = 1 - x - x^2, \quad h_2 = 3x - 2x^2, \quad h_3 = 1 - 2x^2.$$

(i) Prove that B_1 and B_2 are bases of $\mathbb{R}[x]_3$.
(ii) Find the transition matrix from B_1 to B_2.
(iii) Given that $[p(x)]_{B_1} = (1, 4, 3)^{\mathrm{T}}$, find $[p(x)]_{B_2} = (y_1, y_2, y_3)^{\mathrm{T}}$.

Solution:

(i) Because $\dim \mathbb{R}[x]_3 = 3$, so we only need to prove that B_1 and B_2 are both linearly independent. Let each element in B_1 be linear combinations of the natural basis of $\mathbb{R}[x]_3$, $B = \{f_0, f_1, f_2\}$. Then,

$$\begin{cases} g_1 = 1 \cdot f_0, \\ g_2 = -1 \cdot f_0 + 1 \cdot f_1, \\ g_3 = 1 \cdot f_0 - 1 \cdot f_1 + 1 \cdot f_2. \end{cases} \qquad ①$$

Equation ① written in a matrix form is

$$(g_1, g_2, g_3) = (f_0, f_1, f_2) \begin{pmatrix} 1 & -1 & 1 \\ 0 & 1 & -1 \\ 0 & 0 & 1 \end{pmatrix}. \qquad ②$$

The determinant of the matrix on the *RHS* of equation ②

$$\boldsymbol{A} = \begin{pmatrix} 1 & -1 & 1 \\ 0 & 1 & -1 \\ 0 & 0 & 1 \end{pmatrix} \qquad ③$$

is nonzero. So, $B_1 = \{g_1, g_2, g_3\}$ is linearly independent according to Theorem 4.1. With a similar argument, $B_2 = \{h_1, h_2, h_3\}$ is also linearly independent.

(ii) Similar to equation ①, the relationship between B_2 and the natural basis B is

$$(h_1, h_2, h_3) = (f_0, f_1, f_2) \begin{pmatrix} 1 & 0 & 1 \\ -1 & 3 & 0 \\ -1 & -2 & -2 \end{pmatrix}. \qquad ④$$

Denote the matrix on the *RHS* of equation ④ by

$$\boldsymbol{P} = \begin{pmatrix} 1 & 0 & 1 \\ -1 & 3 & 0 \\ -1 & -2 & -2 \end{pmatrix}.$$

The foregoing matrices $\boldsymbol{A}$ and $\boldsymbol{P}$ are transition matrices from basis B to B_1 and B to B_2, respectively. As $\boldsymbol{A}$ is invertible, we obtain from equation ②

$$(f_0, f_1, f_2) = (g_1, g_2, g_3)\boldsymbol{A}^{-1}. \qquad ⑤$$

Substituting equation ⑤ into equation ④ yields

$$(h_1, h_2, h_3) = (g_1, g_2, g_3)(\boldsymbol{A}^{-1}\boldsymbol{P}).$$

Hence, the transition matrix from B_1 to B_2 is

$$\boldsymbol{C} = \boldsymbol{A}^{-1}\boldsymbol{P} = \begin{pmatrix} 1 & 1 & 0 \\ 0 & 1 & 1 \\ 0 & 0 & 1 \end{pmatrix} \begin{pmatrix} 1 & 0 & 1 \\ -1 & 3 & 0 \\ -1 & -2 & -2 \end{pmatrix} = \begin{pmatrix} 0 & 3 & 1 \\ -2 & 1 & -2 \\ -1 & -2 & -2 \end{pmatrix}.$$

(iii) Given that $[p(x)]_{B_1} = (1, 4, 3)^{\mathrm{T}}$,

$$\begin{pmatrix} y_1 \\ y_2 \\ y_3 \end{pmatrix} = \boldsymbol{C}^{-1} \begin{pmatrix} 1 \\ 4 \\ 3 \end{pmatrix} = \begin{pmatrix} 6 & -4 & 7 \\ 2 & -1 & 2 \\ -5 & 3 & -6 \end{pmatrix} \begin{pmatrix} 1 \\ 4 \\ 3 \end{pmatrix} = \begin{pmatrix} 11 \\ 4 \\ -11 \end{pmatrix}.$$

■

*4.6 Linear Transformation of Vector Space

First, we give the definition of a mapping.

Definition 4.17

Let X and Y be two nonempty sets. $\boldsymbol{\sigma}$ is said to be a **mapping** from X to Y if there is a uniquely defined element $\boldsymbol{\beta}$ in Y corresponding to an element $\boldsymbol{\alpha}$ in X, which is denoted as

$$\sigma : \alpha \in X \to \beta \in Y.$$

$\boldsymbol{\beta}$ is called the **image** of $\boldsymbol{\alpha}$ under $\boldsymbol{\sigma}$, and $\boldsymbol{\alpha}$ is called an **inverse image** of $\boldsymbol{\beta}$ under $\boldsymbol{\sigma}$. For convenience, it is denoted as

$$\sigma : \alpha \to \beta \quad \text{or} \quad \sigma(\alpha) = \beta.$$

Do note that the image of $\boldsymbol{\alpha}$ is unique, whereas the inverse image of $\boldsymbol{\beta}$ may or may not be so.

The mapping from X to itself is called a **transformation**.

If $\forall \boldsymbol{\alpha}_1, \boldsymbol{\alpha}_2 \in X$, $\boldsymbol{\alpha}_1 \neq \boldsymbol{\alpha}_2$ implies $\boldsymbol{\sigma}(\boldsymbol{\alpha}_1) \neq \boldsymbol{\sigma}(\boldsymbol{\alpha}_2)$, then $\boldsymbol{\sigma}$ is called an injective mapping or an **injection**.

If $\forall \boldsymbol{\beta} \in Y$, there is $\boldsymbol{\alpha} \in X$ such that $\sigma(\boldsymbol{\alpha}) = \boldsymbol{\beta}$, σ is called a **surjection.**

If σ is not only an injection but a surjection as well, then we call it a **bijective mapping** (or **one-to-one correspondence**).

Example 4.17 Difference between surjection and injection

$f(x) = \sin x$ is a mapping of $\mathbb{R} \to [-1, 1]$, it is a surjection but not an injection.
$f(x) = \mathrm{e}^x$ is a mapping of $\mathbb{R} \to \mathbb{R}$ and it is an injection but not a surjection. ■

The well-known unary linear function in the calculus,

$$y = f(x) = ax \tag{4.21}$$

(its general form $y = ax + b$ can be transformed to the above form by shifting) is a mapping of $\mathbb{R} \to \mathbb{R}$ (or $\mathbb{R}^1 \to \mathbb{R}^1$), it is obviously a bijective mapping if $a \neq 0$. The mapping has the following properties:

(i) $f(x_1 + x_2) = f(x_1) + f(x_2)$. (4.22)

(ii) $f(\lambda x) = \lambda f(x)$, λ is a constant.

Now, we try to generalize the unary linear function to an n-dimensional vector space. Let $\boldsymbol{A} \in \mathbb{R}^{n \times n}$. If the mapping

$$\sigma : \boldsymbol{x} \to \boldsymbol{A}\boldsymbol{x},$$

that is,

$$\sigma(\boldsymbol{x}) = \boldsymbol{A}\boldsymbol{x}$$

(a mapping of $\mathbb{R}^n \to \mathbb{R}^n$) satisfies the following properties:

(iii) $\sigma(\boldsymbol{x}_1 + \boldsymbol{x}_2) = \boldsymbol{A}(\boldsymbol{x}_1 + \boldsymbol{x}_2) = \boldsymbol{A}\boldsymbol{x}_1 + \boldsymbol{A}\boldsymbol{x}_2 = \sigma(\boldsymbol{x}_1) + \sigma(\boldsymbol{x}_2)$,

(iv) $\sigma(\lambda \boldsymbol{x}) = \boldsymbol{A}(\lambda \boldsymbol{x}) = \lambda \boldsymbol{A}\boldsymbol{x} = \lambda \sigma(\boldsymbol{x})$, $\lambda \in \mathbb{R}$,

for every column vector $\boldsymbol{x} \in \mathbb{R}^n$, then the mapping σ is called a **linear mapping** of $\mathbb{R}^n \to \mathbb{R}^n$ (also called a **linear transformation**).

More generally, if $\boldsymbol{A} \in \mathbb{R}^{m \times n}$, $\boldsymbol{x} \in \mathbb{R}^n$, then the mapping

$$\sigma : \boldsymbol{x} \to \boldsymbol{A}\boldsymbol{x} \in \mathbb{R}^m$$

is a **linear mapping** from $\mathbb{R}^n$ to $\mathbb{R}^m$. For instance, a binary linear function,

$$y = f(x_1, x_2) = a_1 x_1 + a_2 x_2 = (a_1, a_2) \begin{pmatrix} x_1 \\ x_2 \end{pmatrix},$$

is a linear mapping from $\mathbb{R}^2 \to \mathbb{R}^1$.

In this section, we mainly focus on the linear mapping of $\mathbb{R}^n \to \mathbb{R}^n$ (we also call it a linear transformation of $\mathbb{R}^n$). We first generalize the domain and range (both are real number sets) of the unary function (4.21) to the general vector space $V(F)$, with the properties in (4.22). This leads to the linear transformation that will be considered next.

4.6.1 Definition of linear transformation and its properties

Definition 4.18

Let $V(F)$ be a vector space. If a transformation $\boldsymbol{\sigma}$ of $V(F)$ satisfies:

(i) $\sigma(\boldsymbol{\alpha}+\boldsymbol{\beta}) = \sigma(\boldsymbol{\alpha}) + \sigma(\boldsymbol{\beta})$, (4.23)

(ii) $\sigma(\lambda\boldsymbol{\alpha}) = \lambda\sigma(\boldsymbol{\alpha})$,

$\forall \boldsymbol{\alpha}, \boldsymbol{\beta} \in V$ and $\lambda \in F$, then σ is called a linear transformation of $V(F)$, and $\sigma(\boldsymbol{\alpha})$ is called the image of $\boldsymbol{\alpha}$, $\boldsymbol{\alpha}$ an inverse image of $\sigma(\boldsymbol{\alpha})$.

Equation (4.23) is equivalently expressed as

$$\sigma(\lambda\boldsymbol{\alpha} + \mu\boldsymbol{\beta}) = \lambda\sigma(\boldsymbol{\alpha}) + \mu\sigma(\boldsymbol{\beta}) \tag{4.23$'$}$$

$\forall \boldsymbol{\alpha}, \boldsymbol{\beta} \in V$ and $\lambda, \mu \in F$.

Example 4.18 Rotation of vectors in $\mathbb{R}^2$

Rotation—the transformation $\mathbf{R}_\theta$ rotating the vectors in $\mathbb{R}^2$ (all the vectors emanating from the origin within the plane Oxy) counterclockwise through an angle θ around the origin (as shown in Figure 4.4) is a linear transformation in $\mathbb{R}^2$.

$$\mathbf{R}_\theta(x, y) = \mathbf{R}_\theta(\alpha) = \alpha' = (x', y'), \tag{4.24}$$

where $|\boldsymbol{\alpha}| = r$, and

$$x' = r\cos(\beta+\theta) = r\cos\beta\cos\theta - r\sin\beta\sin\theta = x\cos\theta - y\sin\theta,$$
$$y' = r\sin(\beta+\theta) = r\sin\beta\cos\theta + r\cos\beta\sin\theta = y\cos\theta + x\sin\theta.$$

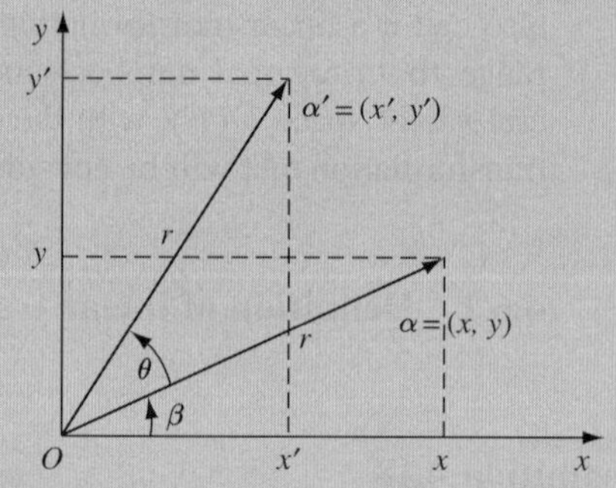

Figure 4.4 Transformation of $\mathbf{R}_\theta$ on vectors in $\mathbb{R}^2$

$\forall \boldsymbol{\alpha}_1 = (x_1, y_1), \boldsymbol{\alpha}_2 = (x_2, y_2) \in \mathbb{R}^2$ and $\forall \lambda, \mu \in \mathbb{R}$, by applying equation (4.24), yield

$$\begin{aligned}\mathbf{R}_\theta(\lambda\boldsymbol{\alpha}_1 + \mu\boldsymbol{\alpha}_2) &= \mathbf{R}_\theta(\lambda x_1 + \mu x_2, \lambda y_1 + \mu y_2)\\ &= ((\lambda x_1 + \mu x_2)\cos\theta - (\lambda y_1 + \mu y_2)\sin\theta,\\ &\quad (\lambda x_1 + \mu x_2)\sin\theta + (\lambda y_1 + \mu y_2)\cos\theta)\\ &= \lambda(x_1\cos\theta - y_1\sin\theta, x_1\sin\theta + y_1\cos\theta)\\ &\quad +\mu(x_2\cos\theta - y_2\sin\theta, x_2\sin\theta + y_2\cos\theta)\\ &= \lambda\mathbf{R}_\theta(x_1, y_1) + \mu\mathbf{R}_\theta(x_2, y_2)\\ &= \lambda\mathbf{R}_\theta(\boldsymbol{\alpha}_1) + \mu\mathbf{R}_\theta(\boldsymbol{\alpha}_2).\end{aligned}$$

So, $\mathbf{R}_\theta$ is a linear transformation of $\mathbb{R}^2$. ■

Example 4.19 Reflection of vectors in $\mathbb{R}^2$

The reflection (the involution transformation) φ that maps a vector in $\mathbb{R}^2$ to its reflection over a straight line L (known as a mirror) passing through the origin is a linear transformation of $\mathbb{R}^2$, namely (as shown in Figure 4.5, L is the perpendicular bisector of AB.)

$$\varphi(\boldsymbol{\alpha}) = \boldsymbol{\alpha}'.$$

Let $\boldsymbol{\omega}$ be a unit vector in a straight line L. Then,

$$\overrightarrow{OC} = (\boldsymbol{\alpha}, \boldsymbol{\omega})\boldsymbol{\omega},$$

where $(\boldsymbol{\alpha}, \boldsymbol{\omega})$ is the inner product of $\boldsymbol{\alpha}$ and $\boldsymbol{\omega}$. Thus,

$$\boldsymbol{\alpha}' = \boldsymbol{\alpha} + \overrightarrow{AB} = \boldsymbol{\alpha} + 2\overrightarrow{AC} = \boldsymbol{\alpha} + 2(\overrightarrow{OC} - \boldsymbol{\alpha}) = -\boldsymbol{\alpha} + 2(\boldsymbol{\alpha}, \boldsymbol{\omega})\boldsymbol{\omega}$$

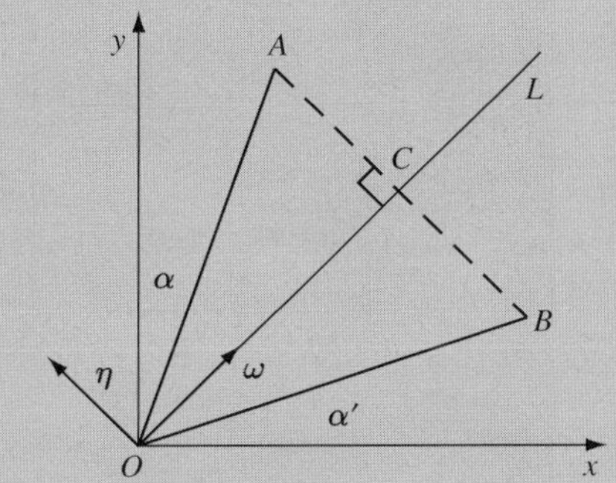

Figure 4.5 Transformation of φ on vectors in $\mathbb{R}^2$

and

$$\varphi(\boldsymbol{\alpha}) = -\boldsymbol{\alpha} + 2(\boldsymbol{\alpha}, \boldsymbol{\omega})\boldsymbol{\omega}. \tag{4.25}$$

Now, we need to verify that φ is a linear transformation. $\forall \boldsymbol{\alpha}_1, \boldsymbol{\alpha}_2 \in \mathbb{R}^2$ and $\forall \lambda, \mu \in \mathbb{R}$, we have

$$\begin{aligned}\varphi(\lambda\boldsymbol{\alpha}_1 + \mu\boldsymbol{\alpha}_2) &= -(\lambda\boldsymbol{\alpha}_1 + \mu\boldsymbol{\alpha}_2) + 2(\lambda\boldsymbol{\alpha}_1 + \mu\boldsymbol{\alpha}_2, \boldsymbol{\omega})\boldsymbol{\omega}\\ &= \lambda(-\boldsymbol{\alpha}_1 + 2(\boldsymbol{\alpha}_1, \boldsymbol{\omega})\boldsymbol{\omega}) + \mu(-\boldsymbol{\alpha}_2 + 2(\boldsymbol{\alpha}_2, \boldsymbol{\omega})\boldsymbol{\omega})\\ &= \lambda\varphi(\boldsymbol{\alpha}_1) + \mu\varphi(\boldsymbol{\alpha}_2).\end{aligned}$$

Therefore, reflection φ is a linear transformation of $\mathbb{R}^2$. ■

The linear transformations of the above two examples are injective. An injective linear transformation is also called monomorphism.

Example 4.20 Projection in $\mathbb{R}^3$ is a linear transformation

The projection $\mathbf{P}(\boldsymbol{\alpha}) = \boldsymbol{\beta}$ (see Figure 4.6) that projects a vector $\boldsymbol{\alpha} = (x_1, x_2, x_3)$ in $\mathbb{R}^3$ on plane Oxy to $\boldsymbol{\beta} = (x_1, x_2, 0)$, that is,

$$\mathbf{P}(x_1, x_2, x_3) = (x_1, x_2, 0) \tag{4.26}$$

is a linear transformation of $\mathbb{R}^3$. Readers can verify that

$$\forall \boldsymbol{\alpha}_1 = (a_1, b_1, c_1), \boldsymbol{\alpha}_2 = (a_2, b_2, c_2) \in \mathbb{R}^3, \forall k \in \mathbb{R},$$

$$\mathbf{P}(\boldsymbol{\alpha}_1 + \boldsymbol{\alpha}_2) = \mathbf{P}(\boldsymbol{\alpha}_1) + \mathbf{P}(\boldsymbol{\alpha}_2), \quad \mathbf{P}(k\boldsymbol{\alpha}_1) = k\mathbf{P}(\boldsymbol{\alpha}_1).$$

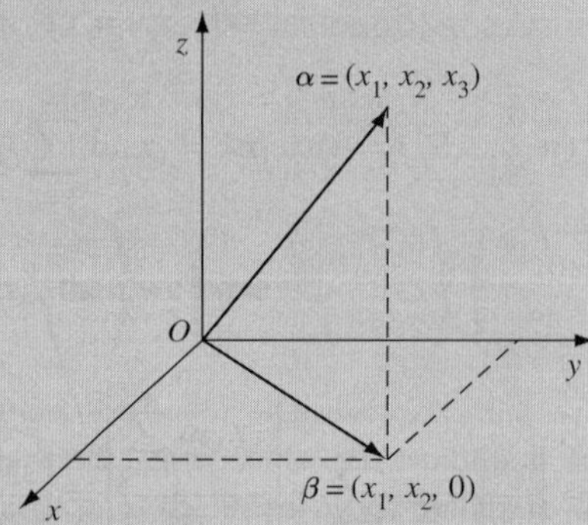

Figure 4.6 Projection of $\mathbf{P}(\boldsymbol{\alpha}) = \boldsymbol{\beta}$ in $\mathbb{R}^3$

Let $(x_1, x_2, x_3)^{\mathrm{T}}$ and $(y_1, y_2, y_3)^{\mathrm{T}}$ denote the coordinate vectors of $\boldsymbol{\alpha}$ and $\boldsymbol{\beta}$ under a basis, respectively. It is obvious that

$$\begin{pmatrix} y_1 \\ y_2 \\ y_3 \end{pmatrix} = \begin{pmatrix} 1 & & \\ & 1 & \\ & & 0 \end{pmatrix} \begin{pmatrix} x_1 \\ x_2 \\ x_3 \end{pmatrix}. \tag{4.27}$$

This is a matrix representation of projection. ■

Example 4.21 Some other transformations in $\mathbb{R}^n$

The following are transformations in $\mathbb{R}^n$

Identity transformation: $\boldsymbol{\sigma}(\boldsymbol{\alpha}) = \boldsymbol{\alpha}$, $(\forall \boldsymbol{\alpha} \in \mathbb{R}^n)$,

Zero transformation: $\boldsymbol{\sigma}(\boldsymbol{\alpha}) = \mathbf{0}$, $(\forall \boldsymbol{\alpha} \in \mathbb{R}^n)$,

Transformation of scalar multiplication: $\boldsymbol{\sigma}(\boldsymbol{\alpha}) = \lambda \boldsymbol{\alpha}$, $(\forall \boldsymbol{\alpha} \in \mathbb{R}^n$, λ a real number),

are all linear transformations of $\mathbb{R}^n$ (the proof is left to readers as exercises). ■

Example 4.22 A linear transformation in $\mathbb{R}^3$

If a transformation in $\mathbb{R}^3$ is given by

$$\boldsymbol{\sigma}(x_1, x_2, x_3) = (x_1 + x_2, x_2 - 4x_3, 2x_3),$$

then $\boldsymbol{\sigma}$ is a linear transformation of $\mathbb{R}^3$.

For any $\boldsymbol{\alpha} = (a_1, a_2, a_3)$, $\boldsymbol{\beta} = (b_1, b_2, b_3) \in \mathbb{R}^3$, we have

$$\begin{aligned}\sigma(\boldsymbol{\alpha}+\boldsymbol{\beta}) &= (a_1+b_1, a_2+b_2, a_3+b_3)\\ &= (a_1+a_2+b_1+b_2, a_2+b_2-4a_3-4b_3, 2a_3+2b_3)\\ &= (a_1+a_2, a_2-4a_3, 2a_3)+(b_1+b_2, b_2-4b_3, 2b_3)\\ &= \sigma(\boldsymbol{\alpha})+\sigma(\boldsymbol{\beta}).\end{aligned}$$

Similarly, for any $\boldsymbol{\alpha} \in \mathbb{R}^3$, $k \in \mathbb{R}$, we also have $\sigma(k\boldsymbol{\alpha}) = k\sigma(\boldsymbol{\alpha})$. Hence, σ is a linear transformation. ■

Example 4.23 Transformation in $\mathbb{R}^3$ which is not linear

If a transformation in $\mathbb{R}^3$ is given by

$$\sigma(x_1, x_2, x_3) = (x_1^2, x_2+x_3, x_2),$$

then σ is not a linear transformation.

There exist $\boldsymbol{\alpha} = (1, 0, 0)$ and $\boldsymbol{\beta} = (2, 0, 0)$ with,

$$\sigma(\boldsymbol{\alpha}+\boldsymbol{\beta}) = \sigma(3,0,0) = (9,0,0) \neq (1,0,0)+(4,0,0) = \sigma(\boldsymbol{\alpha})+\sigma(\boldsymbol{\beta})$$

So σ is not a linear transformation of $\mathbb{R}^3$. It can also be shown by verifying $\sigma(\lambda\boldsymbol{\alpha}) \neq \lambda\sigma(\boldsymbol{\alpha})$. ■

From Example 4.22, a transformation of $\mathbb{R}^n$

$$\sigma(x_1, x_2, \ldots, x_n) = (y_1, y_2, \ldots, y_n),$$

is a linear transformation of $\mathbb{R}^n$ when $y_i\,(i = 1, 2, \ldots, n)$ are all linear combinations of $x_1, x_2, \ldots, x_n$. In Example 4.23, an item $y_1 = x_1^2$ indicates it is not a linear transformation. With the same judgment, it is not a linear transformation if $y_1 = x_1x_2$.

Properties of linear transformations are listed below. A linear transformation σ of a vector space V over a field F has the following properties.

Property 4.9

$$\sigma(\mathbf{0}) = \mathbf{0},\ \sigma(-\boldsymbol{\alpha}) = -\sigma(\boldsymbol{\alpha})(\forall\boldsymbol{\alpha} \in V).$$

The property is true by letting $\lambda = 0, -1$ in $\sigma(\lambda\boldsymbol{\alpha}) = \lambda\sigma(\boldsymbol{\alpha})$.

Property 4.10

If $\alpha = k_1\alpha_1 + k_2\alpha_2 + \cdots + k_n\alpha_n$ $(k_i \in F,\ \alpha_i \in V,\ i = 1, 2, \ldots, n)$, then

$$\sigma(\alpha) = k_1\sigma(\alpha_1) + k_2\sigma(\alpha_2) + \cdots + k_n\sigma(\alpha_n).$$

Proof: By mathematical induction and the definition of linear transformation, we can get

$$\sigma(\beta_1 + \beta_2 + \cdots + \beta_n) = \sigma(\beta_1) + \sigma(\beta_2) + \cdots + \sigma(\beta_n),\ n \geq 2.$$

Then,

$$\begin{aligned}\sigma(\alpha) &= \sigma(k_1\alpha_1 + k_2\alpha_2 + \cdots + k_n\alpha_n)\\ &= \sigma(k_1\alpha_1) + \sigma(k_2\alpha_2) + \cdots + \sigma(k_n\alpha_n)\\ &= k_1\sigma(\alpha_1) + k_2\sigma(\alpha_2) + \cdots + k_n\sigma(\alpha_n).\end{aligned}$$ □

Property 4.11

If $\alpha_1, \alpha_2, \ldots, \alpha_n$ are linearly dependent, then their image vectors $\sigma(\alpha_1), \sigma(\alpha_2), \ldots, \sigma(\alpha_n)$ are also linearly dependent.

There exist $k_1, k_2, \ldots, k_n$, not all 0 such that

$$k_1\alpha_1 + k_2\alpha_2 + \cdots + k_n\alpha_n = \mathbf{0}.$$

Together with $\sigma(\mathbf{0}) = \mathbf{0}$, then $k_1\sigma(\alpha_1) + k_2\sigma(\alpha_2) + \cdots + k_n\sigma(\alpha_n) = \mathbf{0}$.

Notice that the converse of Property 4.11 is not true. For instance, in Example 4.20,

$$\alpha_1 = (1, 1, 2), \quad \alpha_2 = (2, 2, 2)$$

are linearly independent, while $\mathbf{P}(\alpha_1) = (1, 1, 0)$ and $\mathbf{P}(\alpha_2) = (2, 2, 0)$ are linearly dependent.

4.6.2 Matrix representation of a linear transformation

At the beginning of §4.6, we pointed out that the transformation of $\boldsymbol{x}$ into $\boldsymbol{Ax}$ by an nth-order $\boldsymbol{A}$ is a linear transformation of $\mathbb{R}^n$. We have also mentioned in Example 4.19 that the projection $\mathbf{P}(\alpha) = \beta$ is illustrated by a matrix form (4.27). The matrix

$$\boldsymbol{P} = \begin{pmatrix} 1 & & \\ & 1 & \\ & & 0 \end{pmatrix}$$

plays the role of the projection, that is, $\boldsymbol{Px}$ gives the projection vector of $\boldsymbol{x}$ in the plane Oxy. These suggest that there are certain relationships between a linear transformation σ in an n-dimensional vector space $V(F)$ and an nth-order matrix. We shall now focus on this relationship and represent a linear transformation by a matrix.

Let $\{\boldsymbol{\alpha}_1, \boldsymbol{\alpha}_2, \ldots, \boldsymbol{\alpha}_n\}$ be a basis of $V(F)$ and σ be a linear transformation of $V(F)$. If $\boldsymbol{\alpha} \in V(F)$ and

$$\boldsymbol{\alpha} = x_1\boldsymbol{\alpha}_1 + x_2\boldsymbol{\alpha}_2 + \cdots + x_n\boldsymbol{\alpha}_n, \tag{4.28}$$

then, from Property 4.10, we have

$$\sigma(\boldsymbol{\alpha}) = x_1\sigma(\boldsymbol{\alpha}_1) + x_2\sigma(\boldsymbol{\alpha}_2) + \cdots + x_n\sigma(\boldsymbol{\alpha}_n). \tag{4.29}$$

So we know the image $\sigma(\boldsymbol{\alpha})$ of any vector $\boldsymbol{\alpha}$ if images $\sigma(\boldsymbol{\alpha}_1), \sigma(\boldsymbol{\alpha}_2), \ldots, \sigma(\boldsymbol{\alpha}_n)$ are known. The following theorem further explains the effect that a linear transformation is determined by its image of a basis.

THEOREM 4.16

Let $\{\boldsymbol{\alpha}_1, \boldsymbol{\alpha}_2, \ldots, \boldsymbol{\alpha}_n\}$ be a basis of $V(F)$. If two linear transformations σ and τ of V(F) have the same image under this basis, that is,

$$\sigma(\boldsymbol{\alpha}_i) = \tau(\boldsymbol{\alpha}_i), \quad i = 1, 2, \ldots, n,$$

then $\sigma = \tau$.

Proof: $\sigma = \tau$ means that any vector has the same image under their mappings. Namely, $\sigma(\boldsymbol{\alpha}) = \tau(\boldsymbol{\alpha})$ for any $\boldsymbol{\alpha} \in V$. For a vector $\boldsymbol{\alpha}$ as shown in (4.28), we thus get

$$\begin{aligned}\sigma(\boldsymbol{\alpha}) &= x_1\sigma(\boldsymbol{\alpha}_1) + x_2\sigma(\boldsymbol{\alpha}_2) + \cdots + x_n\sigma(\boldsymbol{\alpha}_n)\\ &= x_1\tau(\boldsymbol{\alpha}_1) + x_2\tau(\boldsymbol{\alpha}_2) + \cdots + x_n\tau(\boldsymbol{\alpha}_n) = \tau(\boldsymbol{\alpha}).\end{aligned}$$

□

As $\sigma(\boldsymbol{\alpha}_i) \in V(F)(i = 1, 2, \ldots, n)$, so they can be written as a linear combination of the basis $\{\boldsymbol{\alpha}_1, \boldsymbol{\alpha}_2, \ldots, \boldsymbol{\alpha}_n\}$. That is,

$$\begin{cases}\sigma(\boldsymbol{\alpha}_1) = a_{11}\boldsymbol{\alpha}_1 + a_{21}\boldsymbol{\alpha}_2 + \cdots + a_{n1}\boldsymbol{\alpha}_n,\\ \sigma(\boldsymbol{\alpha}_2) = a_{12}\boldsymbol{\alpha}_1 + a_{22}\boldsymbol{\alpha}_2 + \cdots + a_{n2}\boldsymbol{\alpha}_n,\\ \cdots\cdots\cdots\cdots\cdots\cdots\\ \sigma(\boldsymbol{\alpha}_n) = a_{1n}\boldsymbol{\alpha}_1 + a_{2n}\boldsymbol{\alpha}_2 + \cdots + a_{nn}\boldsymbol{\alpha}_n.\end{cases} \tag{4.30}$$

Denote

$$\sigma(\boldsymbol{\alpha}_1, \boldsymbol{\alpha}_2, \ldots, \boldsymbol{\alpha}_n) = (\sigma(\boldsymbol{\alpha}_1), \sigma(\boldsymbol{\alpha}_2), \ldots, \sigma(\boldsymbol{\alpha}_n)). \tag{4.31}$$

Express (4.30) in matrix form

$$\sigma(\boldsymbol{\alpha}_1, \boldsymbol{\alpha}_2, \ldots, \boldsymbol{\alpha}_n) = (\boldsymbol{\alpha}_1, \boldsymbol{\alpha}_2, \ldots, \boldsymbol{\alpha}_n)\begin{pmatrix} a_{11} & a_{12} & \cdots & a_{1n} \\ a_{21} & a_{22} & \cdots & a_{2n} \\ \vdots & \vdots & & \vdots \\ a_{n1} & a_{n2} & \cdots & a_{nn} \end{pmatrix}, \tag{4.32}$$

or

$$\sigma(\boldsymbol{\alpha}_1, \boldsymbol{\alpha}_2, \ldots, \boldsymbol{\alpha}_n) = (\boldsymbol{\alpha}_1, \boldsymbol{\alpha}_2, \ldots, \boldsymbol{\alpha}_n)\, \boldsymbol{A}. \tag{4.32$'$}$$

Definition 4.19

A matrix **A** is called the **matrix representation** or the **(corresponding) matrix** of σ under a basis $\{\boldsymbol{\alpha}_1, \boldsymbol{\alpha}_2, \ldots, \boldsymbol{\alpha}_n\}$, if there exists a linear transformation σ in $V(F)$ such that the basis $\{\boldsymbol{\alpha}_1, \boldsymbol{\alpha}_2, \ldots, \boldsymbol{\alpha}_n\}$ of $V(F)$ and their images $\sigma(\boldsymbol{\alpha}_1), \sigma(\boldsymbol{\alpha}_2), \ldots, \sigma(\boldsymbol{\alpha}_n)$ have the relationship of (4.30) [that is, (4.32)].

Obviously, the matrix of σ is uniquely determined under a given basis. If $\boldsymbol{\alpha}_j$ and $\sigma(\boldsymbol{\alpha}_j)(j = 1, 2, \ldots, n)$ are vectors in $\mathbb{R}^n$ and in column form, then (4.32) is an equation of three nth-order matrices. In this case, we get the matrix $\boldsymbol{A}$ of (4.32) by computing $(\boldsymbol{\alpha}_1, \boldsymbol{\alpha}_2, \ldots, \boldsymbol{\alpha}_n)^{-1}\sigma(\boldsymbol{\alpha}_1, \boldsymbol{\alpha}_2, \ldots, \boldsymbol{\alpha}_n)$. How can we compute $\sigma(\boldsymbol{\alpha})$ when a matrix $\boldsymbol{A}$ and a basis are given?

THEOREM 4.17

Suppose **A** *is the matrix of a linear transformation* σ *of* $V(F)$ *under a basis* $\{\boldsymbol{\alpha}_1, \boldsymbol{\alpha}_2, \ldots, \boldsymbol{\alpha}_n\}$, *let* $\boldsymbol{x} = (x_1, x_2, \ldots, x_n)^{\mathrm{T}}$ *and* $\boldsymbol{y} = (y_1, y_2, \ldots, y_n)^{\mathrm{T}}$ *be the coordinate vectors of a vector* $\boldsymbol{\alpha}$ *and* $\sigma(\boldsymbol{\alpha})$ *under this basis, then*

$$\boldsymbol{y} = \boldsymbol{A}\boldsymbol{x}. \tag{4.33}$$

Proof:
As

$$\sigma(\boldsymbol{\alpha}_1, \boldsymbol{\alpha}_2, \ldots, \boldsymbol{\alpha}_n) = (\boldsymbol{\alpha}_1, \boldsymbol{\alpha}_2, \ldots, \boldsymbol{\alpha}_n)\boldsymbol{A} \qquad ①$$

$$\boldsymbol{\alpha} = x_1\boldsymbol{\alpha}_1 + x_2\boldsymbol{\alpha}_2 + \cdots + x_n\boldsymbol{\alpha}_n = (\boldsymbol{\alpha}_1, \boldsymbol{\alpha}_2, \ldots, \boldsymbol{\alpha}_n)\begin{pmatrix} x_1 \\ x_2 \\ \vdots \\ x_n \end{pmatrix},$$

then

$$\sigma(\boldsymbol{\alpha}) = x_1\sigma(\boldsymbol{\alpha}_1) + x_2\sigma(\boldsymbol{\alpha}_2) + \cdots + x_n\sigma(\boldsymbol{\alpha}_n)$$

$$= (\sigma(\boldsymbol{\alpha}_1), \sigma(\boldsymbol{\alpha}_2), \ldots, \sigma(\boldsymbol{\alpha}_n)) \begin{pmatrix} x_1 \\ x_2 \\ \vdots \\ x_n \end{pmatrix} \xlongequal{\text{(by ①)}} (\boldsymbol{\alpha}_1, \boldsymbol{\alpha}_2, \ldots, \boldsymbol{\alpha}_n)\boldsymbol{A} \begin{pmatrix} x_1 \\ x_2 \\ \vdots \\ x_n \end{pmatrix}.$$

So, the coordinate vector of $\sigma(\boldsymbol{\alpha})$ under $\{\boldsymbol{\alpha}_1, \boldsymbol{\alpha}_2, \ldots, \boldsymbol{\alpha}_n\}$ is

$$\begin{pmatrix} y_1 \\ y_2 \\ \vdots \\ y_n \end{pmatrix} = \boldsymbol{A} \begin{pmatrix} x_1 \\ x_2 \\ \vdots \\ x_n \end{pmatrix},$$

that is,

$$\boldsymbol{y} = \boldsymbol{A}\boldsymbol{x}. \qquad \square$$

Solving a nonhomogeneous system of linear equations $\boldsymbol{Ax} = \boldsymbol{y}$ in previous sections, that is, determining $\boldsymbol{x}$ for any given $\boldsymbol{y}$, is equivalent to finding an inverse image $\boldsymbol{\alpha}$ (its coordinate vector is $\boldsymbol{x}$) for a given image $\sigma(\boldsymbol{\alpha})$ (its coordinate vector is $\boldsymbol{y}$) of a given linear transformation σ (its corresponding matrix is $\boldsymbol{A}$).

Example 4.24 Matrix representation for a rotation transformation

Find a matrix representation for the rotation transformation $\mathbf{R}_\theta$ in Example 4.19 under the orthonormal basis $\boldsymbol{\varepsilon}_1 = (1, 0)^{\mathrm{T}}$, $\boldsymbol{\varepsilon}_2 = (0, 1)^{\mathrm{T}}$ of $\mathbb{R}^2$ (see Figure 4.7).

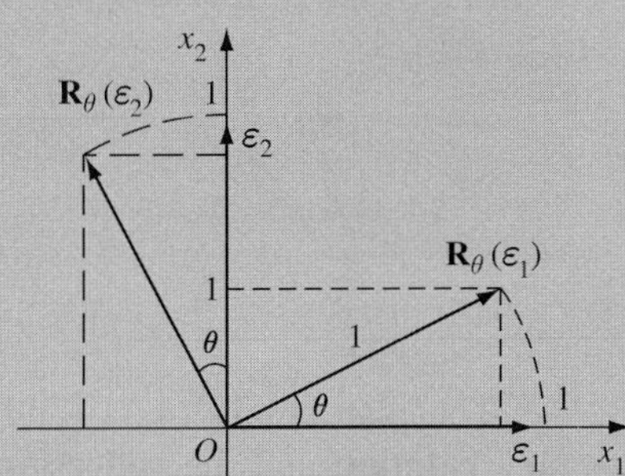

Figure 4.7 Rotation transformation $\mathbf{R}_\theta$ under the orthonormal basis of $\mathbb{R}^2$

Solution: Firstly, we find the images $\mathbf{R}_\theta(\boldsymbol{\varepsilon}_1), \mathbf{R}_\theta(\boldsymbol{\varepsilon}_2)$ of the basis vectors. Due to the fact that they are vectors with length 1, we have

$$\begin{cases} \mathbf{R}_\theta(\boldsymbol{\varepsilon}_1) = (\cos\theta)\boldsymbol{\varepsilon}_1 + (\sin\theta)\boldsymbol{\varepsilon}_2, \\ \mathbf{R}_\theta(\boldsymbol{\varepsilon}_2) = (-\sin\theta)\boldsymbol{\varepsilon}_1 + (\cos\theta)\boldsymbol{\varepsilon}_2, \end{cases}$$

that is,

$$\mathbf{R}_\theta(\boldsymbol{\varepsilon}_1, \boldsymbol{\varepsilon}_2) = (\mathbf{R}_\theta(\boldsymbol{\varepsilon}_1), \mathbf{R}_\theta(\boldsymbol{\varepsilon}_2)) = (\boldsymbol{\varepsilon}_1, \boldsymbol{\varepsilon}_2)\begin{pmatrix} \cos\theta & -\sin\theta \\ \sin\theta & \cos\theta \end{pmatrix}.$$

The matrix of the rotation transformation $\mathbf{R}_\theta$ under the orthonormal basis $\{\boldsymbol{\varepsilon}_1, \boldsymbol{\varepsilon}_2\}$ is

$$R = \begin{pmatrix} \cos\theta & -\sin\theta \\ \sin\theta & \cos\theta \end{pmatrix}. \tag{4.34}$$

■

Example 4.25 Matrix representation for a reflection transformation

Find the matrix H of the reflection φ in Example 4.20 under the orthonormal basis $\{\boldsymbol{\omega}, \boldsymbol{\eta}\}$ (see Figure 4.5) of $\mathbb{R}^3$.

Solution: According to the definition of the reflection, obviously we get

$$\begin{cases} \varphi(\boldsymbol{\omega}) = \boldsymbol{\omega}, \\ \varphi(\boldsymbol{\eta}) = -\boldsymbol{\eta}, \end{cases} \quad \text{that is,} \quad \varphi(\boldsymbol{\omega}, \boldsymbol{\eta}) = (\boldsymbol{\omega}, \boldsymbol{\eta})\begin{pmatrix} 1 & 0 \\ 0 & -1 \end{pmatrix}.$$

So, the matrix of φ under the orthonormal basis $\{\boldsymbol{\omega}, \boldsymbol{\eta}\}$ is

$$H = \begin{pmatrix} 1 & 0 \\ 0 & -1 \end{pmatrix}. \tag{4.35}$$

■

The matrices in (4.34) and (4.35) are respectively called the **elementary rotation** and the **reflection matrices**, which are both orthogonal. The corresponding elementary rotation transformation and mirror reflection transformation are called **orthogonal transformations**, under which the length of any vector remains invariant.

Example 4.26 Some other matrix representations of transformations

Under any basis, the matrices of identity transformation, zero transformation, and scalar multiplication transformation (see Example 4.21) of $\mathbb{R}^n$ are $\boldsymbol{I}_n, \boldsymbol{0}_{n\times n}, \lambda\boldsymbol{I}_n$ respectively (the proofs are left to readers as exercises). ■

Example 4.27 Computation and application of matrix of a transformation

Let σ be a linear transformation and $B=\{\boldsymbol{\alpha}_1,\boldsymbol{\alpha}_2,\boldsymbol{\alpha}_3\}$ be a basis of $\mathbb{R}^3$. Given that

$$\boldsymbol{\alpha}_1=(1,0,0)^{\mathrm{T}},\quad \boldsymbol{\alpha}_2=(1,1,0)^{\mathrm{T}},\quad \boldsymbol{\alpha}_3=(1,1,1)^{\mathrm{T}},$$

$$\sigma(\boldsymbol{\alpha}_1)=(1,-1,0)^{\mathrm{T}},\quad \sigma(\boldsymbol{\alpha}_2)=(-1,1,-1)^{\mathrm{T}},\quad \sigma(\boldsymbol{\alpha}_3)=(1,-1,2)^{\mathrm{T}}.$$

(i) Give the corresponding matrix of σ under the basis B.
(ii) Calculate $\sigma^2(\boldsymbol{\alpha}_1), \sigma^2(\boldsymbol{\alpha}_2)$, and $\sigma^2(\boldsymbol{\alpha}_3)$.
(iii) The coordinate vector of $\sigma(\boldsymbol{\beta})$ under the basis B is $(2,1,-2)^{\mathrm{T}}$; determine whether the inverse image $\boldsymbol{\beta}$ of $\sigma(\boldsymbol{\beta})$ is unique. Give the coordinates of $\boldsymbol{\beta}$ under the basis B.

Solution:

(i) **Method 1**: Substitute $\boldsymbol{\alpha}_1,\boldsymbol{\alpha}_2,\boldsymbol{\alpha}_3$ and $\sigma(\boldsymbol{\alpha}_1),\sigma(\boldsymbol{\alpha}_2),\sigma(\boldsymbol{\alpha}_3)$ into (4.32) as column vectors, then

$$\begin{pmatrix}1&-1&1\\-1&1&-1\\0&-1&2\end{pmatrix}=\begin{pmatrix}1&1&1\\0&1&1\\0&0&1\end{pmatrix}\boldsymbol{A}.$$

So, the matrix of σ under the basis B is

$$\boldsymbol{A}=\begin{pmatrix}1&1&1\\0&1&1\\0&0&1\end{pmatrix}^{-1}\begin{pmatrix}1&-1&1\\-1&1&-1\\0&-1&2\end{pmatrix}=\begin{pmatrix}2&-2&2\\-1&2&-3\\0&-1&2\end{pmatrix}.$$

Method 2: $\sigma(\boldsymbol{\alpha}_1),\sigma(\boldsymbol{\alpha}_2),\sigma(\boldsymbol{\alpha}_3)$ are written as a linear combination of the basis $B=\{\boldsymbol{\alpha}_1,\boldsymbol{\alpha}_2,\boldsymbol{\alpha}_3\}$,

$$\begin{cases}\sigma(\boldsymbol{\alpha}_1)=2\boldsymbol{\alpha}_1-\boldsymbol{\alpha}_2,\\ \sigma(\boldsymbol{\alpha}_2)=-2\boldsymbol{\alpha}_1+2\boldsymbol{\alpha}_2-\boldsymbol{\alpha}_3,\\ \sigma(\boldsymbol{\alpha}_3)=2\boldsymbol{\alpha}_1-3\boldsymbol{\alpha}_2+2\boldsymbol{\alpha}_3.\end{cases}$$

The transpose of the coefficient matrix on the *RHS* of the above equation system is the matrix of σ under basis B.

(ii) **Method 1**: First, we get $(\sigma(\alpha_1))_B = (x_1, x_2, x_3)^{\mathrm{T}} = (2, -1, 0)^{\mathrm{T}}$. Then, the coordinate vector of $\sigma^2(\alpha) = \sigma(\sigma(\alpha))$ under the basis B from (4.33) is

$$\begin{pmatrix} y_1 \\ y_2 \\ y_3 \end{pmatrix} = \begin{pmatrix} 2 & -2 & 2 \\ -1 & 2 & -3 \\ 0 & -1 & 2 \end{pmatrix} \begin{pmatrix} 2 \\ -1 \\ 0 \end{pmatrix} = \begin{pmatrix} 6 \\ -4 \\ 1 \end{pmatrix}.$$

So, $\sigma^2(\alpha_1) = 6\alpha_1 - 4\alpha_2 + \alpha_3 = (3, -3, 1)^{\mathrm{T}}$. We can get $\sigma^2(\alpha_2)$ and $\sigma^2(\alpha_3)$ similarly.

Method 2: According to

$$\sigma(\alpha_1, \alpha_2, \alpha_3) = [\sigma(\alpha_1), \sigma(\alpha_2), \sigma(\alpha_3)] = (\alpha_1, \alpha_2, \alpha_3)A,$$

Applying σ to both sides of the expression yields

$$\begin{aligned}(\sigma^2(\alpha_1), \sigma^2(\alpha_2), \sigma^2(\alpha_3)) &= \sigma((\alpha_1, \alpha_2, \alpha_3)A) = (\sigma(\alpha_1, \alpha_2, \alpha_3))A \\ &= (\alpha_1, \alpha_2, \alpha_3)A^2 \\ &= (\alpha_1, \alpha_2, \alpha_3) \begin{pmatrix} 6 & -10 & 14 \\ -4 & 9 & -14 \\ 1 & -4 & 7 \end{pmatrix},\end{aligned}$$

which provides

$$\begin{aligned}\sigma^2(\alpha_1) &= \sigma(\sigma(\alpha_1)) = 6\alpha_1 - 4\alpha_2 + \alpha_3 = (3, -3, 1)^{\mathrm{T}}, \\ \sigma^2(\alpha_2) &= \sigma(\sigma(\alpha_2)) = -10\alpha_1 + 9\alpha_2 - 4\alpha_3 = (-5, 5, -4)^{\mathrm{T}}, \\ \sigma^2(\alpha_3) &= \sigma(\sigma(\alpha_3)) = 14\alpha_1 - 14\alpha_2 + 7\alpha_3 = (7, -7, 7)^{\mathrm{T}}.\end{aligned}$$

(iii) Let $(\boldsymbol{\beta})_B = (x_1, x_2, x_3)^{\mathrm{T}}$. By virtue of (4.33), we get

$$\begin{pmatrix} 2 & -2 & 2 \\ -1 & 2 & -3 \\ 0 & -1 & 2 \end{pmatrix} \begin{pmatrix} x_1 \\ x_2 \\ x_3 \end{pmatrix} = \begin{pmatrix} 2 \\ 1 \\ -2 \end{pmatrix}$$

and a solution of the linear equations is

$$(x_1, x_2, x_3)^{\mathrm{T}} = (3, 2, 0)^{\mathrm{T}} + k(1, 2, 1)^{\mathrm{T}},$$

where k is an arbitrary constant. So the inverse image $\boldsymbol{\beta}$ of $\sigma(\boldsymbol{\beta})$ is not unique. ■

The matrix representation of a linear transformation is connected to a basis of the space. There are generally different matrix representations of a linear transformation under different bases. We now show the relationship between matrices of a linear transformation under different bases.

THEOREM 4.18

Suppose $\boldsymbol{A}$ and $\boldsymbol{B}$ are the matrices of a linear transformation σ under two bases $B_1=\{\boldsymbol{\alpha}_1, \boldsymbol{\alpha}_2, \ldots, \boldsymbol{\alpha}_n\}$ and $B_2=\{\boldsymbol{\beta}_1, \boldsymbol{\beta}_2, \ldots, \boldsymbol{\beta}_n\}$, and $\boldsymbol{C}$ is the transition matrix from basis B_1 to B_2. Then,

$$\boldsymbol{B}=\boldsymbol{C}^{-1}\boldsymbol{A}\boldsymbol{C}. \tag{4.36}$$

Proof:
As

$$\sigma(\boldsymbol{\alpha}_1, \boldsymbol{\alpha}_2, \ldots, \boldsymbol{\alpha}_n)=(\boldsymbol{\alpha}_1, \boldsymbol{\alpha}_2, \ldots, \boldsymbol{\alpha}_n)\boldsymbol{A}, \qquad ①$$

$$(\boldsymbol{\beta}_1, \boldsymbol{\beta}_2, \ldots, \boldsymbol{\beta}_n)=(\boldsymbol{\alpha}_1, \boldsymbol{\alpha}_2, \ldots, \boldsymbol{\alpha}_n)\boldsymbol{C}, \qquad ②$$

we get from ②

$$(\boldsymbol{\alpha}_1, \boldsymbol{\alpha}_2, \ldots, \boldsymbol{\alpha}_n)=(\boldsymbol{\beta}_1, \boldsymbol{\beta}_2, \ldots, \boldsymbol{\beta}_n)\boldsymbol{C}^{-1}, \qquad ③$$

then

$$\begin{aligned}\sigma(\boldsymbol{\beta}_1, \boldsymbol{\beta}_2, \ldots, \boldsymbol{\beta}_n)&=\sigma(\boldsymbol{\alpha}_1, \boldsymbol{\alpha}_2, \ldots, \boldsymbol{\alpha}_n)\boldsymbol{C}\\ &=(\boldsymbol{\alpha}_1, \boldsymbol{\alpha}_2, \ldots, \boldsymbol{\alpha}_n)\boldsymbol{A}\boldsymbol{C} \text{ (from ①)}\\ &=(\boldsymbol{\beta}_1, \boldsymbol{\beta}_2, \ldots, \boldsymbol{\beta}_n)\boldsymbol{C}^{-1}\boldsymbol{A}\boldsymbol{C} \text{ (from ②)}.\end{aligned}$$

Hence,

$$\boldsymbol{B}=\boldsymbol{C}^{-1}\boldsymbol{A}\boldsymbol{C}. \qquad \square$$

Theorem 4.18 indicates that matrices $\boldsymbol{A}$ and $\boldsymbol{B}$ of a same linear transformation under different bases are similar (see Definition 5.3 in Chapter 5). On the other hand, $\boldsymbol{B}$ must be the matrix of σ under another basis if two matrices $\boldsymbol{A}$ and $\boldsymbol{B}$ enjoy the relation (4.36), that is, they are similar and $\boldsymbol{A}$ is the matrix of σ under a basis $\boldsymbol{\alpha}_1, \boldsymbol{\alpha}_2, \ldots, \boldsymbol{\alpha}_n$ (readers are encouraged to compute the basis on their own).

Example 4.28 Matrix of linear transformation under multiple bases

The matrix of a linear transformation σ of $\mathbb{R}^3$ under the natural basis $(\boldsymbol{\varepsilon}_1, \boldsymbol{\varepsilon}_2, \boldsymbol{\varepsilon}_3)$ is

$$A = \begin{pmatrix} 2 & -1 & -1 \\ -1 & 2 & -1 \\ -1 & -1 & 2 \end{pmatrix}.$$

(i) Find the matrix of σ under a basis $\{\boldsymbol{\beta}_1, \boldsymbol{\beta}_2, \boldsymbol{\beta}_3\}$, where $\boldsymbol{\beta}_1 = (1, 1, 1,)^{\mathrm{T}}$, $\boldsymbol{\beta}_2 = (-1, 1, 0)^{\mathrm{T}}$, and $\boldsymbol{\beta}_3 = (-1, 0, 1)^{\mathrm{T}}$.

(ii) For $\boldsymbol{\alpha} = (1, 2, 3)^{\mathrm{T}}$, find the coordinate $(y_1, y_2, y_3)^{\mathrm{T}}$ of $\sigma(\boldsymbol{\alpha})$ under the basis $\{\boldsymbol{\beta}_1, \boldsymbol{\beta}_2, \boldsymbol{\beta}_3\}$ and find $\sigma(\boldsymbol{\alpha})$.

Solution:

(i) First, we find the transition matrix $\boldsymbol{C}$ from the natural basis $\{\boldsymbol{\varepsilon}_1, \boldsymbol{\varepsilon}_2, \boldsymbol{\varepsilon}_3\}$ to $\{\boldsymbol{\beta}_1, \boldsymbol{\beta}_2, \boldsymbol{\beta}_3\}$.

$$(\boldsymbol{\beta}_1, \boldsymbol{\beta}_2, \boldsymbol{\beta}_3) = (\boldsymbol{\varepsilon}_1, \boldsymbol{\varepsilon}_2, \boldsymbol{\varepsilon}_3)\boldsymbol{C},$$

yields

$$\boldsymbol{C} = (\boldsymbol{\beta}_1, \boldsymbol{\beta}_2, \boldsymbol{\beta}_3) = \begin{pmatrix} 1 & -1 & -1 \\ 1 & 1 & 0 \\ 1 & 0 & 1 \end{pmatrix},$$

$$\boldsymbol{C}^{-1} = \frac{1}{3}\begin{pmatrix} 1 & 1 & 1 \\ -1 & 2 & -1 \\ -1 & -1 & 2 \end{pmatrix}.$$

Then, from (4.36) in Theorem 4.18, the matrix of σ under the basis $\{\boldsymbol{\beta}_1, \boldsymbol{\beta}_2, \boldsymbol{\beta}_3\}$ is

$$\boldsymbol{B} = \boldsymbol{C}^{-1}\boldsymbol{A}\boldsymbol{C} = \frac{1}{3}\begin{pmatrix} 1 & 1 & 1 \\ -1 & 2 & -1 \\ -1 & -1 & 2 \end{pmatrix}\begin{pmatrix} 2 & -1 & -1 \\ -1 & 2 & -1 \\ -1 & -1 & 2 \end{pmatrix}\begin{pmatrix} 1 & -1 & -1 \\ 1 & 1 & 0 \\ 1 & 0 & 1 \end{pmatrix}$$

$$= \begin{pmatrix} 0 & 0 & 0 \\ 0 & 3 & 0 \\ 0 & 0 & 3 \end{pmatrix}.$$

(ii) First, find the coordinate vector $(x_1, x_2, x_3)^{\mathrm{T}}$ of $\boldsymbol{\alpha}$ under a basis $\{\boldsymbol{\beta}_1, \boldsymbol{\beta}_2, \boldsymbol{\beta}_3\}$. Because the coordinate vector of $\boldsymbol{\alpha}$ under the natural basis $\{\boldsymbol{\varepsilon}_1, \boldsymbol{\varepsilon}_2, \boldsymbol{\varepsilon}_3\}$ is itself, that is, $(1, 2, 3)^{\mathrm{T}}$, then from the coordinate transformation formula

$$\begin{pmatrix} x_1 \\ x_2 \\ x_3 \end{pmatrix} = \boldsymbol{C}^{-1}\begin{pmatrix} 1 \\ 2 \\ 3 \end{pmatrix} = \frac{1}{3}\begin{pmatrix} 1 & 1 & 1 \\ -1 & 2 & -1 \\ -1 & -1 & 2 \end{pmatrix}\begin{pmatrix} 1 \\ 2 \\ 3 \end{pmatrix} = \begin{pmatrix} 2 \\ 0 \\ 1 \end{pmatrix}.$$

According to (4.33) in Theorem 4.17, the coordinate vector of $\boldsymbol{\sigma}(\boldsymbol{\alpha})$ under the basis $\{\boldsymbol{\beta}_1, \boldsymbol{\beta}_2, \boldsymbol{\beta}_3\}$ is

$$\begin{pmatrix} y_1 \\ y_2 \\ y_3 \end{pmatrix} = \boldsymbol{B} \begin{pmatrix} x_1 \\ x_2 \\ x_3 \end{pmatrix} = \begin{pmatrix} 0 & 0 & 0 \\ 0 & 3 & 0 \\ 0 & 0 & 3 \end{pmatrix} \begin{pmatrix} 2 \\ 0 \\ 1 \end{pmatrix} = \begin{pmatrix} 0 \\ 0 \\ 3 \end{pmatrix}.$$

So, $\boldsymbol{\sigma}(\boldsymbol{\alpha}) = 0\boldsymbol{\beta}_1 + 0\boldsymbol{\beta}_2 + 3\boldsymbol{\beta}_3 = (-3, 0, 3)^{\mathrm{T}}$. ■

To arrive at a conclusion from the above discussion, a linear transformation $\boldsymbol{\sigma}$ is solely determined by the image $\boldsymbol{\sigma}(\boldsymbol{\alpha}_1), \boldsymbol{\sigma}(\boldsymbol{\alpha}_2), \ldots, \boldsymbol{\sigma}(\boldsymbol{\alpha}_n)$ for a basis $\{\boldsymbol{\alpha}_1, \boldsymbol{\alpha}_2, \ldots, \boldsymbol{\alpha}_n\}$ of $\mathbb{R}^n$ and $\boldsymbol{\sigma}$ is uniquely represented by a matrix under this basis. We now turn to its reverse problem: Is any nth-order matrix $\boldsymbol{A}$ a matrix of certain linear transformation under a given basis? From (4.32),

$$\begin{aligned} &(\boldsymbol{\sigma}(\boldsymbol{\alpha}_1), \boldsymbol{\sigma}(\boldsymbol{\alpha}_2), \ldots, \boldsymbol{\sigma}(\boldsymbol{\alpha}_n)) \\ &= (\boldsymbol{\alpha}_1, \boldsymbol{\alpha}_2, \ldots, \boldsymbol{\alpha}_n) \begin{pmatrix} a_{11} & a_{12} & \cdots & a_{1n} \\ a_{21} & a_{22} & \cdots & a_{2n} \\ \vdots & \vdots & & \vdots \\ a_{n1} & a_{n2} & \cdots & a_{nn} \end{pmatrix} \overset{\text{say}}{=\!=} (\boldsymbol{\beta}_1, \boldsymbol{\beta}_2, \ldots, \boldsymbol{\beta}_n), \end{aligned}$$

(where $\boldsymbol{\beta}_j = \sum_{i=1}^{n} a_{ij}\boldsymbol{\alpha}_i$, $j = 1, 2, \ldots, n$), we see that vectors $\boldsymbol{\beta}_1, \boldsymbol{\beta}_2, \ldots, \boldsymbol{\beta}_n$ in $\mathbb{R}^n$ and a matrix $\boldsymbol{A}$ are in one-to-one correspondence for a given basis $\{\boldsymbol{\alpha}_1, \boldsymbol{\alpha}_2, \ldots, \boldsymbol{\alpha}_n\}$ of $\mathbb{R}^n$. So an inverse problem is stated as: Is there a unique linear transformation $\boldsymbol{\sigma}$ such that $\boldsymbol{\sigma}(\boldsymbol{\alpha}_j) = \boldsymbol{\beta}_j$, $j = 1, 2, \ldots, n$, for a given basis $\{\boldsymbol{\alpha}_1, \boldsymbol{\alpha}_2, \ldots, \boldsymbol{\alpha}_n\}$ of $\mathbb{R}^n$ and n vectors $\boldsymbol{\beta}_1, \boldsymbol{\beta}_2, \ldots, \boldsymbol{\beta}_n$? The following theorem gives an affirmative answer to this question.

THEOREM 4.19

Let $\{\boldsymbol{\alpha}_1, \boldsymbol{\alpha}_2, \ldots, \boldsymbol{\alpha}_n\}$ be a basis of $\mathbb{R}^n$ and $\boldsymbol{\beta}_1, \boldsymbol{\beta}_2, \ldots, \boldsymbol{\beta}_n$ be n vectors in $\mathbb{R}^n$. Then, there exists a unique linear transformation $\boldsymbol{\sigma}$ such that

$$\boldsymbol{\sigma}(\boldsymbol{\alpha}_j) = \boldsymbol{\beta}_j, \; j = 1, 2, \ldots, n. \tag{4.37}$$

Proof: The existence can be proved by firstly constructing a transformation that satisfies (4.37), and then establishing its linearity.
For

$$\boldsymbol{\zeta} = x_1\boldsymbol{\alpha}_1 + x_2\boldsymbol{\alpha}_2 + \cdots + x_n\boldsymbol{\alpha}_n \; (x_j \in \mathbb{R}, \; j = 1, 2, \ldots, n),$$

define a transformation

$$\boldsymbol{\sigma}(\boldsymbol{\zeta}) = x_1\boldsymbol{\beta}_1 + x_2\boldsymbol{\beta}_2 + \cdots + x_n\boldsymbol{\beta}_n. \qquad ①$$

It is obvious that

$$\sigma(\boldsymbol{\alpha}_j)=\boldsymbol{\beta}_j,\ j=1,2,\ldots,n, \qquad ②$$

when

$$\boldsymbol{\zeta}=\boldsymbol{\alpha}_j=0\boldsymbol{\alpha}_1+\cdots+0\boldsymbol{\alpha}_{j-1}+1\boldsymbol{\alpha}_j+0\boldsymbol{\alpha}_{j+1}+\cdots+0\boldsymbol{\alpha}_n.$$

For any two given vectors $\boldsymbol{\zeta}_1=\sum\limits_{j=1}^{n}a_j\boldsymbol{\alpha}_j$ and $\boldsymbol{\zeta}_2=\sum\limits_{j=1}^{n}b_j\boldsymbol{\alpha}_j$ in $\mathbb{R}^n$ and any $k\in\mathbb{R}$, we have

$$\begin{aligned}\sigma(\boldsymbol{\zeta}_1+\boldsymbol{\zeta}_2)&=\sigma\Big(\sum_{j=1}^{n}(a_j+b_j)\boldsymbol{\alpha}_j\Big)\\&=\sum_{j=1}^{n}(a_j+b_j)\boldsymbol{\beta}_j=\sum_{j=1}^{n}a_j\boldsymbol{\beta}_j+\sum_{j=1}^{n}b_j\boldsymbol{\beta}\ \text{(from ①)}\\&=\sigma(\boldsymbol{\zeta}_1)+\sigma(\boldsymbol{\zeta}_2),\ \text{(from ①)} \qquad ③\end{aligned}$$

$$\sigma(k\boldsymbol{\zeta})=\sigma\left(\sum_{j=1}^{n}kx_j\boldsymbol{\alpha}_j\right)=\sum_{j=1}^{n}kx_j\boldsymbol{\beta}_j=k\sum_{j=1}^{n}x_j\boldsymbol{\beta}_j=k\sigma(\boldsymbol{\zeta}).\ \text{(from ①)} \qquad ④$$

②, ③, and ④ show that the transformation σ defined by ① is a linear transformation. The existence thus holds and the uniqueness is guaranteed by Theorem 4.16. □

One conclusion is that a linear transformation and a matrix are in one-to-one correspondence under a pre-chosen basis of $\mathbb{R}^n$.

4.6.3 Operations of linear transformations

Definition 4.20

Let σ and τ be two linear transformations of a linear space $V(F)$, $\lambda\in F$. The sum $(\sigma+\tau)$ of σ and τ, the scalar multiplication $\lambda\sigma$ of σ by λ, and the multiplication $\sigma\tau$ of σ by τ are defined as follows, respectively:

$\forall\boldsymbol{\alpha}\in V$,

$$(\sigma+\tau)(\boldsymbol{\alpha})=\sigma(\boldsymbol{\alpha})+\tau(\boldsymbol{\alpha}), \tag{4.38}$$

$$(\lambda\sigma)(\boldsymbol{\alpha})=\lambda\sigma(\boldsymbol{\alpha}), \tag{4.39}$$

$$(\sigma\tau)(\boldsymbol{\alpha})=\sigma(\tau(\boldsymbol{\alpha})). \tag{4.40}$$

$(\sigma+\tau)$, $\lambda\sigma$ and $\sigma\tau$ defined above are still linear transformations in $V(F)$, which can be verified from the definition of a linear transformation. To prove that $\sigma\tau$ is a linear transformation, $\forall \alpha_1, \alpha_2 \in V$, $k_1, k_2 \in F$, we have

$$\begin{aligned}(\sigma\tau)(k_1\alpha_1+k_2\alpha_2) &= \sigma\tau(k_1\alpha_1+k_2\alpha_2) = \sigma(k_1\tau(\alpha_1)+k_2\tau(\alpha_2)) \\ &= k_1\sigma(\tau(\alpha_1))+k_2\sigma(\tau(\alpha_2)) \\ &= k_1(\sigma\tau)(\alpha_1)+k_2(\sigma\tau)(\alpha_2).\end{aligned}$$

So, $\sigma\tau$ is also a linear transformation. The proofs of $(\sigma+\tau)$ and $\lambda\sigma$ being linear transformations are left as exercises for readers.

In addition, it is easily proved that the addition and multiplication of the linear transformations satisfy the associative law. That is, for any linear transformations σ, τ, and φ, on a linear space $V(F)$

$$(\sigma+\tau)+\varphi = \sigma+(\tau+\varphi); \quad (\sigma\tau)\varphi = \sigma(\tau\varphi).$$

The addition of transformations also satisfies the commutative law, that is,

$$\sigma+\tau = \tau+\sigma.$$

However the multiplication operation generally does not satisfy the commutative law, namely,

$$\sigma\tau \neq \tau\sigma.$$

The multiplication satisfies left distributive law and right distributive law with respect to addition, that is,

$$\sigma(\tau+\varphi) = \sigma\tau+\sigma\varphi; \quad (\tau+\varphi)\sigma = \tau\sigma+\varphi\sigma.$$

We know that the linear transformations of $V(F)$ are in one-to-one correspondence with the elements in $F^{n\times n}$ (i.e., nth-order matrix over number field F) when a basis $\{\alpha_1, \alpha_2, \ldots, \alpha_n\}$ of $V(F)$ is given. This correspondence remains true for operations as stated in the following theorem.

THEOREM 4.20

Let $\mathbf{A}$ and $\mathbf{B}$ be matrices of linear transformations σ and τ in a linear space $V(F)$ under a basis $\{\alpha_1, \alpha_2, \ldots, \alpha_n\}$ of V, respectively. Then the matrices of $(\sigma+\tau)$, $\lambda\sigma$, and $\sigma\tau$ are $\mathbf{A}+\mathbf{B}$, $\lambda\mathbf{A}$, and $\mathbf{AB}$, respectively.

Proof: Let $\boldsymbol{A}=(a_{ij})_{n\times n}$, $\boldsymbol{B}=(b_{ij})_{n\times n}$. According to (4.32), we have

$$\sigma(\boldsymbol{\alpha}_j)=\sum_{i=1}^{n} a_{ij}\boldsymbol{\alpha}_i, \quad \tau(\boldsymbol{\alpha}_j)=\sum_{i=1}^{n} b_{ij}\boldsymbol{\alpha}_i,\ j=1,2,\ldots,n.$$

Thus,

$$\begin{aligned}(\sigma+\tau)(\boldsymbol{\alpha}_j)&=\sigma(\boldsymbol{\alpha}_j)+\tau(\boldsymbol{\alpha}_j)=\sum_{i=1}^{n} a_{ij}\boldsymbol{\alpha}_i+\sum_{i=1}^{n} b_{ij}\boldsymbol{\alpha}_i\\&=\sum_{i=1}^{n}(a_{ij}+b_{ij})\boldsymbol{\alpha}_i,\ j=1,2,\ldots,n.\end{aligned}$$

This implies that the row i column j entry of the matrix of $(\sigma+\tau)$ is $a_{ij}+b_{ij}=$ the row i column j entry of $(\boldsymbol{A}+\boldsymbol{B})$. So the matrix of $(\sigma+\tau)$ is $(\boldsymbol{A}+\boldsymbol{B})$. Moreover,

$$\begin{aligned}(\sigma\tau)(\boldsymbol{\alpha}_j)&=\sigma(\tau(\boldsymbol{\alpha}_j))=\sigma\left(\sum_{i=1}^{n} b_{ij}\boldsymbol{\alpha}_i\right)=\sum_{i=1}^{n} b_{ij}\sigma(\boldsymbol{\alpha}_i)\\&=\sum_{i=1}^{n} b_{ij}\left(\sum_{k=1}^{n} a_{ki}\boldsymbol{\alpha}_k\right)\\&=\sum_{k=1}^{n}\left(\sum_{i=1}^{n} a_{ki}b_{ij}\right)\boldsymbol{\alpha}_k\\&=\sum_{k=1}^{n} c_{kj}\boldsymbol{\alpha}_k.\end{aligned}$$

So, the row k column j entry of the matrix of $\sigma\tau$ is $c_{kj}=\sum_{i=1}^{n} a_{ki}b_{ij}=(\boldsymbol{AB})_{kj}$ (i.e., the row k, column j entry of $\boldsymbol{AB}$). Therefore, the matrix of $\sigma\tau$ is $\boldsymbol{AB}$. □

Similarly, the matrix of $\lambda\sigma$ is $\lambda\boldsymbol{A}$ (the proof is left as exercises for readers).

If the matrix $\boldsymbol{A}$ of a linear transformation σ is invertible, then σ is called an **invertible linear transformation**. An invertible σ may also be defined as: There exists a linear transformation τ such that

$$\sigma\tau=\tau\sigma=\boldsymbol{I} \quad \text{(identity transformation).}$$

Because a linear transformation and a matrix are in one-to-one correspondence under a given basis, we can therefore use matrices to study linear transformations and vice versa.

4.6.4 Image (range) and kernel of a linear transformation

Definition 4.21

For a linear transformation σ of a linear space $V(F)$, the set generated by the image of V under σ,

$$\sigma(V)=\{\boldsymbol{\beta}|\boldsymbol{\beta}=\sigma(\boldsymbol{\alpha}),\boldsymbol{\alpha}\in V\} \tag{4.41}$$

is called the **image** (or range) of σ, the set of all inverse-images of the zero element $\boldsymbol{\theta}$ in V

$$\sigma^{-1}(\boldsymbol{\theta})=\{\boldsymbol{\alpha}|\sigma(\boldsymbol{\alpha})=\boldsymbol{\theta},\boldsymbol{\alpha}\in V\} \tag{4.42}$$

is called the **kernel** of σ. Here, σ(V) and $\sigma^{-1}(\boldsymbol{\theta})$ are generally denoted as Im σ and Ker σ.

Example 4.29 Image and kernel of the rotation and reflection matrices

For rotation $\mathbf{R}_\theta$ and reflection φ on $\mathbb{R}^2$, their images (ranges) are both $\mathbb{R}^2$ themselves, and their kernels contain only zero vector $\{\mathbf{0}\}$. ■

Example 4.30 Image and kernel of the projection

The image and kernel of the projection $\mathbf{P}$ on $\mathbb{R}^3$ in Example 4.21 are

$$\text{Im}\,\mathbf{P}=L(\boldsymbol{\varepsilon}_1,\boldsymbol{\varepsilon}_2),\ \text{Ker}\,\mathbf{P}=L(\boldsymbol{\varepsilon}_3).$$ ■

We have the following conclusions for the image and kernel of a linear transformation σ.

(i) $\sigma(V)$ is a subspace of linear space $V(F)$.

Actually, $\sigma(V)$ is a nonempty set as $\sigma(\boldsymbol{\theta})=\boldsymbol{\theta}$, and $\forall\boldsymbol{\beta}_1,\boldsymbol{\beta}_2\in\sigma(V)$, $\exists\boldsymbol{\alpha}_1,\boldsymbol{\alpha}_2\in V$ such that $\sigma(\boldsymbol{\alpha}_1)=\boldsymbol{\beta}_1,\sigma(\boldsymbol{\alpha}_2)=\boldsymbol{\beta}_2$. Then, $\forall\lambda_1,\lambda_2\in F$, we have

$$\lambda_1\boldsymbol{\beta}_1+\lambda_2\boldsymbol{\beta}_2=\lambda_1\sigma(\boldsymbol{\alpha}_1)+\lambda_2\sigma(\boldsymbol{\alpha}_2)=\sigma(\lambda_1\boldsymbol{\alpha}_1+\lambda_2\boldsymbol{\alpha}_2)\in\sigma(V).$$

So, $\sigma(V)$ is a subspace of $V(F)$.

(ii) Ker σ (i.e., $\sigma^{-1}(\theta)$) is also a subspace of $V(F)$.
$\sigma^{-1}(\theta)$ is not an empty set as $\theta \in \sigma^{-1}(\theta)$ and $\forall \alpha_1, \alpha_2 \in \sigma^{-1}(\theta)$ and $\forall \lambda_1, \lambda_2 \in F$,

$$\sigma(\lambda_1\alpha_1 + \lambda_2\alpha_2) = \lambda_1\sigma(\alpha_1) + \lambda_2\sigma(\alpha_2) = \lambda_1\theta + \lambda_2\theta = \theta,$$

that is, $\lambda_1\alpha_1 + \lambda_2\alpha_2 \in \sigma^{-1}(\theta)$. So, $\sigma^{-1}(\theta)$ is a subspace of $V(F)$.

(iii) Linear transformation σ is a monomorphism, if and only if, $\sigma^{-1}(\theta) = \{\theta\}$.

To prove the 'only if' direction, let σ be a monomorphism. $\forall \alpha \in V$ with $\sigma(\alpha) = \theta = \sigma(\theta)$, we get $\alpha = \theta$, which means $\sigma^{-1}(\theta) = \{\theta\}$.

To prove the 'if' direction, let $\sigma^{-1}(\theta) = \{\theta\}$. We get $\alpha_1 = \alpha_2$ if $\sigma(\alpha_1) = \sigma(\alpha_2)$ $\forall \alpha_1, \alpha_2 \in V$ due to $\sigma(\alpha_1) - \sigma(\alpha_2) = \sigma(\alpha_1 - \alpha_2) = \theta$, and $\alpha_1 - \alpha_2 = \theta$. So, σ is a monomorphism.

The two subspaces, $\sigma(V)$ and $\sigma^{-1}(\theta)$ have their dimensions. The dimension of the range $\dim\sigma(V)$ is called the rank of σ, denoted as $\mathrm{r}(\sigma)$. The dimension of the kernel, $\dim\sigma^{-1}(\theta)$ is called the nullity of σ, denoted as $N(\sigma)$. Their relation is characterized in the following theorem.

THEOREM 4.21

Suppose the dimension of a linear space $V(F)$ be n and σ be a linear transformation. Then,

$$\mathrm{r}(\sigma) + N(\sigma) = \dim\sigma(V) + \dim\sigma^{-1}(\theta) = n. \tag{4.43}$$

Proof: Let $\dim\sigma^{-1}(\theta) = k$, $B_1 = \{\alpha_1, \ldots, \alpha_k\}$ be a basis of kernel $\sigma^{-1}(\theta)$ and $B = \{\alpha_1, \ldots, \alpha_k, \alpha_{k+1}, \ldots, \alpha_n\}$ be a basis of V expanded from B_1.

$$\sigma(\alpha) = x_1\sigma(\alpha_1) + x_2\sigma(\alpha_2) + \cdots + x_n\sigma(\alpha_n),$$
$$\forall \alpha = x_1\alpha_1 + x_2\alpha_2 + \cdots + x_n\alpha_n \in V.$$

The range of σ is therefore the subspace generated by the image of σ with respect to the basis of V, that is, $\sigma(V) = L\{\sigma(\alpha_1), \ldots, \sigma(\alpha_k), \sigma(\alpha_{k+1}), \ldots, \sigma(\alpha_n)\}$. Together with $\sigma(\alpha_i) = \theta (i = 1, \ldots, k)$, we get $\sigma(V) = L\{\sigma(\alpha_{k+1}), \ldots, \sigma(\alpha_n)\}$.

So, we only need to prove that $\{\sigma(\alpha_{k+1}), \ldots, \sigma(\alpha_n)\}$ is linearly independent to get $\mathrm{r}(\sigma) = \dim\sigma(V) = n - k$.

Let

$$c_{k+1}\sigma(\alpha_{k+1}) + \cdots + c_n\sigma(\alpha_n) = \theta.$$

Then,

$$\sigma(c_{k+1}\alpha_{k+1} + \cdots + c_n\alpha_n) = \theta,$$

and $c_{k+1}\alpha_{k+1} + \cdots + c_n\alpha_n \in \sigma^{-1}(\theta)$. Hence, it can be linearly expressed by B_1 with

$$c_1\alpha_1 + \cdots + c_k\alpha_k - c_{k+1}\alpha_{k+1} - \cdots - c_n\alpha_n = \theta.$$

We get $c_1 = \cdots = c_k = c_{k+1} = \cdots = c_n = 0$ and then $\{\sigma(\boldsymbol{\alpha}_{k+1}), \ldots, \sigma(\boldsymbol{\alpha}_n)\}$ is linearly independent. □

Because the range $\sigma(V)$ of a linear transformation σ is a subspace generated by the image $\sigma(\boldsymbol{\alpha}_1), \sigma(\boldsymbol{\alpha}_2), \ldots, \sigma(\boldsymbol{\alpha}_n)$ with respect to the basis $B = \{\boldsymbol{\alpha}_1, \boldsymbol{\alpha}_2, \ldots, \boldsymbol{\alpha}_n\}$ of V, so

$$\mathrm{r}(\sigma) = \dim \sigma(V) = \operatorname{rank}\{\sigma(\boldsymbol{\alpha}_1), \sigma(\boldsymbol{\alpha}_2), \ldots, \sigma(\boldsymbol{\alpha}_n)\}. \tag{4.44}$$

The *RHS* of (4.44) is equal to the rank of the transformation matrix $\boldsymbol{A} = (a_{ij})_{n\times n}$ under the basis B, that is,

$$\mathrm{r}(\sigma) = \mathrm{r}(\boldsymbol{A}). \tag{4.45}$$

This result is proved as follows.

Proof: From (4.32), we have

$$[\sigma(\boldsymbol{\alpha}_1), \sigma(\boldsymbol{\alpha}_2), \ldots, \sigma(\boldsymbol{\alpha}_n)] = (\boldsymbol{\alpha}_1, \boldsymbol{\alpha}_2, \ldots, \boldsymbol{\alpha}_n)\begin{pmatrix} a_{11} & a_{12} & \cdots & a_{1n} \\ a_{21} & a_{22} & \cdots & a_{2n} \\ \vdots & \vdots & & \vdots \\ a_{n1} & a_{n2} & \cdots & a_{nn} \end{pmatrix}.$$

Then, we have to prove that image set $\{\sigma(\boldsymbol{\alpha}_1), \sigma(\boldsymbol{\alpha}_2), \ldots, \sigma(\boldsymbol{\alpha}_n)\}$ and the column vector set $\{\boldsymbol{\beta}_1, \boldsymbol{\beta}_2, \ldots, \boldsymbol{\beta}_n\}$ of $\boldsymbol{A}$ have the same linear dependence. For

$$x_1\sigma(\boldsymbol{\alpha}_1) + x_2\sigma(\boldsymbol{\alpha}_2) + \cdots + x_n\sigma(\boldsymbol{\alpha}_n) = \boldsymbol{\theta}, \tag*{①}$$

that is,

$$\sum_{j=1}^{n} x_j\sigma(\boldsymbol{\alpha}_j) = \sum_{j=1}^{n} x_j \sum_{i=1}^{n} a_{ij}\boldsymbol{\alpha}_i = \sum_{i=1}^{n}\left(\sum_{j=1}^{n} a_{ij}x_j\right)\boldsymbol{\alpha}_i = \boldsymbol{\theta},$$

we get

$$\sum_{j=1}^{n} a_{ij}x_j = 0, \quad i = 1, 2, \ldots, n,$$

as $\boldsymbol{\alpha}_1, \boldsymbol{\alpha}_2, \ldots, \boldsymbol{\alpha}_n$ are linearly independent. It is a homogeneous system of linear equations with coefficient matrix $\boldsymbol{A} = (a_{ij})_{n\times n}$, that is,

$$\boldsymbol{A}\boldsymbol{x} = \boldsymbol{0}. \tag*{②}$$

If $\{\sigma(\boldsymbol{\alpha}_1), \sigma(\boldsymbol{\alpha}_2), \ldots, \sigma(\boldsymbol{\alpha}_n)\}$ are linearly independent, then ① is true only for $x_1, x_x, \ldots, x_n$ all 0. So, ② only has the zero solution, and therefore the column vectors of $\boldsymbol{A}$ are linearly independent, and vice versa. If $\{\sigma(\boldsymbol{\alpha}_1), \sigma(\boldsymbol{\alpha}_2), \ldots, \sigma(\boldsymbol{\alpha}_n)\}$ are linearly dependent, then ② has nonzero solutions, that is, the column vectors are linearly dependent, and vice versa. With the same argument, we can prove that any subset of the image set

$\{\sigma(\boldsymbol{\alpha}_1), \sigma(\boldsymbol{\alpha}_2), \ldots, \sigma(\boldsymbol{\alpha}_n)\}$ and the corresponding subset of column vectors in $\boldsymbol{A}$ have the same linear dependence, which induces $\mathrm{r}(\sigma) = \mathrm{r}(\boldsymbol{A})$. □

If the corresponding matrix of a linear transformation $\boldsymbol{\sigma}$ is $\boldsymbol{A}$ under a basis $B = \{\boldsymbol{\alpha}_1, \boldsymbol{\alpha}_2, \ldots, \boldsymbol{\alpha}_n\}$, then the coordinate vector $(x_1, x_2, \ldots, x_n)^{\mathrm{T}}$ of any vector $\boldsymbol{\alpha} = x_1\boldsymbol{\alpha}_1 + x_2\boldsymbol{\alpha}_2 + \cdots + x_n\boldsymbol{\alpha}_n$ in kernel $\boldsymbol{\sigma}^{-1}(\boldsymbol{\theta})$ is a solution vector of the homogeneous system of linear equations $\boldsymbol{Ax} = \boldsymbol{0}$. The dimension of the solution space $N(\boldsymbol{A})$ is equal to the dimension of kernel $\boldsymbol{\sigma}^{-1}(\boldsymbol{\theta})$, that is,

$$\dim \boldsymbol{\sigma}^{-1}(\boldsymbol{\theta}) = \dim N(\boldsymbol{A}). \tag{4.46}$$

Exercises and Supplementary Exercises

Exercises

4.1 Prove that $\boldsymbol{\alpha}_1 = (1, 1, 1, 1)^{\mathrm{T}}, \boldsymbol{\alpha}_2 = (1, 1, -1, -1)^{\mathrm{T}}$, $\boldsymbol{\alpha}_3 = (1, -1, 1, -1)^{\mathrm{T}}$, and $\boldsymbol{\alpha}_4 = (1, -1, -1, 1)^{\mathrm{T}}$ make a basis of $\mathbb{R}^4$ and find the coordinates of $\boldsymbol{\beta} = (1, 2, 1, 1)^{\mathrm{T}}$ under this basis.

4.2 Two bases of $\mathbb{R}^3$ are

$$\boldsymbol{\alpha}_1 = (1, 2, 1)^{\mathrm{T}},\ \boldsymbol{\alpha}_2 = (2, 3, 3)^{\mathrm{T}},\ \boldsymbol{\alpha}_3 = (3, 7, 1)^{\mathrm{T}},$$

$$\boldsymbol{\beta}_1 = (3, 1, 4)^{\mathrm{T}},\ \boldsymbol{\beta}_2 = (5, 2, 1)^{\mathrm{T}},\ \boldsymbol{\beta}_3 = (1, 1, -6)^{\mathrm{T}}.$$

(i) Find the coordinates of vector $\boldsymbol{\gamma} = (3, 6, 2)^{\mathrm{T}}$ under $\{\boldsymbol{\alpha}_1, \boldsymbol{\alpha}_2, \boldsymbol{\alpha}_3\}$,

(ii) Compute the transition matrix from $\{\boldsymbol{\alpha}_1, \boldsymbol{\alpha}_2, \boldsymbol{\alpha}_3\}$ to $\{\boldsymbol{\beta}_1, \boldsymbol{\beta}_2, \boldsymbol{\beta}_3\}$,

(iii) Calculate the coordinates of $\boldsymbol{\gamma}$ under $\{\boldsymbol{\beta}_1, \boldsymbol{\beta}_2, \boldsymbol{\beta}_3\}$ using equation (4.7) from Theorem 4.2.

4.3 Two bases of $\mathbb{R}^4$ are

$$\boldsymbol{\alpha}_1 = (1, 2, -1, 0)^{\mathrm{T}}, \boldsymbol{\alpha}_2 = (1, -1, 1, 1)^{\mathrm{T}},$$

$$\boldsymbol{\alpha}_3 = (-1, 2, 1, 1)^{\mathrm{T}}, \boldsymbol{\alpha}_4 = (-1, -1, 0, 1)^{\mathrm{T}},$$

and

$$\boldsymbol{\beta}_1 = (2, 1, 0, 1)^{\mathrm{T}}, \boldsymbol{\beta}_2 = (0, 1, 2, 2)^{\mathrm{T}},$$

$$\boldsymbol{\beta}_3 = (-2, 1, 1, 2)^{\mathrm{T}}, \boldsymbol{\beta}_4 = (1, 3, 1, 2)^{\mathrm{T}}.$$

(i) Find the transition matrix from $\{\boldsymbol{\alpha}_1, \boldsymbol{\alpha}_2, \boldsymbol{\alpha}_3, \boldsymbol{\alpha}_4\}$ to $\{\boldsymbol{\beta}_1, \boldsymbol{\beta}_2, \boldsymbol{\beta}_3, \boldsymbol{\beta}_4\}$ and the coordinates of $\boldsymbol{\gamma}$ under $\{\boldsymbol{\beta}_1, \boldsymbol{\beta}_2, \boldsymbol{\beta}_3, \boldsymbol{\beta}_4\}$ if the coordinates of $\boldsymbol{\gamma}$ under $\{\boldsymbol{\alpha}_1, \boldsymbol{\alpha}_2, \boldsymbol{\alpha}_3, \boldsymbol{\alpha}_4\}$ is $(1, 0, 0, 0)^{\mathrm{T}}$.

(ii) Find the transition matrix from $\{\boldsymbol{\beta}_1, \boldsymbol{\beta}_2, \boldsymbol{\beta}_3, \boldsymbol{\beta}_4\}$ to $\{\boldsymbol{\alpha}_1, \boldsymbol{\alpha}_2, \boldsymbol{\alpha}_3, \boldsymbol{\alpha}_4\}$ and the coordinates of $\boldsymbol{\xi}$ under $\{\boldsymbol{\alpha}_1, \boldsymbol{\alpha}_2, \boldsymbol{\alpha}_3, \boldsymbol{\alpha}_4\}$ if the coordinates of $\boldsymbol{\xi}$ under $\{\boldsymbol{\beta}_1, \boldsymbol{\beta}_2, \boldsymbol{\beta}_3, \boldsymbol{\beta}_4\}$ is $(1, 2, -1, 0)^T$.

(iii) If the coordinates of $\boldsymbol{\alpha}$ under $\{\boldsymbol{\alpha}_1, \boldsymbol{\alpha}_2, \boldsymbol{\alpha}_3, \boldsymbol{\alpha}_4\}$ is $(1, 2, -1, 0)^{\mathrm{T}}$, find its coordinates under $\{\boldsymbol{\beta}_1, \boldsymbol{\beta}_2, \boldsymbol{\beta}_3, \boldsymbol{\beta}_4\}$.

4.4 Find a vector $\boldsymbol{\gamma}$ in $\mathbb{R}^4$ which has the same coordinates under the natural basis $\{\boldsymbol{\varepsilon}_1, \boldsymbol{\varepsilon}_2, \boldsymbol{\varepsilon}_3, \boldsymbol{\varepsilon}_4\}$ and basis $\boldsymbol{\beta}_1 = (2, 1, -1, 1)^{\mathrm{T}}$, $\boldsymbol{\beta}_2 = (0, 3, 1, 0)^{\mathrm{T}}$, $\boldsymbol{\beta}_3 = (5, 3, 2, 1)^{\mathrm{T}}$, and $\boldsymbol{\beta}_4 = (6, 6, 1, 3)^{\mathrm{T}}$.

4.5 For given $\boldsymbol{\alpha} = (1, 2, -1, 1), \boldsymbol{\beta} = (2, 3, 1, -1)$, and $\boldsymbol{\gamma} = (-1, -1, -2, 2)$,

(i) calculate the lengths of $\boldsymbol{\alpha}$, $\boldsymbol{\beta}$, and $\boldsymbol{\gamma}$ and draw figures for $\langle \boldsymbol{\alpha}, \boldsymbol{\beta} \rangle$, $\langle \boldsymbol{\alpha}, \boldsymbol{\gamma} \rangle$,

(ii) find all vectors orthogonal to $\boldsymbol{\alpha}$, $\boldsymbol{\beta}$, and $\boldsymbol{\gamma}$.

4.6 Find the unit vector orthogonal to all the three vectors $(1, 1, -1, 1)$, $(1, -1, -1, 1)$, $(2, 1, 1, 3)$.

4.7 Prove that $\boldsymbol{\beta}$ is orthogonal to any linear combination of $\boldsymbol{\alpha}_1, \boldsymbol{\alpha}_2, \ldots, \boldsymbol{\alpha}_m$ if the vector $\boldsymbol{\beta}$ is orthogonal to each vector of $\boldsymbol{\alpha}_1, \boldsymbol{\alpha}_2, \ldots, \boldsymbol{\alpha}_m$.

4.8 Construct an orthonormal vector set from each of the following sets by Gram-Schmidt orthogonalization.

(i) $(1, 2, 2, -1)$, $(1, 1, -5, 3)$, $(3, 2, 8, -7)$.

(ii) $(1, 1, -1, -2)$, $(5, 8, -2, -3)$, $(3, 9, 3, 8)$.

(iii) $(2, 1, 3, -1)$, $(7, 4, 3, -3)$, $(1, 1, -6, 0)$, $(5, 7, 7, 8)$.

4.9 Construct an orthonormal basis of $\mathbb{R}^3$ from each of following bases by Gram-Schmidt orthogonalization, and find the coordinates of $\boldsymbol{\alpha} = (1, -1, 0)$ under the orthonormal basis.

(i) $(1, 1, 1)$, $(0, 1, 1)$, $(1, 0, 1)$.

(ii) $(1, -1, 1)$, $(-1, 1, 1)$, $(1, 1, -1)$.

4.10 Let $\{\boldsymbol{\alpha}_1, \boldsymbol{\alpha}_2, \boldsymbol{\alpha}_3\}$ be an orthonormal basis of $\mathbb{R}^3$. Prove that

$$\boldsymbol{\beta}_1 = \frac{1}{3}(2\boldsymbol{\alpha}_1 + 2\boldsymbol{\alpha}_2 - \boldsymbol{\alpha}_3),$$

$$\boldsymbol{\beta}_2 = \frac{1}{3}(2\boldsymbol{\alpha}_1 - \boldsymbol{\alpha}_2 + 2\boldsymbol{\alpha}_3),$$

$$\boldsymbol{\beta}_3 = \frac{1}{3}(\boldsymbol{\alpha}_1 - 2\boldsymbol{\alpha}_2 - 2\boldsymbol{\alpha}_3)$$

is another orthonormal basis of $\mathbb{R}^3$.

4.11 Suppose

$$Q=\begin{pmatrix} a & -\frac{3}{7} & \frac{2}{7} \\ b & c & d \\ -\frac{3}{7} & \frac{2}{7} & e \end{pmatrix}$$

is orthogonal. Find the values of $a, b, c, d,$ and e.

4.12 Prove that the adjoint matrix A^* of A is also an orthogonal matrix if A is an orthogonal matrix.

4.13 Prove that if A is an orthogonal matrix, then

(i) the determinant of A is equal to 1 or -1,

(ii) $A^{-1} = A^{\mathrm{T}}$.

4.14 A matrix A is called involutary if $A^2 = I$. Prove that if a square matrix has any two of the three properties (symmetric matrix, orthogonal matrix, involutary matrix), then it must have the third property.

4.15 Verify that the following matrices are symmetric, orthogonal, and involutary.

(i) $\begin{pmatrix} \frac{1}{3} & -\frac{2}{3} & -\frac{2}{3} \\ -\frac{2}{3} & \frac{1}{3} & -\frac{2}{3} \\ -\frac{2}{3} & -\frac{2}{3} & \frac{1}{3} \end{pmatrix}$.

(ii) $\begin{pmatrix} \frac{1}{2} & \frac{1}{2} & \frac{1}{2} & \frac{1}{2} \\ \frac{1}{2} & \frac{1}{2} & -\frac{1}{2} & -\frac{1}{2} \\ \frac{1}{2} & -\frac{1}{2} & \frac{1}{2} & -\frac{1}{2} \\ \frac{1}{2} & -\frac{1}{2} & -\frac{1}{2} & \frac{1}{2} \end{pmatrix}$.

***4.16** Prove that a triangular matrix is a diagonal matrix with main diagonal entries $+1$ or -1 if it is orthogonal.

4.17 Choose the linear spaces over the real number field with the given addition and scalar multiplication from the following instances.

(i) All the real symmetric (antisymmetric, triangular) matrices with the matrix addition and scalar multiplication.

(ii) All nth-order invertible matrices (orthogonal matrices) with matrix addition and scalar multiplication.

(iii) All the nth-degree polynomial ($n \geq 1$) with real coefficients along with polynomial addition and scalar multiplication.

(iv) In a plane, all vectors not parallel to a given vector, with vector addition and scalar multiplication.

(v) All vectors in a plane with the usual addition of vectors and scalar multiplication defined as the following

$$k \cdot \boldsymbol{\alpha} = \mathbf{0},$$

where $k \in \mathbb{R}$, $\boldsymbol{\alpha}$ is a vector, $\mathbf{0}$ is the zero vector.

(vi) All solutions of the homogeneous linear differential equation $y''' + 3y'' + 3y' + y = 0$ with the function addition and scalar multiplication.

(vii) All solutions of nonhomogenous linear differential equations. $y''' + 3y'' + 3y' + y = x + 1$ with the function addition and scalar multiplication.

(viii) All the positive number $\mathbb{R}^+$, the addition and scalar multiplication defined as $a \oplus b = ab$, $k \cdot a = a^k$, where $a, b \in \mathbb{R}^+$, $k \in \mathbb{R}$.

4.18 Determine if the set of all complex numbers is a linear space over the real number field and the complex field with the usual number addition and scalar multiplication? If it is a linear space, compute its dimension, and a basis.

4.19 Are the following subsets of $\mathbb{R}^3$ subspaces of $\mathbb{R}^3$? Verify your conclusions by definition and explain their geometrical meanings.

(i) $W_1 = \{(x_1, x_2, x_3) \,|\, x_1 - x_2 + x_3 = 0\}$,

(ii) $W_2 = \{(x_1, x_2, x_3) \,|\, x_1 + x_2 = 1\}$,

(iii) $W_3 = \{(1, x_2, x_3) \,|\, x_2, x_3 \in R\}$,

(iv) $W_4 = \{(x_1, 0, x_3) \,|\, x_1, x_3 \in R\}$,

(v) $W_5 = \left\{(x_1, x_2, x_3) \,\middle|\, x_1 = \frac{x_2}{2} = \frac{x_3}{3}\right\}$,

(vi) $W_6 = \left\{ (x_1, x_2, x_3) \,\middle|\, \frac{x_1 - 1}{2} = \frac{x_2}{3} = \frac{x_3 - 5}{-1} \right\}.$

4.20 Find the bases and dimensions of subspaces W_1, W_4, and W_5 in **Q4.19**.

4.21 Let

$$A = \begin{pmatrix} 2 & 1 & -1 & 1 & -3 \\ 1 & 1 & 1 & 0 & 1 \\ 3 & 2 & -1 & 1 & -2 \end{pmatrix}.$$

Find the dimension of the solution space of linear system of homogenous equations $\boldsymbol{Ax} = \boldsymbol{0}$, and an orthonormal basis of the solution space.

4.22 Let V be a 2-dimensional subspace of $\mathbb{R}^5$ with a basis $\boldsymbol{\alpha}_1$=(1, 1, 1, 1, 1) and $\boldsymbol{\alpha}_2 = (1, 1, 0, 1, 1)$. Extend this basis to a basis of $\mathbb{R}^5$.

4.23 Find the dimension and a basis of a subspace of $\mathbb{R}^4$,

$$V = \{(x_1, x_2, x_3, x_4) | x_1 - x_2 + x_3 - x_4 = 0\}$$

and extend it into a basis of $\mathbb{R}^4$.

4.24 In $\mathbb{R}^4$, find a basis and dimension of subspaces generated by the following vector sets, and find an orthonormal basis for them,

(i) $\begin{cases} \boldsymbol{\alpha}_1 = (2, 1, 3, 1), \\ \boldsymbol{\alpha}_2 = (1, 2, 0, 1), \\ \boldsymbol{\alpha}_3 = (-1, 1, -3, 0), \\ \boldsymbol{\alpha}_4 = (1, 1, 1, 1). \end{cases}$

(ii) $\begin{cases} \boldsymbol{\beta}_1 = (2, 1, 3, -1), \\ \boldsymbol{\beta}_2 = (-1, 1, -3, 1), \\ \boldsymbol{\beta}_3 = (4, 5, 3, -1), \\ \boldsymbol{\beta}_4 = (1, 5, -3, 1). \end{cases}$

4.25 Let $\boldsymbol{\alpha}, \boldsymbol{\beta}, \boldsymbol{\gamma} \in \mathbb{R}^n$, $c_1, c_2, c_3 \in \mathbb{R}$ and $c_1, c_3 \neq 0$. Prove that $L(\boldsymbol{\alpha}, \boldsymbol{\beta}) = L(\boldsymbol{\beta}, \boldsymbol{\gamma})$ if $c_1\boldsymbol{\alpha} + c_2\boldsymbol{\beta} + c_3\boldsymbol{\gamma} = \boldsymbol{0}$.

4.26 Find the dimension and a basis for each linear space in **Q4.17**.

4.27 Let $\{\boldsymbol{\alpha}_1, \boldsymbol{\alpha}_2, \ldots, \boldsymbol{\alpha}_n\}$ be a basis of an n-dimensional linear space V, and none of the coordinates $(x_1, x_2, \ldots, x_n)$ of vector $\boldsymbol{\alpha}_{n+1}$ in V is zero. Prove that any n vectors in $\boldsymbol{\alpha}_1, \boldsymbol{\alpha}_2, \ldots, \boldsymbol{\alpha}_n, \boldsymbol{\alpha}_{n+1}$ must form a basis of V. Find the coordinates of $\boldsymbol{\alpha}_1$ under the basis $\{\boldsymbol{\alpha}_2, \ldots, \boldsymbol{\alpha}_n, \boldsymbol{\alpha}_{n+1}\}$.

4.28 An old basis of $\mathbb{R}[x]_5$ is $\{1, x, x^2, x^3, x^4\}$ and a new basis is

$\{1, 1+x, 1+x+x^2, 1+x+x^2+x^3, 1+x+x^2+x^3+x^4\}$.

(i) Find the transition matrix from the old basis to the new basis.

(ii) Find the coordinates of the polynomial $1 + 2x + 3x^2 + 4x^3 + 5x^4$ under the new basis.

(iii) If the coordinates of the polynomial $f(x)$ is (1, 2, 3, 4, 5) under the new basis, find its coordinates under the old basis.

4.29 Let

$$\boldsymbol{E}_{11} = \begin{pmatrix} 1 & 0 \\ 0 & 0 \end{pmatrix}, \quad \boldsymbol{E}_{12} = \begin{pmatrix} 0 & 1 \\ 0 & 0 \end{pmatrix},$$

$$\boldsymbol{E}_{21} = \begin{pmatrix} 0 & 0 \\ 1 & 0 \end{pmatrix}, \quad \boldsymbol{E}_{22} = \begin{pmatrix} 0 & 0 \\ 0 & 1 \end{pmatrix},$$

$$\boldsymbol{G}_1 = \begin{pmatrix} 0 & 1 \\ 1 & 1 \end{pmatrix}, \quad \boldsymbol{G}_2 = \begin{pmatrix} 1 & 0 \\ 1 & 1 \end{pmatrix},$$

$$\boldsymbol{G}_3 = \begin{pmatrix} 1 & 1 \\ 0 & 1 \end{pmatrix}, \quad \boldsymbol{G}_4 = \begin{pmatrix} 1 & 1 \\ 1 & 0 \end{pmatrix}.$$

(i) Prove that $\boldsymbol{G}_1, \boldsymbol{G}_2, \boldsymbol{G}_3$, and $\boldsymbol{G}_4$ form a basis of $\mathbb{R}^{2\times 2}$;

(ii) Find the transition matrix from $\{\boldsymbol{E}_{11}, \boldsymbol{E}_{12}, \boldsymbol{E}_{21}, \boldsymbol{E}_{22}\}$ to $\{\boldsymbol{G}_1, \boldsymbol{G}_2, \boldsymbol{G}_3, \boldsymbol{G}_4\}$.

(iii) Find the coordinate vectors of the matrix

$$\boldsymbol{A} = \begin{pmatrix} 0 & 1 \\ 2 & -3 \end{pmatrix}$$

under the two bases respectively.

4.30 Let

$$\boldsymbol{A} = \begin{pmatrix} 1 & 0 & 0 \\ 0 & \omega & 0 \\ 0 & 0 & \omega^2 \end{pmatrix},$$

where $\omega = \dfrac{-1 + \sqrt{3}\mathrm{i}}{2}$.

(i) Prove that all real coefficient polynomials of $\boldsymbol{A}$ form a linear space under addition and scalar multiplication of matrix polynomial over the real number field.

(ii) Find the dimension and a basis of this linear space.

4.31 Let $\boldsymbol{A} \in \mathbb{R}^{n\times n}$.

(i) Prove that the matrices commutative with $\boldsymbol{A}$ form a subspace of $\mathbb{R}^{n\times n}$, which is denoted as $C(\boldsymbol{A})$.

(ii) Find $C(\boldsymbol{A})$ if $\boldsymbol{A}=\boldsymbol{I}$.

(iii) Find the dimension and a basis of $C(\boldsymbol{A})$ if $\boldsymbol{A}=\mathrm{diag}(1,2,\ldots,n)$.

4.32 Let

$$\boldsymbol{A}=\begin{pmatrix}1&0&0\\0&1&0\\3&2&2\end{pmatrix}.$$

Find the dimension and a basis of the subspace made up by all matrices commutative with $\boldsymbol{A}$ in $\mathbb{R}^{3\times 3}$.

***4.33** Let V_1 and V_2 be subspaces generated by vectors $\{\boldsymbol{\alpha}_i\}$ and $\{\boldsymbol{\beta}_i\}$ respectively. Find the dimension and a basis of $V_1\cap V_2$ and V_1+V_2 respectively. Let

(i) $\begin{cases}\boldsymbol{\alpha}_1=(1,2,1,0),\\ \boldsymbol{\alpha}_2=(-1,1,1,1).\end{cases}$ $\begin{cases}\boldsymbol{\beta}_1=(2,-1,0,1),\\ \boldsymbol{\beta}_2=(1,-1,3,7).\end{cases}$

(ii) $\begin{cases}\boldsymbol{\alpha}_1=(1,2,-1,-2),\\ \boldsymbol{\alpha}_2=(3,1,1,1),\\ \boldsymbol{\alpha}_3=(-1,0,1,-1),\end{cases}$ $\begin{cases}\boldsymbol{\beta}_1=(2,5,-6,-5),\\ \boldsymbol{\beta}_2=(-1,2,-7,3).\end{cases}$

4.34 Let V_1 and V_2 be the solution spaces of homogeneous systems $x_1+x_2+\cdots+x_n=0$ and $x_1=x_2=\cdots=x_n$, respectively. Prove that $\mathbb{R}^n=V_1\oplus V_2$.

4.35 Let W_1 be a subspace of $\mathbb{R}^n$ and $W_1\neq\mathbb{R}^n$. Prove that there exists another subspace W_2 such that $\mathbb{R}^n=W_1\oplus W_2$.

***4.36** Let

$$\boldsymbol{A}=\begin{pmatrix}1&1&-1&0&1\\2&3&1&-1&0\\0&1&3&-1&-2\\4&1&-13&3&10\end{pmatrix}.$$

(i) Find dimensions and bases of the row space and column space of the matrix $\boldsymbol{A}$.

(ii) Find the dimension and a basis of the zero space of matrix $\boldsymbol{A}$.

(iii) Find the dimension of the orthogonal complement of the row space of $\boldsymbol{A}$.

***4.37** In $\mathbb{R}^3$, which of the following subspaces are orthogonal? Which two are orthogonal complements of each other? Explain your answers.

(i) $W_1=\{(x,y,z)|3x-y+2z=0\}$,

(ii) $W_2=\{(x,y,z)|x-y-2z=0\}$,

(iii) $W_3=\left\{(x,y,z)\,\middle|\,\dfrac{x}{3}=\dfrac{y}{-1}=\dfrac{z}{2}\right\}$,

(iv) $W_4=\left\{(x,y,z)\,\middle|\,\dfrac{x}{3}=\dfrac{y}{5}=\dfrac{z}{-2}\right\}$.

4.38 Let $\boldsymbol{\alpha}=(x_1,x_2,x_3)\in\mathbb{R}^3$. Are the following transformations linear? Justify your answers.

(i) $\sigma(\boldsymbol{\alpha})=(2x_1,0,0)$;

(ii) $\sigma(\boldsymbol{\alpha})=(x_1x_2,0,x_2)$;

(iii) $\sigma(\boldsymbol{\alpha})=(x_1,x_2,-x_3)$;

(iv) $\sigma(\boldsymbol{\alpha})=(1,1,x_3)$;

(v) $\sigma(\boldsymbol{\alpha})=\boldsymbol{\alpha}_0+\boldsymbol{\alpha}$; where $\boldsymbol{\alpha}_0\in\mathbb{R}^3$ is a fixed vector.

4.39 If (x_1,x_2,x_3) in **Q4.38** are the coordinates of $\boldsymbol{\alpha}$ under the natural basis $\{\boldsymbol{\varepsilon}_1,\boldsymbol{\varepsilon}_2,\boldsymbol{\varepsilon}_3\}$, explain the geometric meaning of the two linear transformations (i) and (iii).

4.40 For the two linear transformations (i) and (iii) in **Q4.38**, find, respectively, the matrix of σ under

(i) the natural basis $\{\boldsymbol{\varepsilon}_1,\boldsymbol{\varepsilon}_2,\boldsymbol{\varepsilon}_3\}$,

(ii) a basis $\boldsymbol{\alpha}_1=(1,0,0)$, $\boldsymbol{\alpha}_2=(-1,1,0)$, and $\boldsymbol{\alpha}_3=(1,-1,1)$.

4.41 A linear transformation in $\mathbb{R}^3$ is defined as

$$\sigma(x_1,x_2,x_3)^{\mathrm{T}}=(x_1+x_2,x_1-x_2,x_3)^{\mathrm{T}}.$$

(i) Find the matrix of σ under the natural basis $\{\boldsymbol{\varepsilon}_1,\boldsymbol{\varepsilon}_2,\boldsymbol{\varepsilon}_3\}$.

(ii) Find the matrix of σ under a basis $\boldsymbol{\beta}_1=(1,0,0)^{\mathrm{T}}$, $\boldsymbol{\beta}_2=(1,1,0)^{\mathrm{T}}$, and $\boldsymbol{\beta}_3=(1,1,1)^{\mathrm{T}}$.

4.42 Let $\{\boldsymbol{\alpha}_1, \boldsymbol{\alpha}_2, \boldsymbol{\alpha}_3\}$ and $\{\boldsymbol{\beta}_1, \boldsymbol{\beta}_2, \boldsymbol{\beta}_3\}$ be two bases of $\mathbb{R}^3$, where $\boldsymbol{\beta}_1 = 2\boldsymbol{\alpha}_1 + \boldsymbol{\alpha}_2 + 3\boldsymbol{\alpha}_3$, $\boldsymbol{\beta}_2 = \boldsymbol{\alpha}_1 + \boldsymbol{\alpha}_2 + 2\boldsymbol{\alpha}_3$, $\boldsymbol{\beta}_3 = -\boldsymbol{\alpha}_1 + \boldsymbol{\alpha}_2 + \boldsymbol{\alpha}_3$, and the matrix of $\boldsymbol{\sigma}$ under the basis $\{\boldsymbol{\alpha}_1, \boldsymbol{\alpha}_2, \boldsymbol{\alpha}_3\}$ is

$$\boldsymbol{A} = \begin{pmatrix} 5 & 7 & -5 \\ 0 & 4 & -1 \\ 2 & 8 & 3 \end{pmatrix}.$$

Find the matrix of $\boldsymbol{\sigma}$ under the basis:
(i) $\{-\boldsymbol{\alpha}_2, 2\boldsymbol{\alpha}_1, \boldsymbol{\alpha}_3\}$, (ii) $\{\boldsymbol{\beta}_1, \boldsymbol{\beta}_2, \boldsymbol{\beta}_3\}$.

4.43 The image of a linear transformation $\boldsymbol{\sigma}$ of $\mathbb{R}^3$ under a basis

$$\boldsymbol{\alpha}_1 = (-1, 0, 2)^{\mathrm{T}}, \boldsymbol{\alpha}_2 = (0, 1, 1)^{\mathrm{T}},$$
$$\boldsymbol{\alpha}_3 = (3, -1, -6)^{\mathrm{T}},$$

is $\boldsymbol{\sigma}(\boldsymbol{\alpha}_1) = \boldsymbol{\beta}_1 = (-1, 0, 1)^{\mathrm{T}}, \boldsymbol{\sigma}(\boldsymbol{\alpha}_2) = \boldsymbol{\beta}_2 = (0, -1, 2)^{\mathrm{T}}$, and $\boldsymbol{\sigma}(\boldsymbol{\alpha}_3) = \boldsymbol{\beta}_3 = (-1, -1, 3)^{\mathrm{T}}$.

(i) Find the matrix of $\boldsymbol{\sigma}$ under the basis $\{\boldsymbol{\alpha}_1, \boldsymbol{\alpha}_2, \boldsymbol{\alpha}_3\}$;
(ii) Find $\boldsymbol{\sigma}(\boldsymbol{\beta}_1)$, $\boldsymbol{\sigma}(\boldsymbol{\beta}_2)$, and $\boldsymbol{\sigma}(\boldsymbol{\beta}_3)$;
(iii) The coordinate vector of $\boldsymbol{\alpha}$ under the basis $\{\boldsymbol{\alpha}_1, \boldsymbol{\alpha}_2, \boldsymbol{\alpha}_3\}$ is $(5, 1, 1)^{\mathrm{T}}$. Find the coordinate vector of $\boldsymbol{\sigma}(\boldsymbol{\alpha})$ under the basis $\{\boldsymbol{\alpha}_1, \boldsymbol{\alpha}_2, \boldsymbol{\alpha}_3\}$;
(iv) Find $\boldsymbol{\sigma}(\boldsymbol{\beta})$ if $\boldsymbol{\beta} = (1, 1, 1)^{\mathrm{T}}$;
(v) The coordinate vector of $\boldsymbol{\sigma}(\boldsymbol{\gamma})$ under the basis $\{\boldsymbol{\alpha}_1, \boldsymbol{\alpha}_2, \boldsymbol{\alpha}_3\}$ is $(2, -4, -2)^{\mathrm{T}}$. Is the inverse image of $\boldsymbol{\gamma}$ unique or not? If not, find all the inverse images of $\boldsymbol{\gamma}$.

4.44 Define a transformation $\boldsymbol{\sigma}(\boldsymbol{X}) = \boldsymbol{BXC}$ in $\mathbb{R}^{n\times n}$, where $\boldsymbol{B}, \boldsymbol{C} \in \mathbb{R}^{n\times n}$ are two fixed matrices. Prove that $\boldsymbol{\sigma}$ is a linear transformation of $\mathbb{R}^{n\times n}$.

4.45 Find the matrix representation of the differentiation transformation $D(f(x)) = f'(x)$ of $\mathbb{R}[x]_4$ under a basis $\left\{1, 1+x, 1+x+x^2, 1+x+x^2+x^3\right\}$.

4.46 Use $\boldsymbol{B} = \boldsymbol{C} = \begin{pmatrix} a & b \\ c & d \end{pmatrix}$ in **Q4.44**. Find the matrix of $\boldsymbol{\sigma}(\boldsymbol{X}) = \boldsymbol{BXC}$ under the basis $\boldsymbol{E}_{11}, \boldsymbol{E}_{12}, \boldsymbol{E}_{21}$, and $\boldsymbol{E}_{22}$ (provided in **Q4.29**).

4.47 Let $\boldsymbol{\sigma}$ be a linear transformation on linear space V. If $\boldsymbol{\sigma}^{k-1}(\boldsymbol{\xi}) \neq \boldsymbol{\theta}$, but $\boldsymbol{\sigma}^{k}(\boldsymbol{\xi}) = \boldsymbol{\theta}$, prove that $\boldsymbol{\xi}, \boldsymbol{\sigma}(\boldsymbol{\xi}), \boldsymbol{\sigma}^2(\boldsymbol{\xi}), \ldots, \boldsymbol{\sigma}^{k-1}(\boldsymbol{\xi})$ are linearly independent ($k > 0$).

4.48 Find the image (range) and kernel of a linear transformation $\boldsymbol{\sigma}$ as well as the rank of $\boldsymbol{\sigma}$.

(i) $\boldsymbol{\sigma}(x_1, x_2, x_3) = (x_1 + x_2 + x_3, -x_1 - 2x_3, x_2 - x_3)$.
(ii) $\boldsymbol{\sigma}$ is the zero transformation of n-dimensional linear space.
(iii) $\boldsymbol{\sigma}$ is the identity transformation of n-dimension linear space V.
(iv) $\boldsymbol{\sigma} : \mathbb{R}^n \to \mathbb{R}^n$ and $\boldsymbol{\sigma}(x_1, x_2, \ldots, x_n) = (x_1, 0, \ldots, 0)$.

4.49 Find a linear transformation $\boldsymbol{\sigma}$ such that the image of $\boldsymbol{\sigma}$ is $\boldsymbol{\sigma}(\mathbb{R}^3) = L(\boldsymbol{\alpha}_1, \boldsymbol{\alpha}_2)$, where $\boldsymbol{\alpha}_1 = (1, 0, -1), \boldsymbol{\alpha}_2 = (1, 2, 2)$.

4.50 A linear transformation of $\mathbb{R}^2$ is $\boldsymbol{\sigma}(x_1, x_2) = (x_1 - x_2, x_1 + x_2)$.

(i) Find $\boldsymbol{\sigma}^2(x_1, x_2) = ?$
(ii) Is $\boldsymbol{\sigma}$ invertible? If yes, compute $\boldsymbol{\sigma}^{-1}(x_1, x_2) = ?$

Supplementary exercises

4.51 Let $\boldsymbol{A}$ be an orthogonal matrix, $\boldsymbol{I} + \boldsymbol{A}$ is invertible. Prove that

(i) $(\boldsymbol{I} - \boldsymbol{A})(\boldsymbol{I} + \boldsymbol{A})^{-1}$ is commutative,
(ii) $(\boldsymbol{I} - \boldsymbol{A})(\boldsymbol{I} + \boldsymbol{A})^{-1}$ is an antisymmetric matrix.

4.52 Prove the following:

(i) If $\det\boldsymbol{A} = 1$, then $\boldsymbol{A}$ is an orthogonal matrix, if and only if, each entry of $\boldsymbol{A}$ is equal to its algebraic complement.
(ii) If $\det\boldsymbol{A} = -1$, then $\boldsymbol{A}$ is an orthogonal matrix, if and only if, each entry of $\boldsymbol{A}$ is equal to its algebraic complement multiplied by -1.

4.53 Let $\boldsymbol{\alpha}_1, \boldsymbol{\alpha}_2, \ldots, \boldsymbol{\alpha}_m \in \mathbb{R}^n$. Prove that $\boldsymbol{\alpha}_1, \boldsymbol{\alpha}_2, \ldots, \boldsymbol{\alpha}_m$ are linearly independent, if and only if,

$$\det\begin{pmatrix} (\boldsymbol{\alpha}_1, \boldsymbol{\alpha}_1) & (\boldsymbol{\alpha}_1, \boldsymbol{\alpha}_2) & \cdots & (\boldsymbol{\alpha}_1, \boldsymbol{\alpha}_m) \\ (\boldsymbol{\alpha}_2, \boldsymbol{\alpha}_1) & (\boldsymbol{\alpha}_2, \boldsymbol{\alpha}_2) & \cdots & (\boldsymbol{\alpha}_2, \boldsymbol{\alpha}_m) \\ \vdots & \vdots & & \vdots \\ (\boldsymbol{\alpha}_m, \boldsymbol{\alpha}_1) & (\boldsymbol{\alpha}_m, \boldsymbol{\alpha}_2) & \cdots & (\boldsymbol{\alpha}_m, \boldsymbol{\alpha}_m) \end{pmatrix} \neq 0.$$

4.54 Prove that $L(\boldsymbol{\alpha}_1, \boldsymbol{\alpha}_2, \ldots, \boldsymbol{\alpha}_s) + L(\boldsymbol{\beta}_1, \boldsymbol{\beta}_2, \ldots, \boldsymbol{\beta}_t) = L(\boldsymbol{\alpha}_1, \boldsymbol{\alpha}_2, \ldots, \boldsymbol{\alpha}_s, \boldsymbol{\beta}_1, \boldsymbol{\beta}_2, \ldots, \boldsymbol{\beta}_t)$.

4.55 Suppose that rank $\{\boldsymbol{\alpha}_1, \boldsymbol{\alpha}_2, \ldots, \boldsymbol{\alpha}_s\} = r$, rank $\{\boldsymbol{\beta}_1, \boldsymbol{\beta}_2, \ldots, \boldsymbol{\beta}_t\} = p$. Prove that $\dim L(\boldsymbol{\alpha}_1, \boldsymbol{\alpha}_2, \ldots, \boldsymbol{\alpha}_s, \boldsymbol{\beta}_1, \boldsymbol{\beta}_2, \ldots, \boldsymbol{\beta}_t) \leq r + p$.

4.56 Let $\boldsymbol{A}$ be an $m \times n$ matrix and $\boldsymbol{B}$ be an $s \times n$ matrix. Consider the intersection of solution spaces of two linear homogeneous systems $\boldsymbol{Ax} = \boldsymbol{0}$ and $\boldsymbol{Bx} = \boldsymbol{0}$. How do we compute their intersection?

4.57 How do we compute the intersection of the two subspaces $L(\boldsymbol{\alpha}_1, \boldsymbol{\alpha}_2, \ldots, \boldsymbol{\alpha}_s)$ and $L(\boldsymbol{\beta}_1, \boldsymbol{\beta}_2, \ldots, \boldsymbol{\beta}_t)$?

4.58 Let $\boldsymbol{A}$ be an nth-order real matrix. Determine if the following propositions are true and provide your reasons.

(i) If $\dim R(\boldsymbol{A}) = n$, then $R(\boldsymbol{A}) = R(\boldsymbol{A}^{\mathrm{T}})$.

(ii) If $\dim R(\boldsymbol{A}) = m < n$, then $R(\boldsymbol{A}) \neq R(\boldsymbol{A}^{\mathrm{T}})$.

4.59 Let V_1 and V_2 be two non-trivial subspaces of $\mathbb{R}^n$. Prove that there exists a vector $\boldsymbol{\alpha}$ in $\mathbb{R}^n$ such that $\boldsymbol{\alpha} \notin V_1$ and $\boldsymbol{\alpha} \notin V_2$. Give an example to explain this result in $\mathbb{R}^3$.

5 Eigenvalues and Eigenvectors, Diagonalization of Matrices

CHAPTER OVERVIEW

Thus far, we have learnt to solve almost every problem that originated or evolved from the consideration of systems of linear equations, where elementary transformations are employed to reduce a matrix into an echelon form. However, we have come to a turning point, and henceforth the emphasis will be on a new tool. Elementary transformation will no longer be our main tool, although we may still be required to reduce a matrix to a diagonal or upper triangular form. Elementary transformation would only be expected to play an auxiliary role here. The major focus of this chapter would be: eigenvalues and eigenvectors of matrices, diagonalization of matrices in the sense of similarity transformation, and diagonalization of real symmetric matrices.

5.1 Eigenvalues and Eigenvectors of Matrices, Similar Matrices

Eigenvalues and eigenvectors are two related concepts in linear algebra. In linear algebra, we study linear transformations, represented by matrices acting on vectors. Eigenvalues and eigenvectors are properties of a matrix and the prefix "eigen" comes from the German word for "innate." They provide vital information about the matrix and are commonly used in matrix factorization. Eigenvalues and eigenvectors have applications in areas such as finance, optimization, and quantum mechanics.

5.1.1 Concepts of eigenvalues and eigenvectors

Definition 5.1

Let $\boldsymbol{A}$ be an nth-order matrix over the complex number field $\mathbb{C}$. If there exists a number $\lambda \in \mathbb{C}$ and nonzero n-dimensional vector $\boldsymbol{x}$ such that

$$\boldsymbol{A}\boldsymbol{x} = \lambda \boldsymbol{x}, \tag{5.1}$$

then λ is called an **eigenvalue** of the matrix $\boldsymbol{A}$, $\boldsymbol{x}$ is called an **eigenvector** belonging to (corresponding to) eigenvalue λ of $\boldsymbol{A}$.

We require that the eigenvector $\boldsymbol{x} \neq \boldsymbol{0}$ and the matrices here are square, which are default assumptions in this chapter, unless otherwise stated. According to the definition, an eigenvalue of an nth-order matrix $\boldsymbol{A}$ is the value of λ such that the system of homogeneous linear equations

$$(\lambda \boldsymbol{I} - \boldsymbol{A})\boldsymbol{x} = \boldsymbol{0}$$

has a nonzero solution. That is, any λ satisfying the equation

$$\det(\lambda \boldsymbol{I} - \boldsymbol{A}) = 0 \tag{5.2}$$

is an eigenvalue of the matrix $\boldsymbol{A}$. Therefore, eigenvalues are the roots of the polynomial $\det(\lambda \boldsymbol{I} - \boldsymbol{A})$ in λ.

Definition 5.2

For an nth-order matrix $\boldsymbol{A} = (a_{ij})$,

$$f(\lambda) = \det(\lambda \boldsymbol{I} - \boldsymbol{A}) = \begin{vmatrix} \lambda - a_{11} & -a_{12} & \cdots & -a_{1n} \\ -a_{21} & \lambda - a_{22} & \cdots & -a_{2n} \\ \vdots & \vdots & & \vdots \\ -a_{n1} & -a_{n2} & \cdots & \lambda - a_{nn} \end{vmatrix} \tag{5.3}$$

is called the **characteristic polynomial** of $\boldsymbol{A}$, $\lambda\boldsymbol{I}-\boldsymbol{A}$ is called the **eigenmatrix** of matrix $\boldsymbol{A}$, (5.2) is called the **characteristic equation** of $\boldsymbol{A}$.

Obviously, the characteristic polynomial of an nth-order matrix $\boldsymbol{A}$ is an nth-degree polynomial in λ. A root of multiplicity k of the characteristic polynomial is also called an **eigenvalue of multiplicity k**. When $n \geq 5$, the characteristic polynomial has no general formula to analytically express its roots in terms of its coefficients. Sometimes, it is even difficult to find roots of the characteristic polynomial of a third-order matrix. Therefore, numerical methods are commonly used to find eigenvalues of matrices, which are typical issues in the numerical analysis.

Example 5.1 Eigenvalues and eigenvectors of a matrix

Find the eigenvalues and eigenvectors of the matrix

$$\boldsymbol{A}=\begin{pmatrix}5 & -1 & -1\\ 3 & 1 & -1\\ 4 & -2 & 1\end{pmatrix}.$$

Solution: The characteristic polynomial of matrix $\boldsymbol{A}$ is

$$\det(\lambda\boldsymbol{I}-\boldsymbol{A})=\begin{vmatrix}\lambda-5 & 1 & 1\\ -3 & \lambda-1 & 1\\ -4 & 2 & \lambda-1\end{vmatrix}=0.$$

Adding each column to the first column, we get $(\lambda-3)$ in each position of the first column of the determinant, which we factor out and then adding the first row multiplied by -1 to the second and third rows, we have

$$\det(\lambda\boldsymbol{I}-\boldsymbol{A})=(\lambda-3)\begin{vmatrix}1 & 1 & 1\\ 0 & \lambda-2 & 0\\ 0 & 1 & \lambda-2\end{vmatrix}=(\lambda-3)(\lambda-2)^2=0.$$

So, the eigenvalues of $\boldsymbol{A}$ are $\lambda_1=3$ and $\lambda_2=2$ (of multiplicity 2).

To find the eigenvectors belonging to the eigenvalue $\lambda_1=3$, we consider the homogeneous system, which is called the eigenvalue equation of $\boldsymbol{A}$ for $\lambda_1=3$,

$$\begin{pmatrix}-2 & 1 & 1\\ -3 & 2 & 1\\ -4 & 2 & 2\end{pmatrix}\begin{pmatrix}x_1\\ x_2\\ x_3\end{pmatrix}=\begin{pmatrix}0\\ 0\\ 0\end{pmatrix}.$$

We get a solution basis $\boldsymbol{x}_1=(1,1,1)^{\mathrm{T}}$, and so all $k_1\boldsymbol{x}_1$ (k_1 is a nonzero arbitrary constant) are eigenvectors corresponding to $\lambda_1=3$ of $\boldsymbol{A}$.

The eigenvalue equation of $\boldsymbol{A}$ for $\lambda_2 = 2$ is

$$\begin{pmatrix} -3 & 1 & 1 \\ -3 & 1 & 1 \\ -4 & 2 & 1 \end{pmatrix} \begin{pmatrix} x_1 \\ x_2 \\ x_3 \end{pmatrix} = \begin{pmatrix} 0 \\ 0 \\ 0 \end{pmatrix},$$

which has a solution basis $\boldsymbol{x}_2 = (1, 1, 2)^{\mathrm{T}}$. So, $k_2\boldsymbol{x}_2$ (k_2 is a nonzero arbitrary constant) are eigenvectors corresponding to $\lambda_2 = 2$ of $\boldsymbol{A}$. ■

Example 5.2 Characteristic polynomial of a diagonal matrix

The characteristic polynomials of a diagonal matrix $\boldsymbol{A}$ or an upper (lower) triangular matrix $\boldsymbol{B}$ with main diagonal entries $a_{11}, a_{22}, \ldots, a_{nn}$, are

$$|\lambda \boldsymbol{I} - \boldsymbol{A}| = |\lambda \boldsymbol{I} - \boldsymbol{B}| = (\lambda - a_{11})(\lambda - a_{22}) \cdots (\lambda - a_{nn}).$$

The n eigenvalues of $\boldsymbol{A}$ and $\boldsymbol{B}$ therefore are just the n main diagonal entries. ■

5.1.2 Properties of eigenvalues and eigenvectors

THEOREM 5.1

If $\boldsymbol{x}_1$ and $\boldsymbol{x}_2$ are both eigenvectors corresponding to an eigenvalue λ_0 of $\boldsymbol{A}$, then $k_1\boldsymbol{x}_1 + k_2\boldsymbol{x}_2 \neq \boldsymbol{0}$ is also an eigenvector belonging to λ_0 (k_1, k_2 are arbitrary constants).

Proof: Since $\boldsymbol{x}_1$ and $\boldsymbol{x}_2$ are solutions to the system of homogeneous linear equations

$$(\lambda_0 \boldsymbol{I} - \boldsymbol{A})\boldsymbol{x} = \boldsymbol{0},$$

so $k_1\boldsymbol{x}_1 + k_2\boldsymbol{x}_2$ is also a solution to the above system of equations (k_1, k_2 are arbitrary constants), and therefore is an eigenvector belong to λ_0, provided $k_1\boldsymbol{x}_1 + k_2\boldsymbol{x}_2$ is nonzero. □

In the solution space of $(\lambda \boldsymbol{I} - \boldsymbol{A})\boldsymbol{x} = 0$, all solution vectors except the zero vector are eigenvectors belonging to the eigenvalue λ of matrix $\boldsymbol{A}$. Any eigenvector of λ is a solution of $(\lambda \boldsymbol{I} - \boldsymbol{A})\boldsymbol{x} = 0$. Therefore, the solution space of

$(\lambda \boldsymbol{I} - \boldsymbol{A})\boldsymbol{x} = 0$ is called the **eigensubspace** corresponding to λ of the matrix $\boldsymbol{A}$ and is denoted as V_λ. An eigensubspace of an nth-order matrix $\boldsymbol{A}$ is a subspace of an n-dimensional vector space with dimension

$$\dim V_\lambda = n - \mathrm{r}(\lambda \boldsymbol{I} - \boldsymbol{A}).$$

Note that the eigenvalue of an nth-order real matrix may be a complex number, so an eigensubspace is usually a subspace of the complex vector space $\mathbb{C}^n$ of dimension n (see Appendix A).

Two eigensubspaces of a matrix $\boldsymbol{A}$ in Example 5.1 are:

$$V_{\lambda_1} = \{k\boldsymbol{x}\,|\boldsymbol{x} = (1, 1, 1)^{\mathrm{T}}, k \in \mathbb{C}\}, \quad V_{\lambda_2} = \{k\boldsymbol{x}\,|\boldsymbol{x} = (1, 1, 2)^{\mathrm{T}}, k \in \mathbb{C}\}.$$

THEOREM 5.2

Suppose the n eigenvalues of an nth-order matrix $\boldsymbol{A} = (a_{ij})$ *is* $\lambda_1, \lambda_2, \ldots, \lambda_n$. *Then,*

$$\text{(i) } \sum_{i=1}^{n} \lambda_i = \sum_{i=1}^{n} a_{ii}, \qquad \text{(ii) } \prod_{i=1}^{n} \lambda_i = \det \boldsymbol{A},$$

where $\sum_{i=1}^{n} a_{ii}$ *is the sum of the main diagonal entries, which is called the* ***trace*** *of a matrix* $\boldsymbol{A}$ *and written as* tr($\boldsymbol{A}$).

***Proof:** Suppose

$$\det(\lambda \boldsymbol{I} - \boldsymbol{A}) = \begin{vmatrix} \lambda - a_{11} & 0 - a_{12} & \cdots & 0 - a_{1n} \\ 0 - a_{21} & \lambda - a_{22} & \cdots & 0 - a_{2n} \\ \vdots & \vdots & \ddots & \vdots \\ 0 - a_{n1} & 0 - a_{n2} & \cdots & \lambda - a_{nn} \end{vmatrix} \tag{5.4}$$

$$= \lambda^n + c_1\lambda^{n-1} + c_2\lambda^{n-2} + \cdots + c_{n-1}\lambda + c_n. \tag{5.5}$$

Equation (5.4) can be written as the sum of 2^n determinants. The determinants containing λ^{n-1} after expansion are

$$\begin{vmatrix} -a_{11} & 0 & 0 & \cdots & 0 \\ -a_{21} & \lambda & 0 & \cdots & 0 \\ -a_{31} & 0 & \lambda & \ddots & 0 \\ \vdots & \vdots & \vdots & \ddots & \vdots \\ -a_{n1} & 0 & 0 & \cdots & \lambda \end{vmatrix}, \quad \begin{vmatrix} \lambda & -a_{12} & 0 & \cdots & 0 \\ 0 & -a_{22} & 0 & \cdots & 0 \\ 0 & -a_{32} & \lambda & \cdots & 0 \\ \vdots & \vdots & \vdots & \ddots & \vdots \\ 0 & -a_{n2} & 0 & \cdots & \lambda \end{vmatrix}, \ldots, \begin{vmatrix} \lambda & 0 & 0 & \cdots & -a_{1n} \\ 0 & \lambda & 0 & \cdots & -a_{2n} \\ 0 & 0 & \lambda & \cdots & -a_{3n} \\ \vdots & \vdots & \vdots & \ddots & \vdots \\ 0 & 0 & 0 & \cdots & -a_{nn} \end{vmatrix}.$$

The sum of all of them is equal to

$$-(a_{11} + a_{22} + \cdots + a_{nn})\lambda^{n-1} = \left(-\sum_{i=1}^{n} a_{ii}\right)\lambda^{n-1}.$$

That is, $c_1 = -\sum_{i=1}^{n} a_{ii}$ in (5.5).

The constant term without λ in (5.4) is

$$\begin{vmatrix} -a_{11} & -a_{12} & \cdots & -a_{1n} \\ -a_{21} & -a_{22} & \cdots & -a_{2n} \\ \vdots & \vdots & & \vdots \\ -a_{n1} & -a_{n2} & \cdots & -a_{nn} \end{vmatrix} = (-1)^n \det \boldsymbol{A},$$

that is, $c_n = (-1)^n \det \boldsymbol{A}$.

Suppose $\boldsymbol{A}$ has n eigenvalues $\lambda_1, \lambda_2, \ldots, \lambda_n$. According to the relation between roots and coefficients of an nth-degree polynomial, we have

$$\sum_{i=1}^{n} \lambda_i = -c_1 = \sum_{i=1}^{n} a_{ii}, \quad (-1)^n \prod_{i=1}^{n} \lambda_i = c_n = (-1)^n \det \boldsymbol{A},$$

the second term is equal to $\prod_{i=1}^{n} \lambda_i = \det \boldsymbol{A}$. □

From Theorem 5.2 (ii), any eigenvalue of $\boldsymbol{A}$ is nonzero when $\det \boldsymbol{A} \neq 0$ (that is, $\boldsymbol{A}$ is an invertible matrix). On the other hand, a singular matrix $\boldsymbol{A}$ has at least one zero eigenvalue.

If interested, readers may prove that c_k in (5.5) is

$$c_k = (-1)^k s_k, \quad k = 1, 2, \ldots, n,$$

where s_k is the sum of all kth-order principal minors of the nth-order matrix $\boldsymbol{A}$.

Eigenvectors of a matrix always correspond to eigenvalues of the matrix. One eigenvector cannot belong to different eigenvalues. Indeed, suppose if possible, an eigenvector $\boldsymbol{x}$ corresponds to eigenvalues λ_1, λ_2 $(\lambda_1 \neq \lambda_2)$ of $\boldsymbol{A}$ simultaneously, that is,

$$\boldsymbol{A}\boldsymbol{x} = \lambda_1 \boldsymbol{x} \text{ and } \boldsymbol{A}\boldsymbol{x} = \lambda_2 \boldsymbol{x},$$

then

$$\lambda_1 \boldsymbol{x} = \lambda_2 \boldsymbol{x}, \text{ that is, } (\lambda_1 - \lambda_2)\boldsymbol{x} = \boldsymbol{0}.$$

Thus, $\boldsymbol{x} = \boldsymbol{0}$ for $\lambda_1 - \lambda_2 \neq 0$, which is contradictory to the fact that $\boldsymbol{x} \neq \boldsymbol{0}$.

Now, some additional properties of eigenvalues and eigenvectors of matrices are listed below.

Property 5.1

If λ is an eigenvalue of $\boldsymbol{A}$, $\boldsymbol{x}$ is an eigenvector belonging to λ, then

(i) $k\lambda$ is an eigenvalue of $k\boldsymbol{A}$ (k is an arbitrary constant),
(ii) λ^m is an eigenvalue of $\boldsymbol{A}^m$ (m is a positive integer),

(iii) if A is invertible, λ^{-1} is an eigenvalue of A^{-1}, and x remains an eigenvector of matrices kA, A^m, A^{-1} corresponding to eigenvalues $k\lambda$, λ^m, λ^{-1}, respectively.

Proof:

(i) The proof is left as an exercise for readers.

(ii) According to $Ax = \lambda x$, we have:

$$A(Ax) = A(\lambda x) = \lambda(Ax) = \lambda(\lambda x),$$

that is,

$$A^2 x = \lambda^2 x.$$

Repeating the above steps $(m - 2)$ times, we have

$$A^m x = \lambda^m x.$$

So, λ^m is an eigenvalue of matrix A^m and x is also an eigenvector corresponding to λ^m of A^m.

(iii) $\lambda \neq 0$ when A is invertible, from $Ax = \lambda x$, we have

$$A^{-1}(Ax) = A^{-1}(\lambda x) = \lambda A^{-1} x, \text{ where } A^{-1} x = \lambda^{-1} x.$$

Therefore, λ^{-1} is an eigenvalue of A^{-1} and x is also an eigenvector associated with λ^{-1} of A^{-1}. □

Property 5.2

Matrices A and A^{T} have the same eigenvalues.

Proof: Since $(\lambda I - A)^{\mathrm{T}} = (\lambda I)^{\mathrm{T}} - A^{\mathrm{T}} = \lambda I - A^{\mathrm{T}}$, so

$$\det(\lambda I - A) = \det(\lambda I - A^{\mathrm{T}}).$$

Then, A and A^{T} have the same set of eigenvalues. □

*THEOREM 5.3

Suppose $A = (a_{ij})$ is an nth-order matrix. If any one of the two conditions

(i) $\sum_{j=1}^{n} |a_{ij}| < 1 \ (i = 1, 2, \ldots, n)$; (ii) $\sum_{i=1}^{n} |a_{ij}| < 1 \ (j = 1, 2, \ldots, n)$

holds, then the modulus $|\lambda_k|$ (referring to the absolute value when λ is a real number) of every eigenvalue $\lambda_k (k = 1, 2, \ldots, n)$ of A is less than 1.

Proof: Suppose λ is an arbitrary eigenvalue of $\boldsymbol{A}$ and $\boldsymbol{x}$ is the eigenvector belonging to λ. Then, $\boldsymbol{Ax} = \lambda\boldsymbol{x}$, that is,

$$\sum_{j=1}^{n} a_{ij}x_j = \lambda x_i \quad (i = 1, 2, \ldots, n).$$

Denote $\max\limits_{1\le j\le n} |x_j| = x_k$, then we have

$$|\lambda| = \left|\frac{\lambda x_k}{x_k}\right| = \left|\frac{\sum\limits_{j=1}^{n} a_{kj}x_j}{x_k}\right| \le \sum_{j=1}^{n} |a_{kj}| \left|\frac{x_j}{x_k}\right| \le \sum_{j=1}^{n} |a_{kj}|.$$

Thus, $|\lambda| < 1$ if condition (i) is true, which implies $|\lambda_k| < 1$ $(k = 1, 2, \ldots, n)$. With a similar argument, the result holds when condition (ii) is true. □

Example 5.3 A diagonalizable matrix

Suppose $\boldsymbol{A} = \begin{pmatrix} 1 & -1 & 1 \\ 2 & -2 & 2 \\ -1 & 1 & -1 \end{pmatrix}$.

(i) Find eigenvalues and eigenvectors of $\boldsymbol{A}$.
(ii) Find an invertible matrix $\boldsymbol{P}$ such that $\boldsymbol{P}^{-1}\boldsymbol{AP}$ is a diagonal matrix.

Solution:

(i) $$|\lambda\boldsymbol{I} - \boldsymbol{A}| = \begin{vmatrix} \lambda-1 & 1 & -1 \\ -2 & \lambda+2 & -2 \\ 1 & -1 & \lambda+1 \end{vmatrix}$$

$$= \begin{vmatrix} \lambda-1 & 0 & -1 \\ -2 & \lambda & -2 \\ 1 & \lambda & \lambda+1 \end{vmatrix} = \begin{vmatrix} \lambda-1 & 0 & -1 \\ -2 & \lambda & -2 \\ 3 & 0 & \lambda+3 \end{vmatrix}$$

$$= \lambda[(\lambda-1)(\lambda+3)+3] = \lambda^2(\lambda+2).$$

The eigenvalues of $\boldsymbol{A}$ are $\lambda_1 = \lambda_2 = 0$ (of multiplicity 2) and $\lambda_3 = -2$.

When $\lambda_1 = 0$, $\boldsymbol{x}_1 = (1, 1, 0)^{\mathrm{T}}$ and $\boldsymbol{x}_2 = (-1, 0, 1)^{\mathrm{T}}$ make a solution basis of $(\lambda_1\boldsymbol{I} - \boldsymbol{A})\boldsymbol{x} = \boldsymbol{0}$. So, eigenvectors corresponding to $\lambda_1 = 0$ of $\boldsymbol{A}$ are $k_1\boldsymbol{x}_1 + k_2\boldsymbol{x}_2 = k_1(1, 1, 0)^{\mathrm{T}} + k_2(-1, 0, 1)^{\mathrm{T}}$ (where k_1, k_2 are arbitrary constants not both zero).

For $\lambda_3 = -2$,

$$\begin{pmatrix} -3 & 1 & -1 \\ -2 & 0 & -2 \\ 1 & -1 & -1 \end{pmatrix} \begin{pmatrix} x_1 \\ x_2 \\ x_3 \end{pmatrix} = \begin{pmatrix} 0 \\ 0 \\ 0 \end{pmatrix},$$

$\boldsymbol{x}_3 = (-1, -2, 1)^{\mathrm{T}}$ is a solution basis. All eigenvectors corresponding to $\lambda_3 = -2$ are $k_3\boldsymbol{x}_3 = k_3(-1, -2, 1)^{\mathrm{T}}$ (k_3 is a nonzero arbitrary constant).

(ii) Arranging $\boldsymbol{A}\boldsymbol{x}_i = \lambda_i \boldsymbol{x}_i$ $(i = 1, 2, 3)$ as a matrix equality

$$\boldsymbol{A}(\boldsymbol{x}_1, \boldsymbol{x}_2, \boldsymbol{x}_3) = (\boldsymbol{x}_1, \boldsymbol{x}_2, \boldsymbol{x}_3) \begin{pmatrix} \lambda_1 & 0 & 0 \\ 0 & \lambda_2 & 0 \\ 0 & 0 & \lambda_3 \end{pmatrix},$$

taking

$$\boldsymbol{P} = (\boldsymbol{x}_1, \boldsymbol{x}_2, \boldsymbol{x}_3) = \begin{pmatrix} 1 & -1 & -1 \\ 1 & 0 & -2 \\ 0 & 1 & 1 \end{pmatrix}, \quad \boldsymbol{\Lambda} = \begin{pmatrix} 0 & & \\ & 0 & \\ & & -2 \end{pmatrix},$$

then $\boldsymbol{A}\boldsymbol{P} = \boldsymbol{P}\boldsymbol{\Lambda}$ and $|\boldsymbol{P}| = 2 \neq 0$. Thus, we conclude that $\boldsymbol{P}^{-1}\boldsymbol{A}\boldsymbol{P} = \boldsymbol{\Lambda}$ is a diagonal matrix. ■

5.1.3 Similar matrices and their properties

Definition 5.3

If there exists an invertible matrix $\boldsymbol{P}$ such that $\boldsymbol{P}^{-1}\boldsymbol{A}\boldsymbol{P} = \boldsymbol{B}$ for matrices $\boldsymbol{A}$ and $\boldsymbol{B}$, then $\boldsymbol{A}$ is said to be **similar** to $\boldsymbol{B}$, denoted as $\boldsymbol{A} \sim \boldsymbol{B}$.

The similarity among matrices is an equivalence relation. Indeed, one may easily verify the following three properties.

(i) Reflexivity: $\boldsymbol{A} \sim \boldsymbol{A}$, for every $\boldsymbol{A}$.
(ii) Symmetry: if $\boldsymbol{A} \sim \boldsymbol{B}$, then $\boldsymbol{B} \sim \boldsymbol{A}$.
(iii) Transitivity: if $\boldsymbol{A} \sim \boldsymbol{B}$, $\boldsymbol{B} \sim \boldsymbol{C}$, then $\boldsymbol{A} \sim \boldsymbol{C}$.

Similar matrices have the following properties.

(i) $\boldsymbol{P}^{-1}(k_1\boldsymbol{A}_1 + k_2\boldsymbol{A}_2)\boldsymbol{P} = k_1\boldsymbol{P}^{-1}\boldsymbol{A}_1\boldsymbol{P} + k_2\boldsymbol{P}^{-1}\boldsymbol{A}_2\boldsymbol{P}$ (where k_1, k_2 are arbitrary constants).
(ii) $\boldsymbol{P}^{-1}(\boldsymbol{A}_1\boldsymbol{A}_2)\boldsymbol{P} = (\boldsymbol{P}^{-1}\boldsymbol{A}_1\boldsymbol{P})(\boldsymbol{P}^{-1}\boldsymbol{A}_2\boldsymbol{P})$.
(iii) If $\boldsymbol{A} \sim \boldsymbol{B}$, then $\boldsymbol{A}^m \sim \boldsymbol{B}^m$ (m is a positive integer).

Proof: Since $\boldsymbol{A} \sim \boldsymbol{B}$, there exists an invertible matrix $\boldsymbol{P}$ such that

$$\boldsymbol{P}^{-1}\boldsymbol{A}\boldsymbol{P} = \boldsymbol{B}.$$

Then,

$$\boldsymbol{B}^m = (\boldsymbol{P}^{-1}\boldsymbol{A}\boldsymbol{P})(\boldsymbol{P}^{-1}\boldsymbol{A}\boldsymbol{P})\cdots(\boldsymbol{P}^{-1}\boldsymbol{A}\boldsymbol{P}) = \boldsymbol{P}^{-1}\boldsymbol{A}^m\boldsymbol{P}.$$

So,

$$\boldsymbol{A}^m \sim \boldsymbol{B}^m. \qquad \square$$

*(iv) If $\boldsymbol{A} \sim \boldsymbol{B}$, then $f(\boldsymbol{A}) \sim f(\boldsymbol{B})$, where

$$f(x) = a_n x^n + a_{n-1}x^{n-1} + \cdots + a_1 x + a_0,$$
$$f(\boldsymbol{A}) = a_n\boldsymbol{A}^n + a_{n-1}\boldsymbol{A}^{n-1} + \cdots + a_1\boldsymbol{A} + a_0\boldsymbol{I},$$
$$f(\boldsymbol{B}) = a_n\boldsymbol{B}^n + a_{n-1}\boldsymbol{B}^{n-1} + \cdots + a_1\boldsymbol{B} + a_0\boldsymbol{I}.$$

Using the conclusions of properties (i) and (iii), one may prove property (iv) easily.

Property (v) is important so it has been classified as a theorem.

THEOREM 5.4

Similar matrices have the same eigenvalues.

Proof: We only need to prove that similar matrices have the same characteristic polynomial. Suppose $\boldsymbol{A} \sim \boldsymbol{B}$. Then, there exists an invertible matrix $\boldsymbol{P}$ such that

$$\boldsymbol{P}^{-1}\boldsymbol{A}\boldsymbol{P} = \boldsymbol{B},$$

and

$$\begin{aligned}|\lambda\boldsymbol{I} - \boldsymbol{B}| &= \left|\lambda\boldsymbol{I} - \boldsymbol{P}^{-1}\boldsymbol{A}\boldsymbol{P}\right| = \left|\boldsymbol{P}^{-1}(\lambda\boldsymbol{I} - \boldsymbol{A})\boldsymbol{P}\right| \\ &= \left|\boldsymbol{P}^{-1}\right| |\lambda\boldsymbol{I} - \boldsymbol{A}| \, |\boldsymbol{P}| = |\lambda\boldsymbol{I} - \boldsymbol{A}|. \text{ (since } \left|\boldsymbol{P}^{-1}\right| |\boldsymbol{P}| = 1) \end{aligned} \qquad \square$$

Readers must note that the converse proposition of Theorem 5.4 is not true. For example,

$$\boldsymbol{I} = \begin{pmatrix} 1 & 0 \\ 0 & 1 \end{pmatrix}, \quad \boldsymbol{A} = \begin{pmatrix} 1 & 1 \\ 0 & 1 \end{pmatrix},$$

both have 1 as an eigenvalue of multiplicity 2, but for any invertible matrix $\boldsymbol{P}$, we have $\boldsymbol{P}^{-1}\boldsymbol{I}\boldsymbol{P} = \boldsymbol{I} \neq \boldsymbol{A}$, so $\boldsymbol{A}$ and $\boldsymbol{I}$ are not similar.

5.2 Diagonalization Conditions

A matrix is called diagonalizable if the matrix is similar to a diagonal matrix. In this section, we shall discuss the conditions for a matrix to be diagonalizable. The main conclusion is that an nth-order matrix is diagonalizable, if and only if, it has n linearly independent eigenvectors, or the (algebraic) multiplicity of each eigenvalue of the matrix is equal to the (geometric) dimension of the corresponding eigensubspace.

From now on, we shall denote the diagonal matrix with main diagonal entries $a_1, a_2, \ldots, a_n$ by diag $(a_1, a_2, \ldots, a_n)$ or by $\boldsymbol{\Lambda}$.

From Example 5.3 in §5.1, when a third-order matrix $\boldsymbol{A}$ has three linearly independent eigenvectors $\boldsymbol{x}_1, \boldsymbol{x}_2, \boldsymbol{x}_3$, then we have

$$\boldsymbol{P}^{-1}\boldsymbol{A}\boldsymbol{P} = \text{diag}\,(\lambda_1, \lambda_2, \lambda_3),$$

where $\boldsymbol{P} = (\boldsymbol{x}_1, \boldsymbol{x}_2, \boldsymbol{x}_3)$ and $\lambda_1, \lambda_2, \lambda_3$ are eigenvalues corresponding to eigenvectors $\boldsymbol{x}_1, \boldsymbol{x}_2, \boldsymbol{x}_3$, respectively. This indicates that, for a third-order matrix $\boldsymbol{A}$, having three linearly independent eigenvectors is a sufficient condition under which $\boldsymbol{A}$ and a diagonal matrix are similar. Indeed, it is also a necessary condition. A general conclusion is stated in the following.

THEOREM 5.5

*An nth-order matrix **A** is similar to a diagonal matrix, if and only if, **A** has n linearly independent eigenvectors.*

Proof: To prove the 'only if' part, suppose

$$\boldsymbol{P}^{-1}\boldsymbol{A}\boldsymbol{P} = \text{diag}\,(\lambda_1, \lambda_2, \ldots, \lambda_n) \xlongequal{\text{written as}} \boldsymbol{\Lambda},$$

that is,

$$\boldsymbol{A}\boldsymbol{P} = \boldsymbol{P}\boldsymbol{\Lambda}.$$

The column partition of matrix $\boldsymbol{P}$ is

$$\boldsymbol{P} = (\boldsymbol{x}_1, \boldsymbol{x}_2, \ldots, \boldsymbol{x}_n).$$

Then,

$$\boldsymbol{A}(\boldsymbol{x}_1, \boldsymbol{x}_2, \ldots, \boldsymbol{x}_n) = (\boldsymbol{x}_1, \boldsymbol{x}_2, \ldots, \boldsymbol{x}_n)\begin{pmatrix} \lambda_1 & & & \\ & \lambda_2 & & \\ & & \ddots & \\ & & & \lambda_n \end{pmatrix},$$

that is,

$$(\boldsymbol{A}\boldsymbol{x}_1, \boldsymbol{A}\boldsymbol{x}_2, \ldots, \boldsymbol{A}\boldsymbol{x}_n) = (\lambda_1\boldsymbol{x}_1, \lambda_2\boldsymbol{x}_2, \ldots, \lambda_n\boldsymbol{x}_n).$$

So,

$$Ax_i = \lambda_i x_i \quad (i = 1, 2, \ldots, n).$$

So, $x_1, x_2, \ldots, x_n$ are eigenvectors of A belonging to the eigenvalues $\lambda_1, \lambda_2, \ldots, \lambda_n$, respectively. They are linearly independent as P is invertible. The necessity condition is proved.

The 'if' part (sufficiency condition) is also true, since the steps described above are obviously invertible. □

There are only two linearly independent eigenvectors for A in Example 5.1 of §5.1, so it cannot be diagonalized.

From Theorem 5.4, we know that if A is similar to a diagonal matrix Λ, then the main diagonal of matrix Λ are all eigenvalues of A. Up to the order of arrangement of the eigenvalues, Λ is unique. We call Λ the **similar canonical form** of A.

THEOREM 5.6

The eigenvectors corresponding to different eigenvalues of matrix A *are linearly independent.*

Proof: Suppose m distinct eigenvalues of A is $\lambda_1, \lambda_2, \ldots, \lambda_m$ and the corresponding eigenvectors are $x_1, x_2, \ldots, x_m$, respectively.

We prove that $x_1, x_2, \ldots, x_m$ are linearly independent by induction on m.

When $m = 1$, the conclusion holds obviously (eigenvector $x_1 \neq 0$).

Suppose the eigenvectors $x_1, x_2, \ldots, x_k$ belonging to k distinct eigenvalues $\lambda_1, \lambda_2, \ldots, \lambda_k$ are linearly independent. Now consider the case of eigenvectors belonging to $(k+1)$ distinct eigenvalues.

Suppose

$$a_1 x_1 + a_2 x_2 + \cdots + a_k x_k + a_{k+1} x_{k+1} = 0. \quad ①$$

Then,

$$A(a_1 x_1 + a_2 x_2 + \cdots + a_k x_k + a_{k+1} x_{k+1}) = 0,$$

that is,

$$a_1 \lambda_1 x_1 + a_2 \lambda_2 x_2 + \cdots + a_k \lambda_k x_k + a_{k+1} \lambda_{k+1} x_{k+1} = 0. \quad ②$$

Subtracting expression ② from expression ① multiplied by λ_{k+1} yields

$$a_1(\lambda_{k+1} - \lambda_1)x_1 + a_2(\lambda_{k+1} - \lambda_2)x_2 + \cdots + a_k(\lambda_{k+1} - \lambda_k)x_k = 0.$$

According to the hypothesis, $x_1, x_2, \ldots, x_k$ are linearly independent, so

$$a_i(\lambda_{k+1} - \lambda_i) = 0, \quad i = 1, 2, \ldots, k.$$

Since $\lambda_{k+1} \neq \lambda_i, i = 1, 2, \ldots, k$, hence,

$$a_i = 0, \quad i = 1, 2, \ldots, k. \qquad ③$$

Substituting the outcome of ③ into ①, we have

$$a_{k+1}\boldsymbol{x}_{k+1} = \mathbf{0}.$$

As eigenvector $\boldsymbol{x}_{k+1} \neq \mathbf{0}$, so $a_{k+1} = 0$, where $\boldsymbol{x}_1, \boldsymbol{x}_2, \ldots, \boldsymbol{x}_{k+1}$ are linearly independent. □

COROLLARY 5.1

If an nth-order matrix $\boldsymbol{A}$ *has n distinct eigenvalues, then* $\boldsymbol{A}$ *is similar to a diagonal matrix.*

However, the converse proposition of the corollary is not true. For instance, in Example 5.3 in §5.1, $\boldsymbol{A}$ is similar to a diagonal matrix, but the eigenvalue 0 is of multiplicity 2.

***THEOREM 5.7**

Suppose $\lambda_1, \lambda_2, \ldots, \lambda_m$ *are m distinct eigenvalues of an nth-order matrix* $\boldsymbol{A}$ *and linearly independent eigenvectors belonging to* λ_i *are* $\boldsymbol{x}_{i_1}, \boldsymbol{x}_{i_2}, \ldots, \boldsymbol{x}_{i_{r_i}}$ $(i = 1, 2, \ldots, m)$, *then the set of vectors made up by all the eigenvectors (the total number is* $r_1 + r_2 + \cdots + r_m$*)* $\{\boldsymbol{x}_{i_1}, \boldsymbol{x}_{i_2}, \ldots, \boldsymbol{x}_{i_{r_i}} \mid i = 1, 2, \ldots, m\}$ *are linearly independent.*

Proof:
Suppose

$$\sum_{i=1}^{m} (k_{i_1}\boldsymbol{x}_{i_1} + k_{i_2}\boldsymbol{x}_{i_2} + \cdots + k_{i_{r_i}}\boldsymbol{x}_{i_{r_i}}) = \mathbf{0}, \qquad ①$$

and let

$$\boldsymbol{y}_i = k_{i_1}\boldsymbol{x}_{i_1} + k_{i_2}\boldsymbol{x}_{i_2} + \cdots + k_{i_{r_i}}\boldsymbol{x}_{i_{r_i}}. \qquad ②$$

Then, ① gives

$$\boldsymbol{y}_1 + \boldsymbol{y}_2 + \cdots + \boldsymbol{y}_m = \mathbf{0}, \qquad ③$$

where $\boldsymbol{y}_i$ is a zero vector or an eigenvector corresponding to λ_i $(i = 1, 2, \ldots, m)$. According to Theorem 5.6, none of $\boldsymbol{y}_1, \boldsymbol{y}_2, \ldots, \boldsymbol{y}_m$ in ③ is an eigenvector (because if there are one or several eigenvectors, then we know that the sum cannot be equal to zero by virtue of their linear independence). So,

$$\boldsymbol{y}_i = \mathbf{0}, \quad (i = 1, 2, \ldots, m). \qquad ④$$

Since $\boldsymbol{x}_{i_1}, \boldsymbol{x}_{i_2}, \dots, \boldsymbol{x}_{i_{r_i}}$ are linearly independent, then from ④ and ②, we have

$$k_{i_1} = k_{i_2} = \cdots = k_{i_{r_i}} = 0, \quad i = 1, 2, \dots, m.$$

Hence, the theorem follows. □

From Theorem 5.7, we know that eigenvectors $\boldsymbol{x}_1, \boldsymbol{x}_2, \boldsymbol{x}_3$ of $\boldsymbol{A}$ in Example 5.3 of §5.1 are linearly independent.

*THEOREM 5.8

Suppose λ_0 is an eigenvalue of multiplicity k of an nth-order matrix and the largest number of linearly independent eigenvectors corresponding to λ_0 is l, then $k \geq l$.

Proof: Suppose $l > k$. As

$$\boldsymbol{A}\boldsymbol{x}_i = \lambda_0 \boldsymbol{x}_i, \quad \boldsymbol{x}_i \neq \boldsymbol{0}, \quad i = 1, 2, \dots, l, \qquad ①$$

we extend $\boldsymbol{x}_1, \boldsymbol{x}_2, \dots, \boldsymbol{x}_l$ to a basis of the n-dimensional vector space $\mathbb{C}^n$,

$$\boldsymbol{x}_1, \boldsymbol{x}_2, \dots, \boldsymbol{x}_l, \boldsymbol{x}_{l+1}, \dots, \boldsymbol{x}_n,$$

where $\boldsymbol{x}_{l+1}, \dots, \boldsymbol{x}_n$ are generally not eigenvectors corresponding to λ_0, but $\boldsymbol{A}\boldsymbol{x}_m \in \mathbb{C}^n (m = l+1, \dots, n)$ which can be linearly expressed by the above basis, that is,

$$\boldsymbol{A}\boldsymbol{x}_m = a'_{1m}\boldsymbol{x}_1 + a'_{2m}\boldsymbol{x}_2 + \cdots + a'_{lm}\boldsymbol{x}_l + a'_{l+1\,m}\boldsymbol{x}_{l+1} + \cdots + a'_{nm}\boldsymbol{x}_n$$
$$(m = l+1, \dots, n). \qquad ②$$

The n equalities in ① and ② are written into a matrix equality

$$\boldsymbol{A}(\boldsymbol{x}_1, \dots, \boldsymbol{x}_l, \boldsymbol{x}_{l+1}, \dots, \boldsymbol{x}_n)$$

$$= (\boldsymbol{x}_1, \dots, \boldsymbol{x}_l, \boldsymbol{x}_{l+1}, \dots, \boldsymbol{x}_n) \left(\begin{array}{ccc|ccc} \lambda_0 & & & a'_{1l+1} & \cdots & a'_{1n} \\ & \ddots & & \vdots & \ddots & \vdots \\ & & \lambda_0 & a'_{ll+1} & \cdots & a'_{ln} \\ \hline & & & a'_{l+1l+1} & \cdots & a'_{l+1n} \\ & \boldsymbol{0} & & \vdots & \ddots & \vdots \\ & & & a'_{nl+1} & \cdots & a'_{nn} \end{array} \right), \qquad ③$$

where the number of λ_0's is l.

Denoting $\boldsymbol{P}=(\boldsymbol{x}_1,\ldots,\boldsymbol{x}_l,\boldsymbol{x}_{l+1},\ldots,\boldsymbol{x}_n)$ and expressing the matrix on the *RHS* of ③ as a block matrix, we have

$$\boldsymbol{P}^{-1}\boldsymbol{AP}=\begin{pmatrix}\lambda_0\boldsymbol{I}_l & \boldsymbol{A}_1\\ \boldsymbol{0} & \boldsymbol{A}_2\end{pmatrix}.$$

According to the fact that similar matrices have the same characteristic polynomial, we have

$$\begin{aligned}|\lambda\boldsymbol{I}_n-\boldsymbol{A}|&=\left|\lambda\boldsymbol{I}_n-\boldsymbol{P}^{-1}\boldsymbol{AP}\right|\\&=\begin{vmatrix}(\lambda-\lambda_0)\boldsymbol{I}_l & -\boldsymbol{A}_1\\ \boldsymbol{0} & \lambda\boldsymbol{I}_{n-l}-\boldsymbol{A}_2\end{vmatrix}\\&=|(\lambda-\lambda_0)\boldsymbol{I}_l|\,|\lambda\boldsymbol{I}_{n-l}-\boldsymbol{A}_2|\\&=(\lambda-\lambda_0)^l g(\lambda).\end{aligned} \qquad ④$$

This indicates that λ_0 is an eigenvalue with $\boldsymbol{A}$ with multiplicity of at least l $(l>k)$, which is contradictory to λ_0 being an eigenvalue of multiplicity k, so we conclude that $l\le k$. □

Since any solution basis of $(\lambda_0\boldsymbol{I}-\boldsymbol{A})\boldsymbol{x}=\boldsymbol{0}$ contains $l=n-\mathrm{r}(\lambda_0\boldsymbol{I}-\boldsymbol{A})$ vectors, then the dimension of the eigensubspace is $\dim V_{\lambda_0}=l$. Theorem 5.8 can now be stated as $\dim V_{\lambda_0}\le$ the multiplicity of eigenvalue λ_0.

***THEOREM 5.9**

*An nth-order matrix **A** is similar to a diagonal matrix, if and only if, the largest number of linearly independent eigenvectors corresponding to each eigenvalue of **A** is equal to the multiplicity of the eigenvalue (i.e., the dimension of each eigensubspace V_{λ_i} of **A** is equal to the multiplicity of eigenvalue λ_i).*

Proof: Suppose $|\lambda\boldsymbol{I}-\boldsymbol{A}|=\prod_{i=1}^{m}(\lambda-\lambda_i)^{r_i}$, where $\lambda_1,\ldots,\lambda_m\in\mathbb{C}$ are different from each other and $\sum_{i=1}^{m}r_i=n$.

Sufficiency: Under the assumption, there are r_i linearly independent eigenvectors corresponding to λ_i and m different eigenvalues. From Theorem 5.7, $\boldsymbol{A}$ has n linearly independent eigenvectors. Thus, $\boldsymbol{A}$ is similar to a diagonal matrix according to Theorem 5.5.

Use reduction to absurdity to prove the necessity condition. Suppose there is an eigenvalue λ_i such that $l_i<r_i$, where l_i is the number of linearly independent eigenvectors belonging to λ_i and r_i is the multiplicity of λ_i. We know from Theorem 5.8 that the number of linearly independent eigenvectors of $\boldsymbol{A}$ is less than n, so $\boldsymbol{A}$ is not similar to a diagonal matrix. □

Example 5.4 Diagonalization of a matrix

Consider a real symmetric matrix

$$A=\begin{pmatrix} 1 & -1 & -1 & -1 \\ -1 & 1 & -1 & -1 \\ -1 & -1 & 1 & -1 \\ -1 & -1 & -1 & 1 \end{pmatrix}.$$

Is A similar to a diagonal matrix? If yes, give the diagonal matrix Λ and an invertible matrix P such that $P^{-1}AP=\Lambda$, and compute A^k (k is a positive integer).

Solution: The characteristic polynomial of A is

$$|\lambda I-A|=\begin{vmatrix} \lambda-1 & 1 & 1 & 1 \\ 1 & \lambda-1 & 1 & 1 \\ 1 & 1 & \lambda-1 & 1 \\ 1 & 1 & 1 & \lambda-1 \end{vmatrix}$$

$$=(\lambda+2)\begin{vmatrix} 1 & 1 & 1 & 1 \\ 1 & \lambda-1 & 1 & 1 \\ 1 & 1 & \lambda-1 & 1 \\ 1 & 1 & 1 & \lambda-1 \end{vmatrix}$$

$$=(\lambda+2)\begin{vmatrix} 1 & 1 & 1 & 1 \\ 0 & \lambda-2 & 0 & 0 \\ 0 & 0 & \lambda-2 & 0 \\ 0 & 0 & 0 & \lambda-2 \end{vmatrix}$$

$$=(\lambda+2)(\lambda-2)^3.$$

So, eigenvalues of A are $\lambda_1=-2$ (single root) and $\lambda_1=2$ (triple roots).

Upon solving $(\lambda_1 I-A)x=\mathbf{0}$, that is,

$$\begin{pmatrix} -3 & 1 & 1 & 1 \\ 1 & -3 & 1 & 1 \\ 1 & 1 & -3 & 1 \\ 1 & 1 & 1 & -3 \end{pmatrix}\begin{pmatrix} x_1 \\ x_2 \\ x_3 \\ x_4 \end{pmatrix}=\begin{pmatrix} 0 \\ 0 \\ 0 \\ 0 \end{pmatrix},$$

we get eigenvectors $\{k_1 x_1 \,|\, x_1=(1,1,1,1)^{\mathrm{T}}, k_1\neq 0\}$ corresponding to λ_1.

Again, upon solving $(\lambda_2 I-A)x=\mathbf{0}$, that is

$$x_1+x_2+x_3+x_4=0,$$

we get a solution basis $x_{21}=(1,-1,0,0)^{\mathrm{T}}$, $x_{22}=(1,0,-1,0)^{\mathrm{T}}$, and $x_{23}=(1,0,0,-1)^{\mathrm{T}}$.

So, A is similar to a diagonal matrix as it has four linearly independent eigenvectors.

Let

$$P = (x_1, x_{21}, x_{22}, x_{23}) = \begin{pmatrix} 1 & 1 & 1 & 1 \\ 1 & -1 & 0 & 0 \\ 1 & 0 & -1 & 0 \\ 1 & 0 & 0 & -1 \end{pmatrix}.$$

Then,

$$P^{-1}AP = \begin{pmatrix} -2 & & & \\ & 2 & & \\ & & 2 & \\ & & & 2 \end{pmatrix} = \Lambda.$$

The four diagonal entries are eigenvalues corresponding to the four eigenvectors. Since eigenvectors [or a solution basis of $(\lambda I - A)x = 0$] are not unique, neither P is unique.

From $A = P\Lambda P^{-1}$, we have

$$A^k = (P\Lambda P^{-1})^k = P\Lambda P^{-1} P\Lambda P^{-1} \cdots P\Lambda P^{-1} = P\Lambda^k P^{-1}$$

$$= \begin{pmatrix} 1 & 1 & 1 & 1 \\ 1 & -1 & 0 & 0 \\ 1 & 0 & -1 & 0 \\ 1 & 0 & 0 & -1 \end{pmatrix} \begin{pmatrix} (-2)^k & & & \\ & 2^k & & \\ & & 2^k & \\ & & & 2^k \end{pmatrix} \frac{1}{4} \begin{pmatrix} 1 & 1 & 1 & 1 \\ 1 & -3 & 1 & 1 \\ 1 & 1 & -3 & 1 \\ 1 & 1 & 1 & -3 \end{pmatrix}$$

$$= \begin{cases} 2^k I_4, & k \text{ is even}, \\ 2^{k-1} A, & k \text{ is odd}. \end{cases}$$ ■

Example 5.5 Non-diagonalizability of an upper triangular matrix with equal diagonal entries

Suppose $A = (a_{ij})_{n\times n}$ is an upper triangular matrix with all main diagonal entries 2 and there exist $a_{ij} \neq 0\,(i < j)$. Is A similar to a diagonal matrix?

Solution: Let $A = \begin{pmatrix} 2 & * & \cdots & * \\ 0 & 2 & \cdots & * \\ \vdots & \vdots & \ddots & \vdots \\ 0 & 0 & \cdots & 2 \end{pmatrix}$ (where $*$ are arbitrary constants not all zero). Then, $|\lambda I - A| = (\lambda - 2)^n$ and $\lambda = 2$ is an eigenvalue of multiplicity n of $\boldsymbol{A}$. Moreover, $\mathrm{r}(2\boldsymbol{I}-\boldsymbol{A}) \geq 1$ implies that the number of vectors in a solution basis of $(2\boldsymbol{I}-\boldsymbol{A})\boldsymbol{x}=\boldsymbol{0}$ is less than or equal to $(n-1)$, that is, the number of linearly independent eigenvectors of $\boldsymbol{A}$ is less than or equal to $(n-1)$. Therefore, $\boldsymbol{A}$ is not similar to a diagonal matrix. ■

*Example 5.6 Diagonalizing a matric polynomial

Let $f(x)=x^3-2x+5$ and a matrix $\boldsymbol{A}$ be defined as in Example 5.4. Give an invertible matrix $\boldsymbol{P}$ and a diagonal matrix $\boldsymbol{\Lambda}_1$ such that $\boldsymbol{P}^{-1}f(\boldsymbol{A})\boldsymbol{P}=\boldsymbol{\Lambda}_1$, where $f(\boldsymbol{A})=\boldsymbol{A}^3-2\boldsymbol{A}+5\boldsymbol{I}$.

Solution: Using Property 5.1 in §5.1 for $\boldsymbol{A}\boldsymbol{x}=\lambda\boldsymbol{x}$ ($\boldsymbol{x}\neq\boldsymbol{0}$), then

$$(k\boldsymbol{A})\boldsymbol{x}=(k\lambda)\boldsymbol{x}, \quad \boldsymbol{A}^m\boldsymbol{x}=\lambda^m\boldsymbol{x} \quad (k\in\mathbb{C}, m\in\mathbb{N}).$$

So, we have $f(\boldsymbol{A})\boldsymbol{x}=f(\lambda)\boldsymbol{x}$ (readers are invited to prove this) for arbitrary polynomial $f(x)$. Then, $f(\lambda)$ is an eigenvalue of $f(\boldsymbol{A})$ and $\boldsymbol{x}$ is the corresponding eigenvector of $f(\lambda)$. With $\lambda_1=-2$ and $\lambda_2=2$, we have $f(\lambda_1)=1$, $f(\lambda_2)=9$ and $\boldsymbol{x}_1, \boldsymbol{x}_{21}, \boldsymbol{x}_{22}, \boldsymbol{x}_{23}$ in Example 5.4 are eigenvectors corresponding to eigenvalues $f(\lambda_1)$ and $f(\lambda_2)$ of $f(\boldsymbol{A})$.

Let $\boldsymbol{P}=(\boldsymbol{x}_1, \boldsymbol{x}_{21}, \boldsymbol{x}_{22}, \boldsymbol{x}_{23})$, then

$$\boldsymbol{P}^{-1}f(\boldsymbol{A})\boldsymbol{P}=\mathrm{diag}(f(\lambda_1), f(\lambda_2), f(\lambda_2), f(\lambda_2))=\mathrm{diag}(1,9,9,9)=\boldsymbol{\Lambda}_1.$$

There is another approach to this problem. In Example 5.4,

$$\boldsymbol{P}^{-1}\boldsymbol{A}\boldsymbol{P}=\mathrm{diag}(-2,2,2,2)=\boldsymbol{\Lambda},$$

that is,

$$\boldsymbol{A}=\boldsymbol{P}\boldsymbol{\Lambda}\boldsymbol{P}^{-1}.$$

Then,

$$\begin{aligned} f(\boldsymbol{A})&=\boldsymbol{A}^3-2\boldsymbol{A}+5\boldsymbol{I}=(\boldsymbol{P}\boldsymbol{\Lambda}\boldsymbol{P}^{-1})^3-2(\boldsymbol{P}\boldsymbol{\Lambda}\boldsymbol{P}^{-1})+5\boldsymbol{I}\\ &=\boldsymbol{P}(\boldsymbol{\Lambda}^3-2\boldsymbol{\Lambda}+5\boldsymbol{I})\boldsymbol{P}^{-1}=\boldsymbol{P}f(\boldsymbol{\Lambda})\boldsymbol{P}^{-1}. \end{aligned}$$

Hence,

$$\begin{aligned} \boldsymbol{P}^{-1}f(\boldsymbol{A})\boldsymbol{P}=f(\boldsymbol{\Lambda})&=\mathrm{diag}(f(-2), f(2), f(2), f(2))\\ &=\mathrm{diag}(1,9,9,9)=\boldsymbol{\Lambda}_1. \end{aligned}$$

■

***Example 5.7 Diagonalizing an idempotent matrix**

Suppose the rank of an idempotent matrix $\boldsymbol{A}$ (i.e., $\boldsymbol{A}^2 = \boldsymbol{A}$) of order n is $r (0 < r \leq n)$. Prove that, $\boldsymbol{A} \sim \text{diag}(1, 1, \ldots, 1, 0, 0, \ldots, 0)$, in which there are r 1's.

Proof: Let $\boldsymbol{Ax} = \lambda\boldsymbol{x}$ ($\boldsymbol{x} \neq \boldsymbol{0}$). Then, $\lambda\boldsymbol{x} = \boldsymbol{Ax} = \boldsymbol{A}^2\boldsymbol{x} = \lambda^2\boldsymbol{x}$ ($\boldsymbol{x} \neq \boldsymbol{0}$) implies $\lambda^2 = \lambda$. This indicates that each eigenvalue of the idempotent matrix is 0 or 1.

When $\lambda_1 = 1$, the rank of the eigenmatrix $\boldsymbol{I} - \boldsymbol{A}$ is $(n - r)$. This is due to $\boldsymbol{A} - \boldsymbol{A}^2 = \boldsymbol{A}(\boldsymbol{I} - \boldsymbol{A}) = \boldsymbol{0}$ and

$$\text{r}(\boldsymbol{A}) + \text{r}(\boldsymbol{I} - \boldsymbol{A}) \leq n.$$

Moreover,

$$\text{r}(\boldsymbol{A}) + \text{r}(\boldsymbol{I} - \boldsymbol{A}) \geq \text{r}(\boldsymbol{A} + (\boldsymbol{I} - \boldsymbol{A})) = \text{r}(\boldsymbol{I}) = n,$$

So,

$$\text{r}(\boldsymbol{A}) + \text{r}(\boldsymbol{I} - \boldsymbol{A}) = n.$$

Then,

$$\text{r}(\boldsymbol{I} - \boldsymbol{A}) = n - r, \quad \dim V_{\lambda_1} = n - (n - r) = r.$$

For $(\boldsymbol{I} - \boldsymbol{A})\boldsymbol{x} = \boldsymbol{0}$, we can choose r linearly independent eigenvectors $\boldsymbol{x}_1, \boldsymbol{x}_2, \ldots, \boldsymbol{x}_r$ from the column vectors of $\boldsymbol{A}$ corresponding to $\lambda_1 = 1$.

For $(\lambda_2\boldsymbol{I} - \boldsymbol{A})\boldsymbol{x} = \boldsymbol{0}$ and $\lambda_2 = 0$, that is, $\boldsymbol{Ax} = \boldsymbol{0}$, we can choose $(n - r)$ linearly independent eigenvectors $\boldsymbol{x}_{r+1}, \ldots, \boldsymbol{x}_n$ from the column vectors of $(\boldsymbol{I} - \boldsymbol{A})$. Take

$$\boldsymbol{P} = (\boldsymbol{x}_1, \boldsymbol{x}_2, \ldots, \boldsymbol{x}_r, \boldsymbol{x}_{r+1}, \ldots, \boldsymbol{x}_n).$$

Then,

$$\boldsymbol{P}^{-1}\boldsymbol{AP} = \text{diag}(1, 1, \ldots, 1, 0, \cdots, 0),$$

where the number of 1's is $r = \text{r}(\boldsymbol{A})$ and the number of 0's is $n - r$. If $\text{r}(\boldsymbol{I} - \boldsymbol{A}) = 0$, then $\boldsymbol{A} = \boldsymbol{I}$. The proposition is also true when $r = n$. ■

5.3 Diagonalization of Real Symmetric Matrices

There exist matrices not similar to diagonal matrices as pointed out in the last section. However, a real symmetric matrix, which plays an important part in applications, can always be diagonalized. The eigenvalues of these symmetric matrices are all real. Furthermore, there exists an orthogonal matrix $\boldsymbol{T}$ such that $\boldsymbol{T}^{-1}\boldsymbol{A}\boldsymbol{T}$ is a diagonal matrix for a real symmetric matrix $\boldsymbol{A}$. In order to prove these important properties, we first introduce the relevant concepts and properties for complex matrices and complex vectors.

Definition 5.4

A matrix or a vector with complex entries is called a **complex matrix** or a **complex vector**, respectively.

Definition 5.5

Let $\boldsymbol{A} = (a_{ij})_{m\times n}$ with complex number a_{ij}. Then, $\overline{\boldsymbol{A}} = (\overline{a}_{ij})_{m\times n}$ is called the **conjugate matrix** of $\boldsymbol{A}$, where $\overline{a}_{ij}$ is the conjugate complex number of a_{ij}.

From Definition 5.5, we have $\overline{\overline{\boldsymbol{A}}} = \boldsymbol{A}$, $\overline{\boldsymbol{A}}^{\mathrm{T}} = \overline{(\boldsymbol{A}^{\mathrm{T}})}$, $\overline{\boldsymbol{A}}^{\mathrm{T}} = \boldsymbol{A}$ when $\boldsymbol{A}$ is a real symmetric matrix.

By the definition and computational properties of conjugate complex numbers, the conjugate matrices have the following properties.

(i) $\overline{k\boldsymbol{A}} = \overline{k}\,\overline{\boldsymbol{A}}$ (k is a complex number),
(ii) $\overline{\boldsymbol{A}+\boldsymbol{B}} = \overline{\boldsymbol{A}} + \overline{\boldsymbol{B}}$,
(iii) $\overline{\boldsymbol{A}\boldsymbol{B}} = \overline{\boldsymbol{A}}\,\overline{\boldsymbol{B}}$,
(iv) $\overline{(\boldsymbol{A}\boldsymbol{B})}^{\mathrm{T}} = \overline{\boldsymbol{B}}^{\mathrm{T}}\overline{\boldsymbol{A}}^{\mathrm{T}}$,
(v) If $\boldsymbol{A}$ is invertible, then $\overline{\boldsymbol{A}^{-1}} = \left(\overline{\boldsymbol{A}}\right)^{-1}$,
(vi) $\det \overline{\boldsymbol{A}} = \overline{\det \boldsymbol{A}}$.

The n-dimensional complex vector (written in column form) $\boldsymbol{x}$ satisfies the following property.

$\overline{\boldsymbol{x}}^{\mathrm{T}}\boldsymbol{x} \geq 0$, the equality sign holds, if and only if, $\boldsymbol{x} = \boldsymbol{0}$. This is because, if $\boldsymbol{x} = (x_1, x_2, \cdots, x_n)^{\mathrm{T}}$, $x_i \in \mathbb{C}$ $(i = 1, 2, \ldots, n)$, then

$$\overline{\boldsymbol{x}}^{\mathrm{T}}\boldsymbol{x} = \overline{x_1}x_1 + \overline{x_2}x_2 + \cdots + \overline{x_n}x_n = |x_1|^2 + |x_2|^2 + \cdots + |x_n|^2 \geq 0,$$

where $|x_i|$ is the modulus of the complex number x_i. Hence, $\overline{\boldsymbol{x}}^{\mathrm{T}}\boldsymbol{x} = 0$, if and only if, $x_i = 0$ $(i = 1, 2, \ldots, n)$, that is, $\boldsymbol{x} = \boldsymbol{0}$.

5.3.1 Eigenvalues and eigenvectors of real symmetric matrices

Although the coefficients of the characteristic polynomials of real matrices are all real, their roots may be complex. Eigenvectors corresponding to these roots could then be complex as well. However, eigenvalues of a real symmetric matrix are all real, the corresponding eigenvectors are real over the real number field, and the eigenvectors corresponding to different eigenvalues are orthogonal. We shall prove these results separately in the following text.

THEOREM 5.10

Any eigenvalue of a real symmetric matrix $\boldsymbol{A}$ is real.

Proof: Suppose λ is an arbitrary eigenvalue of $\boldsymbol{A}$. From $\overline{\boldsymbol{A}}^{\mathrm{T}} = \boldsymbol{A}$ and $\boldsymbol{A}\boldsymbol{x} = \lambda\boldsymbol{x}$, we have

$$\overline{(\boldsymbol{A}\boldsymbol{x})}^{\mathrm{T}} = \overline{(\lambda\boldsymbol{x})^{\mathrm{T}}}, \text{ that is, } \overline{\boldsymbol{x}^{\mathrm{T}}\boldsymbol{A}^{\mathrm{T}}}\boldsymbol{x} = \overline{\lambda}\,\overline{\boldsymbol{x}}^{\mathrm{T}}\boldsymbol{x}, \text{ that is, } \overline{\boldsymbol{x}}^{\mathrm{T}}\boldsymbol{A}\boldsymbol{x} = \lambda\overline{\boldsymbol{x}}^{\mathrm{T}}\boldsymbol{x} = \overline{\lambda}\,\overline{\boldsymbol{x}}^{\mathrm{T}}\boldsymbol{x}.$$

Together with $\boldsymbol{x} \neq \boldsymbol{0}$, $\overline{\boldsymbol{x}}^{\mathrm{T}}\boldsymbol{x} > 0$, this gives $\lambda = \overline{\lambda}$, that is, λ is a real number. □

THEOREM 5.11

The real eigenvectors of a real symmetric matrix $\boldsymbol{A}$ corresponding to different eigenvalues are orthogonal.

Proof: As $\boldsymbol{A}\boldsymbol{x}_i = \lambda_i\boldsymbol{x}_i$, $(\boldsymbol{x}_i \neq \boldsymbol{0}, i = 1, 2)$, $\lambda_1 \neq \lambda_2$, $\boldsymbol{A}^{\mathrm{T}} = \boldsymbol{A}$, then

$$\lambda_1\boldsymbol{x}_2{}^{\mathrm{T}}\boldsymbol{x}_1 = \boldsymbol{x}_2{}^{\mathrm{T}}\boldsymbol{A}\boldsymbol{x}_1 = \boldsymbol{x}_2{}^{\mathrm{T}}\boldsymbol{A}^{\mathrm{T}}\boldsymbol{x}_1 = (\boldsymbol{A}\boldsymbol{x}_2)^{\mathrm{T}}\boldsymbol{x}_1 = (\lambda_2\boldsymbol{x}_2)^{\mathrm{T}}\boldsymbol{x}_1 = \lambda_2\boldsymbol{x}_2{}^{\mathrm{T}}\boldsymbol{x}_1.$$

Since $\lambda_1 \neq \lambda_2$, then $\boldsymbol{x}_2{}^{\mathrm{T}}\boldsymbol{x}_1 = 0$, that is, $(\boldsymbol{x}_2, \boldsymbol{x}_1) = 0$. Then, $\boldsymbol{x}_1$ and $\boldsymbol{x}_2$ are orthogonal. □

5.3.2 Diagonalization of a real symmetric matrix

In this section, we discuss the diagonalization of real symmetric matrices with the related theorems and also illustrate their applications in the examples that follow.

THEOREM 5.12

For any nth-order real symmetric matrix $\boldsymbol{A}$, there exists an nth-order orthogonal matrix $\boldsymbol{T}$, such that

$$\boldsymbol{T}^{-1}\boldsymbol{A}\boldsymbol{T} = \mathrm{diag}(\lambda_1, \lambda_2, \ldots, \lambda_n).$$

Proof: This is proved by using mathematical induction on n. It holds obviously when $n = 1$.

Suppose it holds for any $(n-1)$th-order real symmetric matrix $\boldsymbol{B}$, that is, there exists an $(n-1)$th-order orthogonal matrix $\boldsymbol{Q}$ such that $\boldsymbol{Q}^{-1}\boldsymbol{B}\boldsymbol{Q}=\boldsymbol{\Lambda}_1$. Here, we will prove that it also holds for any nth-order real symmetric matrix.

Let $\boldsymbol{A}\boldsymbol{x}_1=\lambda_1\boldsymbol{x}_1$ and $\boldsymbol{x}_1$ be an eigenvector of unit length. Now, extend $\boldsymbol{x}_1$ to an orthonomal basis

$$\boldsymbol{x}_1, \boldsymbol{x}_2, \dots, \boldsymbol{x}_n,$$

where $\boldsymbol{x}_2, \dots, \boldsymbol{x}_n$ are not necessarily the eigenvectors of $\boldsymbol{A}$. Hence, we have

$$\begin{aligned}\boldsymbol{A}(\boldsymbol{x}_1, \boldsymbol{x}_2, \dots, \boldsymbol{x}_n) &= (\boldsymbol{A}\boldsymbol{x}_1, \boldsymbol{A}\boldsymbol{x}_2, \dots, \boldsymbol{A}\boldsymbol{x}_n)\\ &= (\boldsymbol{x}_1, \boldsymbol{x}_2, \dots, \boldsymbol{x}_n)\left(\begin{array}{c|ccc}\lambda_1 & b_{12} & \cdots & b_{1n}\\ \hline 0 & b_{22} & \cdots & b_{2n}\\ \vdots & \vdots & & \vdots\\ 0 & b_{n2} & \cdots & b_{nn}\end{array}\right). \qquad ①\end{aligned}$$

Denote

$$\boldsymbol{P}=(\boldsymbol{x}_1, \boldsymbol{x}_2, \dots, \boldsymbol{x}_n) \ (\boldsymbol{P} \text{ is an orthogonal matrix})$$

and express the matrix on the *RHS* of ① in a partitioned form. Equation ① is expressed as

$$\boldsymbol{P}^{-1}\boldsymbol{A}\boldsymbol{P}=\begin{pmatrix}\lambda_1 & \boldsymbol{b}\\ \boldsymbol{0} & \boldsymbol{B}\end{pmatrix}. \qquad ②$$

Since $\boldsymbol{P}^{-1}=\boldsymbol{P}^{\mathrm{T}}$, $(\boldsymbol{P}^{-1}\boldsymbol{A}\boldsymbol{P})^{\mathrm{T}}=\boldsymbol{P}^{\mathrm{T}}\boldsymbol{A}^{\mathrm{T}}(\boldsymbol{P}^{-1})^{\mathrm{T}}=\boldsymbol{P}^{-1}\boldsymbol{A}\boldsymbol{P}$, so

$$\begin{pmatrix}\lambda_1 & \boldsymbol{0}\\ \boldsymbol{b}^{\mathrm{T}} & \boldsymbol{B}^{\mathrm{T}}\end{pmatrix}=\begin{pmatrix}\lambda_1 & \boldsymbol{b}\\ \boldsymbol{0} & \boldsymbol{B}\end{pmatrix}.$$

Therefore, $\boldsymbol{b}=\boldsymbol{0}$, $\boldsymbol{B}^{\mathrm{T}}=\boldsymbol{B}$ (where $\boldsymbol{B}$ is a real symmetric matrix with order $n-1$), and ② becomes

$$\boldsymbol{P}^{-1}\boldsymbol{A}\boldsymbol{P}=\begin{pmatrix}\lambda_1 & \boldsymbol{0}\\ \boldsymbol{0} & \boldsymbol{B}\end{pmatrix}.$$

According to the inductive assumption, we now construct an orthogonal matrix

$$\boldsymbol{S}=\begin{pmatrix}1 & \boldsymbol{0}\\ \boldsymbol{0} & \boldsymbol{Q}\end{pmatrix},$$

(readers can easily verify that $\boldsymbol{S}^{\mathrm{T}}\boldsymbol{S}=\boldsymbol{I}_n$), and we have

$$\begin{aligned}\boldsymbol{S}^{-1}(\boldsymbol{P}^{-1}\boldsymbol{A}\boldsymbol{P})\boldsymbol{S} &= \begin{pmatrix}1 & \boldsymbol{0}\\ \boldsymbol{0} & \boldsymbol{Q}^{-1}\end{pmatrix}\begin{pmatrix}\lambda_1 & \boldsymbol{0}\\ \boldsymbol{0} & \boldsymbol{B}\end{pmatrix}\begin{pmatrix}1 & \boldsymbol{0}\\ \boldsymbol{0} & \boldsymbol{Q}\end{pmatrix}\\ &= \begin{pmatrix}\lambda_1 & \boldsymbol{0}\\ \boldsymbol{0} & \boldsymbol{Q}^{-1}\boldsymbol{B}\boldsymbol{Q}\end{pmatrix}=\begin{pmatrix}\lambda_1 & \boldsymbol{0}\\ \boldsymbol{0} & \boldsymbol{\Lambda}_1\end{pmatrix}\\ &= \operatorname{diag}(\lambda_1, \lambda_2, \dots, \lambda_n).\end{aligned}$$

Take $\boldsymbol{T} = \boldsymbol{P}\boldsymbol{S}$ (the product of two orthogonal matrices is also orthogonal), so that $\boldsymbol{T}^{-1} = \boldsymbol{S}^{-1}\boldsymbol{P}^{-1}$, where

$$\boldsymbol{T}^{-1}\boldsymbol{A}\boldsymbol{T} = \mathrm{diag}(\lambda_1, \lambda_2, \ldots, \lambda_n),$$

where $\lambda_1, \lambda_2, \ldots, \lambda_n$ are the eigenvalues of $\boldsymbol{A}$. □

For a given real symmetric matrix $\boldsymbol{A}$ of order n, determine an orthogonal matrix $\boldsymbol{T}$ such that $\boldsymbol{T}^{-1}\boldsymbol{A}\boldsymbol{T} = \boldsymbol{\Lambda}$. First, get all the distinct eigenvalues $\lambda_1, \lambda_2, \ldots, \lambda_m$ from the characteristic polynomial $|\lambda\boldsymbol{I} - \boldsymbol{A}| = \prod_{i=1}^{m} (\lambda - \lambda_i)^{r_i}$. Then, get r_i linearly independent eigenvectors $\boldsymbol{x}_{i_1}, \boldsymbol{x}_{i_2}, \ldots, \boldsymbol{x}_{i_{r_i}}$ belonging to λ_i of multiplicity r_i according to Theorem 5.9, since $\boldsymbol{A}$ can be diagonalized. Next, get r_i orthogonal unit vectors $\boldsymbol{y}_{i_1}, \ldots, \boldsymbol{y}_{i_{r_i}}$ by applying Gram-Schmidt orthogonalization procedure. $\left\{\boldsymbol{y}_{i_1}, \ldots, \boldsymbol{y}_{i_{r_i}} \mid i = 1, \ldots, m\right\}$ corresponding to different eigenvalues are n mutually orthogonal unit eigenvectors from Theorem 5.11. Finally, arranging them in columns as an nth-order matrix, we get the desired orthogonal matrix $\boldsymbol{T}$.

Example 5.8 Orthogonal diagonalization of a symmetric matrix

Suppose

$$\boldsymbol{A} = \begin{pmatrix} 1 & -2 & 2 \\ -2 & -2 & 4 \\ 2 & 4 & -2 \end{pmatrix}.$$

Get an orthogonal matrix $\boldsymbol{T}$ such that $\boldsymbol{T}^{-1}\boldsymbol{A}\boldsymbol{T}$ is a diagonal matrix.

Solution:

$$|\lambda\boldsymbol{I} - \boldsymbol{A}| = \begin{vmatrix} \lambda-1 & 2 & -2 \\ 2 & \lambda+2 & -4 \\ -2 & -4 & \lambda+2 \end{vmatrix} = \begin{vmatrix} 0 & -2(\lambda-2) & (\lambda+3)(\lambda-2)/2 \\ 0 & \lambda-2 & \lambda-2 \\ -2 & -4 & \lambda+2 \end{vmatrix}$$

$$= (-2)(\lambda-2)^2 \begin{vmatrix} -2 & (\lambda+3)/2 \\ 1 & 1 \end{vmatrix}$$

$$= (\lambda-2)^2(\lambda+7).$$

Then, $\lambda_1 = 2$ (multiplicity 2) and $\lambda_2 = -7$.

For $\lambda_1 = 2$ and $(\lambda_1\boldsymbol{I} - \boldsymbol{A})\boldsymbol{x} = \boldsymbol{0}$,

$$\begin{pmatrix} 1 & 2 & -2 \\ 2 & 4 & -4 \\ -2 & -4 & 4 \end{pmatrix} \begin{pmatrix} x_1 \\ x_2 \\ x_3 \end{pmatrix} = \begin{pmatrix} 0 \\ 0 \\ 0 \end{pmatrix},$$

we get linearly independent eigenvectors $\boldsymbol{x}_1 = (2, -1, 0)^{\mathrm{T}}$ and $\boldsymbol{x}_2 = (2, 0, 1)^{\mathrm{T}}$.

By applying Gram-Schmidt orthogonalization procedure, we have

$$\boldsymbol{\beta}_1 = \boldsymbol{x}_1,$$

$$\boldsymbol{\beta}_2 = \boldsymbol{x}_2 - \frac{(\boldsymbol{x}_2, \boldsymbol{\beta}_1)}{(\boldsymbol{\beta}_1, \boldsymbol{\beta}_1)}\boldsymbol{\beta}_1 = \begin{pmatrix} 2 \\ 0 \\ 1 \end{pmatrix} - \frac{4}{5}\begin{pmatrix} 2 \\ -1 \\ 0 \end{pmatrix} = \frac{1}{5}\begin{pmatrix} 2 \\ 4 \\ 5 \end{pmatrix}.$$

Then, to normalize $\boldsymbol{\beta}_1$ and $\boldsymbol{\beta}_2$ to unit length, we have

$$\boldsymbol{y}_1 = \left(\frac{2\sqrt{5}}{5}, -\frac{\sqrt{5}}{5}, 0\right)^{\mathrm{T}}, \quad \boldsymbol{y}_2 = \left(\frac{2\sqrt{5}}{15}, \frac{4\sqrt{5}}{15}, \frac{\sqrt{5}}{3}\right)^{\mathrm{T}}.$$

For $\lambda_2 = -7$ and $(\lambda_2 \boldsymbol{I} - \boldsymbol{A})\boldsymbol{x} = \boldsymbol{0}$,

$$\begin{pmatrix} -8 & 2 & -2 \\ 2 & -5 & -4 \\ -2 & -4 & -5 \end{pmatrix}\begin{pmatrix} x_1 \\ x_2 \\ x_3 \end{pmatrix} = \begin{pmatrix} 0 \\ 0 \\ 0 \end{pmatrix},$$

we get an eigenvector $\boldsymbol{x}_3 = (1, 2, -2)^{\mathrm{T}}$, which is normalized to $\boldsymbol{y}_3 = \left(\frac{1}{3}, \frac{2}{3}, -\frac{2}{3}\right)^{\mathrm{T}}$. The orthogonal matrix is

$$\boldsymbol{T} = (\boldsymbol{y}_1, \boldsymbol{y}_2, \boldsymbol{y}_3) = \begin{pmatrix} \frac{2\sqrt{5}}{5} & \frac{2\sqrt{5}}{15} & \frac{1}{3} \\ -\frac{\sqrt{5}}{5} & \frac{4\sqrt{5}}{15} & \frac{2}{3} \\ 0 & \frac{\sqrt{5}}{3} & -\frac{2}{3} \end{pmatrix},$$

and

$$\boldsymbol{T}^{-1}\boldsymbol{A}\boldsymbol{T} = \mathrm{diag}(\lambda_1, \lambda_1, \lambda_2) = \mathrm{diag}(2, 2, -7).$$ ■

Example 5.9 Similar real symmetric matrices are orthogonally similar

For similar real symmetric matrices $\boldsymbol{A}$ and $\boldsymbol{B}$, prove that there exists an orthogonal matrix $\boldsymbol{T}$ such that $\boldsymbol{T}^{-1}\boldsymbol{A}\boldsymbol{T} = \boldsymbol{B}$.

Proof: Similar matrices $\boldsymbol{A}$ and $\boldsymbol{B}$ have the same eigenvalues $\lambda_1, \lambda_2, \ldots, \lambda_n$. According to Theorem 5.12, there exist orthogonal matrices $\boldsymbol{T}_1$ and $\boldsymbol{T}_2$ for $\boldsymbol{A}$ and $\boldsymbol{B}$, respectively, such that

$$\boldsymbol{T}_1^{-1}\boldsymbol{A}\boldsymbol{T}_1 = \mathrm{diag}(\lambda_1, \lambda_1, \ldots, \lambda_n) = \boldsymbol{T}_2^{-1}\boldsymbol{B}\boldsymbol{T}_2.$$

Then,

$$T_2 T_1^{-1} A T_1 T_2^{-1} = B.$$

Take $T = T_1 T_2^{-1}$ (an orthogonal matrix), then $T^{-1}AT = B$. ■

Example 5.10 Real symmetric matrices with same eigenvalues

If two nth-order real symmetric matrices A and B have the same n eigenvalues, prove that there exists an orthogonal matrix T and an nth-order matrix Q, such that $A = QT$ and $B = TQ$ hold simultaneously.

Proof: Since a real symmetric matrix is similar to a diagonal matrix Λ with n eigenvalues in its diagonal, then $A \sim \Lambda \sim B$, that is, $A \sim B$. From Example 5.9, we know there exists orthogonal matrix T_1, such that $T_1^{-1} A T_1 = B$, that is, $AT_1 = T_1 B$. Let $Q = AT_1$ and $T = T_1^{-1}$, then $A = QT_1^{-1} = QT$, $B = T_1^{-1} Q = TQ$ (where $T = T_1^{-1}$ is still orthogonal) hold simultaneously. ■

Example 5.11 Product of two real symmetric matrices and diagonalization

Suppose both A and B are real symmetric matrices of order n. If there exists an orthogonal matrix T such that $T^{-1}AT$ and $T^{-1}BT$ are both diagonal matrices, then AB is a real symmetric matrix.

Proof: From $(AB)^{\mathrm{T}} = B^{\mathrm{T}} A^{\mathrm{T}} = BA$, we know that AB is symmetric, if and only if, AB is commutative. So, we only need to prove $AB = BA$. According to the assumption, we have

$$T^{-1}AT = \operatorname{diag}(\lambda_1, \lambda_2, \ldots, \lambda_n), \quad T^{-1}BT = \operatorname{diag}(\mu_1, \mu_2, \cdots, \mu_n).$$

Thus,

$$(T^{-1}AT)(T^{-1}BT) = (T^{-1}BT)(T^{-1}AT) = \operatorname{diag}(\lambda_1\mu_1, \ldots, \lambda_n\mu_n),$$

therefore,

$$AB = BA, \text{ where } (AB)^{\mathrm{T}} = BA = AB.$$ ■

Exercises and Supplementary Exercises

Exercises

5.1 Find eigenvalues and eigenvectors of the following matrices:

$$\text{(i)} \begin{pmatrix} 2 & -3 \\ -3 & 1 \end{pmatrix}, \quad \text{(ii)} \begin{pmatrix} 3 & -1 & 1 \\ 2 & 0 & 1 \\ 1 & -1 & 2 \end{pmatrix},$$

$$\text{(iii)} \begin{pmatrix} 2 & 0 & 0 \\ 1 & 1 & 1 \\ 1 & -1 & 3 \end{pmatrix}, \quad \text{(iv)} \begin{pmatrix} 1 & 2 & 3 & 4 \\ 0 & 1 & 2 & 3 \\ 0 & 0 & 1 & 2 \\ 0 & 0 & 0 & 1 \end{pmatrix},$$

$$\text{(v)} \begin{pmatrix} 4 & 5 & -2 \\ -2 & -2 & 1 \\ -1 & -1 & 1 \end{pmatrix}, \quad \text{(vi)} \begin{pmatrix} 2 & -2 & 0 \\ -2 & 1 & -2 \\ 0 & -2 & 0 \end{pmatrix}.$$

5.2 $\lambda_1 = 3$ (multiplicity 2) and $\lambda_2 = 12$ are eigenvalues of the matrix

$$A = \begin{pmatrix} 7 & 4 & -1 \\ 4 & 7 & -1 \\ -4 & -4 & x \end{pmatrix}.$$

Determine x and then find its eigenvectors.

5.3 If $\boldsymbol{x}_1$ and $\boldsymbol{x}_2$ are different eigenvectors belonging to different eigenvalues of matrix $\boldsymbol{A}$, prove that $\boldsymbol{x}_1 + \boldsymbol{x}_2$ is not an eigenvector of $\boldsymbol{A}$.

5.4 If $\boldsymbol{x}_1, \boldsymbol{x}_2, \boldsymbol{x}_3$ are eigenvectors belonging to different eigenvalues $\lambda_1, \lambda_2, \lambda_3$ of $\boldsymbol{A}$, prove that $\boldsymbol{x}_1 + \boldsymbol{x}_2 + \boldsymbol{x}_3$ is not an eigenvector of $\boldsymbol{A}$.

5.5 Prove that the eigenvalues of the involutary matrix $\boldsymbol{A}$ (Hint: $\boldsymbol{A}^2 = \boldsymbol{I}$) are only 1 or -1.

5.6 If $\boldsymbol{A}$ is invertible, discuss the relation between eigenvalues (eigenvectors) of $\boldsymbol{A}$ and $\boldsymbol{A}^*$.

5.7 If $\boldsymbol{P}^{-1}\boldsymbol{A}\boldsymbol{P} = \boldsymbol{B}$, does $\boldsymbol{P}^{-1}(\boldsymbol{A} - 2\boldsymbol{I})\boldsymbol{P} = \boldsymbol{B} - 2\boldsymbol{I}$ hold?

5.8 If $\boldsymbol{A} \sim \boldsymbol{\Lambda} = \begin{pmatrix} -1 & 0 \\ 0 & 2 \end{pmatrix}$, compute $\det(\boldsymbol{A} - \boldsymbol{I})$.

5.9 If $\boldsymbol{P} = \begin{pmatrix} 2 & -1 \\ 3 & -2 \end{pmatrix}$ and $\boldsymbol{P}^{-1}\boldsymbol{A}\boldsymbol{P} = \begin{pmatrix} -1 & 0 \\ 0 & 2 \end{pmatrix}$, compute $\boldsymbol{A}^n$.

***5.10** If $\boldsymbol{B} = \boldsymbol{P}^{-1}\boldsymbol{A}\boldsymbol{P}$ and $\boldsymbol{x}$ is an eigenvector belonging to the eigenvalue λ_0 of matrix $\boldsymbol{A}$, prove that $\boldsymbol{P}^{-1}\boldsymbol{x}$ is an eigenvector belonging to the eigenvalue λ_0 of matrix $\boldsymbol{B}$.

***5.11** If $\boldsymbol{A}$ is a nonsingular matrix, prove that $\boldsymbol{AB}$ and $\boldsymbol{BA}$ are similar.

***5.12** If $\boldsymbol{A} \sim \boldsymbol{B}$, $\boldsymbol{C} \sim \boldsymbol{D}$, prove that

$$\begin{pmatrix} \boldsymbol{A} & \boldsymbol{0} \\ \boldsymbol{0} & \boldsymbol{C} \end{pmatrix} \sim \begin{pmatrix} \boldsymbol{B} & \boldsymbol{0} \\ \boldsymbol{0} & \boldsymbol{D} \end{pmatrix}.$$

***5.13** Prove that the mth-order matrix

$$\boldsymbol{J} = \begin{pmatrix} 0 & 1 & & \\ & 0 & \ddots & \\ & & \ddots & 1 \\ & & & 0 \end{pmatrix}$$

has only the zero eigenvalue, and the eigensubspace is a one-dimensional subspace of $\mathbb{R}^m$. Provide a basis for it.

5.14 If $\boldsymbol{I} + \boldsymbol{A}$ is invertible, $(\boldsymbol{I} - \boldsymbol{A})$ is not invertible, then what can we conclude about the eigenvalues of $\boldsymbol{A}$?

5.15 If $\det(\boldsymbol{I} - \boldsymbol{A}^2) = 0$, prove that at least one of 1 and -1 is an eigenvalue of $\boldsymbol{A}$.

5.16 In **Q5.1**, which matrices are diagonalizable? For the diagonalizable matrix $\boldsymbol{A}$, find an invertible matrix $\boldsymbol{P}$ and a diagonal matrix $\boldsymbol{\Lambda}$, such that $\boldsymbol{P}^{-1}\boldsymbol{A}\boldsymbol{P} = \boldsymbol{\Lambda}$.

5.17 Can an upper (lower) triangular matrix with distinct main diagonal entries be similar to a diagonal matrix? Please provide the reason(s).

5.18 If n^2 entries of an nth-order matrix are all 1, find an invertible matrix $\boldsymbol{P}$ such that $\boldsymbol{P}^{-1}\boldsymbol{A}\boldsymbol{P}$ is a diagonal matrix, and write the diagonal matrix that is similar to $\boldsymbol{A}$.

5.19 For a 4th-order matrix A with eigenvalues $\lambda_1 = 1$ (multiplicity 3) and $\lambda_2 = -3$, eigenvectors $x_1 = (1, -1, 0, 0)^T$, $x_2 = (-1, 1, -1, 0)^T$, $x_3 = (0, -1, 1, -1)^T$ belonging to λ_1 and the eigenvector $x_4 = (0, 0, -1, 1)^T$ belonging to λ_2, can A be diagonalized? If it can be diagonalized, determine A and A^n (n is a positive integer).

5.20 Suppose a 3rd-order matrix A has an eigenvalue λ_1 of multiplicity 2 and $x_1 = (1, 0, 1)^T$, $x_2 = (-1, 0, -1)^T$, $x_3 = (1, 1, 0)^T$, $x_4 = (0, 1, -1)^T$ are all eigenvectors corresponding to λ_1. Can A be diagonalized?

5.21 Let $A = \begin{pmatrix} -3 & 2 \\ -2 & 2 \end{pmatrix}$.

(i) Find A^4, A^5, A^k (k is a positive number).

(ii) If $f(x) = \begin{vmatrix} x^4 - 1 & x \\ x^3 & x^6 + 1 \end{vmatrix}$, determine $f(A)$.

5.22 Suppose $A = \begin{pmatrix} 3 & 4 & 0 & 0 \\ 4 & -3 & 0 & 0 \\ 0 & 0 & 2 & 4 \\ 0 & 0 & 0 & 2 \end{pmatrix}$, Compute A^k (k is a positive number).

(Hint: Compute A^k by diagonal block matrix.)

5.23 For the matrix A of Example 5.4 in §5.2, find an orthogonal matrix T such that $T^{-1}AT$ is a diagonal matrix.

5.24 For each of the following real symmetric matrices A, find an orthogonal matrix T and diagonal matrix Λ, such that $T^{-1}AT = \Lambda$:

(i) $\begin{pmatrix} 3 & 2 & 4 \\ 2 & 0 & 2 \\ 4 & 2 & 3 \end{pmatrix}$, (ii) $\begin{pmatrix} 1 & 3 & 0 \\ 3 & 4 & -1 \\ 0 & -1 & 1 \end{pmatrix}$,

(iii) $\begin{pmatrix} 1 & 0 & 2 \\ 0 & 1 & 2 \\ 2 & 2 & -1 \end{pmatrix}$, (iv) $\begin{pmatrix} 0 & 0 & 4 & 1 \\ 0 & 0 & 1 & 4 \\ 4 & 1 & 0 & 0 \\ 1 & 4 & 0 & 0 \end{pmatrix}$,

(v) $\begin{pmatrix} -1 & -3 & 3 & -3 \\ -3 & -1 & -3 & 3 \\ 3 & -3 & -1 & -3 \\ -3 & 3 & -3 & -1 \end{pmatrix}$.

5.25 Suppose A is an nth-order real symmetric matrix and $A^2 = A$. Prove that there exists an orthogonal matrix T such that $T^{-1}AT = \text{diag}(1, 1, \cdots, 1, 0, \ldots, 0)$.

***5.26** Suppose an nth-order real symmetric matrix A have the eigenvalues $\lambda_i \geq 0$ ($i = 1, 2, \ldots, n$). Prove that there exists a real symmetric matrix B with nonnegative eigenvalues such that $A = B^2$.

***5.27** Suppose A is a real symmetric idempotent matrix of order $n (A^2 = A)$ and $r(A) = r$. Find $\det(A - 2I)$.

Supplementary exercises

5.28 The polynomial is

$$f(x) = a_n x^n + a_{n-1} x^{n-1} + \cdots + a_1 x + a_0,$$

λ_0 is an eigenvalue of matrix A, and x is the eigenvector belonging to λ_0. Prove that $f(\lambda_0)$ is the eigenvalue of $f(A)$ while x is still an eigenvector belonging to $f(\lambda_0)$.

5.29 Suppose $A \sim B$ and $f(A) = A^3 + 2A - 3I$. Prove that $f(A) \sim f(B)$.

5.30 Let $A = (a_{ij})_{4\times 4}$, 0 is a double eigenvalue of A and 1 is another eigenvalue of A. Find the characteristic polynomial $\det(\lambda I - A)$ of A.

5.31 Suppose the sum of the entries of each row is 1. Can we find at least one eigenvalue of A?

5.32 Suppose that $\lambda_1, \lambda_2, \cdots, \lambda_n$ are n eigenvalues of a matrix $A = (a_{ij})_{n\times n}$. Prove that

$$\sum_{i=1}^{n} \lambda_i^2 = \sum_{j=1}^{n} \sum_{i=1}^{n} a_{ij} a_{ji}.$$

5.33 Suppose $AB = BA$ and x is an eigenvector belonging to the eigenvalue λ_0 of A. Prove that

$$Bx \in V_{\lambda_0} \text{ (eigensubspace of } A).$$

5.34 If an nth-order matrix $\boldsymbol{A}$ has n different eigenvalues, prove that $\boldsymbol{AB} = \boldsymbol{BA}$, if and only if, the eigenvectors of $\boldsymbol{A}$ are also the eigenvectors of $\boldsymbol{B}$.

5.35 Suppose $\boldsymbol{A}$ and $\boldsymbol{B}$ are both nth-order matrices and $\varphi(\lambda) = |\lambda \boldsymbol{I} - \boldsymbol{B}|$. Prove that $\varphi(\boldsymbol{A})$ is invertible, if and only if, none of the eigenvalues of $\boldsymbol{B}$ is an eigenvalue of $\boldsymbol{A}$.
(Hint: Consider the necessary and sufficient condition of $|\varphi(\boldsymbol{A})| \neq 0$ using $\varphi(\lambda) = |\lambda \boldsymbol{I} - \boldsymbol{B}| = (\lambda - \lambda_1)(\lambda - \lambda_2)\cdots(\lambda - \lambda_n)$, and $|\mu \boldsymbol{I} - \boldsymbol{A}| \neq 0$ where μ is not an eigenvalue of $\boldsymbol{A}$.)

5.36 Prove that eigenvalues of an antisymmetric real matrix is 0 or a pure imaginary number.

5.37 For two nonzero orthogonal vectors in $\mathbb{R}^n$,

$$\boldsymbol{\alpha} = (a_1, a_2, \ldots, a_n), \boldsymbol{\beta} = (b_1, b_2, \ldots, b_n),$$

prove that the eigenvalues of the matrix $\boldsymbol{A} = \boldsymbol{\alpha}^{\mathrm{T}}\boldsymbol{\beta}$ are all 0, and $\boldsymbol{A}$ cannot be diagonalized.

5.38 Suppose $\boldsymbol{\alpha} = (a_1, a_2, \ldots, a_n) \in \mathbb{R}^n$, and $a_i \neq 0$ $(i = 1, 2, \cdots, n)$. Find the eigenvalues of the matrix $\boldsymbol{A} = \boldsymbol{\alpha}^{\mathrm{T}}\boldsymbol{\alpha}$ and find an invertible matrix $\boldsymbol{P}$ such that $\boldsymbol{P}^{-1}\boldsymbol{AP}$ is in diagonal form.

5.39 $\boldsymbol{\xi} = (1, 1, -1)^{\mathrm{T}}$ is one eigenvector of

$$\boldsymbol{A} = \begin{pmatrix} 2 & -1 & 2 \\ 5 & a & 3 \\ -1 & b & -2 \end{pmatrix}.$$

(i) Detemine a, b and the eigenvalue corresponding to $\boldsymbol{\xi}$.
(ii) Can $\boldsymbol{A}$ be similar to a diagonal matrix? Provide a reason in favor of your answer.

5.40 Suppose $\boldsymbol{A} = \begin{pmatrix} a & -1 & c \\ 5 & b & 3 \\ 1-c & 0 & -a \end{pmatrix}$. It is known that $|\boldsymbol{A}| = 1$ and $\boldsymbol{A}^*$ has an eigenvalue λ_0 with eigenvector $\boldsymbol{x} = (-1, -1, 1)^{\mathrm{T}}$. Find a, b, c, and λ_0.

5.41 Suppose $\boldsymbol{A} = \begin{pmatrix} 1 & -1 & 1 \\ x & 4 & y \\ -3 & -3 & 5 \end{pmatrix}$. It is known that $\boldsymbol{A}$ has three linearly independent eigenvectors, and $\lambda_1 = 2$ is a double eigenvalue. Find $\boldsymbol{P}$ such that $\boldsymbol{P}^{-1}\boldsymbol{AP} = \boldsymbol{\Lambda}$ (diagonal matrix).

5.42 Suppose $\boldsymbol{\alpha} = (a_1, a_2, \ldots, a_n)^{\mathrm{T}}$, $\boldsymbol{\beta} = (b_1, b_2, \ldots, b_n)^{\mathrm{T}}$ are both nonzero vectors and $\boldsymbol{\alpha}^{\mathrm{T}}\boldsymbol{\beta} = 0$, $\boldsymbol{A} = \boldsymbol{\alpha}\boldsymbol{\beta}^{\mathrm{T}}$. Find: (i) $\boldsymbol{A}^2$, (ii) the eigenvalues and eigenvectors of $\boldsymbol{A}$.

Q5.43–5.46: Select the correct answer for these multiple choice questions.

5.43 It is known that $2, 4, 6, \ldots, 2n$ are n eigenvalues of nth-order matrix $\boldsymbol{A}$, then the determinant $|\boldsymbol{A} - 3\boldsymbol{I}| = (\quad)$.

(a) $2 \cdot n! - 3^n$.
(b) $(2n-3)!! = 1 \cdot 3 \cdot 5 \cdots (2n-3)$.
(c) $-[(2n-3)!!]$.
(d) $5 \cdot 7 \cdot 9 \cdots (2n+3)$.

5.44 The determinant of an nth-order matrix $\boldsymbol{A}$, $|\boldsymbol{A}| \neq 0$ and λ_1 is an eigenvalue of $\boldsymbol{A}$. Then, $(\boldsymbol{A}^*)^2 + \boldsymbol{I}$ has an eigenvalue _______.

(a) $(\lambda_1 |\boldsymbol{A}|)^2 + 1$. (b) $\left(\frac{|\boldsymbol{A}|}{\lambda_1}\right)^2 + 1$.
(c) $(1 + \lambda_1 |\boldsymbol{A}|)^2$. (d) $\left(1 + \frac{|\boldsymbol{A}|}{\lambda_1}\right)^2$.

5.45 If $\boldsymbol{A}$ and $\boldsymbol{B}$ are both nth-order matrices and $\boldsymbol{A} \sim \boldsymbol{B}$, then _______.

(a) $\lambda \boldsymbol{I} - \boldsymbol{A} = \lambda \boldsymbol{I} - \boldsymbol{B}$.
(b) $\boldsymbol{A}$ and $\boldsymbol{B}$ have identical eigenvalues and eigenvectors.
(c) $\boldsymbol{AB} \sim \boldsymbol{B}^2$.
(d) for any constant t, $t\boldsymbol{I} - \boldsymbol{A} \sim t\boldsymbol{I} - \boldsymbol{B}$.

5.46 If $A = \begin{pmatrix} 2 & 0 & 0 \\ 0 & 0 & 1 \\ 0 & 1 & x \end{pmatrix}$ and $B = \begin{pmatrix} 2 & 0 & 0 \\ 0 & y & 0 \\ 0 & 0 & -1 \end{pmatrix}$ are similar, then _______.
(a) $x = 0, y = 1$. (b) $x = -1, y = 0$.
(c) $x = y = 0$. (d) $x = y = 1$.

6 Quadratic Forms

CHAPTER OVERVIEW

A quadratic form is a homogeneous polynomial of degree two. The general equation of a central quadratic curve discussed in analytic geometry, when its center is the coordinate origin, is

$$ax^2 + 2bxy + cy^2 = f. \qquad ①$$

The *LHS* of the equation is a quadratic homogeneous polynomial in x and y. To study the geometric properties of the quadratic curve, we will deform equation ① to a standard equation without cross (product) term of x and y, as

$$a'x'^2 + c'y'^2 = f', \qquad ②$$

by basis transformation (coordinate transformation). There are also similar problems in the study of quadratic surfaces. Quadratic homogeneous polynomials are encountered not only in geometric problems but also in other branches of mathematics as well as physics, mechanics, and network computation.

A basic problem of quadratic forms is to deform the general quadratic homogeneous polynomial to a sum of squares, in a manner similar to the transformation of the general equations of quadratic curves centered at origin, to standard equations. In this chapter, besides this basic question, we will also focus on properties and discrimination of the definite quadratic form (mainly the positive definite quadratic form), which is necessary for many important applications. While some of these applications will also be introduced, we will study quadratic form through the tool of matrices, and therefore, let us first discuss the matrix representation of the quadratic form.

6.1 Definition and Matrix Representation of Quadratic Forms, Congruent Matrices

Definition 6.1

A quadratic homogenous polynomial in n variables $x_1, x_2, \cdots, x_n$,

$$\begin{aligned} &f(x_1, x_2, \cdots, x_n) \\ &= a_{11}x_1{}^2 + 2a_{12}x_1x_2 + 2a_{13}x_1x_3 + \ldots + 2a_{1n}x_1x_n \\ &\quad + a_{22}x_2{}^2 + 2a_{23}x_2x_3 + \ldots + 2a_{2n}x_2x_n + \cdots + a_{nn}x_n{}^2, \end{aligned} \tag{6.1}$$

is called a **quadratic form** of n-variable over the number field F, if the coefficients belong to a number field F. In this chapter, we will briefly discuss quadratic forms over the field of real numbers, which are called **quadratic forms**.

Since $x_ix_j = x_jx_i$ is symmetric, if we suppose

$$a_{ij} = a_{ji}, \ i < j, \tag{6.2}$$

then $2a_{ij}x_ix_j = a_{ij}x_ix_j + a_{ji}x_jx_i (i < j)$, and thus (6.1) can be rewritten in a symmetric form

$$\begin{aligned} &f(x_1, x_2, \ldots, x_n) \\ &= a_{11}x_1{}^2 + a_{12}x_1x_2 + \cdots + a_{1n}x_1x_n + a_{21}x_2x_1 + a_{22}x_2{}^2 + \cdots + a_{2n}x_2x_n \\ &\quad + \cdots + a_{n1}x_nx_1 + a_{n2}x_nx_2 + \cdots + a_{nn}x_n{}^2 \\ &= \sum_{i=1}^{n} x_i(a_{i1}x_1 + a_{i2}x_2 + \cdots + a_{in}x_n) = \sum_{i=1}^{n} x_i \sum_{j=1}^{n} a_{ij}x_j = \sum_{i=1}^{n}\sum_{j=1}^{n} a_{ij}x_ix_j \end{aligned} \tag{6.3}$$

$$= (x_1, x_2, \ldots, x_n)\begin{pmatrix} a_{11}x_1 + a_{12}x_2 + \cdots + a_{1n}x_n \\ a_{21}x_1 + a_{22}x_2 + \cdots + a_{2n}x_n \\ \vdots \\ a_{n1}x_1 + a_{n2}x_2 + \cdots + a_{nn}x_n \end{pmatrix}$$

$$= (x_1, x_2, \ldots, x_n)\begin{pmatrix} a_{11} & a_{12} & \cdots & a_{1n} \\ a_{21} & a_{22} & \cdots & a_{2n} \\ \vdots & \vdots & & \vdots \\ a_{n1} & a_{n2} & \cdots & a_{nn} \end{pmatrix}\begin{pmatrix} x_1 \\ x_2 \\ \vdots \\ x_n \end{pmatrix} = \boldsymbol{x}^{\mathrm{T}}\boldsymbol{A}\boldsymbol{x}, \qquad (6.4)$$

where, $\boldsymbol{x} = (x_1, x_2, \ldots, x_n)^{\mathrm{T}}$, $\boldsymbol{A} = (a_{ij})_{n\times n}$, and $\boldsymbol{A}$ is called the **corresponding matrix** of the quadratic form (6.3). Any quadratic form (6.1) can always be written in the symmetric form (6.3) by (6.2), and corresponds to a matrix $\boldsymbol{A}$. From (6.2), we know $\boldsymbol{A}$ is a symmetric matrix. If $\boldsymbol{A}$ and $\boldsymbol{B}$ are symmetric square matrices of order n, and

$$f(x_1, x_2, \ldots, x_n) = \boldsymbol{x}^{\mathrm{T}}\boldsymbol{A}\boldsymbol{x} = \boldsymbol{x}^{\mathrm{T}}\boldsymbol{B}\boldsymbol{x},$$

then we must have $\boldsymbol{A} = \boldsymbol{B}$ (the proof is left as an exercise for readers). Therefore, a quadratic form and its matrix are uniquely determined by each other. So, studying the properties of the quadratic form can be transformed to studying the properties of $\boldsymbol{A}$.

Example 6.1 Corresponding matrix of a quadratic form

Let $f(x_1, x_2, x_3, x_4) = 2{x_1}^2 + x_1x_2 + 2x_1x_3 + 4x_2x_4 + {x_3}^2 + 5{x_4}^2$.

Then, its matrix is

$$\boldsymbol{A} = \begin{pmatrix} 2 & \frac{1}{2} & 1 & 0 \\ \frac{1}{2} & 0 & 0 & 2 \\ 1 & 0 & 1 & 0 \\ 0 & 2 & 0 & 5 \end{pmatrix}.$$

■

A quadratic form $\boldsymbol{x}^{\mathrm{T}}\boldsymbol{A}\boldsymbol{x}$ can also be considered as a function of an n-dimensional vector $\boldsymbol{\alpha}$, that is,

$$f(\boldsymbol{\alpha}) = \boldsymbol{x}^{\mathrm{T}}\boldsymbol{A}\boldsymbol{x},$$

where $\boldsymbol{x} = (x_1, x_2, \ldots, x_n)^{\mathrm{T}}$ is the coordinate vector of $\boldsymbol{\alpha}$ with respect to a basis in $\mathbb{R}^n$. Therefore, the quadratic form $\boldsymbol{x}^{\mathrm{T}}\boldsymbol{A}\boldsymbol{x}$ is a quadratic homogenous function of n coordinates of the vector $\boldsymbol{\alpha}$. So, as a function of n-dimensional vector $\boldsymbol{\alpha}$, the matrix of a quadratic form is related to a basis.

If the coordinate vectors of the n-dimensional vector $\boldsymbol{\alpha}$ under two bases $\{\boldsymbol{\varepsilon}_1, \boldsymbol{\varepsilon}_2, \ldots, \boldsymbol{\varepsilon}_n\}$ and $\{\boldsymbol{\eta}_1, \boldsymbol{\eta}_2, \ldots, \boldsymbol{\eta}_n\}$ of $\mathbb{R}^n$ are respectively

$$\boldsymbol{x} = (x_1, x_2, \ldots, x_n)^{\mathrm{T}} \text{ and } \boldsymbol{y} = (y_1, y_2, \ldots, y_n)^{\mathrm{T}},$$

and

$$(\boldsymbol{\eta}_1, \boldsymbol{\eta}_2, \ldots, \boldsymbol{\eta}_n) = (\boldsymbol{\varepsilon}_1, \boldsymbol{\varepsilon}_2, \ldots, \boldsymbol{\varepsilon}_n)\boldsymbol{C},$$

then

$$\boldsymbol{x} = \boldsymbol{C}\boldsymbol{y},$$

where the basis transformation matrix $\boldsymbol{C}$ is an invertible matrix. Thus, we have a quadratic form

$$f(\boldsymbol{\alpha}) = \boldsymbol{x}^{\mathrm{T}}\boldsymbol{A}\boldsymbol{x} = \boldsymbol{y}^{\mathrm{T}}(\boldsymbol{C}^{\mathrm{T}}\boldsymbol{A}\boldsymbol{C})\boldsymbol{y},$$

that is, the matrices of quadratic form $f(\boldsymbol{\alpha})$ under the two bases $\{\boldsymbol{\varepsilon}_1, \boldsymbol{\varepsilon}_2, \ldots, \boldsymbol{\varepsilon}_n\}$ and $\{\boldsymbol{\eta}_1, \boldsymbol{\eta}_2, \ldots, \boldsymbol{\eta}_n\}$ are

$$\boldsymbol{A} \text{ and } \boldsymbol{C}^{\mathrm{T}}\boldsymbol{A}\boldsymbol{C},$$

where $\boldsymbol{C}^{\mathrm{T}}\boldsymbol{A}\boldsymbol{C}$ is still a symmetric matrix, $\boldsymbol{y}^{\mathrm{T}}(\boldsymbol{C}^{\mathrm{T}}\boldsymbol{A}\boldsymbol{C})\boldsymbol{y}$ is a quadratic form in $y_1, y_2, \ldots, y_n$.

Example 6.2 Transformation of a quadratic form to a standard equation under counterclockwise rotation

Assume the coordinates $(x_1, x_2)^{\mathrm{T}}$ of vector $\boldsymbol{\alpha}$ under the standard basis $\{\boldsymbol{\varepsilon}_1, \boldsymbol{\varepsilon}_2\}$ satisfy the equation

$$5{x_1}^2 + 5{x_2}^2 - 6x_1x_2 = 4. \qquad ①$$

If a basis transformation is performed, $\boldsymbol{\varepsilon}_1$ and $\boldsymbol{\varepsilon}_2$ can be transformed to $\boldsymbol{\eta}_1$ and $\boldsymbol{\eta}_2$ by counterclockwise rotation of 45°, that is,

$$(\boldsymbol{\eta}_1, \boldsymbol{\eta}_2) = (\boldsymbol{\varepsilon}_1, \boldsymbol{\varepsilon}_2)\begin{pmatrix} \cos 45^\circ & -\sin 45^\circ \\ \sin 45^\circ & \cos 45^\circ \end{pmatrix}. \qquad ②$$

Then, the coordinate $(y_1, y_2)^{\mathrm{T}}$ of $\boldsymbol{\alpha}$ under basis $\{\boldsymbol{\eta}_1, \boldsymbol{\eta}_2\}$ satisfy

$$\boldsymbol{x} = \begin{pmatrix} x_1 \\ x_2 \end{pmatrix} = \begin{pmatrix} \cos 45^\circ & -\sin 45^\circ \\ \sin 45^\circ & \cos 45^\circ \end{pmatrix}\begin{pmatrix} y_1 \\ y_2 \end{pmatrix} = \boldsymbol{C}\boldsymbol{y}. \qquad ③$$

Equation ② can be expressed in matrix form as

$$\boldsymbol{x}^{\mathrm{T}}\boldsymbol{A}\boldsymbol{x} = (x_1, x_2)\begin{pmatrix} 5 & -3 \\ -3 & 5 \end{pmatrix}\begin{pmatrix} x_1 \\ x_2 \end{pmatrix} = 4.$$

Substituting ③ into the above expression, we get

$$\begin{aligned}\boldsymbol{x}^{\mathrm{T}}\boldsymbol{A}\boldsymbol{x} &= \boldsymbol{y}^{\mathrm{T}}(\boldsymbol{C}^{\mathrm{T}}\boldsymbol{A}\boldsymbol{C})\boldsymbol{y}\\ &= (y_1, y_2)\begin{pmatrix}\frac{\sqrt{2}}{2} & \frac{\sqrt{2}}{2}\\ -\frac{\sqrt{2}}{2} & \frac{\sqrt{2}}{2}\end{pmatrix}\begin{pmatrix}5 & -3\\ -3 & 5\end{pmatrix}\begin{pmatrix}\frac{\sqrt{2}}{2} & -\frac{\sqrt{2}}{2}\\ \frac{\sqrt{2}}{2} & \frac{\sqrt{2}}{2}\end{pmatrix}\begin{pmatrix}y_1\\ y_2\end{pmatrix}\\ &= (y_1, y_2)\begin{pmatrix}2 & 0\\ 0 & 8\end{pmatrix}\begin{pmatrix}y_1\\ y_2\end{pmatrix} = 2{y_1}^2 + 8{y_2}^2 = 4,\end{aligned}$$

that is,

$$\frac{1}{2}{y_1}^2 + 2{y_2}^2 = 1. \qquad ④$$

Thus, we have transformed ① to a standard equation ④ under the coordinate system of the basis $\{\boldsymbol{\eta}_1, \boldsymbol{\eta}_2\}$, its graph is an ellipse (see Figure 6.1).

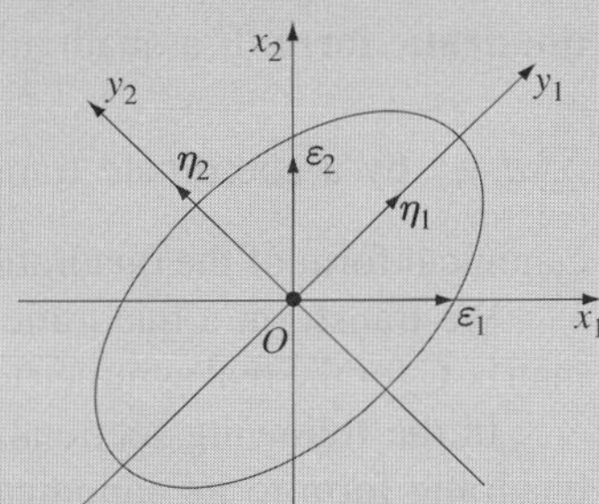

Figure 6.1 Graph of $5{x_1}^2 + 5{x_2}^2 - 6x_1x_2 = 4$ ■

The basic method used to transform a general quadratic form $f(x_1, x_2, \ldots, x_n)$ to an algebraic sum of perfect squared terms (or squared sum in short) in $y_1, y_2, \ldots, y_n$ is to make a transformation of coordinates (or a nonsingular linear transformation)

$$\boldsymbol{x} = \boldsymbol{C}\boldsymbol{y},$$

where $\boldsymbol{C}$ is an invertible matrix, such that

$$\boldsymbol{x}^{\mathrm{T}}\boldsymbol{A}\boldsymbol{x} = \boldsymbol{y}^{\mathrm{T}}(\boldsymbol{C}^{\mathrm{T}}\boldsymbol{A}\boldsymbol{C})\boldsymbol{y} = d_1{y_1}^2 + \cdots + d_n{y_n}^2.$$

From the matrix viewpoint, the basic problem is to seek an invertible matrix $\boldsymbol{C}$ for a real symmetric matrix $\boldsymbol{A}$ such that $\boldsymbol{C}^{\mathrm{T}}\boldsymbol{A}\boldsymbol{C}$ is in diagonal form.

Definition 6.2

For two matrices $\boldsymbol{A}$ and $\boldsymbol{B}$, if there exists an invertible matrix $\boldsymbol{C}$, such that $\boldsymbol{C}^{\mathrm{T}}\boldsymbol{A}\boldsymbol{C} = \boldsymbol{B}$, then $\boldsymbol{A}$ is said to be **congruent** to $\boldsymbol{B}$, denoted by $\boldsymbol{A} \simeq \boldsymbol{B}$.

It is easy to prove that the **congruence** relation between matrices satisfies reflexivity, symmetry, and transitivity. Since the congruence relation has symmetry, so $\boldsymbol{A}$ being congruent to $\boldsymbol{B}$ is equivalent to $\boldsymbol{B}$ being congruent to $\boldsymbol{A}$, where $\boldsymbol{A}$ and $\boldsymbol{B}$ are **congruent matrices**.

6.2 Transformation of a Quadratic Form to Its Canonical Form

The problem discussed in this section is how to deform a quadratic form $f(x_1, x_2, \ldots, x_n) = \boldsymbol{x}^{\mathrm{T}}\boldsymbol{A}\boldsymbol{x}$ to a squared sum $d_1 y_1{}^2 + \cdots + d_n y_n{}^2$ of $y_1, y_2, \ldots, y_n$ by transformation of coordinates, $\boldsymbol{x} = \boldsymbol{C}\boldsymbol{y}$. A quadratic form that contains squared terms only and no cross term is called the **canonical quadratic form**. If a quadratic form $\boldsymbol{x}^{\mathrm{T}}\boldsymbol{A}\boldsymbol{x} = \sum_{i=1}^{n}\sum_{j=1}^{n} a_{ij}x_i x_j$ is transformed to $\sum_{i=1}^{n} d_i y_i{}^2$ by a coordinate transformation $\boldsymbol{x} = \boldsymbol{C}\boldsymbol{y}$, we call $\sum_{i=1}^{n} d_i y_i{}^2$ to be the canonical form of the quadratic form $\boldsymbol{x}^{\mathrm{T}}\boldsymbol{A}\boldsymbol{x}$.

We transform a quadratic form to a canonical form to seek an invertible matrix $\boldsymbol{C}$ for the real symmetric matrix $\boldsymbol{A}$, such that $\boldsymbol{C}^{\mathrm{T}}\boldsymbol{A}\boldsymbol{C}$ is a diagonal matrix.

In the following sections, we introduce three methods of transforming a quadratic form to its canonical form, and prove that, for any real symmetric matrix $\boldsymbol{A}$, there must exist an invertible matrix $\boldsymbol{C}$, such that $\boldsymbol{C}^{\mathrm{T}}\boldsymbol{A}\boldsymbol{C}$ is a diagonal matrix.

6.2.1 Orthogonal transformation method

It has been discussed in §5.3 that, for any nth-order real symmetric matrix $\boldsymbol{A}$, there must exist an orthogonal matrix $\boldsymbol{Q}$, such that $\boldsymbol{Q}^{-1}\boldsymbol{A}\boldsymbol{Q} = \boldsymbol{\Lambda}$. Since $\boldsymbol{Q}^{-1} = \boldsymbol{Q}^{\mathrm{T}}$, we get

$$\boldsymbol{Q}^{\mathrm{T}}\boldsymbol{A}\boldsymbol{Q} = \operatorname{diag}(\lambda_1, \lambda_2, \ldots, \lambda_n).$$

Therefore, for any quadratic form $f(x_1, x_2, \ldots, x_n) = \boldsymbol{x}^{\mathrm{T}}\boldsymbol{A}\boldsymbol{x}$, we have the following important theorem.

THEOREM 6.1 ***(Principal axis theorem)***

For any n-variable quadratic form

$$f(x_1, x_2, \ldots, x_n) = \boldsymbol{x}^{\mathrm{T}}\boldsymbol{A}\boldsymbol{x},$$

there exists an orthogonal transformation $\boldsymbol{x} = \boldsymbol{Q}\boldsymbol{y}$ *(*$\boldsymbol{Q}$ *is an nth-order orthogonal matrix) such that*

$$\boldsymbol{x}^{\mathrm{T}}\boldsymbol{A}\boldsymbol{x} = \boldsymbol{y}^{\mathrm{T}}(\boldsymbol{Q}^{\mathrm{T}}\boldsymbol{A}\boldsymbol{Q})\boldsymbol{y} = \lambda_1 {y_1}^2 + \lambda_2 {y_2}^2 + \cdots + \lambda_n {y_n}^2, \tag{6.5}$$

where $\lambda_1, \lambda_2, \ldots, \lambda_n$ *are n eigenvalues of the real symmetric matrix* $\boldsymbol{A}$*; n column vectors* $\boldsymbol{\alpha}_1, \boldsymbol{\alpha}_2, \ldots, \boldsymbol{\alpha}_n$ *of* $\boldsymbol{Q}$ *are orthonormal eigenvectors of* $\boldsymbol{A}$ *belonging to eigenvalues* $\lambda_1, \lambda_2, \ldots, \lambda_n$.

Example 6.3 Tranformation of quadratic form to canonical form

Transform the quadratic form

$$f(x_1, x_2, x_3) = 2{x_1}^2 + 5{x_2}^2 + 5{x_3}^2 + 4x_1x_2 - 4x_1x_3 - 8x_2x_3$$

to canonical form by the orthogonal transformation method.

Solution: The corresponding matrix of the quadratic form is

$$\boldsymbol{A} = \begin{pmatrix} 2 & 2 & -2 \\ 2 & 5 & -4 \\ -2 & -4 & 5 \end{pmatrix},$$

and its characteristic polynomial is

$$|\lambda\boldsymbol{I} - \boldsymbol{A}| = (\lambda - 1)^2(\lambda - 10).$$

The eigenvalues of $\boldsymbol{A}$ are $\lambda_1 = 1$ (multiplicity of 2), $\lambda_2 = 10$. From $(\lambda_1\boldsymbol{I} - \boldsymbol{A})\boldsymbol{x} = \boldsymbol{0}$, that is,

$$\begin{pmatrix} -1 & -2 & 2 \\ -2 & -4 & 4 \\ 2 & 4 & -4 \end{pmatrix}\begin{pmatrix} x_1 \\ x_2 \\ x_3 \end{pmatrix} = \begin{pmatrix} 0 \\ 0 \\ 0 \end{pmatrix},$$

and $(\lambda_2\boldsymbol{I} - \boldsymbol{A})\boldsymbol{x} = \boldsymbol{0}$, that is,

$$\begin{pmatrix} 8 & -2 & 2 \\ -2 & 5 & 4 \\ 2 & 4 & 5 \end{pmatrix}\begin{pmatrix} x_1 \\ x_2 \\ x_3 \end{pmatrix} = \begin{pmatrix} 0 \\ 0 \\ 0 \end{pmatrix},$$

we get linearly independent eigenvectors corresponding to $\lambda_1 = 1$,

$$\boldsymbol{x}_1 = (-2, 1, 0)^{\mathrm{T}}, \qquad \boldsymbol{x}_2 = (2, 0, 1)^{\mathrm{T}},$$

and eigenvector $\boldsymbol{x}_3 = (1, 2, -2)^{\mathrm{T}}$ corresponding to $\lambda_2 = 10$.

Applying Gram-Schmidt orthogonalization procedure on $\boldsymbol{x}_1, \boldsymbol{x}_2$, we get $\boldsymbol{\xi}_1$, $\boldsymbol{\xi}_2$, and then normalize $\boldsymbol{x}_3$ to unit length $\boldsymbol{\xi}_3$, where

$$\boldsymbol{\xi}_1 = \left(-\frac{2\sqrt{5}}{5}, \frac{\sqrt{5}}{5}, 0\right)^{\mathrm{T}}, \ \boldsymbol{\xi}_2 = \left(\frac{2\sqrt{5}}{15}, \frac{4\sqrt{5}}{15}, \frac{\sqrt{5}}{3}\right)^{\mathrm{T}}, \ \boldsymbol{\xi}_3 = \left(\frac{1}{3}, \frac{2}{3}, -\frac{2}{3}\right)^{\mathrm{T}}.$$

Take an orthogonal matrix

$$Q=(\xi_1,\xi_2,\xi_3)=\begin{pmatrix} -\frac{2\sqrt{5}}{5} & \frac{2\sqrt{5}}{15} & \frac{1}{3} \\ \frac{\sqrt{5}}{5} & \frac{4\sqrt{5}}{15} & \frac{2}{3} \\ 0 & \frac{\sqrt{5}}{3} & -\frac{2}{3} \end{pmatrix},$$

then

$$Q^{-1}AQ=Q^{\mathrm{T}}AQ=\mathrm{diag}(1,1,10).$$

Let $\boldsymbol{x}=(x_1,x_2,x_3)^{\mathrm{T}}$, $\boldsymbol{y}=(y_1,y_2,y_3)^{\mathrm{T}}$, and take orthogonal transformation $\boldsymbol{x}=Q\boldsymbol{y}$. Then, the quadratic form becomes a canonical form

$$\boldsymbol{x}^{\mathrm{T}}A\boldsymbol{x}=\boldsymbol{y}^{\mathrm{T}}(Q^{\mathrm{T}}AQ)\boldsymbol{y}=y_1{}^2+y_2{}^2+10y_3{}^2.$$

This example can be explained using geometry. For the quadratic surface under standard coordinate system (basis) $\{\boldsymbol{\varepsilon}_1,\boldsymbol{\varepsilon}_2,\boldsymbol{\varepsilon}_3\}$,

$$2x_1{}^2+5x_2{}^2+5x_3{}^2+4x_1x_2-4x_1x_3-8x_2x_3=1,$$

if we transform the coordinate system $\{\boldsymbol{\varepsilon}_1,\boldsymbol{\varepsilon}_2,\boldsymbol{\varepsilon}_3\}$ to another rectangular coordinate system

$$\xi_1=\begin{pmatrix} -\frac{2\sqrt{5}}{5} \\ \frac{\sqrt{5}}{5} \\ 0 \end{pmatrix},\quad \xi_2=\begin{pmatrix} \frac{2\sqrt{5}}{15} \\ \frac{4\sqrt{5}}{15} \\ \frac{\sqrt{5}}{3} \end{pmatrix},\quad \xi_3=\begin{pmatrix} \frac{1}{3} \\ \frac{2}{3} \\ -\frac{2}{3} \end{pmatrix},$$

that is,

$$(\xi_1,\xi_2,\xi_3)=(\boldsymbol{\varepsilon}_1,\boldsymbol{\varepsilon}_2,\boldsymbol{\varepsilon}_3)\begin{pmatrix} -\frac{2\sqrt{5}}{5} & \frac{2\sqrt{5}}{15} & \frac{1}{3} \\ \frac{\sqrt{5}}{5} & \frac{4\sqrt{5}}{15} & \frac{2}{3} \\ 0 & \frac{\sqrt{5}}{3} & -\frac{2}{3} \end{pmatrix},$$

then the equation of quadratic surface, under coordinate (ξ_1,ξ_2,ξ_3), is

$$y_1{}^2+y_2{}^2+10y_3{}^2=1.$$

From analytic geometry, we know that this is an ellipsoid. The lengths of the three principal axes respectively, are 1, 1, $\frac{1}{\sqrt{10}}$, corresponding to $\frac{1}{\sqrt{|\lambda_1|}}$, $\frac{1}{\sqrt{|\lambda_2|}}$, $\frac{1}{\sqrt{|\lambda_3|}}$ of eigenvalues. The sign of the eigenvalues decides the type of the quadratic surface. ■

***Example 6.4 Transformation of the general equation of a quadratic surface to a standard equation**

Transform the general equation of a quadratic surface

$$x^2 - 2y^2 + 10z^2 + 28xy - 8yz + 20zx - 26x + 32y + 28z - 38 = 0 \quad ①$$

to a standard equation (an equation containing only pure squared terms and constants).

Solution: First we apply the orthogonal transformation, as shown in Example 6.3, to the quadratic part in ①,

$$x^2 - 2y^2 + 10z^2 + 28xy - 8yz + 20zx. \quad ②$$

By computing eigenvalues and eigenvectors of matrix $\boldsymbol{A}$ corresponding to ②, and applying Gram-Schmidt orthogonalization to the eigenvectors, we finally get an orthogonal matrix

$$Q = \begin{pmatrix} \frac{1}{3} & \frac{2}{3} & \frac{2}{3} \\ \frac{2}{3} & \frac{1}{3} & -\frac{2}{3} \\ -\frac{2}{3} & \frac{2}{3} & -\frac{1}{3} \end{pmatrix},$$

such that $\boldsymbol{Q}^{\mathrm{T}}\boldsymbol{A}\boldsymbol{Q} = \operatorname{diag}(9, 18, -18)$. Taking the orthogonal transformation $\boldsymbol{x} = \boldsymbol{Q}\boldsymbol{y}$, where $\boldsymbol{x} = (x, y, z)^{\mathrm{T}}$, $\boldsymbol{y} = (x', y', z')^{\mathrm{T}}$, and putting them into $\boldsymbol{x}^{\mathrm{T}}\boldsymbol{A}\boldsymbol{x}$, we get

$$\boldsymbol{x}^{\mathrm{T}}\boldsymbol{A}\boldsymbol{x} = \boldsymbol{y}^{\mathrm{T}}(\boldsymbol{Q}^{\mathrm{T}}\boldsymbol{A}\boldsymbol{Q})\boldsymbol{y} = (x', y', z')\begin{pmatrix} 9 & 0 & 0 \\ 0 & 18 & 0 \\ 0 & 0 & -18 \end{pmatrix}\begin{pmatrix} x' \\ y' \\ z' \end{pmatrix}$$

$$= 9x'^2 + 18y'^2 - 18z'^2.$$

Let us put $\boldsymbol{x} = \boldsymbol{Q}\boldsymbol{y}$, that is,

$$\begin{cases} x = \dfrac{1}{3}x' + \dfrac{2}{3}y' + \dfrac{2}{3}z', \\ y = \dfrac{2}{3}x' + \dfrac{1}{3}y' - \dfrac{2}{3}z', \\ z = -\dfrac{2}{3}x' + \dfrac{2}{3}y' - \dfrac{1}{3}z', \end{cases} \qquad ③$$

into the linear part of the surface equation ①, where the whole equation of the surface is deformed to

$$x'^2 + 2y'^2 - 2z'^2 - \frac{2}{3}x' + \frac{4}{3}y' - \frac{16}{3}z' - \frac{38}{9} = 0.$$

This yields

$$\left(x' - \frac{1}{3}\right)^2 + 2\left(y' + \frac{1}{3}\right)^2 - 2\left(z' + \frac{4}{3}\right)^2 = 1. \qquad ④$$

Let

$$x'' = x' - \frac{1}{3}, \ y'' = y' + \frac{1}{3}, \ z'' = z' + \frac{4}{3}. \qquad ⑤$$

The standard equation of the surface ① is,

$$x''^2 + 2y''^2 - 2z''^2 = 1, \qquad ⑥$$

which is a one-sheet hyperboloid. ■

In Example 6.4, ① is the equation of a surface under the space rectangular coordinate system $Oxyz$, and x, y, z in the equation are coordinates of a space vector (i.e., space point) under the standard basis $\{\boldsymbol{\varepsilon}_1, \boldsymbol{\varepsilon}_2, \boldsymbol{\varepsilon}_3\}$. When the basis $\{\boldsymbol{\varepsilon}_1, \boldsymbol{\varepsilon}_2, \boldsymbol{\varepsilon}_3\}$ is transformed to $\{\xi_1, \xi_2, \xi_3\}$, that is,

$$\xi_1 = \left(\frac{1}{3}, \frac{2}{3}, -\frac{2}{3}\right)^{\mathrm{T}}, \ \xi_2 = \left(\frac{2}{3}, \frac{1}{3}, \frac{2}{3}\right)^{\mathrm{T}}, \ \xi_3 = \left(\frac{2}{3}, -\frac{2}{3}, -\frac{1}{3}\right)^{\mathrm{T}},$$

coordinate vector $\boldsymbol{x} = (x, y, z)^{\mathrm{T}}$ is transformed to $\boldsymbol{y} = (x', y', z')^{\mathrm{T}}$, with the orthogonal transformation $\boldsymbol{x} = \boldsymbol{Q}\boldsymbol{y}$. Therefore, ④ is a coordinate equation of the surface under basis $\{\xi_1, \xi_2, \xi_3\}$. Equation ⑥ indicates a transformation of translation (not a linear transformation, readers should prove this on their own), so that we may get the standard equation of the surface under a space coordinate system $O''x''y''z''$. The coordinates of the new origin O'' is $\left(\frac{1}{3}, -\frac{1}{3}, -\frac{4}{3}\right)^{\mathrm{T}}$ under the coordinate system $Ox'y'z'$ and is $(-1, 1, 0)^{\mathrm{T}}$ under $Oxyz$.

This example illustrates that, when the center of a quadratic surface is the coordinate origin, we can always transform it to a canonical form by an orthogonal transformation. For a quadratic surface whose center is not at the coordinate origin, we can transform it to a canonical form by an orthogonal transformation followed by a translation.

*6.2.2 Methods of completing the squares and elementary operations

An n-variable quadratic form can also be transformed to a canonical form by a general coordinate transformation $\boldsymbol{x} = \boldsymbol{C}\boldsymbol{y}$, that is,

$$\boldsymbol{x}^{\mathrm{T}}\boldsymbol{A}\boldsymbol{x} = \boldsymbol{y}^{\mathrm{T}}\boldsymbol{C}^{\mathrm{T}}\boldsymbol{A}\boldsymbol{C}\boldsymbol{y} = d_1{y_1}^2 + \cdots + d_n{y_n}^2.$$

That is, for any real symmetric matrix $\boldsymbol{A}$, there exists a transformation matrix $\boldsymbol{C}$, such that

$$\boldsymbol{C}^{\mathrm{T}}\boldsymbol{A}\boldsymbol{C} = \mathrm{diag}(d_1, d_2, \ldots, d_n),$$

where $d_1, d_2, \ldots, d_n$ generally are not eigenvalues of $\boldsymbol{A}$, and the column vector of $\boldsymbol{C}$ are generally not the eigenvectors of $\boldsymbol{A}$ too.

The usual methods here are: "completing the squares" and "elementary operations." We now introduce these two methods through the following examples.

Example 6.5 Transformation of a 3-variable quadratic form to a canonical form by completing the squares

Transform a 3-variable quadratic form

$$f(x_1, x_2, x_3) = 2{x_1}^2 + 3{x_2}^2 + {x_3}^2 + 4x_1x_2 - 4x_1x_3 - 8x_2x_3 \qquad ①$$

to a canonical form by completing the squares and determining the coordinate transformation $\boldsymbol{x} = \boldsymbol{C}\boldsymbol{y}$ and transformation matrix $\boldsymbol{C}$.

Solution: First, complete the square for ${x_1}^2$ and other cross terms containing x_1, that is,

$$\begin{aligned} f(x_1, x_2, x_3) &= 2\left[{x_1}^2 + 2x_1(x_2 - x_3) + (x_2 - x_3)^2\right] \\ &\quad - 2(x_2 - x_3)^2 + 3{x_2}^2 + {x_3}^2 - 8x_2x_3 \\ &= 2(x_1 + x_2 - x_3)^2 + {x_2}^2 - {x_3}^2 - 4x_2x_3. \end{aligned}$$

In the above expression, complete the square for ${x_2}^2 - 4x_2x_3$ to get

$$f(x_1, x_2, x_3) = 2(x_1 + x_2 - x_3)^2 + (x_2 - 2x_3)^2 - 5{x_3}^2. \qquad ②$$

Let

$$\begin{cases} y_1 = x_1 + x_2 - \ x_3, \\ y_2 = x_2 - 2x_3, \\ y_3 = x_3. \end{cases} \qquad ③$$

Substituting ③ into ②, we get the canonical quadratic form,

$$f(x_1, x_2, x_3) = 2{y_1}^2 + {y_2}^2 - 5{y_3}^2. \qquad ④$$

From ③, we get

$$\begin{pmatrix} x_1 \\ x_2 \\ x_3 \end{pmatrix} = \begin{pmatrix} 1 & -1 & -1 \\ 0 & 1 & 2 \\ 0 & 0 & 1 \end{pmatrix} \begin{pmatrix} y_1 \\ y_2 \\ y_3 \end{pmatrix}. \qquad ⑤$$

Equation ⑤ is the coordinate transformation $\boldsymbol{x} = \boldsymbol{C}\boldsymbol{y}$ by which we transform the quadratic form ① to its canonical form ④ (where the transformation matrix $\boldsymbol{C}$ is the third-order matrix in ⑤). ■

For a general n-variable quadratic form $f(x_1, x_2, \ldots, x_n)$, if the coefficient of ${x_1}^2$ is not zero, we can transform it to a canonical form in the same way as shown in Example 6.5. If the coefficient of ${x_1}^2$ is zero but the coefficient of ${x_2}^2$ is not zero, we can start by completing the square for x_2. If all the coefficients of the squared terms are zero, for example, the quadratic form only contains cross terms, we can transform it to a canonical form by the method introduced in the following example.

Example 6.6 Transformation of quadratic form with only cross terms to canonical form

Transform the quadratic form

$$f(x_1, x_2, x_3) = 2x_1x_2 + 4x_1x_3$$

to a canonical form by the method of completing the squares, and find the desired coordinate transformation.

Solution: Since there is no square term in the quadratic form, we cannot complete the squares. So, we first take a coordinate transformation such that a square term may appear.

Let

$$\begin{cases} x_1 = y_1 + y_2, \\ x_2 = y_1 - y_2, \\ x_3 = y_3. \end{cases} \qquad ①$$

Substituting ① into the quadratic form, we get

$$\begin{aligned} f(x_1, x_2, x_3) &= 2(y_1 + y_2)(y_1 - y_2) + 4(y_1 + y_2)y_3 \\ &= 2{y_1}^2 - 2{y_2}^2 + 4y_1y_3 + 4y_2y_3. \end{aligned}$$

Then, use the method in Example 6.5 to complete the squares for terms containing y_1, then for y_2, and we get

$$\begin{aligned} f(x_1, x_2, x_3) &= 2(y_1^2 + 2y_1y_3 + y_3^2) - 2y_3^2 - 2y_2^2 + 4y_2y_3 \\ &= 2(y_1 + y_3)^2 - 2(y_2 - y_3)^2. \end{aligned} \quad ②$$

Let

$$\begin{cases} z_1 = y_1 + y_3, \\ z_2 = y_2 - y_3, \\ z_3 = y_3, \end{cases}$$

that is,

$$\begin{cases} y_1 = z_1 - z_3, \\ y_2 = z_2 + z_3, \\ y_3 = z_3, \end{cases} \quad ③$$

Substituting ③ into ②, the quadratic form is transformed into a canonical form, that is,

$$f(x_1, x_2, x_3) = 2z_1^2 - 2z_2^2. \quad ④$$

Here, we have two successive coordinate transformations to transform a quadratic form $2x_1x_2 + 4x_1x_3$ to a canonical form $2z_1^2 - 2z_2^2$. We denote them respectively as

$$\boldsymbol{x} = \boldsymbol{C}_1\boldsymbol{y} \quad \text{and} \quad \boldsymbol{y} = \boldsymbol{C}_2\boldsymbol{z},$$

where

$$\boldsymbol{C}_1 = \begin{pmatrix} 1 & 1 & 0 \\ 1 & -1 & 0 \\ 0 & 0 & 1 \end{pmatrix}, \qquad \boldsymbol{C}_2 = \begin{pmatrix} 1 & 0 & -1 \\ 0 & 1 & 1 \\ 0 & 0 & 1 \end{pmatrix}, \quad ⑤$$

$$\boldsymbol{x} = (x_1, x_2, x_3)^{\mathrm{T}}, \ \boldsymbol{y} = (y_1, y_2, y_3)^{\mathrm{T}}, \ \boldsymbol{z} = (z_1, z_2, z_3)^{\mathrm{T}}.$$

Then, $\boldsymbol{x} = (\boldsymbol{C}_1\boldsymbol{C}_2)\boldsymbol{z}$ is the coordinate transformation to transform the quadratic form to the canonical form ④ and its matrix representation is

$$\boldsymbol{C} = \boldsymbol{C}_1\boldsymbol{C}_2 = \begin{pmatrix} 1 & 1 & 0 \\ 1 & -1 & -2 \\ 0 & 0 & 1 \end{pmatrix}.$$

The corresponding matrices of the original quadratic form $2x_1x_2 + 4x_1x_3$ and the canonical form $2z_1^2 - 2z_2^2$ are

$$\boldsymbol{A} = \begin{pmatrix} 0 & 1 & 2 \\ 1 & 0 & 0 \\ 2 & 0 & 0 \end{pmatrix}, \qquad \boldsymbol{\Lambda} = \begin{pmatrix} 2 & & \\ & -2 & \\ & & 0 \end{pmatrix},$$

respectively. It is not difficult for readers to verify that $\boldsymbol{C}^{\mathrm{T}}\boldsymbol{A}\boldsymbol{C} = \mathrm{diag}(2, -2, 0)$. ■

According to the methods provided by Examples 6.5 and 6.6, any quadratic form in n variables can be transformed to a canonical form by the method of completing the squares. The corresponding transformation matrix is an upper triangular matrix with main diagonal entries 1 or diagonal block matrix $\boldsymbol{C}_1$ similar to ⑤ in Example 6.6, or a product of the two types of matrices.

Any real symmetric matrix $\boldsymbol{A}$ of order n can also be transformed to a congruent canonical form (diagonal matrix) by a series of identical elementary row and column transformations. The term "identical" means:

(i) If $\boldsymbol{A}$ is multiplied on the *RHS* by column addition elementary matrix $\boldsymbol{E}_{ji}(c)$ (i.e., adding column i multiplied by c to column j of $\boldsymbol{A}$), then correspondingly, we multiply $\boldsymbol{A}$ on the *LHS* by $\boldsymbol{E}_{ji}{}^{\mathrm{T}}(c) = \boldsymbol{E}_{ij}(c)$ (i.e., add row i multiplied by c to row j of $\boldsymbol{A}$ after the column operations). The matrix $\boldsymbol{E}_{ji}{}^{\mathrm{T}}(c)\,\boldsymbol{A}\boldsymbol{E}_{ji}(c)$ after transformation is still a symmetric matrix.

(ii) If $\boldsymbol{A}$ is multiplied on the right by $\boldsymbol{E}_i(c)$, then correspondingly, we multiply $\boldsymbol{A}$ on the left by $\boldsymbol{E}_i{}^{\mathrm{T}}(c) = \boldsymbol{E}_i(c)$, that is, row i and column i of $\boldsymbol{A}$ are both multiplied by a nonzero constant c (the entry a_{ii} thus getting multiplied by c^2), obviously $\boldsymbol{E}_i{}^{\mathrm{T}}(c)\,\boldsymbol{A}\boldsymbol{E}_i(c)$ is also a symmetric matrix.

(iii) If $\boldsymbol{A}$ is multiplied on the right by $\boldsymbol{E}_{ij}$, then correspondingly, we multiply $\boldsymbol{A}$ on the left by $\boldsymbol{E}_{ij}{}^{\mathrm{T}} = \boldsymbol{E}_{ji}$, that is, interchange row i and row j of $\boldsymbol{A}$ after interchanging column i and column j of $\boldsymbol{A}$.

For a real symmetric matrix $\boldsymbol{A} = (a_{ij})_{n\times n}$ with order n,

Case 6.1 If $a_{11} \neq 0$, since $a_{1j} = a_{j1} (j = 1, 2, \ldots, n)$, applying the identical row and column elementary operations of addition to $\boldsymbol{A}$, we can reduce the other elements in the first row and the first column to zero, to obtain

$$\begin{pmatrix} a_{11} & \boldsymbol{0} \\ \boldsymbol{0} & \boldsymbol{A}_1 \end{pmatrix},$$

where $\boldsymbol{A}_1$ is a real symmetric matrix with order $(n-1)$.

Case 6.2 If $a_{11} = 0$ but there exists an $a_{ii} \neq 0$, in this case, we interchange the first column and the ith column and then the first row and the ith row. Then, a_{ii} comes to the location in the first row and first column which could then be reduced to Case 6.1 above.

Case 6.3 If the main diagonal entries a_{ii} are all 0, there must exist an $a_{ij} \neq 0$, where we can add the jth row to the ith row and then add the jth column to the ith column. Thus, the entry of the ith row and the ith column becomes $2a_{ij} \neq 0$ now, and the case is reduced to Case 6.2 above.

Thus, we can use mathematical induction to prove the following theorem (the proof is left to readers who are interested).

THEOREM 6.2

For any real symmetric matrix $\boldsymbol{A}$ *of order n, there exists an invertible matrix* $\boldsymbol{C}$, *such that*

$$\boldsymbol{C}^{\mathrm{T}}\boldsymbol{A}\boldsymbol{C} = \mathrm{diag}(d_1, d_2, \ldots, d_n). \tag{6.6}$$

From Corollary 2.5 (of Theorem 2.4), any invertible matrix can be written as a product of a series of elementary matrices $[\boldsymbol{E}_{ij}(c), \ \boldsymbol{E}_i(c), \ \boldsymbol{E}_{ij}]$, that is, $\boldsymbol{C} = \boldsymbol{P}_1\boldsymbol{P}_2\cdots\boldsymbol{P}_k$ (where $\boldsymbol{P}_1, \boldsymbol{P}_2, \ldots, \boldsymbol{P}_k$ are all elementary matrices). Then, readers only need to prove that there exist $\boldsymbol{P}_1, \boldsymbol{P}_2, \ldots, \boldsymbol{P}_k$, such that

$$\boldsymbol{P}_k{}^{\mathrm{T}}\cdots\boldsymbol{P}_2{}^{\mathrm{T}}\boldsymbol{P}_1{}^{\mathrm{T}}\boldsymbol{A}\boldsymbol{P}_1\boldsymbol{P}_2\cdots\boldsymbol{P}_k = \mathrm{diag}\,(d_1, d_2, \ldots, d_n). \tag{6.7}$$

It can be seen from (6.7) that a real symmetric matrix $\boldsymbol{A}$ can be congruently transformed to a diagonal matrix by the method of elementary operations, that is,

$$\boldsymbol{C}^{\mathrm{T}}\boldsymbol{A}\boldsymbol{C} = \mathrm{diag}(d_1, d_2, \ldots, d_n).$$

The transformation matrix is

$$\boldsymbol{C} = \boldsymbol{P}_1\boldsymbol{P}_2\cdots\boldsymbol{P}_k = \boldsymbol{I}\boldsymbol{P}_1\boldsymbol{P}_2\cdots\boldsymbol{P}_k. \tag{6.8}$$

Therefore, if the column transformations applied to $\boldsymbol{A}$ (that is, multiplied on the right by elementary matrices $\boldsymbol{P}_1, \boldsymbol{P}_2, \ldots, \boldsymbol{P}_k$) are applied to the identity matrix $\boldsymbol{I}$ at the same time, then $\boldsymbol{I}$ is turned into the transformation matrix $\boldsymbol{C}$, while $\boldsymbol{A}$ is converted to the diagonal matrix.

Example 6.7 Transformation of quadratic form to canonical form by elementary transformations (I)

Transform the quadratic form

$$f(x_1, x_2, x_3) = (x_1, x_2, x_3)\begin{pmatrix} 2 & 2 & -2 \\ 2 & 5 & -4 \\ -2 & -4 & 5 \end{pmatrix}\begin{pmatrix} x_1 \\ x_2 \\ x_3 \end{pmatrix}$$

to a canonical form by elementary transformations, and give the coordinate transformation $\boldsymbol{x} = \boldsymbol{C}\boldsymbol{y}$.

Solution: In the following transformation, we use $[i]$ to denote the ith column, ⓘ to denote the ith row, and $[j] + [i] \times k$ to denote the addition of the ith column multiplied by k to the jth column.

$$\begin{pmatrix} \boldsymbol{A} \\ \hline \boldsymbol{I} \end{pmatrix} = \begin{pmatrix} 2 & 2 & -2 \\ 2 & 5 & -4 \\ -2 & -4 & 5 \\ \hline 1 & 0 & 0 \\ 0 & 1 & 0 \\ 0 & 0 & 1 \end{pmatrix} \xrightarrow[\substack{[2]+[1]\times(-1) \\ [3]+[1]}]{\substack{[2]+[1]\times(-1) \\ [3]+[1]}} \begin{pmatrix} 2 & 0 & 0 \\ 2 & 3 & -2 \\ -2 & -2 & 3 \\ \hline 1 & -1 & 1 \\ 0 & 1 & 0 \\ 0 & 0 & 1 \end{pmatrix}$$

$$\xrightarrow[\text{no transformations}]{\substack{\text{row transformations} \\ ②+①\times(-1) \\ ③+①}} \begin{pmatrix} 2 & 0 & 0 \\ 0 & 3 & -2 \\ 0 & -2 & 3 \\ \hline 1 & -1 & 1 \\ 0 & 1 & 0 \\ 0 & 0 & 1 \end{pmatrix} \xrightarrow[{[3]+[2]\times\frac{2}{3}}]{\substack{[3]+[2]\times\frac{2}{3} \\ ③+②\times\frac{2}{3}}}$$

$$\begin{pmatrix} 2 & 0 & 0 \\ 0 & 3 & 0 \\ 0 & 0 & \frac{5}{3} \\ \hline 1 & -1 & \frac{1}{3} \\ 0 & 1 & \frac{2}{3} \\ 0 & 0 & 1 \end{pmatrix} = \begin{pmatrix} \boldsymbol{\Lambda} \\ \hline \boldsymbol{C} \end{pmatrix}.$$

The above process converts $\boldsymbol{A}$ to a diagonal matrix $\boldsymbol{\Lambda}$ by applying a series of identical elementary row and column transformations. While only the same series of column operations is applied to identity matrix $\boldsymbol{I}$, it is converted to the transformation matrix $\boldsymbol{C}$ ($\boldsymbol{C}$ is the product of elementary matrices subjected to elementary column transformations). The above transformations can be depicted by the following block matrix

$$\begin{pmatrix} \boldsymbol{C}^{\mathrm{T}} & \boldsymbol{0} \\ \boldsymbol{0} & \boldsymbol{I} \end{pmatrix} \begin{pmatrix} \boldsymbol{A} \\ \boldsymbol{I} \end{pmatrix} \boldsymbol{C} = \begin{pmatrix} \boldsymbol{C}^{\mathrm{T}}\boldsymbol{A}\boldsymbol{C} \\ \boldsymbol{C} \end{pmatrix} = \begin{pmatrix} \boldsymbol{\Lambda} \\ \boldsymbol{C} \end{pmatrix}.$$

Then, by applying coordinate transformation, the primary quadratic form $\boldsymbol{x}^{\mathrm{T}}\boldsymbol{A}\boldsymbol{x}$ is converted to a canonical form

$$\boldsymbol{y}^{\mathrm{T}}\boldsymbol{C}^{\mathrm{T}}\boldsymbol{A}\boldsymbol{C}\boldsymbol{y} = \boldsymbol{y}^{\mathrm{T}}\boldsymbol{\Lambda}\boldsymbol{y} = 2{y_1}^2 + 3{y_2}^2 + \frac{5}{3}{y_3}^2. \qquad \blacksquare$$

Example 6.8 Transformation of quadratic form to canonical form by elementary transformations (II)

Convert $f(x_1, x_2, x_3) = 2x_1x_2 + 4x_1x_3$ in Example 6.6 to a canonical form by elementary transformations, and provide the coordinate transformation $\boldsymbol{x} = \boldsymbol{C}\boldsymbol{y}$ used here.

Solution:

$$f(x_1, x_2, x_3) = (x_1, x_2, x_3)\begin{pmatrix} 0 & 1 & 2 \\ 1 & 0 & 0 \\ 2 & 0 & 0 \end{pmatrix}\begin{pmatrix} x_1 \\ x_2 \\ x_3 \end{pmatrix}$$

Using the same notations $[i]$ and ⓘ in Example 6.7, elementary transformations can be written as

$$\begin{pmatrix} \boldsymbol{A} \\ \hline \boldsymbol{I} \end{pmatrix} = \begin{pmatrix} 0 & 1 & 2 \\ 1 & 0 & 0 \\ 2 & 0 & 0 \\ \hline 1 & 0 & 0 \\ 0 & 1 & 0 \\ 0 & 0 & 1 \end{pmatrix} \xrightarrow[{[1]+[2]}]{\substack{[1]+[2] \\ ①+②}} \begin{pmatrix} 2 & 1 & 2 \\ 1 & 0 & 0 \\ 2 & 0 & 0 \\ \hline 1 & 0 & 0 \\ 1 & 1 & 0 \\ 0 & 0 & 1 \end{pmatrix} \xrightarrow[{[2]+[1]\times\left(-\frac{1}{2}\right)}]{\substack{[2]+[1]\times\left(-\frac{1}{2}\right) \\ ②+①\times\left(-\frac{1}{2}\right)}}$$

$$\begin{pmatrix} 2 & 0 & 2 \\ 0 & -\frac{1}{2} & -1 \\ 2 & -1 & 0 \\ \hline 1 & -\frac{1}{2} & 0 \\ 1 & \frac{1}{2} & 0 \\ 0 & 0 & 1 \end{pmatrix} \xrightarrow[{[3]+[1]\times(-1)}]{\substack{[3]+[1]\times(-1) \\ ③+①\times(-1)}} \begin{pmatrix} 2 & 0 & 0 \\ 0 & -\frac{1}{2} & -1 \\ 0 & -1 & -2 \\ \hline 1 & -\frac{1}{2} & -1 \\ 1 & \frac{1}{2} & -1 \\ 0 & 0 & 1 \end{pmatrix}$$

$$\xrightarrow[{[3]+[2]\times(-2)}]{\substack{[3]+[2]\times(-2) \\ ③+②\times(-2)}} \begin{pmatrix} 2 & 0 & 0 \\ 0 & -\frac{1}{2} & 0 \\ 0 & 0 & 0 \\ \hline 1 & -\frac{1}{2} & 0 \\ 1 & \frac{1}{2} & -2 \\ 0 & 0 & 1 \end{pmatrix} = \begin{pmatrix} \boldsymbol{\Lambda} \\ \hline \boldsymbol{C} \end{pmatrix}.$$

Thus, with a coordinate transformation $\boldsymbol{x} = \boldsymbol{C}\boldsymbol{y}$, where

$$\boldsymbol{C} = \begin{pmatrix} 1 & -\frac{1}{2} & 0 \\ 1 & \frac{1}{2} & -2 \\ 0 & 0 & 1 \end{pmatrix},$$

the quadratic form $f(x_1, x_2, x_3)$ is converted into the canonical form

$$f(x_1, x_2, x_3) = \boldsymbol{x}^{\mathrm{T}}\boldsymbol{A}\boldsymbol{x} = \boldsymbol{y}^{\mathrm{T}}(\boldsymbol{C}^{\mathrm{T}}\boldsymbol{A}\boldsymbol{C})\boldsymbol{y} = \boldsymbol{y}^{\mathrm{T}}\boldsymbol{A}\boldsymbol{y} = 2{y_1}^2 - \frac{1}{2}y_2^2. \quad \blacksquare$$

From the two approaches pursued in Examples 6.3 and 6.7 on one hand and Examples 6.6 and 6.8 on the other, we see that the canonical forms are generally different when different coordinate transformations are used to convert a quadratic form to a canonical form. The fact confirms that a quadratic form, as a vector function of n variables, its coordinate expression (quadratic homogeneous polynomial of n coordinates) are generally different under different bases. However, in the different canonical forms converted from the same quadratic form, the number of positive perfect square terms and the number of negative perfect square terms remains unchanged. For instance, there are three positive square terms in both Examples 6.3 and 6.7, one positive, one negative and one zero square terms in both Examples 6.6 and 6.8. This is not a coincidence, but an inevitable consequence. The proof will be given in the next section.

*6.3 Inertial Theorem and the Normal Form of a Quadratic Form

THEOREM 6.3 *(Inertial Theorem)*

For an n-variable quadratic form $\boldsymbol{x}^{\mathrm{T}}\boldsymbol{A}\boldsymbol{x}$*, whenever we apply a coordinate transformation to convert it to a canonical form, the number p of the positive square terms and the number q of the negative terms are both uniquely determined. Equivalently in its matrix form, for a real symmetric matrix* $\boldsymbol{A}$ *of order n, we may use any invertible matrix* $\boldsymbol{C}$*, as long as*

$$C^{\mathrm{T}}AC = \begin{pmatrix} d_1 & & & & & & & & \\ & \ddots & & & & & & & \\ & & d_p & & & & & & \\ & & & -d_{p+1} & & & & & \\ & & & & \ddots & & & & \\ & & & & & -d_{p+q} & & & \\ & & & & & & 0 & & \\ & & & & & & & \ddots & \\ & & & & & & & & 0 \end{pmatrix},$$

$d_i > 0\,(i = 1, 2, \ldots, p+q)$, $p+q \le n$, p and q are uniquely determined by $\boldsymbol{A}$.

***Proof:** Since $\operatorname{rank}(\boldsymbol{A}) = \operatorname{rank}(\boldsymbol{C}^{\mathrm{T}}\boldsymbol{A}\boldsymbol{C}) = p+q$, so $p+q$ is determined uniquely by the rank of $\boldsymbol{A}$. Therefore, we only need to verify that p is determined uniquely by $\boldsymbol{A}$. Let $p+q = \mathrm{r}(\boldsymbol{A}) = r$, the quadratic form $f = \boldsymbol{x}^{\mathrm{T}}\boldsymbol{A}\boldsymbol{x}$ can be converted to a canonical form by coordinate transformations

$$\boldsymbol{x} = \boldsymbol{B}\boldsymbol{y} \quad \text{and} \quad \boldsymbol{x} = \boldsymbol{C}\boldsymbol{z}, \tag{①}$$

with canonical forms respectively,

$$f = b_1 {y_1}^2 + b_2 {y_2}^2 + \cdots + b_p {y_p}^2 - b_{p+1} {y_{p+1}}^2 - \cdots - b_r {y_r}^2, \tag{②}$$

$$f = c_1 {z_1}^2 + c_2 {z_2}^2 + \cdots + c_t {z_t}^2 - c_{t+1} {z_{t+1}}^2 - \cdots - c_r {z_r}^2, \tag{③}$$

in which $b_i > 0, c_i > 0\,(i = 1, 2, \ldots, r)$.

The number of positive square terms is uniquely determined if $p = t$.

By contradiction, let $p > t$. From ② and ③, we get

$$\begin{array}{l|l|l} f = b_1 {y_1}^2 + \cdots + b_t {y_t}^2 & + b_{t+1} {y_{t+1}}^2 + \cdots + b_p {y_p}^2 & - b_{p+1} {y^2}_{p+1} - \cdots - b_r {y_r}^2 \\ = c_1 {z_1}^2 + \cdots + c_t {z_t}^2 & - c_{t+1} {z_{t+1}}^2 - \cdots - c_p {z_p}^2 & - c_{p+1} {z_{p+1}}^2 - \cdots - c_r {z_r}^2. \end{array} \tag{④}$$

From expression ①, we have $\boldsymbol{z} = \boldsymbol{C}^{-1}\boldsymbol{B}\boldsymbol{y}$ (denote $\boldsymbol{D} = \boldsymbol{C}^{-1}\boldsymbol{B}$), $\boldsymbol{z} = \boldsymbol{D}\boldsymbol{y}$, that is,

$$\begin{cases} z_1 = d_{11} y_1 + d_{12} y_2 + \cdots + d_{1n} y_n, \\ \cdots \\ z_t = d_{t1} y_1 + d_{t2} y_2 + \cdots + d_{tn} y_n, \\ \cdots \\ z_n = d_{n1} y_1 + d_{n2} y_2 + \cdots + d_{nn} y_n. \end{cases} \tag{⑤}$$

In order to obtain a contradiction from ④, let $z_1 = z_2 = \cdots = z_t = 0$, $y_{p+1} = \cdots = y_n = 0$. By using ⑤ again, we get a system of linear equations in $y_1, y_2, \cdots, y_n$

$$\begin{cases} d_{11}y_1 + d_{12}y_2 + \cdots + d_{1n}y_n = 0, \\ \cdots \\ d_{t1}y_1 + d_{t2}y_2 + \cdots + d_{tn}y_n = 0, \\ y_{p+1} = 0, \\ \cdots \\ y_n = 0. \end{cases} \qquad ⑥$$

The homogeneous system of linear equations ⑥ has n unknowns, but the number of equations is $t + (n - p) = n - (p - t) < n$, where it must have a nonzero solution. Since $y_{p+1} = \cdots = y_n = 0$, so $y_1, y_2, \ldots, y_p$ in the nonzero solution of ⑥ are not all zero. Substituting it into ④, we get

$$f = b_1 y_1{}^2 + b_2 y_2{}^2 + \cdots + b_t y_t{}^2 + \cdots + b_p y_p{}^2 > 0. \qquad ⑦$$

Substituting the nonzero solution of ⑥ into ⑤, we get a group of values of $z_1, z_2, \ldots, z_t, \ldots, z_n$ (this time, $z_1 = z_2 = \cdots = z_t = 0$). Substituting it into ④, we get

$$f = -c_{t+1} z_{t+1}{}^2 - \cdots - c_p z_p{}^2 - \cdots - c_r z_r{}^2 \leq 0. \qquad ⑧$$

Obviously, ⑦ and ⑧ are contradictory. The hypothesis $p > t$ therefore is not true.

Using the same argument, $p < t$ is not true, hence $p = t$. It has been shown that the number of positive square terms in the canonical form has no relationship with the applied nonsingular linear transformation. The positive square terms are determined by the quadratic form (or the matrix $\boldsymbol{A}$ of the quadratic form) itself. Since $q = \mathrm{r}(\boldsymbol{A}) - p$, so q is also determined by $\boldsymbol{A}$ uniquely. □

Definition 6.3

For the canonical form of a quadratic form $\boldsymbol{x}^{\mathrm{T}}\boldsymbol{A}\boldsymbol{x}$, the number of positive square terms (i.e., the number of positive diagonal entries in the diagonal matrix which is congruent to $\boldsymbol{A}$) is called the **index of positive inertia** of the quadratic form (or $\boldsymbol{A}$), while the number of negative square terms (i.e., the number of negative diagonal entries in the diagonal matrix which is congruent to $\boldsymbol{A}$) is called the **index of negative inertia**. The difference between positive index and negative index is called the **sign difference**. The rank of the matrix $\boldsymbol{A}$ is also called the rank of the quadratic form $\boldsymbol{x}^{\mathrm{T}}\boldsymbol{A}\boldsymbol{x}$.

If the rank of a real symmetric matrix $\boldsymbol{A}$ of order n is r and the index of positive inertia is p, then the index of negative inertia is $q = r - p$, the sign difference is $p - q = 2p - r$, and the number of zero diagonal entries in the congruent diagonal matrix is $(n - r)$.

From Inertial Theorem, we can get the following corollary.

COROLLARY 6.1

Let $\boldsymbol{A}$ be a real symmetric matrix of order n. If the positive inertial index and the negative inertial index of $\boldsymbol{A}$ are p and q, respectively, then

$$\boldsymbol{A} \simeq \operatorname{diag}(1, \ldots, 1, -1, \ldots, -1, 0, \ldots, 0), \tag{6.9}$$

where the number of 1's is p, the number of -1's is q, the number of 0's is $n - (p + q)$.

Expressed simply, there exists coordinate transformation $\boldsymbol{x} = \boldsymbol{C}\boldsymbol{y}$ for the quadratic form $\boldsymbol{x}^{\mathrm{T}}\boldsymbol{A}\boldsymbol{x}$, such that

$$\boldsymbol{x}^{\mathrm{T}}\boldsymbol{A}\boldsymbol{x} = {y_1}^2 + \cdots + {y_p}^2 - {y_{p+1}}^2 - \cdots - {y_{p+q}}^2. \tag{6.10}$$

The quadratic form on the *RHS* of (6.10) is called the **normal form** of the quadratic form, the diagonal matrix in (6.9) is called the **congruent normal form** of the matrix $\boldsymbol{A}$.

Proof: According to Theorem 6.2 and the Inertial Theorem, there exists an invertible matrix $\boldsymbol{C}_1$, such that

$$\boldsymbol{C}_1{}^{\mathrm{T}}\boldsymbol{A}\boldsymbol{C}_1 = \operatorname{diag}(d_1, \ldots, d_p, -d_{p+1}, \ldots, -d_{p+q}, 0, \ldots, 0),$$

where $d_i > 0$ $(i = 1, \ldots, p, p+1, \ldots, p+q)$. Denote the invertible matrix as

$$\boldsymbol{C}_2 = \operatorname{diag}\left(\frac{1}{\sqrt{d_1}}, \ldots, \frac{1}{\sqrt{d_p}}, \frac{1}{\sqrt{d_{p+1}}}, \ldots, \frac{1}{\sqrt{d_{p+q}}}, 1, \ldots, 1\right).$$

Then, $\boldsymbol{C}_2^{\mathrm{T}} = \boldsymbol{C}_2$ and

$$\boldsymbol{C}_2{}^{\mathrm{T}}(\boldsymbol{C}_1{}^{\mathrm{T}}\boldsymbol{A}\boldsymbol{C}_1)\boldsymbol{C}_2 = \operatorname{diag}(1, \ldots, 1, -1, \ldots, -1, 0, \ldots, 0),$$

where the numbers of $+1$ and -1 are p and q, respectively; the number of 0 is $[n - (p + q)]$.

Let $\boldsymbol{C} = \boldsymbol{C}_1\boldsymbol{C}_2$. Now, the *RHS* of (6.9) is $\boldsymbol{C}^{\mathrm{T}}\boldsymbol{A}\boldsymbol{C}$. Denote $\boldsymbol{x} = \boldsymbol{C}\boldsymbol{y}$ ($\boldsymbol{C}$ is invertible), then (6.10) is true. □

If two real symmetric matrices $\boldsymbol{A}$ and $\boldsymbol{B}$ with order n are congruent, the quadratic forms $\boldsymbol{x}^{\mathrm{T}}\boldsymbol{A}\boldsymbol{x}$ and $\boldsymbol{y}^{\mathrm{T}}\boldsymbol{B}\boldsymbol{y}$ are also said to be congruent.

According to the above results, readers can easily verify the following conclusions.

(i) Two real symmetric matrices $\boldsymbol{A}$ and $\boldsymbol{B}$ are congruent, if and only if, $\boldsymbol{A}$ and $\boldsymbol{B}$ have the same index of positive inertia and the same index of negative inertia.

(ii) All the real symmetric matrices with order n can be classified into $\frac{1}{2}(n+1)(n+2)$ categories according to their congruent canonical forms (ignoring the order of $+1, -1, 0$).

6.4 Positive Definite Quadratic Forms and Positive Definite Matrices

A positive definite quadratic form of n variables is a quadratic form with positive inertia of n. A positive definite matrix with order n is a real symmetric matrix with positive inertia of n. They have diverse applications in engineering, technology, and optimization. Now, we will introduce the concept of positive definiteness of quadratic form through the determination of an extreme point of a two-variable function. For example,

$$f(x, y) = 2x^2 + 4xy + 5y^2. \qquad ①$$

We have $f(0, 0) = 0$, $f'_x(0, 0) = f'_y(0, 0) = 0$, which, taken as a whole, mean that the origin O(0, 0) is the stationary point of $f(x, y)$. Reformulating ① to

$$f(x, y) = 2(x+y)^2 + 3y^2,$$

we observe that $f(x, y)$ is always positive when x or y is not zero, that is, $\boldsymbol{\alpha} = (x, y)^{\mathrm{T}} \neq \mathbf{0}$. Therefore, O(0, 0) is a local minimum point of $f(x, y)$. The quadratic form ① is a typical positive definite quadratic form that will be discussed in this section. Another example is

$$F(x, y) = 4 + 4xy + 2x\sin x + 5y\sin y. \qquad ②$$

It is easy to verify that

$$F(0, 0) = 4, \ F'_x(0, 0) = F'_y(0, 0) = 0.$$

Therefore, the origin O(0, 0) is a stationary point of $F(x, y)$. From the theory of multivariate calculus, to distinguish whether the origin O(0, 0) is an extreme point of $F(x, y)$, we could expand $F(x, y)$ in Taylor's series of order 2 at the origin,

$$\begin{aligned} F(x, y) = &F(0,0) + F'_x(0,0)x + F'_y(0,0)y + \frac{1}{2!}[F_{xx}''(0,0)x^2 \\ &+ F_{xy}''(0,0)xy + F_{yx}''(0,0)yx + F_{yy}''(0,0)y^2] + o(x^2+y^2), \end{aligned}$$

where $o(x^2 + y^2)$ is an infinitesimal with order higher than $x^2 + y^2$. Based on further calculations, we get

$$F_{xx}''(0,0) = 4, \ F_{xy}''(0,0) = F_{yx}''(0,0) = 4, \ F_{yy}''(0,0) = 10.$$

Thus,

$$F(x, y) = 4 + (2x^2 + 4xy + 5y^2) + o(x^2 + y^2).$$

Hence, to determine whether $F(0, 0) = 4$ is a minimum (maximum) point of $F(x, y)$ is to determine whether

$$(2x^2 + 4xy + 5y^2) + o(x^2 + y^2)$$

is positive (negative) in a neighborhood of the origin O(0,0). As $o(x^2 + y^2)$ is an infinitesimal with order higher than $x^2 + y^2$, so the sign of the above expression is determined by the sign of $2x^2 + 4xy + 5y^2$. According to the previous discussion for ①, we know that the origin O(0,0) is also a local minimum point of the function $F(x, y)$.

For a general n-variable function, the problem of checking whether a stationary point is a local extreme point can be converted to the problem of checking whether an n-variable quadratic form is positive or negative over a neighborhood of that point. It is a problem of checking whether the quadratic form is positive definite or negative definite. Now, we will discuss the positive definite quadratic form first.

Definition 6.4

If

$$\sum_{i=1}^{n}\sum_{j=1}^{n} a_{ij}x_ix_j = \boldsymbol{x}^{\mathrm{T}}\boldsymbol{A}\boldsymbol{x} > 0, \tag{6.11}$$

for any nonzero vector $\boldsymbol{x} = (x_1, x_2, \ldots, x_n)^{\mathrm{T}}$, then $\boldsymbol{x}^{\mathrm{T}}\boldsymbol{A}\boldsymbol{x}$ is called a **positive definite quadratic form,** $\boldsymbol{A}$ is called a **positive definite matrix**.

According to Definition 6.4, we have the following conclusions.

(i) The quadratic form

$$f(y_1, y_2, \ldots, y_n) = d_1{y_1}^2 + d_2{y_2}^2 + \cdots + d_n{y_n}^2$$

is positive definite, if and only if, $d_i > 0$ $(i = 1, 2, \ldots, n)$. The sufficiency is obvious. The necessity is proved by the contradiction. If $d_i \le 0$, then take $y_i = 1$, $y_j = 0$ $(j \ne i)$ and substitute them into the quadratic form to get

$$f(0, \ldots, 0, 1, 0, \ldots, 0) = d_i \le 0,$$

which is contradictory to the fact that the quadratic form $f(y_1, y_2, \ldots, y_n)$ is positive definite.

(ii) The positive definiteness of a quadratic form $\boldsymbol{x}^{\mathrm{T}}\boldsymbol{A}\boldsymbol{x}$ is invariant to the non-degenerate linear transformation $\boldsymbol{x} = \boldsymbol{C}\boldsymbol{y}$ by which it is transformed to $\boldsymbol{y}^{\mathrm{T}}(\boldsymbol{C}^{\mathrm{T}}\boldsymbol{A}\boldsymbol{C})\boldsymbol{y}$, that is,

$$\boldsymbol{x}^{\mathrm{T}}\boldsymbol{A}\boldsymbol{x} \overset{x=Cy}{=\!=} \boldsymbol{y}^{\mathrm{T}}(\boldsymbol{C}^{\mathrm{T}}\boldsymbol{A}\boldsymbol{C})\boldsymbol{y} \ (\boldsymbol{C} \text{ is invertible}),$$

the quadratic forms on both sides have the same nature with respect to positive definiteness. For the transformation $\boldsymbol{x} = \boldsymbol{C}\boldsymbol{y}$ ($\boldsymbol{C}$ is invertible), $\boldsymbol{x}_0 = \boldsymbol{C}\boldsymbol{y}_0 \neq \boldsymbol{0}$ for any $\boldsymbol{y}_0 \neq \boldsymbol{0}$, since $\boldsymbol{C}$ is invertible. If $\boldsymbol{x}^{\mathrm{T}}\boldsymbol{A}\boldsymbol{x}$ is positive definite, then $\boldsymbol{x}_0^{\mathrm{T}}\boldsymbol{A}\boldsymbol{x}_0 > 0$, thus we have

$$\boldsymbol{y}_0{}^{\mathrm{T}}(\boldsymbol{C}^{\mathrm{T}}\boldsymbol{A}\boldsymbol{C})\boldsymbol{y}_0 = \boldsymbol{x}_0{}^{\mathrm{T}}\boldsymbol{A}\boldsymbol{x}_0 > 0, \ \forall \boldsymbol{y}_0 \neq \boldsymbol{0}.$$

So, $\boldsymbol{y}^{\mathrm{T}}(\boldsymbol{C}^{\mathrm{T}}\boldsymbol{A}\boldsymbol{C})\boldsymbol{y}$ is a positive definite quadratic form and vice versa.

Based on the above two conclusions, after converting a quadratic form $\boldsymbol{x}^{\mathrm{T}}\boldsymbol{A}\boldsymbol{x}$ (with a real symmetric matrix $\boldsymbol{A}$) to its canonical form (normal form) $\boldsymbol{y}^{\mathrm{T}}(\boldsymbol{C}^{\mathrm{T}}\boldsymbol{A}\boldsymbol{C})\boldsymbol{y} = \sum_{i=1}^{n} d_i y_i{}^2$ through coordinate transformation $\boldsymbol{x} = \boldsymbol{C}\boldsymbol{y}$ ($\boldsymbol{A}$ is congruent to a diagonal matrix, that is, $\boldsymbol{C}^{\mathrm{T}}\boldsymbol{A}\boldsymbol{C} = \boldsymbol{\Lambda}$), we can easily determine the positive definiteness from the diagonal matrix.

The following important result is about the determination of the positive definiteness of a quadratic form from its canonical form (normal form).

THEOREM 6.4

If $\boldsymbol{A}$ is a real symmetric matrix of order n, then the following statements are equivalent.

(i) $\boldsymbol{x}^{\mathrm{T}}\boldsymbol{A}\boldsymbol{x}$ *is a positive definite quadratic form (or $\boldsymbol{A}$ is a positive definite matrix).*
(ii) *The index of positive inertia of $\boldsymbol{A}$ is n, that is, $\boldsymbol{A} \simeq \boldsymbol{I}$.*
(iii) *There exists an invertible matrix $\boldsymbol{P}$, such that $\boldsymbol{A} = \boldsymbol{P}^{\mathrm{T}}\boldsymbol{P}$.*
(iv) *The n eigenvalues $\lambda_1, \lambda_2, \cdots, \lambda_n$ of $\boldsymbol{A}$ are all positive.*

The four propositions in Theorem 6.4 are equivalent, which means that any two propositions are necessary and sufficient conditions of each other.

To prove the equivalence, we can use the following cyclical scheme of proof.

Proof: (i)→(ii). According to Theorem 6.2, there exists an invertible matrix $\boldsymbol{C}$ for $\boldsymbol{A}$, such that

$$\boldsymbol{C}^{\mathrm{T}}\boldsymbol{A}\boldsymbol{C} = \mathrm{diag}\,(d_1, d_2, \ldots, d_n)\,.$$

Suppose that the index of positive inertia is less than n, then there exists at least one $d_i \leq 0$. Apply the transformation $\boldsymbol{x} = \boldsymbol{C}\boldsymbol{y}$, then

$$\boldsymbol{x}^{\mathrm{T}}\boldsymbol{A}\boldsymbol{x} = \boldsymbol{y}^{\mathrm{T}}\left(\boldsymbol{C}^{\mathrm{T}}\boldsymbol{A}\boldsymbol{C}\right)\boldsymbol{y} = d_1 y_1{}^2 + d_2 y_2{}^2 + \cdots + d_n y_n{}^2$$

is not always positive [see the conclusion (i) after Definition 6.4], which contradicts the proposition (i). So, the index of positive inertia is n and thus $\boldsymbol{A} \simeq \boldsymbol{I}$.

(ii)→(iii). From $C^{\mathrm{T}}AC = I$ (C is invertible), we have $A = (C^{\mathrm{T}})^{-1}C^{-1} = (C^{-1})^{\mathrm{T}}C^{-1}$, then $A = P^{\mathrm{T}}P$ if $P = C^{-1}$.
(iii)→(iv). Suppose $Ax = \lambda x$, that is, $(P^{\mathrm{T}}P)x = \lambda x$, then we have

$$x^{\mathrm{T}}P^{\mathrm{T}}Px = \lambda x^{\mathrm{T}}x, \text{ that is, } (Px, Px) = \lambda(x, x).$$

Since the eigenvector $x \neq 0$, thus $Px \neq 0$. So, the eigenvalue of A,

$$\lambda = \frac{(Px, Px)}{(x, x)} > 0.$$

(iv)→(i). For the real symmetric matrix A of order n, there exists an orthogonal matrix Q, such that

$$Q^{\mathrm{T}}AQ = \mathrm{diag}(\lambda_1, \lambda_2, \ldots, \lambda_n),$$

and

$$x^{\mathrm{T}}Ax = \lambda_1 y_1{}^2 + \lambda_2 y_2{}^2 + \cdots + \lambda_n y_n{}^2,$$

if $x = Qy$. So, $x^{\mathrm{T}}Ax$ is positive definite as the eigenvalues $\lambda_1, \lambda_2, \cdots, \lambda_n$ are all positive. □

Example 6.9 A^{-1} is a positive definite matrix when A is so

Prove that if A is a positive definite matrix, then A^{-1} is also a positive definite matrix.

Proof: As a positive definite matrix, A is of full rank, thus A^{-1} exists. The inverse of a symmetric matrix is still a symmetric matrix and A^{-1} is, therefore, a symmetric matrix. There are many ways to prove that A^{-1} is a positive definite matrix.

Method 1: By definition:

$$\begin{aligned} x^{\mathrm{T}}A^{-1}x &= (x, A^{-1}x) \text{ (apply the transformation } x = Ay) \\ &= (Ay, A^{-1}Ay) = (Ay, y) = y^{\mathrm{T}}Ay. \end{aligned}$$

If $x \neq 0$, then $y \neq 0$ with $x = Ay$ as A is invertible and $y^{\mathrm{T}}Ay > 0$. Consequently, $x^{\mathrm{T}}A^{-1}x > 0\ \forall x \neq 0$. So, A^{-1} is positive definite.

Method 2: By Theorem 6.4, (i) $\rightleftharpoons$ (ii).

There exists an invertible matrix C, such that $C^{\mathrm{T}}AC = I$ for a positive definite A.

Then, $C^{-1}A^{-1}(C^{-1})^{\mathrm{T}} = I$ if both sides of $C^{\mathrm{T}}AC = I$ are inverted. Take $D = (C^{-1})^{\mathrm{T}}$. Then, $D^{\mathrm{T}}A^{-1}D = I$ (D is invertible) and A^{-1} is therefore positive definite.

Method 3: By Theorem 6.4, (i)⇌(iii).

$\boldsymbol{A}$ is positive definite, so there exists an invertible matrix $\boldsymbol{P}$ such that $\boldsymbol{A} = \boldsymbol{P}^{\mathrm{T}}\boldsymbol{P}$. Thus, $\boldsymbol{A}^{-1} = \boldsymbol{P}^{-1}(\boldsymbol{P}^{-1})^{\mathrm{T}} = \boldsymbol{S}^{\mathrm{T}}\boldsymbol{S}$ [where $\boldsymbol{S} = (\boldsymbol{P}^{-1})^{\mathrm{T}}$ is an invertible matrix] and $\boldsymbol{A}^{-1}$ is positive definite.

Method 4: By Theorem 6.4, (i)⇌(iv).

According to $\boldsymbol{A}\boldsymbol{x} = \lambda\boldsymbol{x}$, then $\boldsymbol{A}^{-1}\boldsymbol{x} = \frac{1}{\lambda}\boldsymbol{x}$ (where $\lambda \neq 0$). As the eigenvalues of $\boldsymbol{A}$ are all positive, so the eigenvalues of $\boldsymbol{A}^{-1}$ are also all positive and hence $\boldsymbol{A}^{-1}$ is positive definite. ■

Example 6.10 Determining the nature of the quadratic form $f(x_1, x_2, x_3)$ (I)

Determine whether the quadratic form

$$f(x_1, x_2, x_3) = x_1{}^2 + 2x_2{}^2 + 3x_3{}^2 + 2x_1x_2 - 2x_2x_3$$

is positive definite or not.

Solution: By the method of completing the squares

$$\begin{aligned} f(x_1, x_2, x_3) &= (x_1 + 2x_1x_2 + x_2{}^2) + (x_2{}^2 - 2x_2x_3 + x_3{}^2) + 2x_3{}^2 \\ &= (x_1 + x_2)^2 + (x_2 - x_3)^2 + 2x_3{}^2 \geq 0. \end{aligned}$$

The necessary and sufficient conditions that the equality sign holds are that

$$x_1 + x_2 = 0, \quad x_2 - x_3 = 0, \quad x_3 = 0,$$

that is, $x_1 = x_2 = x_3 = 0$, so $f(x_1, x_2, x_3)$ is positive definite. ■

Example 6.11 Determining the nature of the quadratic form $f(x_1, x_2, x_3)$ (II)

Determine whether the quadratic form

$$f(x_1, x_2, x_3) = 3x_1{}^2 + x_2{}^2 + 3x_3{}^2 - 4x_1x_2 - 4x_1x_3 + 4x_2x_3$$

is positive definite or not.

Solution: The method of completing the squares can be used to determine the positive definiteness for any quadratic form. However, the coefficients in this example are rather complicated for adopting the process of completing the squares. So, we may consider the method of eigenvalues. Calculating the eigenvalues of a quadratic form is usually easy.

For instance, the characteristic polynomial in Example 6.10 is

$$|\lambda \boldsymbol{I} - \boldsymbol{A}| = \lambda^3 - 6\lambda^2 + 9\lambda - 2.$$

It is factorized as $(\lambda - 2)(\lambda^2 - 4\lambda + 1)$ and the eigenvalues are $\lambda = 2$, $2 \pm \sqrt{3}$.

The matrix corresponding to the present quadratic form is

$$\boldsymbol{A} = \begin{pmatrix} 3 & -2 & -2 \\ -2 & 1 & 2 \\ -2 & 2 & 3 \end{pmatrix}.$$

From

$$|\lambda \boldsymbol{I} - \boldsymbol{A}| = \begin{vmatrix} \lambda - 3 & 2 & 2 \\ 2 & \lambda - 1 & -2 \\ 2 & -2 & \lambda - 3 \end{vmatrix} \overset{[1]+[3]}{=\!=\!=} \begin{vmatrix} \lambda - 1 & 2 & 2 \\ 0 & \lambda - 1 & -2 \\ \lambda - 1 & -2 & \lambda - 3 \end{vmatrix}$$

$$\overset{③+①\times(-1)}{=\!=\!=\!=} \begin{vmatrix} \lambda - 1 & 2 & 2 \\ 0 & \lambda - 1 & -2 \\ 0 & -4 & \lambda - 5 \end{vmatrix} = (\lambda - 1)(\lambda^2 - 6\lambda - 3) = 0,$$

we get the eigenvalues of $\boldsymbol{A}$; $\lambda_1 = 1, \lambda_2 = 3 + 2\sqrt{3}, \lambda_3 = 3 - 2\sqrt{3} < 0$. $\boldsymbol{A}$ is, therefore, not a positive definite matrix, thus the quadratic form is not positive definite.

The non-positive definite property of this example can also be determined easily by the definition of the positive definite quadratic form. Because the quadratic form is not positive when $x_1 = 1, x_2 = 1, x_3 = 0$, indeed $f(1, 1, 0) = 0$. ■

We now determine the positive definiteness of a quadratic form $\boldsymbol{x}^{\mathrm{T}}\boldsymbol{A}\boldsymbol{x}$ from the minors of $\boldsymbol{A}$. First, we give two necessary conditions for $\boldsymbol{A}$ being positive definite, then a necessary and sufficient condition.

THEOREM 6.5

If a quadratic form $\boldsymbol{x}^{\mathrm{T}}\boldsymbol{A}\boldsymbol{x}$ is positive definite, then

(i) *the main diagonal entries $a_{ii} > 0$ $(i = 1, 2, \cdots, n)$,*
(ii) *the determinant $|\boldsymbol{A}| > 0$.*

Proof:

(i) Since $x^{\mathrm{T}}Ax = \sum_{i=1}^{n}\sum_{j=1}^{n} a_{ij}x_ix_j$ is positive definite, so take $x_i = (0, \ldots, 0, 1, 0, \ldots, 0)^{\mathrm{T}} \neq \mathbf{0}$ (where the ith component $x_i = 1$), then we have

$$x_i{}^{\mathrm{T}}Ax_i = a_{ii}x_i{}^2 = a_{ii} > 0 \ (i = 1, 2, \cdots, n).$$

(ii) Since A is positive definite, so there exists an invertible matrix P, such that $A = P^{\mathrm{T}}P$. Therefore, $|A| = |P^{\mathrm{T}}||P| = |P^2| > 0$.

Or we can obtain $|A| = \lambda_1\lambda_2\cdots\lambda_n > 0$ from the fact that the eigenvalues of the positive definite matrix A are all positive. □

Theorem 6.5 is a necessary condition under which A is positive definite. Using this theorem, it is easy to verify that none of

$$A = \begin{pmatrix} 1 & 2 \\ 2 & 4 \end{pmatrix}, \quad B = \begin{pmatrix} 4 & 5 \\ 5 & 2 \end{pmatrix}, \quad C = \begin{pmatrix} -3 & 2 \\ 2 & 3 \end{pmatrix},$$

is positive definite as $\det(A) = 0$, $\det(B) < 0$, $c_{11} < 0$ in C. Whereas for

$$A = \begin{pmatrix} 1 & 2 & 0 & 0 \\ 2 & 1 & 0 & 0 \\ 0 & 0 & 1 & 2 \\ 0 & 0 & 2 & 1 \end{pmatrix},$$

it can be verified (verification is left to readers) by Theorem 6.4 that A is not positive definite although $a_{ii} > 0$ and

$$\det(A) = \begin{vmatrix} 1 & 2 \\ 2 & 1 \end{vmatrix}^2 = 9 > 0.$$

Let $A = (a_{ij})_{n\times n}$, then

$$\det A_k = \begin{vmatrix} a_{11} & a_{12} & \cdots & a_{1k} \\ a_{21} & a_{22} & \cdots & a_{2k} \\ \vdots & \vdots & & \vdots \\ a_{k1} & a_{k2} & \cdots & a_{kk} \end{vmatrix}$$

is called the kth-order **sequential** (left upper corner) **principal minor** of the nth-order matrix A. We get n sequential principal minors of A as k takes on $1, 2, \ldots, n$.

THEOREM 6.6

An n-variable quadratic form $\mathbf{x}^{\mathrm{T}}\boldsymbol{A}\boldsymbol{x}$ is positive definite, if and only if, all the n sequential principal minors are positive.

Proof: Here, we prove the necessity condition. Denote $\boldsymbol{x}_k = (x_1, x_2, \ldots, x_k)^{\mathrm{T}}$. For $\boldsymbol{x}_k \neq \mathbf{0}$ and thus $\boldsymbol{x} = (\boldsymbol{x}_k^{\mathrm{T}},\ \mathbf{0})^{\mathrm{T}} \neq \mathbf{0}$, we see that

$$\boldsymbol{x}^{\mathrm{T}}\boldsymbol{A}\boldsymbol{x} = (\boldsymbol{x}_k{}^{\mathrm{T}},\ \ \mathbf{0})\begin{pmatrix} \boldsymbol{A}_k & * \\ * & * \end{pmatrix}\begin{pmatrix} \boldsymbol{x}_k \\ \mathbf{0} \end{pmatrix} = (\boldsymbol{x}_k{}^{\mathrm{T}}\boldsymbol{A}_k,\ \ *)\begin{pmatrix} \boldsymbol{x}_k \\ \mathbf{0} \end{pmatrix} = \boldsymbol{x}_k^{\mathrm{T}}\boldsymbol{A}_k\boldsymbol{x}_k > 0$$

is true for any $\boldsymbol{x}_k \neq \mathbf{0}$. So, the k-variable quadratic form $\boldsymbol{x}_k{}^{\mathrm{T}}\boldsymbol{A}_k\boldsymbol{x}_k$ in $x_1, x_2, \ldots, x_k$ is positive definite and $|\boldsymbol{A}_k| > 0$ according to Theorem 6.5. The necessity is proved.

*The sufficiency condition. Apply mathematical induction on n. When $n = 1$, $a_{11} > 0$, $\boldsymbol{x}^{\mathrm{T}}\boldsymbol{A}\boldsymbol{x} = a_{11}x_1{}^2 > 0$ $(\forall x_1 \neq 0)$, so the sufficiency follows. Suppose the sufficiency is true for $(n-1)$-variable quadratic form. In the following, we will prove that it is true for any n-variable quadratic form. A matrix $\boldsymbol{A}$ is expressed in block form as

$$\boldsymbol{A} = \begin{pmatrix} \boldsymbol{A}_{n-1} & \boldsymbol{\alpha} \\ \boldsymbol{\alpha}^{\mathrm{T}} & a_{nn} \end{pmatrix},$$

where $\boldsymbol{\alpha}^{\mathrm{T}} = (a_{n1}, a_{n2}, \ldots, a_{n,n-1})$.

According to Theorem 6.4, we only need to prove that $\boldsymbol{A}$ is congruent to an identity matrix. Taking

$$\boldsymbol{C}_1{}^{\mathrm{T}} = \begin{pmatrix} \boldsymbol{I}_{n-1} & \mathbf{0} \\ -\boldsymbol{\alpha}^{\mathrm{T}}\boldsymbol{A}_{n-1}^{-1} & 1 \end{pmatrix},$$

and

$$\boldsymbol{C}_1 = \begin{pmatrix} \boldsymbol{I}_{n-1} & -\boldsymbol{A}_{n-1}^{-1}\boldsymbol{\alpha} \\ \mathbf{0} & 1 \end{pmatrix},$$

we get

$$\boldsymbol{C}_1{}^{\mathrm{T}}\boldsymbol{A}\boldsymbol{C}_1 = \begin{pmatrix} \boldsymbol{A}_{n-1} & \mathbf{0} \\ \mathbf{0} & a_{nn} - \boldsymbol{\alpha}^{\mathrm{T}}\boldsymbol{A}_{n-1}^{-1}\boldsymbol{\alpha} \end{pmatrix} \xlongequal{\text{denote}} \begin{pmatrix} \boldsymbol{A}_{n-1} & \mathbf{0} \\ \mathbf{0} & a \end{pmatrix}.$$

According to the sufficiency conditions $|\boldsymbol{A}| > 0$, $|\boldsymbol{A}_{n-1}| > 0$, we get $a > 0$ from the previous expression. According to the inductive assumption, $\boldsymbol{A}_{n-1}$ is positive definite, so there exists $(n-1)$th-order invertible matrix $\boldsymbol{G}$ such that $\boldsymbol{G}^{\mathrm{T}}\boldsymbol{A}_{n-1}\boldsymbol{G} = \boldsymbol{I}_{n-1}$. Again, taking

$$C_2{}^{\mathrm{T}} = \begin{pmatrix} \boldsymbol{G}^{\mathrm{T}} & \boldsymbol{0} \\ \boldsymbol{0} & \frac{1}{\sqrt{a}} \end{pmatrix} \quad \text{and} \quad \boldsymbol{C}_2 = \begin{pmatrix} \boldsymbol{G} & \boldsymbol{0} \\ \boldsymbol{0} & \frac{1}{\sqrt{a}} \end{pmatrix},$$

we get $\boldsymbol{C}_2{}^{\mathrm{T}}(\boldsymbol{C}_1{}^{\mathrm{T}}\boldsymbol{A}\boldsymbol{C}_1)\boldsymbol{C}_2 = \boldsymbol{I}_n$. Hence, $\boldsymbol{A}$ is congruent to a unit matrix, thus is positive definite. □

Using Theorem 6.6, we can determine easily that

$$\boldsymbol{A} = \begin{pmatrix} 1 & 2 & 0 & 0 \\ 2 & 1 & 0 & 0 \\ 0 & 0 & 1 & 2 \\ 0 & 0 & 2 & 1 \end{pmatrix}$$

is not positive definite. This is because

$$\det(\boldsymbol{A}_2) = \begin{vmatrix} 1 & 2 \\ 2 & 1 \end{vmatrix} = -3 < 0.$$

***Example 6.12 Every positive definite matrix is a square of another positive definite matrix**

Prove that if $\boldsymbol{A}$ is a positive definite matrix of order n, then there exists a positive definite matrix $\boldsymbol{B}$, such that $\boldsymbol{A} = \boldsymbol{B}^2$.

Proof: Since a positive definite matrix $\boldsymbol{A}$ is a real symmetric matrix and its eigenvalues are all positive, so there exists an orthogonal matrix $\boldsymbol{Q}$, such that

$$\boldsymbol{A} = \boldsymbol{Q}\,\mathrm{diag}(\lambda_1, \lambda_2, \ldots, \lambda_n)\,\boldsymbol{Q}^{\mathrm{T}}, \qquad ①$$

where $\lambda_i > 0$ $(i = 1, 2, \ldots, n)$. By using $\boldsymbol{Q}^{\mathrm{T}}\boldsymbol{Q} = \boldsymbol{I}$ and

$$\mathrm{diag}(\lambda_1, \lambda_2, \ldots, \lambda_n) = [\mathrm{diag}(\sqrt{\lambda_1}, \sqrt{\lambda_2}, \ldots, \sqrt{\lambda_n})]^2,$$

equation ① can be expressed as

$$\boldsymbol{A} = (\boldsymbol{Q}\,\mathrm{diag}(\sqrt{\lambda_1}, \sqrt{\lambda_2}, \ldots, \sqrt{\lambda_n})\,\boldsymbol{Q}^{\mathrm{T}})^2. \qquad ②$$

So, taking

$$\boldsymbol{B} = \boldsymbol{Q}\,\mathrm{diag}(\sqrt{\lambda_1}, \sqrt{\lambda_2}, \ldots, \sqrt{\lambda_n})\,\boldsymbol{Q}^{\mathrm{T}}, \qquad ③$$

we obtain $\boldsymbol{A} = \boldsymbol{B}^2$.

Since $\sqrt{\lambda_i}$ in ③ is an eigenvalue of $\boldsymbol{B}$ and $\sqrt{\lambda_i} > 0$ $(i = 1, 2, \ldots, n)$, then $\boldsymbol{B}$ is also a positive definite matrix. Hence, the proposition is proved. ■

Here, $\boldsymbol{B}$ is usually denoted as $\boldsymbol{A}^{1/2}$.

★6.5 Other Definite Quadratic Forms

Positive definite, positive semidefinite, negative definite, and negative semidefinite quadratic forms are generally called definite. In this section, we will introduce the properties and discrimination theorems of the definite quadratic forms.

Definition 6.5

For all $\boldsymbol{x} = (x_1, x_2, \ldots, x_n)^{\mathrm{T}} \neq \mathbf{0}$, we have

(i) $\boldsymbol{x}^{\mathrm{T}}\boldsymbol{A}\boldsymbol{x} \geq 0$, but there exists at least one $\boldsymbol{x}_0 \neq \mathbf{0}$, such that $\boldsymbol{x}_0{}^{\mathrm{T}}\boldsymbol{A}\boldsymbol{x}_0 = 0$, then $\boldsymbol{x}^{\mathrm{T}}\boldsymbol{A}\boldsymbol{x}$ is called a **positive semidefinite quadratic form** and $\boldsymbol{A}$ is called a **positive semidefinite matrix**;
(ii) $\boldsymbol{x}^{\mathrm{T}}\boldsymbol{A}\boldsymbol{x} < 0$, then $\boldsymbol{x}^{\mathrm{T}}\boldsymbol{A}\boldsymbol{x}$ is called a **negative definite quadratic form** and $\boldsymbol{A}$ is called a **negative definite matrix**;
(iii) $\boldsymbol{x}^{\mathrm{T}}\boldsymbol{A}\boldsymbol{x} \leq 0$, but there exists at least one $\boldsymbol{x}_0 \neq \mathbf{0}$, such that $\boldsymbol{x}_0{}^{\mathrm{T}}\boldsymbol{A}\boldsymbol{x}_0 = 0$, then $\boldsymbol{x}^{\mathrm{T}}\boldsymbol{A}\boldsymbol{x}$ is called a **negative semidefinite quadratic form** and $\boldsymbol{A}$ is called a **negative semidefinite matrix.**

If a quadratic form is not definite, then it is called **indefinite**.

For example, $\boldsymbol{x}^{\mathrm{T}}\boldsymbol{A}\boldsymbol{x} = d_1x_1{}^2 + d_2x_2{}^2 + \cdots + d_nx_n{}^2$ is negative definite if $d_i < 0$ $(i = 1, 2, \ldots, n)$; positive semidefinite if $d_i \geq 0$ $(i = 1, 2, \ldots, n)$ and at least one of them is zero; negative semidefinite if $d_i \leq 0$ $(i = 1, 2, \ldots, n)$ and at least one of them is zero.

Obviously, if $\boldsymbol{A}$ is a positive definite (positive semidefinite) matrix, then $(-\boldsymbol{A})$ is negative definite (negative semidefinite) and vice versa.

According to the definition and applying the method in the above section, we can prove the following theorems.

THEOREM 6.7

Let $\boldsymbol{A}$ be a real symmetric matrix of order n, then the following propositions are equivalent.

(i) $\boldsymbol{x}^{\mathrm{T}}\boldsymbol{A}\boldsymbol{x}$ *is negative definite.*
(ii) *The index of negative inertia of $\boldsymbol{A}$ is n, that is, $\boldsymbol{A} \simeq -\boldsymbol{I}$.*
(iii) *There exists an invertible matrix $\boldsymbol{P}$, such that $\boldsymbol{A} = -\boldsymbol{P}^{\mathrm{T}}\boldsymbol{P}$.*
(iv) *The eigenvalues of $\boldsymbol{A}$ are all negative.*
(v) *All the odd-order sequential principal minors are negative and all the even-order sequential principal minors are positive.*

THEOREM 6.8

Let $\mathbf{A}$ *be a real symmetric matrix of order n, then the following propositions are equivalent.*

(i) $\mathbf{x}^{\mathrm{T}}\mathbf{A}\mathbf{x}$ *is positive semidefinite.*

(ii) *The index of positive inertia of* $\mathbf{A}$ *is* $\mathrm{r}(\mathbf{A}) = r \ (r < n)$ *or* $\mathbf{A} \simeq \mathrm{diag}(1, 1, \cdots, 1, 0, \cdots, 0)$ *with r 1's.*

(iii) *All the eigenvalues of* $\mathbf{A}$ *are nonnegative but there is at least one equal to zero.*

(iv) *There exists a matrix* $\mathbf{P}$*, not of full rank,* $(\mathrm{r}(\mathbf{P}) < n)$*, such that* $\mathbf{A} = \mathbf{P}^{\mathrm{T}}\mathbf{P}$.

(v) *All the principal minors are nonnegative, but there is at least one principal minor equal to zero (the definition of principal minor is explained in Definition 3.9 of Chapter 3).*

It is not difficult for readers to write the corresponding theorem about the negative semidefinite form.

Regarding the proofs for these theorems, (iv) $\to$ (v) is comparatively difficult if we use the cyclic method to prove Theorem 6.7. The proof of (v) in Theorem 6.8, the necessary and sufficient condition for a positive semidefinite form is comparatively difficult too. Hence, we omit these proofs.

Example 6.13 Determining whether a quadratic form is definite

Determine whether the quadratic form $n\sum_{i=1}^{n} x_i^2 - (\sum_{i=1}^{n} x_i)^2$ is a definite quadratic form.

Solution:

Method 1: The original expression

$$\begin{aligned}
&(n-1)\sum_{i=1}^{n} x_i^2 - \sum_{1\le i<j\le n} 2x_i x_j \qquad ①\\
&= (x_1 - x_2)^2 + (x_1 - x_3)^2 + \cdots + (x_1 - x_n)^2\\
&\quad + (x_2 - x_3)^2 + \cdots + (x_2 - x_n)^2 + \cdots + (x_{n-1} - x_n)^2\\
&= \sum_{1\le i<j\le n} (x_i - x_j)^2 \ge 0.
\end{aligned}$$

The equality holds when $x_1 = x_2 = \cdots = x_n$. So, the original quadratic form is positive semidefinite.

Method 2: Get the eigenvalues of matrix $\boldsymbol{A}$ of quadratic form in ①.

$$|\lambda \boldsymbol{I} - \boldsymbol{A}| = \begin{vmatrix} \lambda-(n-1) & 1 & \cdots & 1 \\ 1 & \lambda-(n-1) & \cdots & 1 \\ \vdots & \vdots & \ddots & \vdots \\ 1 & 1 & \cdots & \lambda-(n-1) \end{vmatrix}$$

$$= \lambda \begin{vmatrix} 1 & 1 & \cdots & 1 \\ 0 & \lambda-n & \cdots & 0 \\ \vdots & \vdots & \ddots & \vdots \\ 0 & 0 & \cdots & \lambda-n \end{vmatrix} = \lambda(\lambda-n)^{n-1}.$$

The eigenvalues of $\boldsymbol{A}$ are $\lambda_1 = 0, \lambda_2 = n$ (multiplicity of $n-1$), so the quadratic form is positive semidefinite. ■

Exercises and Supplementary Exercises

Exercises

Q6.1–6.3: Express the following quadratic forms in matrix form.

6.1 $f(x, y) = 4x^2 - 6xy - 7y^2$.

6.2 $f(x, y, z) = 3x^2 + 4xy - y^2 - 6yz + z^2$.

6.3 $f(x_1, x_2, x_3, x_4) = x_1{}^2 + x_3{}^2 + 2x_4{}^2 + 4x_1x_2 + 2x_1x_4 - 2x_2x_3 - 6x_2x_4 + 4x_3x_4$.

6.4 Let the matrix of an n-variable quadratic form $f(x_1, x_2, \ldots, x_n)$ be the tridiagonal symmetric matrix of order n,

$$\boldsymbol{A} = \begin{pmatrix} 1 & -1 & & & \\ -1 & 1 & -1 & & \\ & -1 & 1 & \ddots & \\ & & \ddots & \ddots & -1 \\ & & & -1 & 1 \end{pmatrix}.$$

Express it in quadratic form (quadratic homogeneous polynomial).

★6.5 If a quadratic form $f(x_1, x_2, x_3, x_4) = \boldsymbol{x}^{\mathrm{T}}\boldsymbol{A}\boldsymbol{x}$ always has $f(x_1, x_2, x_3, x_4) = 0$ for any $\boldsymbol{x} = (x_1, x_2, x_3, x_4)^{\mathrm{T}}$, prove that $\boldsymbol{A}$ is a fourth-order zero matrix.
[Hint: Take some particular $\boldsymbol{x}$, such as $(1, 0, 0, 1)^{\mathrm{T}}$, $(0, 1, 0, 0)^{\mathrm{T}}$, $(1, 0, 1, 0)^{\mathrm{T}}$, $(0, 1, 1, 0)^{\mathrm{T}}$, to prove $\boldsymbol{A} = \boldsymbol{0}$.]

★6.6 Prove that, if $\boldsymbol{A}$ and $\boldsymbol{B}$ are both third-order real symmetric matrices and $\boldsymbol{x}^{\mathrm{T}}\boldsymbol{A}\boldsymbol{x} = \boldsymbol{x}^{\mathrm{T}}\boldsymbol{B}\boldsymbol{x}$ for any $\boldsymbol{x}$, then $\boldsymbol{A} = \boldsymbol{B}$.

★6.7 Let $\boldsymbol{A} \simeq \boldsymbol{B}$, $\boldsymbol{C} \simeq \boldsymbol{D}$, and $\boldsymbol{A}, \boldsymbol{B}, \boldsymbol{C}, \boldsymbol{D}$ be all nth-order real symmetric matrices. Are the following conclusions true? If so, prove it.
(i) $(\boldsymbol{A} + \boldsymbol{C}) \simeq (\boldsymbol{B} + \boldsymbol{D})$.
(ii) $\begin{pmatrix} \boldsymbol{A} & \boldsymbol{0} \\ \boldsymbol{0} & \boldsymbol{C} \end{pmatrix} \simeq \begin{pmatrix} \boldsymbol{B} & \boldsymbol{0} \\ \boldsymbol{0} & \boldsymbol{D} \end{pmatrix}$.

6.8 Transform the following quadratic forms to canonical forms by orthogonal transformation $\boldsymbol{x} = \boldsymbol{Q}\boldsymbol{y}$, and give the orthogonal matrix $\boldsymbol{Q}$.
(i) $f = 2x_1{}^2 + 3x_2{}^2 + 3x_3{}^2 + 4x_2x_3$.
(ii) $f = x_1{}^2 + x_2{}^2 + x_3{}^2 + x_4{}^2 + 2x_1x_2 - 2x_1x_4 - 2x_2x_3 - 2x_3x_4$.

6.9 Let

$$\boldsymbol{A} = \begin{pmatrix} 4 & -2 & & & \\ -2 & 1 & & & \\ & & 5 & & \\ & & & -4 & 6 \\ & & & 6 & 1 \end{pmatrix}.$$

Find an orthogonal matrix $\boldsymbol{Q}$ such that $\boldsymbol{Q}^{\mathrm{T}}\boldsymbol{A}\boldsymbol{Q}$ is a diagonal matrix.

6.10 Transform each of the following quadratic forms to its canonical form by the method of completing the squares and find the coordinate transformation.
(i) $x_1{}^2 + 4x_1x_2 - 3x_2x_3$;
(ii) $x_1x_2 + x_1x_3 - 3x_2x_3$;
(iii) $2x_1{}^2 + 5x_2{}^2 + 4x_3{}^2 + 4x_1x_2 - 8x_2x_3 - 4x_3x_1$.

6.11 Transform each of the following quadratic forms to its canonical forms by elementary transformations and find the corresponding coordinate transformations.
(i) $x_1x_2 + x_2x_3 + x_3x_1$;
(ii) $x_1{}^2 - 2x_2{}^2 + x_3{}^2 + 2x_1x_2 + 4x_1x_3 + 2x_2x_3$;
(iii) $x_1{}^2 + 5x_2{}^2 + 4x_3{}^2 - x_4{}^2 + 6x_1x_2 - 4x_1x_3 - 4x_2x_4 - 8x_3x_4$.

6.12 Let $\boldsymbol{C}$ be an invertible matrix and $\boldsymbol{C}^{\mathrm{T}}\boldsymbol{A}\boldsymbol{C} = \mathrm{diag}(d_1, d_2, \ldots, d_n)$. Are the diagonal entries of the diagonal matrix all eigenvalues of $\boldsymbol{A}$? Why?

6.13 Let the rank of a real symmetric matrix $\boldsymbol{A}$ of order n be r $(r < n)$, prove that
(i) there exists an invertible matrix $\boldsymbol{C}$, such that $\boldsymbol{C}^{\mathrm{T}}\boldsymbol{A}\boldsymbol{C} = \mathrm{diag}(d_1, d_2, \ldots, d_r, 0, \ldots, 0)$, where $d_i \neq 0$ $(i = 1, 2, \ldots, r)$.
★(ii) $\boldsymbol{A}$ can be written as the sum of r symmetric matrices of rank 1.

★**6.14** If the rank of an nth-order real symmetric idempotent matrix $\boldsymbol{A}$ (satisfies $\boldsymbol{A}^2 = \boldsymbol{A}$) is r, then find

(i) a canonical form for the quadratic form $\boldsymbol{x}^{\mathrm{T}}\boldsymbol{A}\boldsymbol{x}$,

(ii) $\det(\boldsymbol{I} + \boldsymbol{A} + \boldsymbol{A}^2 + \cdots + \boldsymbol{A}^n)$.

★**6.15** Let $\boldsymbol{A}$ be an nth-order real symmetric matrix, and neither of the positive or negative inertia indices be zero. Prove that there exist nonzero vectors $\boldsymbol{x}_1, \boldsymbol{x}_2, \boldsymbol{x}_3$, such that

$$\boldsymbol{x}_1^{\mathrm{T}}\boldsymbol{A}\boldsymbol{x}_1 > 0, \quad \boldsymbol{x}_2^{\mathrm{T}}\boldsymbol{A}\boldsymbol{x}_2 = 0,$$

$$\boldsymbol{x}_3^{\mathrm{T}}\boldsymbol{A}\boldsymbol{x}_3 < 0.$$

★**6.16** Let $\boldsymbol{A}$ be an odd-order real symmetric matrix and $\det(\boldsymbol{A}) > 0$. Prove that there exists a nonzero vector x_0, such that $\boldsymbol{x}_0^{\mathrm{T}}\boldsymbol{A}\boldsymbol{x}_0 > 0$.

6.17 Transform the quadratic form in **Q6.11** to its normal form and find the transformation matrix $\boldsymbol{C}$.

6.18 For the matrix $\boldsymbol{A}$ in **Q6.9**, work out an invertible matrix $\boldsymbol{C}$ such that $\boldsymbol{C}^{\mathrm{T}}\boldsymbol{A}\boldsymbol{C}$ is a normal form.

★**6.19** Prove Conclusion (i) at the end of §6.3.

★**6.20** Prove Conclusion (ii) at the end of §6.3.

6.21 Determine whether the following matrices are positive.

(i) $\begin{pmatrix} 2 & -1 & 0 \\ -1 & 2 & -1 \\ 0 & -1 & 2 \end{pmatrix}$. (ii) $\begin{pmatrix} 2 & -1 & -1 \\ -1 & 2 & -1 \\ -1 & -1 & 2 \end{pmatrix}$.

(iii) $\begin{pmatrix} 1 & 1 & 1 \\ 1 & 2 & 2 \\ 1 & 2 & 3 \end{pmatrix}$.

6.22 Determine which of the following quadratic forms are positive definite.

(i) $x_1^2 + 3x_2^2 + 20x_3^2 - 2x_1x_2 - 2x_1x_3 - 10x_2x_3$

(ii) $3x_1^2 + 4x_2^2 + 5x_3^2 + 4x_1x_2 - 4x_2x_3$

(iii) $x_1^2 + 2x_2^2 + 3x_3^2 + 4x_4^2 - 2x_1x_2 + 4x_2x_3 - 8x_3x_4$

6.23 Transform the quadratic form $\sum_{i=1}^{n} x_i^2 + \sum_{i<j}^{n} x_ix_j$ to its canonical form by an orthogonal transformation and explain whether it is a positive definite quadratic form. When $n = 3$, work out the matrix $\boldsymbol{Q}$ of the orthogonal transformation.

★**6.24** For the quadratic form of matrix $\boldsymbol{A}$ in **Q6.23**, work out a positive definite matrix $\boldsymbol{B}$ when $n = 3$, such that $\boldsymbol{A} = \boldsymbol{B}^2$.

6.25 Determine the parameter t in the following quadratic forms, such that the quadratic forms are positive definite.

(i) $5x_1^2 + x_2^2 + tx_3^2 + 4x_1x_2 - 2x_1x_3 - 2x_2x_3$;

(ii) $2x_1^2 + x_2^2 + 3x_3^2 + 2tx_1x_2 + 2x_1x_3$.

6.26 Prove that the eigenvalues of a positive definite matrix are positive by the definitions of eigenvalues and eigenvectors of a matrix, and the definition of a positive definite quadratic form.

6.27 Let $\boldsymbol{P}$ be an invertible matrix. Prove that $\boldsymbol{P}^{\mathrm{T}}\boldsymbol{P}$ is positive definite by definition.

6.28 Let $\boldsymbol{A}$ be a positive definite matrix and $\boldsymbol{C}$ be a real invertible matrix. Prove that $\boldsymbol{C}^{\mathrm{T}}\boldsymbol{A}\boldsymbol{C}$ is a real symmetric positive definite matrix.

6.29 Let $\boldsymbol{A}$ be a positive definite matrix. Prove that the adjoint matrix $\boldsymbol{A}^*$ is also positive definite.

6.30 Let both $\boldsymbol{A}$, $\boldsymbol{B}$ be positive definite matrices of order n and both k, l be positive numbers. Prove that $(k\boldsymbol{A} + l\boldsymbol{B})$ is also positive definite by definition.

6.31 Determine which of the following matrices are negative definite, positive semidefinite, or negative semidefinite.

(i) $\begin{pmatrix} -1 & 1 & 0 \\ 1 & -2 & 1 \\ 0 & 1 & -3 \end{pmatrix}$, (ii) $\begin{pmatrix} 1 & -1 & -1 \\ -1 & 2 & -1 \\ -1 & -1 & 3 \end{pmatrix}$,

(iii) $\begin{pmatrix} 0 & -1 & -1 \\ -1 & 2 & -1 \\ -1 & -1 & 2 \end{pmatrix}$, (iv) $\begin{pmatrix} -2 & 1 & 1 \\ 1 & -2 & 1 \\ 1 & 1 & -2 \end{pmatrix}$.

6.32 Prove that all the main diagonal entries of a negative definite matrix are negative.

★6.33 Let $x^{\mathrm{T}}Ax$ be a negative semidefinite quadratic form.

(i) Is $x^{\mathrm{T}}(-A)x$ positive semidefinite?
(ii) Are all the principal minors of every order less than or equal to zero?

6.34 Determine which of the following quadratic forms are definite.

(i) $-x_1{}^2-2x_2{}^2-5x_3{}^2-2x_1x_2+4x_2x_3$;
(ii) $x_1{}^2+2x_2{}^2+4x_3{}^2+2x_1x_2-4x_2x_3$;
(iii) $x_1{}^2+2x_2{}^2+3x_3{}^2+2x_1x_2-4x_2x_3$.

6.35 Prove that A is a negative definite matrix, if and only if, there exists an invertible matrix P, such that $A=-P^{\mathrm{T}}P$.

6.36 Let B be an nth-order matrix and $\mathrm{r}(B)<n$. Prove that $B^{T}B$ is a positive semidefinite matrix.

6.37 Prove that if the matrix A is positive semidefinite, then there exists a positive semidefinite matrix B, such that $A=B^2$.

Supplementary exercises

6.38 If a quadratic form $f(x_1,x_2,\ldots,x_n)$ is always positive for all nonzero $x_1,x_2,\ldots,x_n$, is f positive definite?

6.39 Let A be a real symmetric matrix. Prove that $(A+tI)$ is a positive definite matrix when t is sufficiently large.

6.40 Assume the eigenvalues of a real symmetric matrix A is $\lambda_1,\lambda_2,\ldots,\lambda_n$. Determine t such that $(A-tI)$ is a positive definite matrix.

6.41 Let A be a real symmetric matrix and B be a positive definite matrix. Prove that there exists an invertible matrix C, such that $C^{\mathrm{T}}AC$ and $C^{\mathrm{T}}BC$ are both diagonal matrices.

6.42 Suppose that A and B are both positive definite matrices and $AB=BA$. Prove that AB is a positive definite matrix [Hint: There exist orthogonal matrices C and D, such that $D^{\mathrm{T}}(C^{\mathrm{T}}AC)D=D^{\mathrm{T}}\Lambda_1D$, $D^{\mathrm{T}}(C^{\mathrm{T}}BC)D=\Lambda_2$, then use the sufficient condition in Theorem 6.6].

6.43 Let $A=(a_{ij})$ be an nth-order definite matrix and $x=(x_1,x_2,\ldots,x_n)^{\mathrm{T}}$. Prove that $f(x)=\det\begin{pmatrix}0 & x^{\mathrm{T}}\\ x & A\end{pmatrix}$ is a negative definite quadratic form.

6.44 Let $A=(a_{ij})$ be an nth-order positive definite matrix. Prove that

$$\det A\le\prod_{i=1}^{n}a_{ii}.$$

6.45 Let $B=(b_{ij})$ be a real invertible matrix of order n. Prove that

$$|B|^2\le\prod_{i=1}^{n}(b_{1i}{}^2+b_{2i}{}^2+\cdots+b_{ni}{}^2).$$

6.46 If $f(x_1,x_2,x_3)=2x_1{}^2+3x_2{}^2+3x_3{}^2+2ax_2x_3$ can be transformed to a canonical form $f=y_1{}^2+2y_2{}^2+5y_3{}^2$ through an orthogonal transformation $x=Qy$, work out the parameter a and the orthogonal matrix Q.

6.47 If the rank of $f(x_1,x_2,x_3)=5x_1{}^2+5x_2{}^2+cx_3{}^2-2x_1x_2+6x_1x_3-6x_2x_3$ is 2,

(i) determine c, and
(ii) what surface does equation $f(x_1,x_2,x_3)=1$ represent?

6.48 Let

$$A=\begin{pmatrix}1&0&1\\0&2&0\\1&0&1\end{pmatrix},\qquad B=(kI+A)^2,$$

where k is a real number. Find a diagonal matrix Λ such that $B\simeq\Lambda$. Determine k such that B is a positive definite matrix.

6.49 Assume

$$f(x_1,x_2,\ldots,x_n)=(x_1+a_1x_2)^2+(x_2+a_2x_3)^2+\cdots+(x_{n-1}+a_{n-1}x_n)^2+(x_n+a_nx_1)^2,$$

where $a_1,a_2,\ldots,a_n$ are all real numbers. Discuss the values of $a_1,a_2,\ldots,a_n$, such that the quadratic form $f(x_1,x_2,\ldots,x_n)$ is positive definite.

6.50 Assume that A is a real symmetric matrix with order m and $B\in\mathbb{R}^{m\times n}$. Prove that $B^{\mathrm{T}}AB$ is positive definite, if and only if, $\mathrm{r}(B)=n$.

7 Applications

CHAPTER OVERVIEW

In this chapter, to deepen readers' understanding of the practical applications of linear algebra, we present seven problems. The solutions of these problems can be obtained using specific concepts of linear algebra learnt in the previous chapters. Although the description of the problem or formulation may be hard for some readers, the applications of linear algebra in these problems can be easily understood and accepted.

7.1 Population Models

Population growth is one of the socio-economic issues that can affect national welfare and citizens' livelihood. Demographists and mathematicians worldwide have, over the last two centuries, proposed and refined population models in their attempt to accurately forecast population growth. A few examples include the deterministic, random, continuous, and discrete models. The Leslie population model (also known as the Leslie Matrix), which we will introduce in this chapter, was proposed in the 1940s. It is a discrete model used to forecast the population variation by age-group.

7.1.1 Leslie population model

This model only considers the variation in female population as the proportion of males and females has minimal fluctuations. Suppose the maximum age of women is s and n is the number of age intervals,

$$\Delta t_i = \left[(i-1)\frac{s}{n}, i\frac{s}{n}\right), i = 1, 2, \cdots, n.$$

We name females with ages belonging to Δt_i as the ith group and define the person(s) with age s belonging to Δ_n. Suppose the number of persons in the ith group is $x_i (i = 1, 2, \cdots, n)$, where $\boldsymbol{x} = (x_1, x_2, \ldots, x_n)^{\mathrm{T}}$ is called the distribution vector of (female) population ages. Suppose that $\boldsymbol{x}$ is observed over time t_k every s/n years and the change within a time interval is ignored. Denote initial time as t_0, $t_k = t_0 + ks/n$, the age distribution vector at time t_k as $\boldsymbol{x}^{(k)} = \left(x_1^{(k)}, x_2^{(k)}, \ldots, x_n^{(k)}\right)^{\mathrm{T}}$. We consider only the population development caused by birth, ageing, and death, and ignore the influence of social phenomena such as migration, war, and disasters.

Suppose the fertility rate of the ith group (the death rate of female infants had been deducted from the fertility rate) is a_i (the average number of female offspring produced by 1 female in the ith group during s/n years, $a_i \geq 0$), survival rate is b_i (the ratio of the number of females still alive after s/n years to the original number of females in ith group, $0 < b_i \leq 1$). The death rate is $1 - b_i$. Let a_i and b_i be constants in the same time interval, which are obtained from population statistical data.

Total number $x_1^{(k)}$ of females in the first group at time t_k is the sum of female offspring produced by females (numbers of which are $x_i^{(k-1)}, i = 1, \ldots, n$) of every group at time t_{k-1}, that is,

$$x_1^{(k)} = a_1 x_1^{(k-1)} + a_2 x_2^{(k-1)} + \cdots + a_n x_n^{(k-1)}. \tag{7.1}$$

The number of females $x_{i+1}^{(k)}$ of the $(i+1)$th group at $t_k (i \geq 1)$ is the sum of females alive after s/n years of the ith group at t_{k-1}, that is,

$$x_{i+1}^{(k)} = b_i x_i^{(k-1)}, i = 1, 2, \ldots, n-1. \tag{7.2}$$

The matrix form of (7.1) and (7.2) is

$$\boldsymbol{x}^{(k)} = \boldsymbol{L}\boldsymbol{x}^{(k-1)}, k = 1, 2, \ldots, n, \tag{7.3}$$

where

$$\boldsymbol{L} = \begin{pmatrix} a_1 & a_2 & \cdots & a_{n-1} & a_n \\ b_1 & 0 & \cdots & 0 & 0 \\ 0 & b_2 & \ddots & \vdots & \vdots \\ \vdots & \ddots & \ddots & 0 & 0 \\ 0 & \cdots & 0 & b_{n-1} & 0 \end{pmatrix}. \tag{7.4}$$

$\boldsymbol{L}$ is called the **Leslie matrix** and (7.3) is deduced to be

$$\boldsymbol{x}^{(k)} = \boldsymbol{L}^k \boldsymbol{x}^{(0)}. \tag{7.5}$$

The total population of each age-group at time t_k, the growth rate of population, and percentage of each age-group in the total population can be obtained from (7.5). This model is suitable for animal populations too.

Example 7.1 The Leslie population model

Suppose that the maximum lifespan of a certain type of animal in a farm is 6 years and the data observed in 1998 is given in Table 7.1. How many animals will be in each age-group in 2006? What is their distribution among these groups? What is the overall growth rate?

Solution: From Table 7.1, we can get the animal age distribution vector $\boldsymbol{x}^{(0)}$ in 1998 and the Leslie matrix $\boldsymbol{L}$,

$$\boldsymbol{x}^{(0)} = \begin{pmatrix} 160 \\ 320 \\ 80 \end{pmatrix}, \quad \boldsymbol{L} = \begin{pmatrix} 0 & 4 & 3 \\ \frac{1}{2} & 0 & 0 \\ 0 & \frac{1}{4} & 0 \end{pmatrix}.$$

Table 7.1 Distribution of animals by age group

Age	**[0, 2)**	**[2, 4)**	**[4, 6)**
Count	160	320	80
Birth rate	0	4	3
Survival rate	$\frac{1}{2}$	$\frac{1}{4}$	0

The age distribution vector in 2006 is

$$x^{(4)} = L^4 x^{(0)} = \begin{pmatrix} 0 & 4 & 3 \\ \frac{1}{2} & 0 & 0 \\ 0 & \frac{1}{4} & 0 \end{pmatrix}^4 \begin{pmatrix} 160 \\ 320 \\ 80 \end{pmatrix}$$

$$= \begin{pmatrix} 4 & 3 & \frac{9}{8} \\ \frac{3}{16} & 4 & 3 \\ \frac{1}{4} & \frac{3}{32} & 0 \end{pmatrix} \begin{pmatrix} 160 \\ 320 \\ 80 \end{pmatrix} = \begin{pmatrix} 1690 \\ 1550 \\ 70 \end{pmatrix}.$$

So, the total number of the animals in 2006 is 3310; among them, 1690 (51.06%) are younger than 2 years, 1550 (46.83%) are 2–4 years old, 70 (2.11%) are 4–6 years old. The overall growth is $3310 - 560 = 2750$. The overall growth rate is 491.07% in 8 years. ■

By using the Leslie model to analyze population growth, we find that after a long time the population growth rate and the distribution in different age intervals tend to stabilize, which are related to the eigenvalues and eigenvectors of the matrix $\boldsymbol{L}$.

7.1.2 Dominant eigenvalue of Leslie matrix and its practical significance

THEOREM 7.1

The matrix $\mathbf{L}$ *of (7.4) has a unique simple positive eigenvalue* λ_1, *whose corresponding eigenvector is*

$$x_1 = (1, b_1/\lambda_1, b_1 b_2/\lambda_1{}^2, \ldots, b_1 b_2 \cdots b_{n-1}/\lambda_1{}^{n-1})^{\mathrm{T}}$$

$$(\textit{each component is positive}).$$

Proof:

$$p_n(\lambda) = |\lambda \boldsymbol{I} - \boldsymbol{L}| = \begin{vmatrix} \lambda - a_1 & -a_2 & \cdots & -a_{n-1} & -a_n \\ -b_1 & \lambda & & & \\ & -b_2 & \ddots & & \\ & & \ddots & \lambda & \\ & & & -b_{n-1} & \lambda \end{vmatrix}$$

$$= \lambda p_{n-1}(\lambda) + b_{n-1} \begin{vmatrix} \lambda - a_1 & -a_2 & \cdots & -a_{n-2} & -a_n \\ -b_1 & \lambda & & & \\ & -b_2 & \ddots & & \\ & & \ddots & \lambda & \\ & & & -b_{n-2} & 0 \end{vmatrix}$$

$$= \lambda p_{n-1}(\lambda) + b_{n-1}(-a_n)(-1)^n(-1)^{n-2} b_1 b_2 \cdots b_{n-2}$$
$$= \lambda p_{n-1}(\lambda) - a_n b_1 b_2 \cdots b_{n-1}. \tag{$*$}$$

Denote $\boldsymbol{\beta}_i = a_i b_1 b_2 \cdots b_{i-1} (b_0 = 1)$. Since $a_i \geq 0$ and not all zero, $b_i > 0$, so $\boldsymbol{\beta}_i \geq 0$ $(i = 1, 2, \cdots, n)$ and not all zero. From the recursion relation $(*)$, we have

$$p_n(\lambda) = \lambda p_{n-1}(\lambda) - \boldsymbol{\beta}_n = \lambda(\lambda p_{n-2}(\lambda) - \boldsymbol{\beta}_{n-1}) - \boldsymbol{\beta}_n$$
$$= \lambda^2 p_{n-2}(\lambda) - \lambda \boldsymbol{\beta}_{n-1} - \boldsymbol{\beta}_n = \cdots$$
$$= \lambda^n - \boldsymbol{\beta}_1 \lambda^{n-1} - \boldsymbol{\beta}_2 \lambda^{n-2} - \cdots - \boldsymbol{\beta}_{n-1}\lambda - \boldsymbol{\beta}_n. \tag{7.6}$$

Let
$$q(\lambda) = \boldsymbol{\beta}_1/\lambda + \boldsymbol{\beta}_2/\lambda^2 + \cdots + \boldsymbol{\beta}_n/\lambda^n \quad (\lambda \neq 0). \tag{7.7}$$

Then, the characteristic equation

$$p_n(\lambda) = 0 \tag{7.8}$$

is equivalent to the equation

$$q(\lambda) = 1. \tag{7.9}$$

We can see that $q(\lambda)$ is a continuous monotonic decreasing function over $\lambda > 0$, and $\lim\limits_{\lambda \to 0^+} q(\lambda) = +\infty$, $\lim\limits_{\lambda \to \infty} q(\lambda) = 0$. Hence, there exists a unique $\lambda_1 \in (0, +\infty)$, such that $q(\lambda_1) = 1$ or $p_n(\lambda_1) = 0$, that is, λ_1 is the unique positive eigenvalue. We prove that λ_1 is a single root in the following using reduction to absurdity.

Suppose λ_1 is a root of multiplicity k of $p_n(\lambda) = 0$ $(k \geq 2)$, then $p_n'(\lambda_1) = 0$, and $p_n'(\lambda_1) = n\lambda_1^{n-1} - (n-1)\boldsymbol{\beta}_1 \lambda_1^{n-2} - \cdots - 2\boldsymbol{\beta}_{n-2}\lambda_1 - \boldsymbol{\beta}_{n-1} = 0$.

We can express it as

$$1 = \frac{n-1}{n}\frac{\boldsymbol{\beta}_1}{\lambda_1} + \frac{n-2}{n}\frac{\boldsymbol{\beta}_2}{\lambda_1^2} + \cdots + \frac{2\boldsymbol{\beta}_{n-2}}{n\lambda_1^{n-2}} + \frac{\boldsymbol{\beta}_{n-1}}{n\lambda^{n-1}} < q(\lambda_1) = 1$$
$$\left(\text{for } \frac{(n-i)}{n} < 1\right),$$

which is a contradiction, so λ_1 is a single root.

Solving the equation system $(\lambda_1 \boldsymbol{I} - \boldsymbol{L})\boldsymbol{x} = \boldsymbol{0}$, we easily know that the eigenvector corresponding to λ_1 is the given $\boldsymbol{x}_1$. □

THEOREM 7.2

If λ_1 is a positive eigenvalue of the matrix **L**, *then any eigenvalue λ (both real and complex) of* **L** *satisfies*

$$|\lambda| \leq \lambda_1 \tag{7.10}$$

Proof: If $\lambda = 0$, it holds obviously. Otherwise, suppose $\lambda = r\mathrm{e}^{\mathrm{i}\theta}$ and $r > 0$. Suppose $|\lambda| = r > \lambda_1$, use reduction to absurdity to prove the theorem. From (7.9) and (7.7),

$$q(\lambda) = q(r\mathrm{e}^{\mathrm{i}\theta}) = \frac{\beta_1}{r}\mathrm{e}^{-\mathrm{i}\theta} + \frac{\beta_2}{r^2}\mathrm{e}^{-2\mathrm{i}\theta} + \cdots + \frac{\beta_n}{r^n}\mathrm{e}^{-n\mathrm{i}\theta} = 1.$$

So, the real part of $q(\lambda)$ is 1, that is,

$$\begin{aligned} 1 &= \frac{\beta_1}{r}\cos\theta + \frac{\beta_2}{r^2}\cos 2\theta + \cdots + \frac{\beta_n}{r^n}\cos n\theta \\ &\leq \frac{\beta_1}{r} + \frac{\beta_2}{r^2} + \cdots + \frac{\beta_n}{r^n} < \frac{\beta_1}{\lambda_1} + \frac{\beta_2}{\lambda_1^2} + \cdots + \frac{\beta_n}{\lambda_1^n} = q(\lambda_1) = 1 \end{aligned}$$

This is a contradiction. So, we have $|\lambda| \leq \lambda_1$ for any eigenvalue λ. □

We call the unique positive eigenvalue λ_1 of matrix L the **dominant eigenvalue**. If $|\lambda| < \lambda_1$ for any eigenvalue $\lambda \neq \lambda_1$, then λ_1 is called a **strictly dominant eigenvalue.**

THEOREM 7.3

If the first row of the matrix **L** *has two sequential entries $a_i, a_{i+1} > 0$, then the positive eigenvalue of* **L** *is a strictly dominant eigenvalue.*

Proof: Using Theorem 7.2, $|\lambda| \leq \lambda_1$. We need only to prove that the equality doesn't hold by using the method of reduction to absurdity. Suppose there exists $\lambda \neq \lambda_1$, such that $|\lambda| = \lambda_1$, then

$$\begin{aligned} 1 = q(\lambda) &= \frac{\beta_1}{\lambda} + \frac{\beta_2}{\lambda^2} + \cdots + \frac{\beta_n}{\lambda^n} \\ &\leq \left|\frac{\beta_1}{\lambda}\right| + \left|\frac{\beta_2}{\lambda^2}\right| + \cdots + \left|\frac{\beta_i}{\lambda^i} + \frac{\beta_{i+1}}{\lambda^{i+1}}\right| + \cdots + \left|\frac{\beta_n}{\lambda_n}\right| \\ &= \frac{\beta_1}{\lambda_1} + \frac{\beta_2}{\lambda_1{}^2} + \cdots + \frac{|\beta_i\lambda_1 + \beta_{i+1}|}{\lambda_1^{i+1}} + \cdots + \frac{\beta_n}{\lambda_1^n} \\ &< \frac{\beta_1}{\lambda_1} + \frac{\beta_2}{\lambda_1^2} + \cdots + \frac{\beta_i}{\lambda_1^i} + \frac{\beta_{i+1}}{\lambda_1^{i+1}} + \cdots + \frac{\beta_n}{\lambda_1^n} = q(\lambda_1) = 1 \end{aligned}$$

This is a contradiction, so $|\lambda| < \lambda_1$ (the above "<" is obtained according to $|\boldsymbol{\beta}_i\lambda + \boldsymbol{\beta}_{i+1}| < \boldsymbol{\beta}_i\,|\lambda| + \boldsymbol{\beta}_{i+1}$ for $\boldsymbol{\beta}_i = a_i b_1 \cdots b_{i-1} > 0$, $\boldsymbol{\beta}_{i+1} > 0$, and λ is a negative real number, where λ_1 is the only positive eigenvalue by Theorem 7.1).
□

The condition of Theorem 7.3 in the population model is satisfied, so the matrix $\boldsymbol{L}$ must have a strictly dominant eigenvalue λ_1.

THEOREM 7.4

If the matrix $\boldsymbol{L}$ *has a strictly dominant eigenvalue* λ_1 *and the corresponding eigenvector is* $\boldsymbol{x}_1$, *then*

$$\lim_{k\to\infty}\frac{1}{\lambda_1^k}\boldsymbol{x}^{(k)} = c\boldsymbol{x}_1, \tag{7.11}$$

where $\boldsymbol{x}^{(k)}$ *is determined by* (7.5) *and* c *is a constant number.*

Proof: Suppose $\boldsymbol{L}$ is diagonalized (you should use the Jordan standard form discussed in Appendix B for the general case), that is, there exist linearly independent eigenvectors $\boldsymbol{x}_1, \boldsymbol{x}_2, \ldots, \boldsymbol{x}_n$ and $\boldsymbol{P} = (\boldsymbol{x}_1, \boldsymbol{x}_2, \ldots, \boldsymbol{x}_n)$, such that

$$\boldsymbol{P}^{-1}\boldsymbol{L}\boldsymbol{P} = \operatorname{diag}(\lambda_1, \lambda_2, \ldots, \lambda_n).$$

Then,

$$\boldsymbol{L}^k = (\boldsymbol{P}\operatorname{diag}(\lambda_1, \ldots, \lambda_n)\boldsymbol{P}^{-1})^k = \boldsymbol{P}\operatorname{diag}(\lambda_1^k, \ldots, \lambda_n^k)\boldsymbol{P}^{-1}.$$

From (7.5), we get

$$\begin{aligned}\frac{1}{\lambda_1^k}\boldsymbol{x}^{(k)} &= \frac{1}{\lambda_1^k}\boldsymbol{L}^{(k)}\boldsymbol{x}^{(0)} = \frac{1}{\lambda_1^k}\boldsymbol{P}\operatorname{diag}(\lambda_1^k, \ldots, \lambda_n^k)\boldsymbol{P}^{-1}\boldsymbol{x}^{(0)}\\ &= \boldsymbol{P}\operatorname{diag}\left(1, \left(\frac{\lambda_2}{\lambda_1}\right)^k, \ldots, \left(\frac{\lambda_n}{\lambda_1}\right)^k\right)\boldsymbol{P}^{-1}\boldsymbol{x}^{(0)}.\end{aligned}$$

Denote $\boldsymbol{P}^{-1}\boldsymbol{x}^{(0)} = (c, d_2, \ldots, d_n)^{\mathrm{T}}$. Since $\lim\limits_{k\to+\infty}\left(\frac{\lambda_i}{\lambda_1}\right)^k = 0$ $(i = 2, \ldots, n)$, so we have

$$\begin{aligned}\lim_{k\to+\infty}\frac{1}{\lambda_1^k}\boldsymbol{x}^{(k)} &= \lim_{k\to+\infty}\boldsymbol{P}\operatorname{diag}\left(1, \left(\frac{\lambda_2}{\lambda_1}\right)^k, \ldots, \left(\frac{\lambda_n}{\lambda_1}\right)^k\right)\boldsymbol{P}^{-1}\boldsymbol{x}^{(0)}\\ &= \boldsymbol{P}\operatorname{diag}(1, 0, \ldots, 0)(c, d_2, \ldots, d_n)^{\mathrm{T}}\\ &= (\boldsymbol{x}_1, \boldsymbol{x}_2, \ldots, \boldsymbol{x}_n)\begin{pmatrix}1 & & & \\ & 0 & & \\ & & \ddots & \\ & & & 0\end{pmatrix}\begin{pmatrix}c\\ d_2\\ \vdots\\ d_n\end{pmatrix} = c\boldsymbol{x}_1.\end{aligned}$$

□

From (7.11), when k is sufficiently large,

$$\boldsymbol{x}^{(k)} \approx c{\lambda_1}^k \boldsymbol{x}_1 = \lambda_1 (c{\lambda_1}^{k-1}\boldsymbol{x}_1) \approx \lambda_1 \boldsymbol{x}^{(k-1)}.$$

This indicates that the age distribution vector tends to a stable state as time t_k is sufficiently large. In other words, the percentage of females in each age-group $x_i^{(k)}$, in proportion to the overall population $\sum_{i=1}^{n} x_i^{(k)}$, is almost equal to the percentage of respective component x_i in proportion to the total sum $\sum_{i=1}^{n} x_i^1$ in the eigenvector $\boldsymbol{x}_1$. Moreover, the growth rate $(x_i^{(k+1)} - s_i^{(k)})/x_i^{(k)}$ tends to $(\lambda_1 - 1)$ as t_k is sufficiently large. We can say that the population is increasing when $\lambda_1 > 1$, decreasing when $\lambda_1 < 1$, and unchanged when $\lambda_1 = 1$. From (7.7) and (7.9), we can get the following theorem.

THEOREM 7.5

*$\lambda_1 = 1$, if and only if, $\boldsymbol{\beta}_1 + \boldsymbol{\beta}_2 + \cdots + \boldsymbol{\beta}_n = 1$, where λ_1 is a positive eigenvalue of matrix **L**.*

Denote $R = \boldsymbol{\beta}_1 + \boldsymbol{\beta}_2 + \cdots + \boldsymbol{\beta}_n = a_1 + a_2 b_1 + \cdots + a_n b_1 \cdots b_{n-1}$, which is called the **net regeneration (reproduction) rate**. It represents the average number of female offspring produced by each female throughout her entire life. We can easily prove that $\lambda > 1, \lambda < 1, \lambda = 1$ when $R > 1, R < 1, R = 1$ correspondingly, which indicate the population to be increasing, decreasing and unchanged, respectively, after t_k is large enough from (7.7) and (7.9).

For instance, the characteristic polynomial of $\boldsymbol{L}$ in Example 7.1 is

$$p(\lambda) = |\lambda \boldsymbol{I} - \boldsymbol{L}| = \lambda^3 - 2\lambda - \frac{3}{8} = 0$$

We get the strictly dominant eigenvalue $\lambda_1 = 3/2$, the corresponding eigenvector $\boldsymbol{x}_1 = \left(1, b_1/\lambda_1, b_1 b_2/\lambda_1^2\right)^{\mathrm{T}} = (1, 1/3, 1/18)^{\mathrm{T}}$, the limit of the growth rate $\lambda_1 - 1 = 50\%$ as t_k is large enough. Then, the ratios of the animal numbers for the three age-groups are 1:1/3:1/18, that is, 18:6:1. It is 72% for those younger than 2 years, 24% for those 2–4 years old, and 4% for those within the 4–6 years old group.

Example 7.2 Net regeneration rate

Table 7.2 shows the 1965 statistical data of Canada (since there are few child-bearing women ≥ 50 years old, we only consider the population growth problem of 0–50 years old).

Table 7.2 Distribution of age groups in Canada in 1965

Age group i	Age interval	a_i	b_i
1	[0, 5)	0. 00000	0. 99651
2	[5, 10)	0. 00024	0. 99820
3	[10, 15)	0. 05861	0. 99802
4	[15, 20)	0. 28608	0. 99729
5	[20, 25)	0. 44791	0. 99694
6	[25, 30)	0. 36399	0. 99621
7	[30, 35)	0. 22259	0. 99460
8	[35, 40)	0. 10459	0. 99184
9	[40, 45)	0. 02826	0. 99870
10	[45, 50)	0. 00240	—

By numerical methods, we get the strictly dominant eigenvalue of matrix $\boldsymbol{L}$ and the corresponding eigenvector $\boldsymbol{x}_1$, respectively:

$$\lambda_1 = 1.07622, \quad \boldsymbol{x}_1 = \begin{pmatrix} 1.00000 \\ 0.92594 \\ 0.85881 \\ 0.79641 \\ 0.73800 \\ 0.68364 \\ 0.63281 \\ 0.58482 \\ 0.53897 \\ 0.49429 \end{pmatrix}.$$

If the fertility rate and survival rate of Canadian women remain the same as in 1965, then the sum of the population younger than 50 years old will increase by 7.622% for every 5 years after a long time. From the eigenvector, we can work out the percentages of every age-group in proportion to the total population: 13.79%, 12.77%, 11.84%, 10.98%, 10.17%, 9.42%, 8.72%, 8.06%, 7.43%, and 6.82% in decreasing order. ■

7.2 Markov Chain

Consider a series of experiments with random factors such that each experimental outcome is exactly one of the $S_1, \ldots, S_m$, which are called states. When S_i appears, the system is in a state S_i. The system at each moment is random over states. If the system's next moment depends only on the current state with a transition probability, not upon the states before the current state, the system is said to have **no aftereffect**. In a more direct way, the future state is only related to the current known state, but not to the past. A **Markov chain** is a random system such that its time and states are both discrete and have no aftereffect. Markov chains are extensively used in economics, social sciences, ecology, and genetics.

We use $x_i(n)(n = 0, 1, 2, \ldots, n)$ to denote the probability of the system in state S_i at moment t_n. The vector $\boldsymbol{x}_n = (x_1(n), \ldots, x_m(n))^{\mathrm{T}}$ is called the **probability vector**, where $0 \le x_i(n) \le 1$ and $x_1(n) + \cdots + x_m(n) = 1$. Here, $p_{ij}(i, j = 1, 2, \ldots, m)$, which is called the **transition probability**, is the probability that the system is currently at state S_j and next at state S_i. The matrix $\boldsymbol{P} = (p_{ij})_{m\times m}$ is called the **transition matrix**. The sum of each column of $\boldsymbol{P}$ is 1 (we write such a matrix $\boldsymbol{P}$ as $\boldsymbol{P} \ge \boldsymbol{0}$ for $p_{ij} \ge 0$). Since there is no aftereffect, we have the fundamental equation of the Markov chain

$$x_i(n+1) = p_{i1}x_1(n) + p_{i2}x_2(n) + \cdots + p_{im}x_m(n) \quad i = 1, 2, \ldots, m. \tag{7.12}$$

Its matrix form is

$$\boldsymbol{x}(n+1) = \boldsymbol{P}\boldsymbol{x}(n), \tag*{(7.12)$'$}$$

$$\boldsymbol{P} \ge \boldsymbol{0}, \ \sum_{i=1}^{m} p_{ij} = 1. \tag{7.13}$$

Using this recursion formula, we have

$$\boldsymbol{x}(n) = \boldsymbol{P}^n\boldsymbol{x}(0), \tag{7.14}$$

where $\boldsymbol{x}(0) = (x_1(0), \ldots, x_m(0))^{\mathrm{T}}$ is the **initial probability vector**.

Example 7.3 Use of Markov chain for business application

There are two states about selling a certain product in a retail shop: good (S_1) or poor (S_2). We observe these states monthly and know the probability $p_{21} = \alpha$ of good state in the current month and poor state in the next month, and probability $p_{12} = \beta$ of poor state in the current month and good state in the next month, where $0 \le \alpha, \beta \le 1$. If the initial state is good (or poor), give the probability of having a good state after n months.

Solution: Since the sum of each column is equal to 1, we have the transition matrix

$$\boldsymbol{P} = \begin{pmatrix} 1-\alpha & \beta \\ \alpha & 1-\beta \end{pmatrix}.$$

From $|\lambda \boldsymbol{I} - \boldsymbol{P}| = 0$, we get eigenvalues $\lambda_1 = 1$ and $\lambda_2 = 1 - \alpha - \beta$.

(i) If α and β are not all 0 and 1, then $|\lambda_2| < \lambda_1 = 1$. The eigenvectors corresponding to λ_1, λ_2 are $\boldsymbol{x}_1 = (\beta, \alpha)^{\mathrm{T}}$, $\boldsymbol{x}_2 = (1, -1)^{\mathrm{T}}$. Let $\boldsymbol{R} = (\boldsymbol{x}_1, \boldsymbol{x}_2)$. We have

$$\boldsymbol{R}^{-1}\boldsymbol{P}\boldsymbol{R} = \mathrm{diag}(\lambda_1, \lambda_2),$$

where

$$\boldsymbol{R}^{-1} = \frac{1}{\alpha+\beta}\begin{pmatrix} 1 & 1 \\ \alpha & -\beta \end{pmatrix}.$$

Then,

$$\boldsymbol{x}(n) = \boldsymbol{P}^n \boldsymbol{x}(0) = \boldsymbol{R}\,\mathrm{diag}(\lambda_1^n, \lambda_2^n)\boldsymbol{R}^{-1}\boldsymbol{x}(0),$$

$$\begin{aligned}\lim_{n\to\infty} \boldsymbol{x}(n) &= \frac{1}{\alpha+\beta}\begin{pmatrix} \beta & 1 \\ \alpha & -1 \end{pmatrix}\begin{pmatrix} 1 & 0 \\ 0 & 0 \end{pmatrix}\begin{pmatrix} 1 & 1 \\ \alpha & -\beta \end{pmatrix}\begin{pmatrix} x_1(0) \\ x_2(0) \end{pmatrix} \\ &= \frac{1}{\alpha+\beta}\begin{pmatrix} \beta & \beta \\ \alpha & \alpha \end{pmatrix}\begin{pmatrix} x_1(0) \\ x_2(0) \end{pmatrix} \\ &= \frac{1}{\alpha+\beta}\begin{pmatrix} \beta \\ \alpha \end{pmatrix} \xlongequal{\text{written as}} \boldsymbol{q}.\end{aligned}$$

The third equality is because $x_1(0) + x_2(0) = 1$. The probability vector $\boldsymbol{x}(n)$ tends to a stable value $\boldsymbol{q}$ independent of $\boldsymbol{x}(0)$ as $n \to \infty$ and $\boldsymbol{q}$ is called the **limit state probability**, where $\boldsymbol{q} = (q_1, q_2)^{\mathrm{T}}$ and $q_1 + q_2 = 1$. Here, q_1 and q_2 respectively represent the probabilities of good and poor states, which are completely determined by the transition matrix $\boldsymbol{P}$.

(ii) If $\alpha = \beta = 0$, then $\boldsymbol{P} = \boldsymbol{I}$, $\boldsymbol{x}(n) = \boldsymbol{x}(0)$.

(iii) If $\alpha = \beta = 1$, then $\boldsymbol{P} = \begin{pmatrix} 0 & 1 \\ 1 & 0 \end{pmatrix}$ and $\lambda_1 = 1, \lambda_2 = -1$. Here, $\lim_{n\to\infty} (-1)^n$ does not exist, and

$$\boldsymbol{x}(n) = \boldsymbol{P}^n \boldsymbol{x}(0) = \begin{cases} \boldsymbol{x}(0), & \text{if } n \text{ is an even number,} \\ \boldsymbol{P}\boldsymbol{x}(0), & \text{if } n \text{ is an odd number.} \end{cases}$$

■

Definition 7.1

If there exists a positive integer N such that $\boldsymbol{P}^N > \boldsymbol{0}$ (i.e., $p_{ij} > 0, i, j = 1, \ldots, m$), then $\boldsymbol{P}$ is called a **regular transition matrix**. The corresponding Markov chain is called a **regular Markov chain**.

THEOREM 7.6

Suppose $\boldsymbol{P}$ *is a regular transition matrix of order* m*, then*

$$\lim_{n\to\infty} \boldsymbol{P}^n = (\boldsymbol{q}, \boldsymbol{q}, \ldots, \boldsymbol{q}) = \boldsymbol{Q}, \tag{7.15}$$

where $\boldsymbol{q} = (q_1, q_2, \ldots, q_m)^{\mathrm{T}}, q_i (i = 1, 2, \ldots, m)$ *are positive numbers and* $q_1 + q_2 + \ldots + q_m = 1$.

The proof is omitted.

The matrix $\boldsymbol{Q}$ has the following property. For arbitrary probability vector $\boldsymbol{x} = (x_1, x_2, \ldots, x_m)^{\mathrm{T}}$, we have

$$\boldsymbol{Q}\boldsymbol{x} = \left(\sum_{i=1}^{m} x_i\right)\boldsymbol{q} = \boldsymbol{q}. \tag{7.16}$$

THEOREM 7.7

A regular Markov chain has a unique limit state probability vector $\boldsymbol{q}$*, such that*

$$\boldsymbol{P}\boldsymbol{q} = \boldsymbol{q}. \tag{7.17}$$

Proof:

$$\lim_{n\to\infty} \boldsymbol{x}(n) = \lim_{n\to\infty} \boldsymbol{P}^n\boldsymbol{x}(0) = \boldsymbol{Q}\boldsymbol{x}(0) = \boldsymbol{q}$$

From $\boldsymbol{P}\boldsymbol{P}^n = \boldsymbol{P}^{n+1}$ and $\lim_{n\to\infty} \boldsymbol{P}^n = \lim_{n\to\infty} \boldsymbol{P}^{n+1} = \boldsymbol{Q}$, we get

$$\boldsymbol{P}\boldsymbol{Q}\boldsymbol{x}(0) = \boldsymbol{Q}\boldsymbol{x}(0),$$

that is,

$$\boldsymbol{P}\boldsymbol{q} = \boldsymbol{q}.$$

We now prove the uniqueness. Suppose an arbitrary limit state probability vector $\boldsymbol{r}$ of the regular Markov chain satisfies $\boldsymbol{P}\boldsymbol{r} = \boldsymbol{r}$. Then, $\boldsymbol{P}^n\boldsymbol{r} = \boldsymbol{r}$ holds for $n = 1, 2, \ldots$. Taking limit and together with (7.15), (7.16) and (7.17), we have

$$\lim_{n\to\infty} \boldsymbol{P}^n\boldsymbol{r} = \boldsymbol{Q}\boldsymbol{r} = \boldsymbol{q} = \boldsymbol{r}. \qquad \square$$

From (7.17), we know that, for a regular Markov chain, the limit state probability vector $\boldsymbol{q}$ is the unique solution to the linear equation system

$$(\boldsymbol{I} - \boldsymbol{P})\boldsymbol{x} = \boldsymbol{0},$$

with $q_1 + \cdots + q_m = 1$, that is, $\boldsymbol{q}$ is the eigenvector corresponding to eigenvalue $\lambda_1 = 1$ of the matrix $\boldsymbol{P}$, such that the sum of all the components is 1.

In Example 7.3, $\boldsymbol{x} = (\boldsymbol{\beta}, \boldsymbol{\alpha})^{\mathrm{T}}$ is a solution of $(\boldsymbol{I} - \boldsymbol{P})\boldsymbol{x} = \boldsymbol{0}$. Then, $\boldsymbol{q} = \frac{1}{\alpha+\beta}(\boldsymbol{\beta}, \boldsymbol{\alpha})^{\mathrm{T}}$ is the required limit probability with the sum of all the components equal to 1.

Example 7.4 Use of Markov chain for scientific application

Consider the chemical reactions of the transfer of the element phosphorus. Suppose that phosphorus only occurs naturally in s_1 (soil), s_2, (organisms such as grass, sheep, cow), and s_3 (other systems such as rivers and lakes). We measure it once a month and obtain the changes as shown in Figure 7.1. Thirty percent of the phosphorus in the soil transfers to organisms, 20% is out of the system, and 50% remains in the soil. Forty percent of the phosphorus in organisms is transferred back to the soil, 40% is transferred out of the system, and 20% remains. If the initial proportion of phosphorus in s_1, s_2, and s_3 is 0.5:0.3:0.2, find the proportion of phosphorus in the three states after n months.

Solution: The initial probability vector is $\boldsymbol{x}(0) = (0.5, 0.3, 0.2)^{\mathrm{T}}$, the transition matrix is

$$\boldsymbol{P} = \begin{pmatrix} 0.5 & 0.4 & 0 \\ 0.3 & 0.2 & 0 \\ 0.2 & 0.4 & 1 \end{pmatrix}.$$

Since $p_{33} = 1$, the system will not transfer out of s_3 once it enters it, s_3 is called an **absorbing state**.

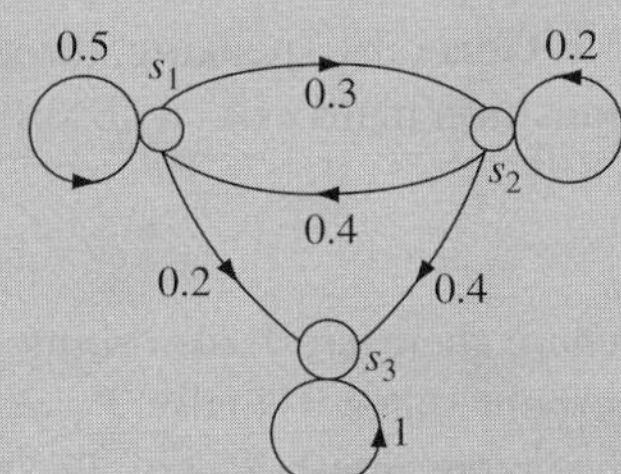

Figure 7.1 Phosphorus transfer between soil, organisms, and other systems

■

Definition 7.2

If a Markov chain has an absorbing state S_i, that is, $p_{ii} = 1$, and the number of transferring steps from any nonabsorbent state to an absorbing state is finite with positive transition probability at each step, then the Markov chain is called an **absorbing chain**, which is different from regular Markov chains.

Denote

$$P_2 = \begin{pmatrix} 0.5 & 0.4 \\ 0.3 & 0.2 \end{pmatrix}, \quad \alpha = (0.2, 0.4).$$

Then,

$$P = \begin{pmatrix} P_2 & \mathbf{0} \\ \alpha & 1 \end{pmatrix}$$

is a lower triangular block matrix and P^n is still a lower triangular block matrix, that is,

$$P^n = \begin{pmatrix} P_2{}^n & \mathbf{0} \\ * & 1 \end{pmatrix}.$$

$(P^n)_{13} = (P^n)_{23} = 0$ for any $n \in \mathbb{N}$, thus $P^n \not> \mathbf{0}$. So, P is not a regular transition matrix. Calculate

$$x(n) = P^n x(0)$$

and we obtain Table 7.3.

Table 7.3 Values of $x(n)$

n		0	1	2	3	$\cdots$	10	$\cdots$	∞
	$x_1(n)$	0.5	0.37	0.27	0.195	$\cdots$	0.020	$\cdots$	$\to 0$
$x(n)$	$x_2(n)$	0.3	0.21	0.15	0.111	$\cdots$	0.011	$\cdots$	$\to 0$
	$x_3(n)$	0.2	0.42	0.58	0.694	$\cdots$	0.969	$\cdots$	$\to 1$

Generally speaking, through rearranging the subscripts of s_i, the mth-order transition matrix of an absorbing chain may be expressed as

$$P = \begin{pmatrix} I_r & P_1 \\ \mathbf{0} & P_2 \end{pmatrix},$$

where there are r absorbing states corresponding to I_r, and $(m-r)$ nonabsorbent states. Usually $P_1 \neq \mathbf{0}$, so the sum of each column of P_2 is < 1, then the eigenvalue $\lambda(P_2)$ of P_2 satisfies $|\lambda(P_2)| < 1$, and $I - P_2$ is invertible. Otherwise, $|I - P_2| = 0$, $\lambda = 1$ is an eigenvalue of P_2, which is contradictory. If we denote

$$M = (I - P_2)^{-1} = I + P_2 + P_2{}^2 + \cdots + P_2{}^n + \cdots = \sum_{n=0}^{\infty} P_2{}^n,$$

$$\alpha = (1, 1, \ldots, 1)^{\mathrm{T}}{}_{1\times(m-r)},$$

then

$$Y = M\alpha = (y_1, \ldots, y_{m-r})^{\mathrm{T}},$$

where y_i is the average time of transitions from the ith nonabsorbent state to an absorbing state.

In Example 7.4, if we modify the order of states as (s_3, s_2, s_1), the corresponding probability transition matrix is

$$\boldsymbol{P} = \begin{pmatrix} 1 & \boldsymbol{\beta} \\ \boldsymbol{0} & \boldsymbol{P}_2 \end{pmatrix}, \quad \boldsymbol{\beta} = (0.4 \quad 0.2), \quad \boldsymbol{P}_2 = \begin{pmatrix} 0.2 & 0.3 \\ 0.4 & 0.5 \end{pmatrix},$$

$$\boldsymbol{I} - \boldsymbol{P}_2 = \begin{pmatrix} 0.8 & -0.3 \\ -0.4 & 0.5 \end{pmatrix}, \quad (\boldsymbol{I} - \boldsymbol{P}_2)^{-1} = \frac{1}{0.28}\begin{pmatrix} 0.5 & 0.3 \\ 0.4 & 0.8 \end{pmatrix}.$$

So,

$$\boldsymbol{Y} = \boldsymbol{M}\boldsymbol{\alpha} = (\boldsymbol{I} - \boldsymbol{P}_2)^{-1}\boldsymbol{\alpha} = \frac{1}{0.28}\begin{pmatrix} 0.5 & 0.3 \\ 0.4 & 0.8 \end{pmatrix}\begin{pmatrix} 1 \\ 1 \end{pmatrix} = \begin{pmatrix} \frac{20}{7} \\ \frac{30}{7} \end{pmatrix}.$$

Namely, it starts from the states s_2 or s_1, and is absorbed by the state s_3 through an average of 3 or 4.3 steps of transitions, respectively.

7.3 Input-Output Mathematical Models

This section introduces an input-output linear model that investigates the balanced relation between "input" and "output" among various departments (enterprises) in any economic system. In an economic system, departments (enterprises) are both producers and consumers. They manufacture products to supply the needs of various internal departments as well as outside the system (including export). They consume resources such as raw materials, equipment, transportation, energy as well as human resources. The purpose of consumption is to produce more products. New value is created through production which is used to reward workers, pay tax, and gain a reasonable profit. Obviously, for each department (enterprise), the sum of the consumed resources and the newly created value is equal to the gross output value. This is the gross balance between "input" and "output."

Here, we discuss only the monetary input-output mathematical model. Input and output are both measured using currencies.

7.3.1 Assignment balance equations

First, we consider the balance between production and sales (including domestic sales and export sales) or the balance of distributed products among all departments in a system.

Suppose an economic system is composed of n enterprises, and let

$x_i \geq 0$ denote the gross output value of the ith enterprise,

$d_i \geq 0$ denote the demand from outside the system for the output of the ith enterprise,

$c_{ij} \geq 0$ denote the output value of the ith enterprise that is consumed by the jth enterprise to produce a unit output value and is called the direct consumption coefficient of the jth enterprise to the ith enterprise.

In order to comprehend the meaning of x_i, d_i, c_{ij} and the mutual relation between production inside and outside the system, we created Table 7.4.

In Table 7.4, the same number of production enterprises indicates an identical enterprise. For example, number "1" represents the coal mine; number "2" represents the power plant, c_{21} represents the output value of the power plant directly consumed by the coal mine to produce 1-unit output value. c_{22} represents the output value of the power plant directly consumed by itself to produce 1-unit output value (i.e., the amount of power consumed by the power plant itself). d_2 represents the external demand of the power plant's output value (i.e., the demand for the electricity generated). x_2 represents the gross output value of the power plant (i.e., the total amount of electricity generated).

According to the above assumption, the output value of the ith enterprise distributed among enterprises inside the system is

$$c_{i1}x_1 + c_{i2}x_2 + \cdots + c_{in}x_n,$$

which is obtained from the ith row. d_i is the output value supplied to external consumers (enterprises). The sum of these two portions is the gross output value x_i of the ith enterprise. Then, we obtain the distribution balance equations

$$x_i = \left(\sum_{j=1}^{n} c_{ij}x_j \right) + d_i, \quad i = 1, 2, \cdots, n. \tag{7.18}$$

Table 7.4 Relations between production and consumption

Direct Consumption Coefficients \ Enterprise		Consumption Enterprises 1	2	$\cdots$	n	External Demand	Gross Output
Production Enterprises	1	c_{11}	c_{12}	$\cdots$	c_{1n}	d_1	x_1
	2	c_{21}	c_{22}	$\cdots$	c_{2n}	d_2	x_2
	$\vdots$	$\vdots$	$\vdots$		$\vdots$	$\vdots$	$\vdots$
	n	c_{n1}	c_{n2}	$\cdots$	c_{nn}	d_n	x_n

Denote

$$C = \begin{pmatrix} c_{11} & c_{12} & \cdots & c_{1n} \\ c_{21} & c_{22} & \cdots & c_{2n} \\ \vdots & \vdots & & \vdots \\ c_{n1} & c_{n2} & \cdots & c_{nn} \end{pmatrix}, \quad x = \begin{pmatrix} x_1 \\ x_2 \\ \vdots \\ x_n \end{pmatrix}, \quad d = \begin{pmatrix} d_1 \\ d_2 \\ \vdots \\ d_n \end{pmatrix}. \tag{7.19}$$

$\boldsymbol{C}$ is called the **matrix of direct consumption coefficients**. $\boldsymbol{C} \geq \boldsymbol{0}$ (i.e., $c_{ij} \geq 0, \forall i, j = 1, 2, \ldots, n$) is called a **nonnegative matrix**, $\boldsymbol{x}$ is called a **production vector**, $\boldsymbol{x} \geq \boldsymbol{0}$ (i.e., $x_i \geq 0, i = 1, 2, \ldots, n$) is called a **nonnegative vector**, $\boldsymbol{d}$ is called the **external demand vector** and $\boldsymbol{d} \geq \boldsymbol{0}$. Then, (7.18) can be written in matrix form as

$$\boldsymbol{x} = \boldsymbol{C}\boldsymbol{x} + \boldsymbol{d}, \tag{7.20}$$

that is,

$$(\boldsymbol{I} - \boldsymbol{C})\boldsymbol{x} = \boldsymbol{d}. \tag{7.21}$$

Equations (7.20) or (7.21) is one of the input-output mathematical models.

Example 7.5 Use of balance equation for economic application

An economic system has three enterprises: coal mine, power plant, and railway. Assume that in one year, the direct consumption coefficients and the external demands are shown in Table 7.5.

In order to balance the output against both the internal and external demands of the system, what should be the gross output values x_1, x_2, x_3 of these enterprises?

Solution: According to the given conditions, the matrix of direct consumption coefficients $\boldsymbol{C}$, external demand vector $\boldsymbol{d}$, production vector $\boldsymbol{x}$

Table 7.5 Demand and consumption of three enterprises

Direct Consumption Coefficients / Enterprises	Enterprises	Consumption Enterprises: Coal	Power	Railway	External Demand	Gross Output
Production Enterprises	Coal	0	0.65	0.55	50000	x_1
	Power	0.25	0.05	0.10	25000	x_2
	Railway	0.25	0.05	0	0	x_3

are, respectively,

$$C = \begin{pmatrix} 0 & 0.65 & 0.55 \\ 0.25 & 0.05 & 0.10 \\ 0.25 & 0.05 & 0 \end{pmatrix}, \quad d = \begin{pmatrix} 50000 \\ 25000 \\ 0 \end{pmatrix}, \quad x = \begin{pmatrix} x_1 \\ x_2 \\ x_3 \end{pmatrix}.$$

Substituting them into (7.21), we have

$$\begin{pmatrix} 1.00 & -0.65 & -0.55 \\ -0.25 & 0.95 & -0.10 \\ -0.25 & -0.05 & 1.00 \end{pmatrix} \begin{pmatrix} x_1 \\ x_2 \\ x_3 \end{pmatrix} = \begin{pmatrix} 50000 \\ 25000 \\ 0 \end{pmatrix},$$

where coefficient matrix $I - C$ is invertible and $(I - C)^{-1} > 0$. We get

$$\begin{pmatrix} x_1 \\ x_2 \\ x_3 \end{pmatrix} = \frac{1}{503} \begin{pmatrix} 756 & 542 & 470 \\ 220 & 690 & 190 \\ 200 & 170 & 630 \end{pmatrix} \begin{pmatrix} 50000 \\ 25000 \\ 0 \end{pmatrix} = \begin{pmatrix} 120087 \\ 56163 \\ 28330 \end{pmatrix}.$$

■

7.3.2 Balance equations for consumption

Here, we consider the balance between the gross output and the new added value of departments (enterprises) by their internally consumed products. For a given direct consumption coefficient matrix $C_{n \times n}$ of a system, the jth enterprise consumes a part of the output itself and that of other enterprises (such as raw materials, transportation, energy, equipment) to obtain the output value x_j. Its value is

$$c_{1j}x_j + c_{2j}x_j + \cdots + c_{nj}x_j.$$

If the net output value from producing output value x_j is z_j, then

$$x_j = c_{1j}x_j + c_{2j}x_j + \cdots + c_{nj}x_j + z_j,$$

that is,

$$x_j = \left(\sum_{i=1}^{n} c_{ij} \right) x_j + z_j, \quad j = 1, 2, \ldots, n. \tag{7.22}$$

The equation system (7.22) is called the **consumption balance equations**. Equation (7.22) can also be expressed as

$$\left(1 - \sum_{i=1}^{n} c_{ij} \right) x_j = z_j, \quad j = 1, 2, \ldots, n. \tag{7.23}$$

Denote

$$D = \operatorname{diag} \left(\sum_{i=1}^{n} c_{i1}, \sum_{i=1}^{n} c_{i2}, \ldots, \sum_{i=1}^{n} c_{in} \right), \tag{7.24}$$

which is called the **enterprise consumption matrix**. z is called the **net output value vector**. Then, (7.22) and (7.23) can be expressed in matrix forms respectively as,

$$\boldsymbol{x} = \boldsymbol{D}\boldsymbol{x} + z, \tag{7.25}$$

and

$$(\boldsymbol{I} - \boldsymbol{D})\boldsymbol{x} = z. \tag{7.26}$$

Equations (7.25) and (7.26) are called the second forms of the input-output model. They reveal the relation among the production vector $\boldsymbol{x}$, the net output vector z, and the enterprise consumption matrix $\boldsymbol{D}$ in an economic system.

From (7.18) and (7.22), we can obtain

$$\sum_{i=1}^{n} x_i = \sum_{i=1}^{n}\left[\left(\sum_{j=1}^{n} c_{ij}x_j\right) + d_i\right] = \sum_{j=1}^{n}\left[\left(\sum_{i=1}^{n} c_{ij}x_j\right) + z_j\right],$$

that is,

$$\sum_{i=1}^{n}\sum_{j=1}^{n} c_{ij}x_j + \sum_{i=1}^{n} d_i = \sum_{i=1}^{n}\sum_{j=1}^{n} c_{ij}x_j + \sum_{j=1}^{n} z_j,$$

so that

$$\sum_{i=1}^{n} d_i = \sum_{j=1}^{n} z_j. \tag{7.27}$$

Equation (7.27) indicates that the total sum of external demands of enterprises is equal to the total sum of the net output value of these enterprises.

7.3.3 Solutions to the balance equations of distribution and consumption

In the consumption balance equations, obviously we have

$$x_j > 0,\ z_j > 0,\ j = 1, 2, \ldots, n. \tag{7.28}$$

From (7.23), we know that the direct consumption coefficients among the enterprises have the following properties:

$$\sum_{i=1}^{n} c_{ij} < 1,\ j = 1, 2, \ldots, n. \tag{7.29}$$

We also have

$$0 \le c_{ij} < 1,\ i, j = 1, 2, \ldots, n. \tag{7.30}$$

So, $\boldsymbol{I}-\boldsymbol{D}=\operatorname{diag}(1-d_1, 1-d_2, \dots, 1-d_n)$ [where $d_j=\sum_{i=1}^{n} c_{ij}$ is given by (7.24), $j=1,2,\dots,n$]. From $1-d_i>0$ $(i=1,2,\dots,n)$, we know that $\boldsymbol{I}-\boldsymbol{D}$ is invertible and $(\boldsymbol{I}-\boldsymbol{D})^{-1}>\boldsymbol{0}$. From (7.26), we have

$$\boldsymbol{x}=(\boldsymbol{I}-\boldsymbol{D})^{-1}\boldsymbol{z}. \tag{7.31}$$

From (7.29) and (7.30), we know that the modulus of all the eigenvalues of $\boldsymbol{C}$ are less than 1 (see Theorem 5.3), so $\boldsymbol{I}-\boldsymbol{C}$ is invertible (otherwise, $\lambda=1$ is an eigenvalue of $\boldsymbol{C}$). That is, the distribution balance equations provide the solution

$$\boldsymbol{x}=(\boldsymbol{I}-\boldsymbol{D})^{-1}\boldsymbol{d}. \tag{7.32}$$

From $\lim\limits_{k\to+\infty}\boldsymbol{C}^k=\boldsymbol{0}$ (Jordan canonical form in Appendix B is used to obtain this result) and the equality

$$(\boldsymbol{I}-\boldsymbol{C})(\boldsymbol{I}+\boldsymbol{C}^2+\cdots+\boldsymbol{C}^{k-1})=\boldsymbol{I}-\boldsymbol{C}^k\to\boldsymbol{I}\,(k\to+\infty),$$

we know that $\sum_{i=0}^{\infty}\boldsymbol{C}^i$ is convergent, and

$$(\boldsymbol{I}-\boldsymbol{C})^{-1}=\sum_{i=0}^{\infty}\boldsymbol{C}^i>\boldsymbol{0}.$$

That is, (7.32) is the only nonnegative solution to the balance equations.

In application problems, we usually use the iterative method (this method is introduced in a course of numerical analysis) to solve the balance equations, however, Gaussian elimination may also be used when n is small.

7.4 Adjacency Matrices of Graph

In graph theory, a set $V=\{v_1,\dots,v_n\}$ is called a vertex set, an ordered pair (v_i,v_j) between vertices is called an arc, and the set of arcs is denoted as A. $G=(V,A)$ is called a **graph**. For example, a tournament among five teams can be described by Figure 7.2, where arc (v_i,v_j) indicates team v_i winning against team v_j. If there is an arc between the two points, then the two points are adjacent. We call $v_1v_2\cdots v_k$ a **path** of length $(k-1)$, if each of the vertices are different and $(v_i,v_{i+1})\in A$ $(i=1,2,\dots,k-1)$. The path $v_4v_5v_1v_2v_3$ in Figure 7.2 is of length 4. If a path begins at v_1 and ends at v_k, then v_1 is called the **initial point** and v_k is the **termination point** of that path. A path whose initial point and termination point are coincident is called a **circuit**; v_1, v_2, v_5, v_1 is a circuit with length 3. If there exists a path between v_1, v_k (written as the path v_1—v_k), then v_1 and v_k are said to be connected. A graph in which any pair of vertices is connected is called a **connected graph**.

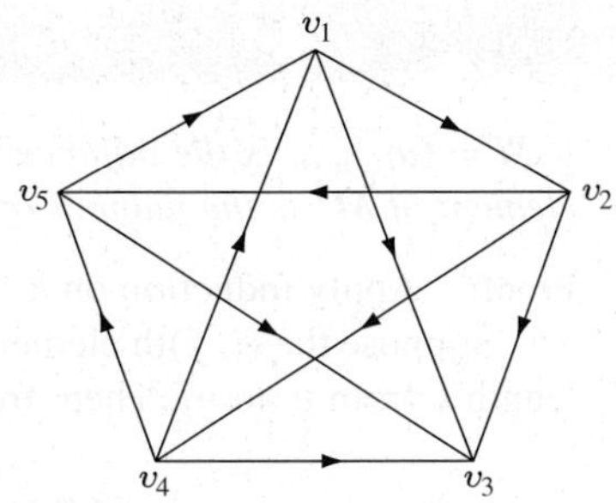

Figure 7.2 Connected graph for five teams in a tournament

A graph can be indicated by a matrix. We call $\boldsymbol{M} = (a_{ij})_{n\times n}$ the **adjacency matrix** of graph $G(V, A)$, if

$$a_{ij} = \begin{cases} 1, & (v_i, v_j) \in A, \\ 0, & \text{otherwise}. \end{cases} \tag{7.33}$$

The adjacency matrix of the graph shown in Figure 7.2 is

$$\boldsymbol{M} = \begin{pmatrix} 0 & 1 & 1 & 0 & 0 \\ 0 & 0 & 1 & 1 & 1 \\ 0 & 0 & 0 & 0 & 0 \\ 1 & 0 & 1 & 0 & 1 \\ 1 & 0 & 1 & 0 & 0 \end{pmatrix}.$$

The sum of row i represents the number of arcs adjacent to v_i, for example, the sum of row ④ is 3, where the number of arcs adjacent to v_4 is 3, which represents that team ④ won 3 games. The sum of column j represents the number of arcs that terminate at the vertex v_j, for example, the sum of column ⑤ is 2, where the number of arcs that terminates at v_5 is 2, which represents that team ⑤ lost 2 games. In the tournament table shown in Figure 7.2, both v_2 and v_4 won more than 3 games. How then do we determine the champion from the tournament table?

We need to calculate the "indirectly winning" teams. If the first team wins against the second, the second wins against the third, then we say that the first wins against the third indirectly. That is, if there exists a path of length k between v_i—v_j ($k \geq 2$), then we say v_i wins against v_j indirectly. To calculate the number of the paths with length 2 from v_i to v_j, calculate

$$a_{i1}a_{1j} + a_{i2}a_{2j} + \cdots + a_{in}a_{nj} = \sum_{k=1}^{n} a_{ik}a_{kj}. \tag{7.34}$$

THEOREM 7.8

If $M = (a_{ij})_{n\times n}$ is the adjacency matrix of a graph $G(V, A)$, then the (i, j)th element of M^k is the number of the paths with length k from v_i to v_j.

Proof: Apply induction on k. It is true for $k = 2$ [see equation (7.34)].

Suppose the (i, j)th element $(M^k)_{ij}$ of M^k is the number of the paths with length k from v_i to v_j. Then, from $M^{k+1} = M^k M$, we have

$$(M^{k+1})_{ij} = \sum_{l=1}^{n} (M^k)_{il}(M)_{lj},$$

where $(M)_{lj} = a_{lj}$ is the number of paths with length 1 from v_l to v_j. So, each term of the sum indicates the number of paths with length $(k+1)$ from v_i to v_l and then from v_l to v_j. The summation over all l is the number of paths with length $(k+1)$ from v_i to v_j. So, the theorem is true for $(k+1)$. □

Example 7.6 Use of adjacency matrices of graph in sports

Regarding a monocycle tournament of five teams, the result of which is shown in Figure 7.2, calculate and list the rank of all five teams.

Solution: If v_i has the largest row sum, then v_i is the champion. Since v_2 and v_4 both have row sum 3, we calculate M^2 and then $M + M^2$

$$M^2 = \begin{pmatrix} 0 & 0 & 1 & 1 & 1 \\ 2 & 0 & 2 & 0 & 1 \\ 0 & 0 & 0 & 0 & 0 \\ 1 & 1 & 2 & 0 & 0 \\ 0 & 1 & 1 & 0 & 0 \end{pmatrix}, \quad M + M^2 = \begin{pmatrix} 0 & 1 & 2 & 1 & 1 \\ 2 & 0 & 3 & 1 & 2 \\ 0 & 0 & 0 & 0 & 0 \\ 2 & 1 & 3 & 0 & 1 \\ 1 & 1 & 2 & 0 & 0 \end{pmatrix}.$$

$(M + M^2)_{ij}$ indicates the number of paths with length ≤ 2 from v_i to v_j, that is, the number of games that v_i beats v_j directly or indirectly $(k = 2)$. In $M + M^2$, the row with the largest row sum (which is 8) is the second row, therefore, v_2 is the champion. Similarly, v_4 is at the second place, v_1, v_5, and v_3 are at the third, fourth, and fifth, respectively.

If $M + M^2$ still includes several rows with identical sums, then calculate $M + M^2 + M^3$ and so on until there are no rows with identical sums. ■

Matrix $P = (p_{ij})_{n\times n}$ is called a **reachability matrix**, if

$$p_{ij} = \begin{cases} 1, & \text{there exists at least one path from } v_i \text{ to } v_j, \\ 0, & \text{the path from } v_i \text{ to } v_j \text{ does not exist.} \end{cases} \tag{7.35}$$

Since the length of any path from v_i to v_j must be less than n if it exists, we can obtain the reachability matrix $\boldsymbol{P}$ by calculating $\boldsymbol{B}_n = \boldsymbol{M} + \boldsymbol{M}^2 + \cdots + \boldsymbol{M}^n$ and changing all the nonzero elements of $\boldsymbol{B}_n$ to 1. If $\boldsymbol{P}$ is a matrix with elements all equal to 1, then any pair of points of G are connected. In this case, G is called a connected graph.

Example 7.7 Use of adjacency matrices of graph for aviation application

Suppose four cities v_1, v_2, v_3, and v_4 have flights as shown in Figure 7.3. Can one reach the other three cities if one takes off from an arbitrary city?

Solution: The adjacency matrix of the graph is

$$\boldsymbol{M} = \begin{pmatrix} 0 & 1 & 0 & 0 \\ 0 & 0 & 1 & 1 \\ 1 & 1 & 0 & 1 \\ 1 & 0 & 0 & 0 \end{pmatrix}.$$

Calculate $\boldsymbol{M}^2$, $\boldsymbol{M}^3$, $\boldsymbol{M}^4$, $\boldsymbol{M} + \boldsymbol{M}^2 + \boldsymbol{M}^3 + \boldsymbol{M}^4 = \boldsymbol{B}_4$, and $\boldsymbol{P}$ as follows.

$$\boldsymbol{M}^2 = \begin{pmatrix} 0 & 0 & 1 & 1 \\ 2 & 1 & 0 & 1 \\ 1 & 1 & 1 & 1 \\ 0 & 1 & 0 & 0 \end{pmatrix}, \quad \boldsymbol{M}^3 = \begin{pmatrix} 2 & 1 & 0 & 1 \\ 1 & 2 & 1 & 1 \\ 2 & 2 & 1 & 2 \\ 0 & 0 & 1 & 1 \end{pmatrix},$$

$$\boldsymbol{M}^4 = \begin{pmatrix} 1 & 2 & 1 & 1 \\ 2 & 2 & 2 & 3 \\ 3 & 3 & 2 & 3 \\ 2 & 1 & 0 & 1 \end{pmatrix}, \quad \boldsymbol{B}_4 = \begin{pmatrix} 3 & 4 & 2 & 3 \\ 5 & 5 & 4 & 6 \\ 7 & 7 & 4 & 7 \\ 3 & 2 & 1 & 2 \end{pmatrix},$$

$$\boldsymbol{P} = \begin{pmatrix} 1 & 1 & 1 & 1 \\ 1 & 1 & 1 & 1 \\ 1 & 1 & 1 & 1 \\ 1 & 1 & 1 & 1 \end{pmatrix}.$$

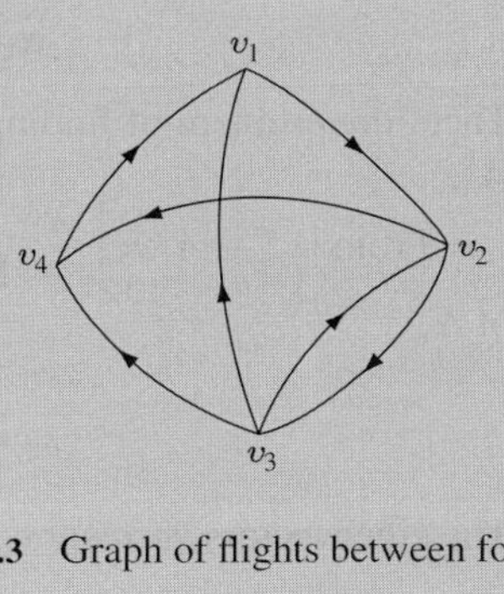

Figure 7.3 Graph of flights between four cities

So, graph G is a connected graph, one can get to the other cities by taking off from any city. The number of connections from v_i to v_j is represented by $(\boldsymbol{B}_4)_{ij}$, for example, $(\boldsymbol{B}_4)_{11}=3$ indicates that there are three connections: $v_1v_2v_4v_1$, $v_1v_2v_3v_1$, and $v_1v_2v_3v_4v_1$ taking off from v_1 and returning to v_1. There are two paths with length 3 $[(\boldsymbol{M}^3)_{11}=2]$ and one path with length 4 $[(\boldsymbol{M}^4)_{11}=1]$. ■

7.5 Matrix Solution of a Recurrence Relation

The method of matrix diagonalization can be exploited to solve some recurrence relations. For example, the famous Fibonacci sequence

$$\{F_n\}: 0, 1, 1, 2, 3, 5, 8, 13, 21, 34, \ldots \text{ satisfies}$$

$$F_{k+2}=F_{k+1}+F_k \ (k=0, 1, 2, \ldots), \qquad ①$$

where $F_0=0$, $F_1=1$.

Here, ① is a difference equation. Now, we use the matrix to find the general term of the sequence. According to the recurrence

$$\begin{cases} F_{k+2}=F_{k+1}+F_k \\ F_{k+1}=F_{k+1} \end{cases} \quad (k=0, 1, 2, \ldots),$$

that is,

$$\boldsymbol{\alpha}_{k+1}=\boldsymbol{A}\boldsymbol{\alpha}_k \quad (k=0, 1, 2, \ldots), \qquad ②$$

where

$$\boldsymbol{A}=\begin{pmatrix}1 & 1\\ 1 & 0\end{pmatrix}, \ \boldsymbol{\alpha}_k=\begin{pmatrix}F_{k+1}\\ F_k\end{pmatrix}, \ \boldsymbol{\alpha}_0=\begin{pmatrix}F_1\\ F_0\end{pmatrix}=\begin{pmatrix}1\\ 0\end{pmatrix},$$

we obtain

$$\boldsymbol{\alpha}_k=\boldsymbol{A}^k\boldsymbol{\alpha}_0, \ (k=1, 2, \ldots). \qquad ③$$

Then, the problem of finding F_k is reduced to find $\boldsymbol{\alpha}_k$, that is, we need to find $\boldsymbol{A}^k$.

From $|\lambda\boldsymbol{I}-\boldsymbol{A}|=\begin{vmatrix}\lambda-1 & -1\\ -1 & \lambda\end{vmatrix}=\lambda^2-\lambda-1=0$, we get the eigenvalues of $\boldsymbol{A}$,

$$\lambda_1=\frac{1+\sqrt{5}}{2}, \ \lambda_2=\frac{1-\sqrt{5}}{2}. \qquad ④$$

The eigenvectors corresponding to λ_1, λ_2 are

$$\boldsymbol{x}_1=(\lambda_1, 1)^{\mathrm{T}}, \quad \boldsymbol{x}_2=(\lambda_2, 1)^{\mathrm{T}}.$$

Let

$$\boldsymbol{P} = (\boldsymbol{x}_1, \boldsymbol{x}_2) = \begin{pmatrix} \lambda_1 & \lambda_2 \\ 1 & 1 \end{pmatrix}.$$

Then,

$$\boldsymbol{P}^{-1} = \frac{1}{\lambda_1 - \lambda_2} \begin{pmatrix} 1 & -\lambda_2 \\ -1 & \lambda_1 \end{pmatrix}.$$

Thus, we have $\boldsymbol{P}^{-1}\boldsymbol{A}\boldsymbol{P} = \mathrm{diag}(\lambda_1, \lambda_2)$ and

$$\boldsymbol{A}^k = \boldsymbol{P} \begin{pmatrix} {\lambda_1}^k & 0 \\ 0 & {\lambda_2}^k \end{pmatrix} \boldsymbol{P}^{-1} = \frac{1}{\lambda_1 - \lambda_2} \begin{pmatrix} {\lambda_1}^{k+1} - {\lambda_2}^{k+1} & \lambda_1 {\lambda_2}^{k+1} - \lambda_2 {\lambda_1}^{k+1} \\ {\lambda_1}^k - {\lambda_2}^k & \lambda_1 {\lambda_2}^k - \lambda_2 {\lambda_1}^k \end{pmatrix},$$

$$\begin{pmatrix} F_{k+1} \\ F_k \end{pmatrix} = \boldsymbol{\alpha}_k = \boldsymbol{A}^k \begin{pmatrix} 1 \\ 0 \end{pmatrix} = \frac{1}{\lambda_1 - \lambda_2} \begin{pmatrix} {\lambda_1}^{k+1} - {\lambda_2}^{k+1} \\ {\lambda_1}^k - {\lambda_2}^k \end{pmatrix}. \qquad ⑤$$

Substituting ④ into ⑤, we have

$$F_k = \frac{1}{\sqrt{5}} \left[\left(\frac{1+\sqrt{5}}{2} \right)^k - \left(\frac{1-\sqrt{5}}{2} \right)^k \right]. \qquad ⑥$$

For any positive integer k, F_k from ⑥ are all positive integers. This may be unexpected from the formulation but is in fact correct.

Denote

$$r_k = \frac{1}{\sqrt{5}} \left| \left(\frac{1-\sqrt{5}}{2} \right)^k \right|. \qquad ⑦$$

Then,

$$F_k = \begin{cases} \dfrac{1}{\sqrt{5}} \left(\dfrac{1+\sqrt{5}}{2} \right)^k + r_k, & k \text{ is an odd number,} \\ \dfrac{1}{\sqrt{5}} \left(\dfrac{1+\sqrt{5}}{2} \right)^k - r_k, & k \text{ is an even number.} \end{cases} \qquad ⑧$$

However, $0 < r_k < \frac{1}{2}$ since $\left| \frac{1-\sqrt{5}}{2} \right| \approx 0.618 < 1$. From ⑧, we know that F_k is equal to the positive integer lying nearest to $\frac{1}{\sqrt{5}} \left(\frac{1+\sqrt{5}}{2} \right)^k$. For example, when $k = 19, 20$,

$$\frac{1}{\sqrt{5}} \left(\frac{1+\sqrt{5}}{2} \right)^k \approx \begin{cases} 4180.999964, & k = 19, \\ 6765.000052, & k = 20, \end{cases}$$

whereas we may directly calculate $F_{19} = 4181$, $F_{20} = 6765$ (you can cross-check the result to be correct using ①).

When k is much larger, $|\lambda_2|^k \approx (0.618)^k << 1$, in this case,

$$F_{k+1}/F_k \approx {\lambda_1}^{k+1}/{\lambda_1}^k = \lambda_1 \approx 1.618, \qquad ⑨$$

or

$$F_k/F_{k+1} \approx \frac{1}{\lambda_1} = |\lambda_2| \approx 0.618. \qquad ⑩$$

Example 7.8 Use of recurrence relation in a practical situation

If one climbs up a flight of stairs, one or two stairs at a time, how many different ways might he have to climb up n stairs?

Solution: Suppose there are F_n different ways to climb up n stairs, then obviously we have $F_1 = 1$ (there is only one way to climb up one stair) and $F_2 = 2$ (there are two ways to climb up two stairs). To climb up n stairs, there are only two ways to take the first step.

(i) If one way is one stair at the first step, then we have F_{n-1} ways to climb up the following $(n-1)$ stairs.
(ii) The other way is to climb two stairs at the first step, then we have F_{n-2} ways to climb up the following $(n-2)$ stairs. Thus,

$$F_n = F_{n-1} + F_{n-2}, \quad n = 3, 4, \ldots. \qquad ⑪$$

Here, we define $F_0 = 1$ and $\boldsymbol{\alpha}_0 = (F_1, F_0)^{\mathrm{T}} = (1, 1)^{\mathrm{T}}$. Then, ⑪ can also be expressed as ②. Similarly, we obtain

$$\boldsymbol{\alpha}_k = \boldsymbol{A}^k \boldsymbol{\alpha}_0, \quad k = 1, 2, 3, \ldots.$$

The number of ways of climbing up the stairs, F_n, are (n begins at 0) 1, 2, 3, 5, 8, 13, 21, 34, ...; a Fibonacci sequence. The difference is, in the former Fibonacci sequence, $F_1 = 1$, $F_2 = 1$, $F_3 = 2$, $F_4 = 3, \ldots$, but here, $F_1 = 1$, $F_2 = 2$, $F_3 = 3, \ldots$.

If someone lives on the second storey (total of 18 steps), then the number of ways of climbing upstairs is $F_{18} = 4181$. Therefore, if he changes his way of climbing up the stairs every day, he can climb to higher stories by different ways every day for 11.45 years. ■

7.6 Systems of Linear Ordinary Differential Equations with Constant Coefficients

Before we introduce a matrix method to solve the systems of linear ordinary differential equations with constant coefficients, we will first provide a brief introduction of the basic concepts required for this method.

7.6.1 Differentials and integrals of matrix-valued functions (a brief introduction)

Suppose $\boldsymbol{A}(t) = (a_{ij}(t))_{m\times n}$. If the limit of every entry $a_{ij}(t)$ of $\boldsymbol{A}$ ($t \to t_0$ or $t \to \infty$) exists, then we say that the limit of $\boldsymbol{A}(t)$ exists, and

$$\lim \boldsymbol{A}(t) = (\lim a_{ij}(t))_{m\times n}.$$

If $a_{ij}(t)$ is differentiable ($i = 1, 2, \ldots, m$; $j = 1, 2, \ldots, n$), then $\boldsymbol{A}(t)$ is said to be differentiable, and

$$\frac{\mathrm{d}}{\mathrm{d}t}\boldsymbol{A}(t) = \left(\frac{\mathrm{d}a_{ij}(t)}{\mathrm{d}t}\right)_{m\times n}.$$

If each element $a_{ij}(t)$ is integrable on $[t_1, t_2]$, then $\boldsymbol{A}(t)$ is said to be integrable on $[t_1, t_2]$, and

$$\int_{t_1}^{t_2} \boldsymbol{A}(t)\,\mathrm{d}t = \left(\int_{t_1}^{t_2} a_{ij}(t)\,\mathrm{d}t\right)_{m\times n}.$$

We also have similar forms to represent matrix-valued series. For example, from

$$\mathrm{e}^x = 1 + x + \frac{x^2}{2!} + \cdots + \frac{x_n}{n!} + \cdots,$$

we have

$$\mathrm{e}^{\boldsymbol{A}} = \boldsymbol{I} + \boldsymbol{A} + \frac{1}{2!}\boldsymbol{A}^2 + \cdots + \frac{1}{n!}\boldsymbol{A}^n + \cdots, \tag{7.36}$$

and

$$\mathrm{e}^{t\boldsymbol{A}} = \boldsymbol{I} + t\boldsymbol{A} + \frac{1}{2!}(t\boldsymbol{A})^2 + \cdots + \frac{1}{n!}(t\boldsymbol{A})^n + \cdots, \tag{7.37}$$

$$\frac{\mathrm{d}}{\mathrm{d}t}(\mathrm{e}^{t\boldsymbol{A}}) = \boldsymbol{A} + t\boldsymbol{A}^2 + \cdots + \frac{1}{(n-1)!}t^{n-1}\boldsymbol{A}^n + \cdots,$$

that is,

$$\frac{\mathrm{d}}{\mathrm{d}t}(\mathrm{e}^{t\boldsymbol{A}}) = \boldsymbol{A}\mathrm{e}^{t\boldsymbol{A}}. \tag{7.38}$$

Here, the definitions of the matrix exponent functions $\mathrm{e}^{\boldsymbol{A}}$ and $\mathrm{e}^{t\boldsymbol{A}}$ as well as their derivatives are similar to their counterparts in calculus.

7.6.2 Matrix solutions to systems of the first-order linear ordinary differential equations with constant coefficients

The system of linear ordinary differential equations with constant coefficients is

$$\begin{cases} \dfrac{\mathrm{d}u_1}{\mathrm{d}t} = a_{11}u_1 + a_{12}u_2 + \cdots + a_{1n}u_n, \\ \dfrac{\mathrm{d}u_2}{\mathrm{d}t} = a_{21}u_1 + a_{22}u_2 + \cdots + a_{2n}u_n, \\ \qquad\qquad \cdots \\ \dfrac{\mathrm{d}u_n}{\mathrm{d}t} = a_{n1}u_1 + a_{n2}u_2 + \cdots + a_{nn}u_n. \end{cases} \tag{7.39}$$

which is denoted in a matrix form

$$\frac{\mathrm{d}\boldsymbol{u}}{\mathrm{d}t} = \boldsymbol{A}\boldsymbol{u}, \tag{7.40}$$

where $\boldsymbol{u} = (u_1, u_2, \ldots, u_n)^{\mathrm{T}}$ and $\boldsymbol{A} = (a_{ij})_{n\times n}$ is the coefficient matrix. Let the solution to the equation (7.40) be

$$\boldsymbol{u} = \mathrm{e}^{\lambda t}\boldsymbol{x}, \tag{7.41}$$

that is,

$$\boldsymbol{u} = (u_1, u_2, \ldots, u_n)^{\mathrm{T}} = \mathrm{e}^{\lambda t}(x_1, x_2, \ldots, x_n)^{\mathrm{T}}.$$

Substituting (7.41) into (7.40), we have

$$\lambda \mathrm{e}^{\lambda t}\boldsymbol{x} = \boldsymbol{A}\mathrm{e}^{\lambda t}\boldsymbol{x} = \mathrm{e}^{\lambda t}\boldsymbol{A}\boldsymbol{x},$$

and consequently $\boldsymbol{A}\boldsymbol{x} = \lambda\boldsymbol{x}$. Then, λ in (7.41) is an eigenvalue of $\boldsymbol{A}$ and $\boldsymbol{x}$ is an eigenvector corresponding to λ. If $\boldsymbol{A}$ can be diagonalized, then there exist n linearly independent eigenvectors $\boldsymbol{x}_1, \boldsymbol{x}_2, \ldots, \boldsymbol{x}_n$. Then, we get n linearly independent particular solutions of (7.40):

$$\boldsymbol{u}_1 = \mathrm{e}^{\lambda_1 t}\boldsymbol{x}_1,\ \boldsymbol{u}_2 = \mathrm{e}^{\lambda_2 t}\boldsymbol{x}_2,\ \ldots,\ \boldsymbol{u}_n = \mathrm{e}^{\lambda_n t}\boldsymbol{x}_n.$$

Their linear combination

$$\boldsymbol{u} = c_1\mathrm{e}^{\lambda_1 t}\boldsymbol{x}_1 + c_2\mathrm{e}^{\lambda_2 t}\boldsymbol{x}_2 + \cdots + c_n\mathrm{e}^{\lambda_n t}\boldsymbol{x}_n \tag{7.42}$$

(where $c_1, c_2, \ldots, c_n$ are arbitrary constant numbers) is the general solution to system (7.39). Now, (7.42) in matrix form is

$$\boldsymbol{u} = (\boldsymbol{x}_1, \boldsymbol{x}_2, \ldots, \boldsymbol{x}_n)\begin{pmatrix} \mathrm{e}^{\lambda_1 t} & & & \\ & \mathrm{e}^{\lambda_2 t} & & \\ & & \ddots & \\ & & & \mathrm{e}^{\lambda_n t} \end{pmatrix}\begin{pmatrix} c_1 \\ c_2 \\ \vdots \\ c_n \end{pmatrix},$$

$\boldsymbol{c} = (c_1, c_2, \ldots, c_n)^{\mathrm{T}}$, $\mathrm{e}^{t\boldsymbol{\Lambda}} = \mathrm{diag}(\mathrm{e}^{\lambda_1 t}, \mathrm{e}^{\lambda_2 t}, \ldots, \mathrm{e}^{\lambda_n t})$, and $\boldsymbol{P} = (\boldsymbol{x}_1, \boldsymbol{x}_2, \ldots, \boldsymbol{x}_n)$. Then system (7.39) or (7.40) has the general solution

$$\boldsymbol{u} = \boldsymbol{P}\mathrm{e}^{t\boldsymbol{\Lambda}}\boldsymbol{c}. \tag{7.43}$$

For the initial value problem,

$$\begin{cases} \dfrac{\mathrm{d}\boldsymbol{u}}{\mathrm{d}t} = \boldsymbol{A}\boldsymbol{u}, \\ \boldsymbol{u}|_{t=0} = \boldsymbol{u}_0. \end{cases} \tag{7.44}$$

Its solution is

$$\boldsymbol{u} = \boldsymbol{P} e^{t\boldsymbol{\Lambda}} \boldsymbol{P}^{-1} \boldsymbol{u}_0, \tag{7.45}$$

where $\boldsymbol{c} = \boldsymbol{P}^{-1}\boldsymbol{u}_0$ from substituting $t = 0$ into (7.43).

Example 7.9 Use of system of linear ordinary differential equations for electrical application

Consider a resistor-capacitor circuit diagram as shown in Figure 7.4. Let $R_1 = 3$, $R_2 = 2$ (unit in Ω), $C_1 = 1$, and $C_2 = 2$ (unit in F). When the switch is turned on, the initial voltage across the capacitor C_1 is 11V, the initial electric current in the *RHS* closed path is 0 and initial voltage across the capacitor C_2 is 0. Find the functional relations between voltages across two capacitors u_1, u_2 over time t after the switch is turned on, that is, find $u_1(t)$ and $u_2(t)$.

Solution: We know from the laws of physics that the relation between current and voltage across a capacitor is

$$i_1 = C_1 \frac{\mathrm{d}u_1}{\mathrm{d}t} = \frac{\mathrm{d}u_1}{\mathrm{d}t}, \quad i_2 = C_2 \frac{\mathrm{d}u_2}{\mathrm{d}t} = 2\frac{\mathrm{d}u_2}{\mathrm{d}t}.$$

Construct voltage equations for the two loops by Kirchhoff's law.

$$\begin{cases} u_1 + 3\left(\dfrac{\mathrm{d}u_1}{\mathrm{d}t} - 2\dfrac{\mathrm{d}u_2}{\mathrm{d}t}\right) = 0, \\ u_2 + 4\dfrac{\mathrm{d}u_2}{\mathrm{d}t} + 3\left(2\dfrac{\mathrm{d}u_2}{\mathrm{d}t} - \dfrac{\mathrm{d}u_1}{\mathrm{d}t}\right) = 0. \end{cases}$$

Convert it to a standard form of system of the first-order nonhomogeneous linear ordinary differential equations with constant coefficients:

$$\begin{cases} \dfrac{\mathrm{d}u_1}{\mathrm{d}t} = -\dfrac{5}{6}u_1 - \dfrac{1}{2}u_2, \\ \dfrac{\mathrm{d}u_2}{\mathrm{d}t} = -\dfrac{1}{4}u_1 - \dfrac{1}{4}u_2. \end{cases} \quad ①$$

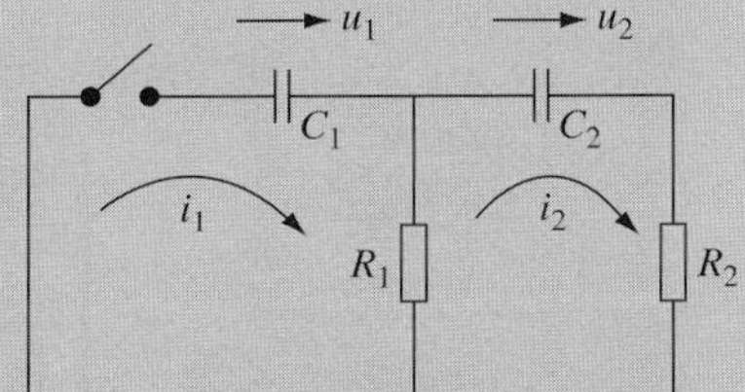

Figure 7.4 A resistor-capacitor circuit

The initial conditions are

$$u_1(0) = 11, \quad u_2(0) = 0. \qquad ②$$

Write ① and ② in matrix form

$$\begin{cases} \dfrac{d\boldsymbol{u}}{dt} = \boldsymbol{A}\boldsymbol{u}, \\ \boldsymbol{u}|_{t=0} = \boldsymbol{u}_0 = (11, 0)^{\mathrm{T}}, \end{cases} \qquad ③$$

where

$$\boldsymbol{A} = \begin{pmatrix} -\dfrac{5}{6} & -\dfrac{1}{2} \\ -\dfrac{1}{4} & -\dfrac{1}{4} \end{pmatrix}.$$

First, find the eigenvalues and eigenvectors. From $|\lambda \boldsymbol{I} - \boldsymbol{A}| = 0$, we obtain $\lambda_1 = -1$, $\lambda_2 = -\frac{1}{12}$. The corresponding eigenvectors are $\boldsymbol{x}_1 = (3, 1)^{\mathrm{T}}$, $\boldsymbol{x}_2 = (2, -3)^{\mathrm{T}}$,

$$\mathrm{e}^{t\Lambda} = \mathrm{diag}(\mathrm{e}^{-t}, \mathrm{e}^{-t/12}), \quad \boldsymbol{P} = (\boldsymbol{x}_1, \boldsymbol{x}_2).$$

From (7.45), we obtain that a solution to the initial-value problem ③,

$$\boldsymbol{u} = \boldsymbol{P}\mathrm{e}^{t\Lambda}\boldsymbol{P}^{-1}\boldsymbol{u}_0,$$

that is,

$$\begin{pmatrix} u_1 \\ u_2 \end{pmatrix} = \begin{pmatrix} 3 & 2 \\ 1 & -3 \end{pmatrix} \begin{pmatrix} \mathrm{e}^{-t} & 0 \\ 0 & \mathrm{e}^{-t/12} \end{pmatrix} \frac{-1}{11} \begin{pmatrix} -3 & -2 \\ -1 & 3 \end{pmatrix} \begin{pmatrix} 11 \\ 0 \end{pmatrix},$$

$$\begin{pmatrix} u_1 \\ u_2 \end{pmatrix} = \begin{pmatrix} 9\mathrm{e}^{-t} + 2\mathrm{e}^{-t/12} \\ 3\mathrm{e}^{-t} - 3\mathrm{e}^{-t/12} \end{pmatrix}.$$

■

Example 7.10 Solving a system of linear ordinary differential equations

Solve a system of linear ordinary differential equations with constant coefficients

$$\begin{cases} \dfrac{dx_1}{dt} = x_1 + x_2, \\ \dfrac{dx_2}{dt} = -4x_1 + 5x_2, \\ \dfrac{dx_3}{dt} = x_1 + 2x_3. \end{cases}$$

The given initial values are

$$x_1(0)=1,\quad x_2(0)=-1,\quad x_3(0)=2.$$

Solution: The initial-value problem is

$$\begin{cases}\dfrac{\mathrm{d}\boldsymbol{x}}{\mathrm{d}t}=\boldsymbol{A}\boldsymbol{x},\\ \boldsymbol{x}(0)=\boldsymbol{x}_0=(1,-1,2)^{\mathrm{T}},\end{cases}$$

where

$$\boldsymbol{A}=\begin{pmatrix}1&1&0\\-4&5&0\\1&0&2\end{pmatrix}.$$

Using the method provided in Appendix B, we can obtain the Jordan canonical form $\boldsymbol{A}$ from an invertible matrix

$$\boldsymbol{P}=\begin{pmatrix}0&1&2\\0&2&5\\1&1&1\end{pmatrix},$$

such that

$$\boldsymbol{P}^{-1}\boldsymbol{A}\boldsymbol{P}=\boldsymbol{J}=\begin{pmatrix}2&0&0\\0&3&1\\0&0&3\end{pmatrix}.$$

According to (7.45), the solution to the initial-value problem is

$$\boldsymbol{x}=\boldsymbol{P}\mathrm{e}^{t\boldsymbol{J}}\boldsymbol{P}^{-1}\boldsymbol{x}_0, \qquad ①$$

where

$$\mathrm{e}^{t\boldsymbol{J}}=\boldsymbol{I}+t\boldsymbol{J}+\frac{(t\boldsymbol{J})^2}{2!}+\cdots+\frac{(t\boldsymbol{J})^n}{n!}+\cdots, \qquad ②$$

$$\boldsymbol{J}^n=\begin{pmatrix}2&0&0\\0&3&1\\0&0&3\end{pmatrix}^n=\begin{pmatrix}2^n&0&0\\0&3^n&3^{n-1}\mathrm{C}_1^n\\0&0&3^n\end{pmatrix}. \qquad ③$$

Substituting ③ into ②, we have

$$\mathrm{e}^{t\boldsymbol{J}}=\begin{pmatrix}\mathrm{e}^{2t}&0&0\\0&\mathrm{e}^{3t}&t\mathrm{e}^{3t}\\0&0&\mathrm{e}^{3t}\end{pmatrix}. \qquad ④$$

Substituting ④ and $\boldsymbol{P}$, $\boldsymbol{P}^{-1}$, $\boldsymbol{x}_0$ into ①, we have

$$\boldsymbol{x}=\begin{pmatrix}x_1(t)\\x_2(t)\\x_3(t)\end{pmatrix}=\begin{pmatrix}0&1&2\\0&2&5\\1&1&1\end{pmatrix}\begin{pmatrix}e^{2t}&0&0\\0&e^{3t}&te^{3t}\\0&0&e^{3t}\end{pmatrix}\begin{pmatrix}-3&1&1\\5&-2&0\\-2&1&0\end{pmatrix}\begin{pmatrix}1\\-1\\2\end{pmatrix}$$

$$=\begin{pmatrix}(1-3t)e^{3t}\\(-1-6t)e^{3t}\\-2e^{2t}+(4-3t)e^{3t}\end{pmatrix}.$$

■

7.7 Least Squares Solutions to a System of Overdetermined Equations

Consider the system of nonhomogeneous linear equations

$$\sum_{j=1}^{m} a_{ij}x_j=b_i \quad (i=1,2,\ldots,n), \tag{7.46}$$

or

$$\boldsymbol{A}_{n\times m}\boldsymbol{x}=\boldsymbol{b}.$$

It is called an **overdetermined system of equations** if $\mathrm{r}(\boldsymbol{A},\boldsymbol{b})\neq\mathrm{r}(\boldsymbol{A})$. It has no solution in the usual sense. That is, $\boldsymbol{Ax}\neq\boldsymbol{b}$ for any $\boldsymbol{x}_{m\times 1}$ or $\|\boldsymbol{Ax}-\boldsymbol{b}\|>0$, where $\|\boldsymbol{Ax}-\boldsymbol{b}\|$ represents the modulus (also called norm) of the n-dimensional vector $\boldsymbol{Ax}-\boldsymbol{b}$. Now, we aim to find $\boldsymbol{x}$ so that $\|\boldsymbol{Ax}-\boldsymbol{b}\|$ reaches its minimum, that is,

$$\|\boldsymbol{b}-\boldsymbol{Ax}\|^2=\sum_{i=1}^{n}\left(b_i-\sum_{j=1}^{m}a_{ij}x_j\right)^2$$

reaches its minimum. The $\boldsymbol{x}$ which achieves this minimum value is called a **least squares solution** of the overdetermined system of equations. In order to find a least squares solution, we first introduce the related concepts of projecting a vector to a subspace and the vertical distance from a vector to a subspace.

Definition 7.3

Suppose $\boldsymbol{\alpha},\boldsymbol{\beta}\in\mathbb{R}^n$, $d(\boldsymbol{\alpha},\boldsymbol{\beta})=\|\boldsymbol{\alpha}-\boldsymbol{\beta}\|$ is called the **distance** between $\boldsymbol{\alpha}$ and $\boldsymbol{\beta}$. If $\boldsymbol{\alpha}=(a_1,a_2,\cdots,a_n)^{\mathrm{T}}$, $\boldsymbol{\beta}=(b_1,b_2,\cdots,b_n)^{\mathrm{T}}$, then

$$d(\boldsymbol{\alpha},\boldsymbol{\beta})=\|\boldsymbol{\alpha}-\boldsymbol{\beta}\|=\sqrt{(a_1-b_1)^2+(a_2-b_2)^2+\cdots+(a_n-b_n)^2}. \tag{7.47}$$

It is not difficult for readers to prove that the distance defined in such a way satisfies three basic properties:

(i) $d(\boldsymbol{\alpha}, \boldsymbol{\beta}) = d(\boldsymbol{\beta}, \boldsymbol{\alpha})$.
(ii) $d(\boldsymbol{\alpha}, \boldsymbol{\beta}) \geq 0$, the equality holds, if and only if, $\boldsymbol{\alpha} = \boldsymbol{\beta}$. (7.48)
(iii) $d(\boldsymbol{\alpha}, \boldsymbol{\beta}) \leq d(\boldsymbol{\alpha}, \boldsymbol{\gamma}) + d(\boldsymbol{\gamma}, \boldsymbol{\beta})$ (triangle inequality).

Definition 7.4

Suppose $\boldsymbol{\alpha} \in \mathbb{R}^n$ and W is a subspace of $\mathbb{R}^n$. If $\boldsymbol{\beta} \in W$ and $(\boldsymbol{\alpha} - \boldsymbol{\beta}) \perp W$, then $\boldsymbol{\beta}$ is called the (orthogonal) projection of $\boldsymbol{\alpha}$ on W, denoted as $(\boldsymbol{\alpha})_W = \boldsymbol{\beta}$. We denote $\|\boldsymbol{\alpha} - \boldsymbol{\beta}\|$ as the vertical distance from $\boldsymbol{\alpha}$ to W.

The geometrical interpretation of Definition 7.4 in three-dimensional geometric space is shown in Figures 7.5 and 7.6 (where W_1 is a one-dimensional subspace of $\mathbb{R}^3$, W_2 is a two-dimensional subspace).

The (orthogonal) projection of a vector on a subspace is unique. If $\boldsymbol{\beta}_1$ and $\boldsymbol{\beta}_2$ are two projections of $\boldsymbol{\alpha}$ on the subspace W, then $(\boldsymbol{\alpha} - \boldsymbol{\beta}_1) \perp W$, $(\boldsymbol{\alpha} - \boldsymbol{\beta}_2) \perp W$. Hence, for any $\boldsymbol{\gamma} \in W$, we have $(\boldsymbol{\alpha} - \boldsymbol{\beta}_1, \boldsymbol{\gamma}) = (\boldsymbol{\alpha} - \boldsymbol{\beta}_2, \boldsymbol{\gamma}) = 0$, then

$$(\boldsymbol{\beta}_1, \boldsymbol{\gamma}) = (\boldsymbol{\beta}_2, \boldsymbol{\gamma}),$$

that is,

$$(\boldsymbol{\beta}_1 - \boldsymbol{\beta}_2, \boldsymbol{\gamma}) = 0.$$

Since the above expression holds for all $\boldsymbol{\gamma} \in W$, then $\boldsymbol{\beta}_1 = \boldsymbol{\beta}_2$ (if $\boldsymbol{\beta}_1 \neq \boldsymbol{\beta}_2$, the above expression does not hold for $\boldsymbol{\gamma} = \boldsymbol{\beta}_1 - \boldsymbol{\beta}_2 \in W$).

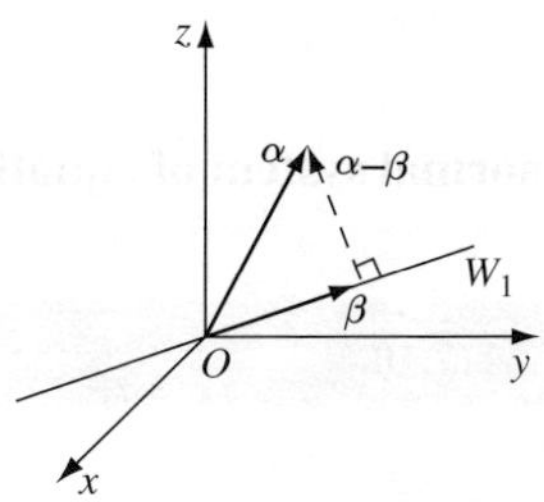

Figure 7.5 Geometrical interpretation of Definition 7.4

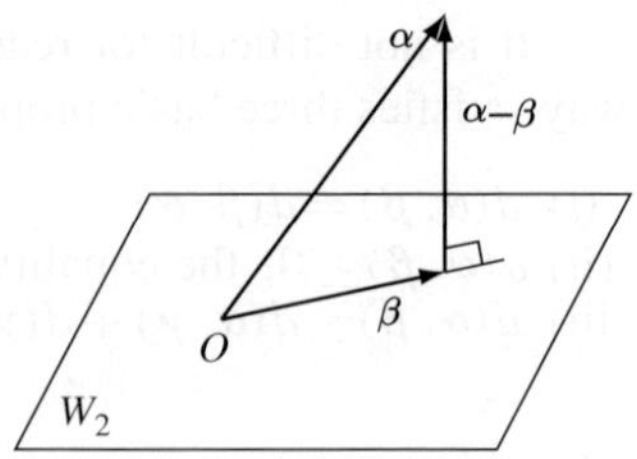

Figure 7.6 Projection of a vector onto a subspace

Suppose $A_{n\times s} = (\boldsymbol{\alpha}_1, \boldsymbol{\alpha}_2, \ldots, \boldsymbol{\alpha}_s)$. $R(A) = L(\boldsymbol{\alpha}_1, \boldsymbol{\alpha}_2, \ldots, \boldsymbol{\alpha}_s)$ is a column space. Obviously,

$$\boldsymbol{\beta} \in R(\boldsymbol{A}) \rightleftharpoons \boldsymbol{\beta} = \sum_{i=1}^{s} x_i \boldsymbol{\alpha}_i = \boldsymbol{A}\boldsymbol{x}.$$

THEOREM 7.9

Suppose $\boldsymbol{b} \in \mathbb{R}^n$ and the (orthogonal) projection of $\boldsymbol{b}$ on $R(A)$ is $(\boldsymbol{b})_{R(A)} = \boldsymbol{A}\boldsymbol{x}^*$, then $\boldsymbol{x}^*$ is a solution to the linear equations $\mathbf{A}^{\mathrm{T}}\mathbf{A}\mathbf{x} = \mathbf{A}^{\mathrm{T}}\boldsymbol{b}$.

Proof: Since $(\boldsymbol{b} - \boldsymbol{A}\boldsymbol{x}^*) \perp R(\boldsymbol{A})$, we have $(\boldsymbol{b} - \boldsymbol{A}\boldsymbol{x}^*) \perp \boldsymbol{A}\boldsymbol{x}$ for arbitrary vector $\boldsymbol{\beta} = \boldsymbol{A}\boldsymbol{x}$ of $R(\boldsymbol{A})$, that is,

$$(\boldsymbol{A}\boldsymbol{x}, \boldsymbol{b} - \boldsymbol{A}\boldsymbol{x}^*) = 0,$$

thus,

$$(\boldsymbol{A}\boldsymbol{x})^{\mathrm{T}}(\boldsymbol{b} - \boldsymbol{A}\boldsymbol{x}^*) = \boldsymbol{x}^{\mathrm{T}}\boldsymbol{A}^{\mathrm{T}}(\boldsymbol{b} - \boldsymbol{A}\boldsymbol{x}^*) = 0.$$

Since $\boldsymbol{x}$ is arbitrary, we have $\boldsymbol{A}^{\mathrm{T}}(\boldsymbol{b} - \boldsymbol{A}\boldsymbol{x}^*) = 0$, where $\boldsymbol{x}^*$ is a solution to $\boldsymbol{A}^{\mathrm{T}}\boldsymbol{A}\boldsymbol{x} = \boldsymbol{A}^{\mathrm{T}}\boldsymbol{b}$. □

We call the system of linear equations

$$\boldsymbol{A}^{\mathrm{T}}\boldsymbol{A}\boldsymbol{x} = \boldsymbol{A}^{\mathrm{T}}\boldsymbol{b} \tag{7.49}$$

a **normal system of equations**.

THEOREM 7.10

(i) *If $\mathrm{r}(\boldsymbol{A}_{n\times s}) = s$, then (7.49) has a unique solution.*
(ii) *If $\mathrm{r}(\boldsymbol{A}) < s$, then (7.49) has infinitely many solutions.*

Proof:

(i) As $A^{\mathrm{T}}A$ is an sth-order matrix, the normal system of equations has a unique solution according to $\mathrm{r}(A^{\mathrm{T}}A) = \mathrm{r}(A) = s$ (refer to Example 3.15, §3.4) and the solution is

$$x^* = (A^{\mathrm{T}}A)^{-1}A^{\mathrm{T}}b. \tag{7.50}$$

(ii) We only need to prove

$$\mathrm{r}(A^{\mathrm{T}}A,\ A^{\mathrm{T}}b) = \mathrm{r}(A^{\mathrm{T}}A) \quad \text{(this is left as an exercise for readers).}$$ □

COROLLARY 7.1

If $\mathrm{r}(A) = s$, *then*

$$(b)_{R(A)} = Ax^* = A(A^{\mathrm{T}}A)^{-1}A^{\mathrm{T}}b = Pb, \tag{7.51}$$

where

$$P = A(A^{\mathrm{T}}A)^{-1}A^{\mathrm{T}}$$

is called the (orthogonal) ***projection matrix***. P *satisfies*

(i) $P^2 = P$ *(P is an idempotent matrix),*
(ii) $P^{\mathrm{T}} = P$ *(P is a symmetric matrix).*

(Proofs are left as exercises for readers).

THEOREM 7.11

If $\mathrm{r}(A_{n\times s}) = s$, *a least squares solution* x^* *of the overdetermined system of linear equations* $Ax = b$ *is the unique solution to (7.49):*

$$x^* = (A^{\mathrm{T}}A)^{-1}A^{\mathrm{T}}b.$$

Proof: A least squares solution x^* satisfies

$$\|b - A\,x^*\| \le \|b - Ax\| \tag{7.52}$$

for all $x \in \mathbb{R}^s$. Since Ax^* and $Ax \in R(A)$, $\|b - Ax^*\|$ is the vertical distance from b to $R(A)$, which satisfies (7.52). If $Ax^* = (b)_{R(A)}$, then $(b - Ax^*) \perp R(A)$. Hence, x^* is a solution to the normal equations according to Theorem 7.9. □

Example 7.11 Data fitting

The following relation holds between solid stretches of a spring and the force exerted on it within the limits of its elasticity.

$$y = a + bx. \quad ①$$

Find the elastic coefficient b according to the data shown in Table 7.6.

Table 7.6 Relation between stretched length and force exerted on a spring

x_i / centimeter	2.6	3.0	3.5	4.3
y_i / netwon	0	1	2	3

Solution: Substituting the data in the table into ①, we have

$$\begin{cases} a + 2.6b = 0, \\ a + 3.0b = 1, \\ a + 3.5b = 2, \\ a + 4.3b = 3. \end{cases} \quad ②$$

This can be written as $\boldsymbol{Ax} = \boldsymbol{y}$ where

$$\boldsymbol{A} = \begin{pmatrix} 1 & 2.6 \\ 1 & 3.0 \\ 1 & 3.5 \\ 1 & 4.3 \end{pmatrix}, \quad \boldsymbol{x} = \begin{pmatrix} a \\ b \end{pmatrix}, \quad \boldsymbol{y} = \begin{pmatrix} 0 \\ 1 \\ 2 \\ 3 \end{pmatrix}.$$

Then, ② is an overdetermined equation system and the least squares solution is

$$\boldsymbol{x}^* = (\boldsymbol{A}^{\mathrm{T}}\boldsymbol{A})^{-1}\boldsymbol{A}^{\mathrm{T}}\boldsymbol{y}, \quad ③$$

in which

$$\boldsymbol{A}^{\mathrm{T}}\boldsymbol{A} = \begin{pmatrix} 1 & 1 & 1 & 1 \\ 2.6 & 3.0 & 3.5 & 4.3 \end{pmatrix} \begin{pmatrix} 1 & 2.6 \\ 1 & 3.0 \\ 1 & 3.5 \\ 1 & 4.3 \end{pmatrix} = \begin{pmatrix} 4 & 13.4 \\ 13.4 & 46.5 \end{pmatrix}.$$

$$(\boldsymbol{A}^{\mathrm{T}}\boldsymbol{A})^{-1} = \frac{1}{6.44}\begin{pmatrix} 46.5 & -13.4 \\ -13.4 & 4 \end{pmatrix}, \quad \boldsymbol{A}^{\mathrm{T}}\boldsymbol{y} = \begin{pmatrix} 6 \\ 22.9 \end{pmatrix}.$$

From ③, we have

$$\boldsymbol{x}^* = \begin{pmatrix} a_0 \\ b_0 \end{pmatrix} = \begin{pmatrix} -4.326 \\ 1.739 \end{pmatrix}.$$

Hence, the elastic coefficient is $b = b_0 = 1.739$ newton/centimeter. ■

In order to fit the data $(x_i, y_i)(i = 1, 2, \ldots, n)$ obtained from experiments with a curve, the least squares method is adopted using the following general steps to finally give a function $y = f(x)$ between x and y.

(i) We could plot the experimental data on a graph paper one by one in order to determine the type of function between x and y through observing the distributive behavior of data points. For example, we may choose to fit a straight line $y = a + bx$ to the data in Figure 7.7, whereas a cubic curve $y = a + bx + cx^2 + dx^3$ could fit the points in Figure 7.8.

(ii) Substitute the experimental data into the selected function model and obtain a nonhomogeneous system of linear equations. For example, choose $y = f(x)$ as an mth-degree polynomial

$$y = a_0 + a_1 x + a_2 x^2 + \cdots + a_m x^m.$$

Substitute the experimental data $(x_i, y_i)(i = 1, 2, \ldots, n)$ into the function and obtain linear equations

$$\boldsymbol{A}\boldsymbol{x} = \boldsymbol{y}, \tag{7.53}$$

in which

$$\boldsymbol{A} = \begin{pmatrix} 1 & x_1 & {x_1}^2 & \cdots & {x_1}^m \\ 1 & x_2 & {x_2}^2 & \cdots & {x_2}^m \\ \vdots & \vdots & \vdots & & \vdots \\ 1 & x_n & {x_n}^2 & \cdots & {x_n}^m \end{pmatrix}, \quad \boldsymbol{x} = \begin{pmatrix} a_0 \\ a_1 \\ \vdots \\ a_m \end{pmatrix}, \quad \boldsymbol{y} = \begin{pmatrix} y_1 \\ y_2 \\ \vdots \\ y_n \end{pmatrix}.$$

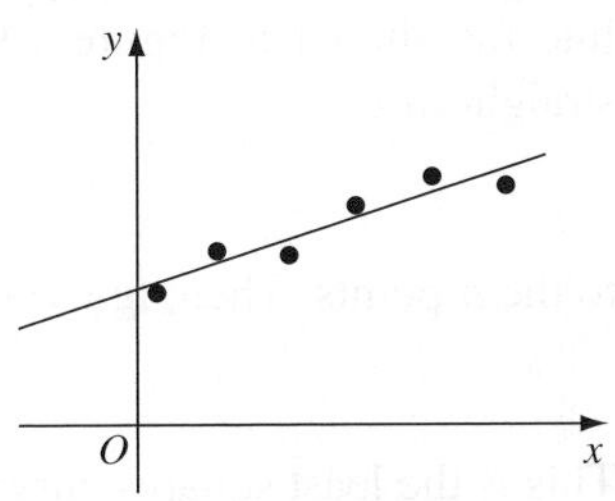

Figure 7.7 Fitting data points to a straight line

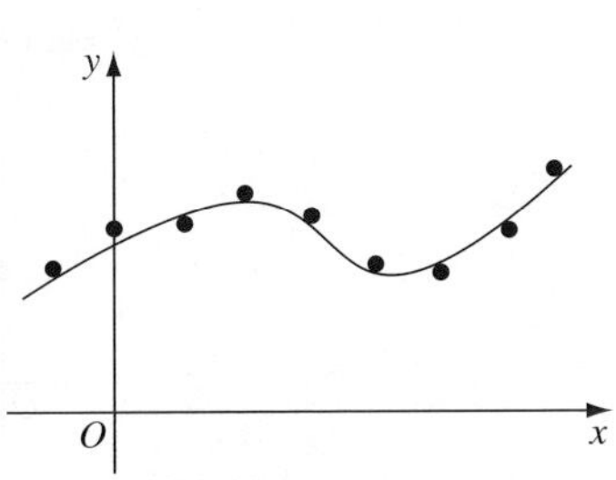

Figure 7.8 Fitting data points to a cubic curve

Usually, $n >> m$ and (7.53) is an overdetermined system of linear equations. However, we can prove that if $x_1, x_2, \ldots, x_n$ are different from each other, then we have $\mathrm{r}(\boldsymbol{A}) = m$, hence the least squares solution to the linear equations (7.53) is

$$\boldsymbol{x}^* = (a_0, a_1, \cdots, a_m)^{\mathrm{T}} = (\boldsymbol{A}^{\mathrm{T}}\boldsymbol{A})^{-1}\boldsymbol{A}^{\mathrm{T}}\boldsymbol{y}.$$

The obtained curvilinear equation $y = a_0 + a_1x + \cdots + a_mx^m$ is usually called an **empirical formula** in engineering research. The coefficients of the empirical formula is worked out by the least squares method as the type of the function is assumed to be correct. Now the question is: whether this assumption about the type of the function reflects the real relation of the experimental data? Also, is it reliable in the general case? Suitable statistical methods should be used for verification.

In fitting a curve to the experimental data with the least squares method, the type of function is not easily determined. Indeed, types of function may be numerous and varied for particular problems, which are not limited to straight lines or polynomial curves. In certain cases, after plotting the experimental data points (x_i, y_i), $i = 1, 2, \cdots, n$, it is not easy to judge the type of function $y = f(x)$.

Sometimes, plotting n points on a logarithmic graph paper, that is,

$$x^* = \lg x,\ y^* = \lg y,$$

may guide us in selecting the right function. For example, the distribution of n logarithmic points (x_i^*, y_i^*) seems to show the characteristics of a straight line (as shown in Figure 7.9). We may use least squares method to fit the straight line

$$y^* = a + bx^*,$$

to the n points. Then, $\lg y = a + b\lg x$ and let $a = \lg c$. Thus, we obtain

$$y = cx^b.$$

This is the least squares curve fitting to n points (x_i, y_i).

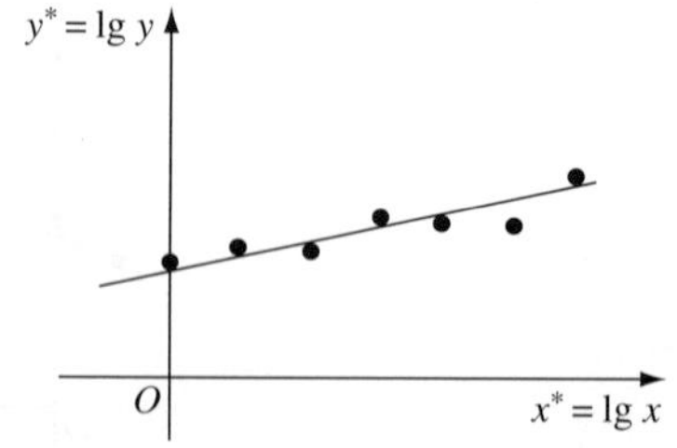

Figure 7.9 Alternative method of line fitting using a logarithmic scaled paper

Exercises and Supplementary Exercise

Exercises

7.1 Suppose a group of animals is divided into two age groups, the span of each age group is 1 year, and the Leslie matrix is

$$L = \begin{pmatrix} 1 & \frac{1}{5} \\ \frac{1}{2} & 0 \end{pmatrix}.$$

If there are 100 animals in group 1 and 10 animals in group 2 at the beginning, how much will the animals in each group increase every year within 5 years? Also, how much will the total number of animals increase every year? Lastly, calculate the ratio of the two groups of animals after 5 years.

7.2 Suppose a group of animals is divided into four age groups and the maximum age is 8 years. Also, we know: fertility rates: $a_1 = 0$, $a_2 = 20/9$, $a_3 = 15/32$, $a_4 = 0$; survival rates: $b_1 = 9/10$, $b_2 = 8/9$, $b_3 = 7/8$, $b_4 = 6/7$; and the number of animals in each group is 100 at the beginning. For the duration of 8 years, how much will the total number of the total number of animals in each age group increase (decrease) in every 2 years? How much will the total number of animals in the group increase (decrease) every 2 years? Also, find the ratio between the four groups of animals.

7.3 Find the eigenvalues and strictly dominant eigenvalues of the following Leslie matrices. Provide the eigenvectors corresponding to the strictly dominant eigenvalues.

$$\text{(i)} \begin{pmatrix} 0 & \frac{5}{4} & 0 \\ \frac{4}{5} & 0 & 0 \\ 0 & \frac{3}{4} & 0 \end{pmatrix}; \quad \text{(ii)} \begin{pmatrix} 0 & \frac{5}{4} & \frac{75}{64} \\ \frac{4}{5} & 0 & 0 \\ 0 & \frac{3}{4} & 0 \end{pmatrix};$$

$$\text{(iii)} \begin{pmatrix} 0 & \frac{20}{9} & \frac{15}{32} & 0 \\ \frac{9}{10} & 0 & 0 & 0 \\ 0 & \frac{8}{9} & 0 & 0 \\ 0 & 0 & \frac{7}{8} & 0 \end{pmatrix}; \quad \text{(iv)} \begin{pmatrix} \frac{1}{6} & \frac{13}{32} & \frac{1}{4} \\ \frac{2}{3} & 0 & 0 \\ 0 & \frac{1}{4} & 0 \end{pmatrix}.$$

7.4 Let the span of each age group be 1 year and the Leslie matrices be (ii), (iii), and (iv) of **Q7.3**.

(i) After k years (k is large enough), by what number will the total number of animals increase each year? What are the percentages in total for each age group? Is there a relation between the ratio of the initial age distribution and that of the current age distribution?

(ii) For the matrix L in **Q7.3** (iii), suppose we know the initial age distribution vector $x^{(0)} = (100, 100, 100, 100)^{\mathrm{T}}$. Find $x^{(1)}$, $x^{(2)}$, $x^{(3)}$, $x^{(4)}$, and calculate their annual rate of increase.

7.5 With respect to the L matrices of **Q7.3**, find the net regeneration rate R.

7.6 Prove that after a sufficiently long period of population growth, (i) the population is increasing, if and only if, the net regeneration rate $R > 1$, (ii) the population is decreasing, if and only if, the net regeneration rate $R < 1$.

7.7 The male and female students in a class like sending e-greeting cards to each other during national holidays. In the first national holiday, only female students sent e-greeting cards, while the male students did not participate. For the rest of the national holidays, a female student receives 70% of their total e-greeting cards from other female students and 30% from male students. Similarly, a male student receives 80% of their total greeting cards from other male students and 20% from female students. What are the probabilities of a male and a female student receiving an e-greeting card respectively, after n holidays ($n = 1, 2, \ldots, 5, n \to \infty$)?

7.8 A library card allows one to borrow books from three libraries. The initial probability vector of a reader borrowing books from these libraries is $\boldsymbol{x}(0) = (0, 1, 0)^{\mathrm{T}}$, the transition probability matrix is

$$\boldsymbol{P} = \begin{pmatrix} 0.8 & 0.3 & 0.2 \\ 0.1 & 0.2 & 0.6 \\ 0.1 & 0.5 & 0.2 \end{pmatrix}.$$

After n loans ($n = 1, 2, \ldots, 5, n \to \infty$), what is the lending probability distribution of that card?

7.9 Suppose $\boldsymbol{P}$ is a probability transition matrix of order n. If the sum of each row is 1, what is its stationary distribution vector?

7.10 Consider m cities $s_1, s_2, \ldots, s_m$. The air pollution rate of s_i at time t ($t = 0, 1, \ldots, n$) is $x_i(t)$, the proportion of the pollution diffused from s_j to s_i from time t to $t+1$ is p_{ij} ($0 \le p_{ij} \le 1$), the pollution that diffuses to $s_1, \ldots, s_r$ ($r = m-3$) is absorbed. The probability transition matrix of s_{m-2}, s_{m-1}, s_m is

$$\boldsymbol{P}_2 = \begin{pmatrix} \frac{1}{3} & \frac{1}{3} & 0 \\ 0 & \frac{1}{3} & \frac{2}{3} \\ \frac{1}{3} & \frac{1}{3} & \frac{1}{3} \end{pmatrix}.$$

Find the average number of transfers starting from city j ($j = m-2, m-1, m$) and to be absorbed by city s_k ($k = 1, \ldots, r$).

7.11 For each direct consumption matrix $\boldsymbol{C}$ of the following, Is $(\boldsymbol{I} - \boldsymbol{C})$ invertible? [Hint: Check if this fulfils the inequality (7.29).]

(i) $\begin{pmatrix} 0.8 & 0.1 \\ 0.3 & 0.6 \end{pmatrix}$, (ii) $\begin{pmatrix} 0.70 & 0.30 & 0.25 \\ 0.10 & 0.25 & 0.25 \\ 0.15 & 0.05 & 0.25 \end{pmatrix}$,

(iii) $\begin{pmatrix} 0.7 & 0.3 & 0.2 \\ 0.1 & 0.4 & 0.3 \\ 0.2 & 0.4 & 0.1 \end{pmatrix}$.

7.12 A builder, an electrician, and a mechanic form a team to start a technical service consultancy. For their profit sharing scheme, when the builder receives a fee of \$10, \$1 will be paid to the electrician and \$3 to the mechanic. Similarly, from a fee of \$10 received by the electrician, \$2 will be paid to the builder while the mechanic gets \$4. Lastly, based on the \$10 received by the mechanic, the builder will receive \$3 while \$4 is paid to the electrician. Each individual need not pay for his own service. If the fees received by the builder is \$5000, the electrician \$7000, and the mechanic \$6000, what is the net income of each partner after profit sharing?

7.13 An economic system is made up of three enterprises where its direct consumption coefficient matrix $\boldsymbol{C}$ and external demand vector $\boldsymbol{d}$ of the system are

$$\boldsymbol{C} = \begin{pmatrix} 0.2 & 0.1 & 0.2 \\ 0.1 & 0.2 & 0.2 \\ 0.1 & 0.1 & 0.1 \end{pmatrix}, \quad \boldsymbol{d} = \begin{pmatrix} 75 \\ 120 \\ 225 \end{pmatrix}.$$

(i) Find the gross output values x_1, x_2, x_3 of these enterprises.

(ii) Find the output value x_{ij} of the ith enterprise consumed by the jth enterprise, ($i, j = 1, 2, 3$). Also, calculate the net output value of each enterprise z_j ($j = 1, 2, 3$).

7.14 An economic system is made up of three departments. The direct consumption coefficient matrix is

$$\boldsymbol{C} = \begin{pmatrix} 0.20 & 0.20 & 0.3125 \\ 0.14 & 0.15 & 0.25 \\ 0.16 & 0.50 & 0.1875 \end{pmatrix}.$$

(i) If the external demand vector $\boldsymbol{d} = (60, 55, 120)^{\mathrm{T}}$, find the gross output value of each department.

(ii) If the external demand vector $\boldsymbol{d} = (70, 55, 120)^{\mathrm{T}}$, find the gross output value of each department.

7.15 A railway line connecting five cities is represented by an adjacency matrix $\boldsymbol{M}$ with $a_{12} = a_{24} = a_{31} = a_{45} = a_{51} = 1$ and the remaining a_{ij} all 0. Draw the graph and state whether any two cities are connected by the railway line.

7.16 Five football teams (A, B, C, D, and E) join a championship series. There is only one match between any two teams. The results are as follows. A defeats B, C and D. B defeats C and E. C defeats D. D defeats B. E defeats A and D.

(i) Draw a graph of the competition results and give the adjacency matrix $\boldsymbol{M}$.

(ii) Rank the positions of the five football teams.

7.17 At the beginning of the year, one pair of young rabbits (1 male, 1 female) matures in one month, and then breeds another two rabbits (1 male, 1 female) in the following month. In short, every matured pair of rabbits can only reproduce another pair of young rabbits every month. How many *pairs* of rabbits (1 male, 1 female) would we have after 18 months?

7.18 A bee crawls from a honeycomb cell A to cell n (see Figure Q7.18). It can only crawl from one cell to an adjacent cell on its right at each step, and cannot crawl back to the cell it was previously in. How many ways can the bee crawl through the cells of the honeycomb?

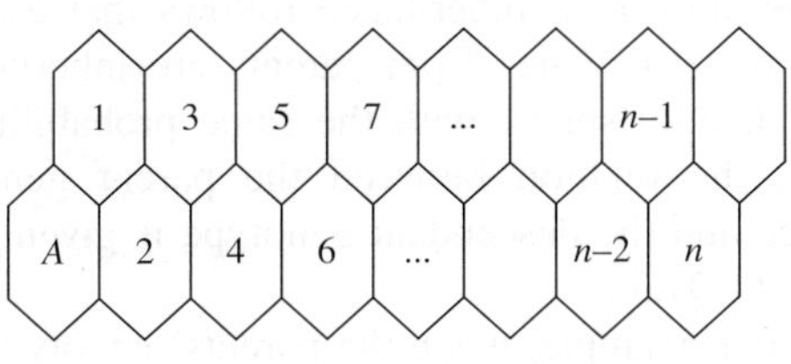

Figure Q7.18

7.19 Solve the following recursion by the matrix method:

$$u_n = u_{n-1} + 2u_{n-2},\ u_1 = 0,\ u_2 = 1.$$

7.20 Suppose the sequence $\{a_k\}$ is $0, \frac{1}{2}, \frac{1}{4}, \frac{3}{8}, \frac{5}{16}, \frac{11}{32}, \ldots$. From the third number, each number is the average of the two numbers just before it, that is,

$$a_{k+2} = \frac{1}{2}(a_{k+1} + a_k), \quad k = 0, 1, 2, \ldots,$$

and $a_0 = 0$, $a_1 = \frac{1}{2}$. Find a_k and the limit of a_k.

7.21 Solve the linear homogeneous system of the first order ordinary differential equations with an initial value.

(i) $\begin{cases} \dfrac{\mathrm{d}\boldsymbol{u}}{\mathrm{d}t} = \boldsymbol{A}\boldsymbol{u}, \\ \boldsymbol{u}(0) = \boldsymbol{u}_0, \end{cases} \quad A = \begin{pmatrix} 1 & 0 & 2 \\ 0 & 1 & 2 \\ 2 & 2 & -1 \end{pmatrix},$

$$\boldsymbol{u}_0 = \begin{pmatrix} 1 \\ 0 \\ 0 \end{pmatrix}.$$

(ii) The equation is the same as equation (i), in which

$$\boldsymbol{A} = \begin{pmatrix} 0 & 0 & 4 & 1 \\ 0 & 0 & 1 & 4 \\ 4 & 1 & 0 & 0 \\ 1 & 4 & 0 & 0 \end{pmatrix}, \quad \boldsymbol{u}(0) = \begin{pmatrix} 1 \\ 0 \\ 1 \\ 0 \end{pmatrix}.$$

7.22 Find the general solution to the following systems of linear ordinary differential equations:

(i) $\begin{cases} \dfrac{\mathrm{d}x_1}{\mathrm{d}t} = 3x_1 + 2x_2 + 4x_3, \\ \dfrac{\mathrm{d}x_2}{\mathrm{d}t} = 2x_1 + 2x_3, \\ \dfrac{\mathrm{d}x_3}{\mathrm{d}t} = 4x_1 + 2x_2 + 3x_3. \end{cases}$

(ii) $\begin{cases} \dfrac{\mathrm{d}x_1}{\mathrm{d}t} = x_1 + 3x_2, \\ \dfrac{\mathrm{d}x_2}{\mathrm{d}t} = 3x_1 + 4x_2 - x_3, \\ \dfrac{\mathrm{d}x_3}{\mathrm{d}t} = -x_2 + x_3. \end{cases}$

7.23 Find the least squares solution to a system of linear equations $\boldsymbol{Ax} = \boldsymbol{b}$, and find the projection vector of vector $\boldsymbol{b}$ on the column space $R(\boldsymbol{A})$ of the matrix $\boldsymbol{A}$, as well as the distance from $\boldsymbol{b}$ to $R(\boldsymbol{A})$.

(i) $\boldsymbol{A} = \begin{pmatrix} 1 & 2 \\ 1 & -1 \\ 0 & 1 \end{pmatrix}, \boldsymbol{b} = \begin{pmatrix} 1 \\ 2 \\ -1 \end{pmatrix}.$

(ii) $\boldsymbol{A} = \begin{pmatrix} 1 & 1 & -1 \\ -1 & -2 & 1 \\ 0 & 1 & -1 \\ 0 & 0 & 1 \end{pmatrix}, \boldsymbol{b} = \begin{pmatrix} 0 \\ 0 \\ 0 \\ 1 \end{pmatrix}.$

7.24 Suppose

$$\boldsymbol{\alpha}_1 = \begin{pmatrix} 1 \\ 1 \\ 1 \\ 1 \end{pmatrix}, \quad \boldsymbol{\alpha}_2 = \begin{pmatrix} 0 \\ 1 \\ 1 \\ 1 \end{pmatrix}, \quad \boldsymbol{\alpha}_3 = \begin{pmatrix} 0 \\ 0 \\ 1 \\ 1 \end{pmatrix},$$

$$\boldsymbol{\alpha}_4 = \begin{pmatrix} 0 \\ 0 \\ 0 \\ 1 \end{pmatrix}, \quad \boldsymbol{b} = \begin{pmatrix} 1 \\ 2 \\ 2 \\ 0 \end{pmatrix}.$$

Find

(i) the projection of vector $\boldsymbol{b}$ on $L(\boldsymbol{\alpha}_1, \boldsymbol{\alpha}_2)$ and the distance from $\boldsymbol{b}$ to $L(\boldsymbol{\alpha}_1, \boldsymbol{\alpha}_2)$,

(ii) the projection of vector $\boldsymbol{b}$ on $L(\boldsymbol{\alpha}_2, \boldsymbol{\alpha}_3, \boldsymbol{\alpha}_4)$ and the distance from $\boldsymbol{b}$ to $L(\boldsymbol{\alpha}_2, \boldsymbol{\alpha}_3, \boldsymbol{\alpha}_4)$.

7.25 Prove that the projection matrix $\boldsymbol{P} = \boldsymbol{A}(\boldsymbol{A}^{\mathrm{T}}\boldsymbol{A})^{-1}\boldsymbol{A}^{\mathrm{T}}$ is an idempotent symmetric matrix, that is, $\boldsymbol{P}^2 = \boldsymbol{P}$, $\boldsymbol{P}^{\mathrm{T}} = \boldsymbol{P}$.

***7.26** Suppose

$$\boldsymbol{A} = \begin{pmatrix} 1 & x_1 & x_1^2 & \cdots & x_1^m \\ 1 & x_2 & x_2^2 & \cdots & x_2^m \\ \vdots & \vdots & \vdots & & \vdots \\ 1 & x_n & x_n^2 & \cdots & x_n^m \end{pmatrix}.$$

Prove that when $n > m$, then $\operatorname{rank}(\boldsymbol{A}) = m + 1$, if and only if, there are at least $(m+1)$ different values of $x_1, x_2, \ldots, x_n$.

7.27 Find a least-squares line that fits these three coordinates: (0, 0), (1, 2), (2, 7).

7.28 Find a least-squares line that fits these four coordinates: (0, 1), (2, 0), (3, 1), (3, 2).

7.29 Apply the least squares method to find a quadratic curve to fit these four coordinates: (2, 0), (3, −10), (5, −48), (6, −76).

7.30 Apply the least squares method to get a cubic polynomial curve to fit the following five coordinates: (−1, −14), (0, −5), (2, −22), (1, −4), (2, 1).

7.31 The turnover curve of a shop in the first five months of a year is a quadratic curve. The turnovers of the first five months are \$40,000, \$44,000, \$52,000, \$64,000 and \$80,000. Use the least squares method to (i) plot the turnover curve, and (ii) forecast the turnover in the month of December.

7.32 Newton's Second Law states that the equation of motion of a body falling vertically near the surface of Earth is

$$s = s_0 + v_0 t + \frac{1}{2} g t^2.$$

Find the approximate value of g by the least squares method through the experimental data given below:

Table Q7.32

t/ second	0.1	0.2	0.3	0.4	0.5
s/ meter	−0.055	0.095	0.314	0.756	1.137

Supplementary exercise

7.33 The external characteristics of an organism is determined by the gene pairs of two genes (A, a), that is, AA, Aa, aa. For example, three gene pairs determine the three colors of a flower. The autosomal inheritance follows that each of the two genes from parents are inherited to their offspring with the same probability. So, the relation between the parent genotype and the descendant genotype is given in Table Q7.33.

For example, when the parents' genotypes are both Aa in column 4, the probability that a descendant is AA, Aa, aa is $\frac{1}{4}, \frac{1}{2}, \frac{1}{4}$ respectively.

Table Q7.33

		Parent (both sides) genotypes					
		AA–AA	*AA–Aa*	*AA–aa*	*Aa–Aa*	*Aa–aa*	*aa–aa*
Descendant genotypes	*AA*	1	$\frac{1}{2}$	0	$\frac{1}{4}$	0	0
	Aa	0	$\frac{1}{2}$	1	$\frac{1}{2}$	$\frac{1}{2}$	0
	Aa	0	0	0	$\frac{1}{4}$	$\frac{1}{2}$	1

(i) Suppose a plant has three genotypes. If they always reproduce with genotype *Aa*, what are the percentages a_n, b_n, c_n of the three genotype plants for n generations? Also, calculate their limits. (Hint: Assume that the initial distribution is a_0, b_0, c_0 and $a_0 + b_0 + c_0 = 1$.)

(ii) If the three genotypes in (i) reproduced with type AA in the beginning, with *Aa* in the first generation, with AA in the second generation, and alternately in the consecutive generations, find a_n, b_n, c_n and their limits.

Appendix

A Inner Product Space, Hermitian Quadratic Form

OVERVIEW

In the linear space $\mathbb{R}^n$, the operations of addition and scalar multiplication as well as the inner product operation provide geometrical metric properties for a vector. The metric properties of a vector are applicable in solving problems of many areas such as analysis and geometry. Therefore, it is necessary for us to define an inner product operation for a general linear space over the real number field or complex number field so that it may become an inner product space. We will discuss real inner product spaces—inner product spaces over the real number field (Euclidean spaces, in particular)—in A.1 and complex inner product spaces (unitary spaces) in A.3.

A.1 Real Inner Product Spaces, Euclidean Spaces

The definition of a linear space consists of abstract concepts such as a set and two kinds of linear operations. The operations are stipulated by properties.

Hence, in an abstract linear space, inner product operations are also stipulated by properties.

Definition A.1

Let V be a linear space over the real number field $\mathbb{R}$. A real number denoted by $(\boldsymbol{\alpha}, \boldsymbol{\beta})$ is defined for any two vectors $\boldsymbol{\alpha}, \boldsymbol{\beta}$ in V, which obeys the following axioms:

(i) $(\boldsymbol{\alpha}, \boldsymbol{\beta}) = (\boldsymbol{\beta}, \boldsymbol{\alpha})$
(ii) $(k\boldsymbol{\alpha}, \boldsymbol{\beta}) = k(\boldsymbol{\alpha}, \boldsymbol{\beta})$
(iii) $(\boldsymbol{\alpha}+\boldsymbol{\beta}, \boldsymbol{\gamma}) = (\boldsymbol{\alpha}, \boldsymbol{\gamma}) + (\boldsymbol{\beta}, \boldsymbol{\gamma})$
(iv) $(\boldsymbol{\alpha}, \boldsymbol{\alpha}) \geq 0$, equality holds, if and only if, $\boldsymbol{\alpha} = 0$,

where $\boldsymbol{\alpha}, \boldsymbol{\beta}, \boldsymbol{\gamma} \in V, k \in \mathbb{R}$. Then, $(\boldsymbol{\alpha}, \boldsymbol{\beta})$ is called an **inner product** of $\boldsymbol{\alpha}$ and $\boldsymbol{\beta}$, while V is called a **real inner product space** if an inner product is defined over the space. A real inner product space with a finite dimension is called an **Euclidean space**.

Property (i) in Definition A.1 shows that the inner product is symmetric. Hence, corresponding to properties (ii) and (iii), we have

$$(\boldsymbol{\alpha}, k\boldsymbol{\beta}) = (k\boldsymbol{\beta}, \boldsymbol{\alpha}) = k(\boldsymbol{\beta}, \boldsymbol{\alpha}) = k(\boldsymbol{\alpha}, \boldsymbol{\beta});$$

$$(\boldsymbol{\alpha}, \boldsymbol{\beta}+\boldsymbol{\gamma}) = (\boldsymbol{\beta}+\boldsymbol{\gamma}, \boldsymbol{\alpha}) = (\boldsymbol{\beta}, \boldsymbol{\alpha}) + (\boldsymbol{\gamma}, \boldsymbol{\alpha}) = (\boldsymbol{\alpha}, \boldsymbol{\beta}) + (\boldsymbol{\alpha}, \boldsymbol{\gamma}).$$

Taking $k = 0$ in property (ii), we get $(\mathbf{0}, \boldsymbol{\beta}) = 0$. Therefore, in a linear space, the inner product of any element with its zero element is zero.

According to property (iv), we can use the inner product to define the length of a vector.

Definition A.2

The length of a vector $\boldsymbol{\alpha}$ in an inner product space V over the real number field $\mathbb{R}$ is defined as

$$\|\boldsymbol{\alpha}\| = \sqrt{(\boldsymbol{\alpha}, \boldsymbol{\alpha})}.$$

Example A.1 Real inner product spaces

Two important real inner product spaces are as follows:

(i) Define an inner product over $\mathbb{R}^n$ *as*

$$(\boldsymbol{\alpha}, \boldsymbol{\beta}) = a_1b_1 + a_2b_2 + \cdots + a_nb_n, \tag{A.1}$$

where $\boldsymbol{\alpha} = (a_1, a_2, \ldots, a_n)^{\mathrm{T}}, \boldsymbol{\beta} = (b_1, b_2, \ldots, b_n)^{\mathrm{T}}$.

It is easy to verify that (A.1) satisfies the four axioms of the inner product. Then, $\mathbb{R}^n$ endowed with this inner product is an n-dimensional Euclidean space. This inner product is called the standard inner product of $\mathbb{R}^n$. The length of a vector $\boldsymbol{\alpha}$ in $\mathbb{R}^n$ is then

$$\|\boldsymbol{\alpha}\| = \sqrt{{a_1}^2 + {a_2}^2 + \cdots + {a_n}^2}.$$

(ii) In the linear space $C[a, b]$ made of all the continuous real-valued functions on interval $[a, b]$, we define $\forall f(x), g(x) \in C[a, b]$,

$$(f, g) = \int_a^b f(x)g(x)\mathrm{d}x. \tag{A.2}$$

(A.2) is an inner product of $f(x)$ and $g(x)$ because it satisfies the four axioms of the inner product. We will now verify that (A.2) satisfies axiom (iii).

Let $f, g, h \in C[a, b]$. Then,

$$\begin{aligned}(f+g, h) &= \int_a^b [f(x)+g(x)]h(x)\,\mathrm{d}x \\ &= \int_a^b [f(x)h(x)+g(x)h(x)]\,\mathrm{d}x \\ &= \int_a^b f(x)h(x)\,\mathrm{d}x + \int_a^b g(x)h(x)\,\mathrm{d}x \\ &= (f, h) + (g, h).\end{aligned}$$

This inner product is called the standard inner product of $C[a, b]$. ■

The methods of defining an inner product in $\mathbb{R}^n$ are not unique; for example, another inner product of $\mathbb{R}^2$ will be given in the following example.

Example A.2 Another inner product of $\mathbb{R}^2$

In $\mathbb{R}^2$, define for any $\boldsymbol{\alpha} = (a_1, a_2)$, $\boldsymbol{\beta} = (b_1, b_2)$,

$$(\boldsymbol{\alpha}, \boldsymbol{\beta}) = a_1b_1 - a_2b_1 - a_1b_2 + 3a_2b_2. \tag{$*$}$$

It is easy to verify that $(\boldsymbol{\alpha}, \boldsymbol{\beta})$ also satisfies the four axioms of the inner product. We verify property (iv) here.

$$(\boldsymbol{\alpha}, \boldsymbol{\alpha}) = a_1{}^2 - 2a_1a_2 + 3a_2{}^2 = (a_1 - a_2)^2 + 2a_2{}^2 \geq 0,$$

and the equality holds, if and only if, $a_1 = a_2 = 0$, that is, $\boldsymbol{\alpha} = \mathbf{0}$.

So, $(*)$ is also an inner product in $\mathbb{R}^2$ with respect to which the length of a vector $\boldsymbol{\alpha}$ is, $\|\boldsymbol{\alpha}\| = \sqrt{(a_1 - a_2)^2 + 2a_2{}^2}$. For this inner product, $\boldsymbol{\alpha} = (1, 0)$ and $\boldsymbol{\beta} = \left(\frac{1}{\sqrt{2}}, \frac{1}{\sqrt{2}}\right)$, we have

$$\|\boldsymbol{\alpha}\| = \|\boldsymbol{\beta}\| = 1, \quad (\boldsymbol{\alpha}, \boldsymbol{\beta}) = 0.$$

■

The distance in a real inner product space has the same geometric interpretation as that in the standard inner product in $\mathbb{R}^n$.

THEOREM A.1

If V is a real inner product space, then the following hold for any $\boldsymbol{\alpha}$, $\boldsymbol{\beta} \in V$, and $\lambda \in \mathbb{R}$.

(i) $\|\lambda\boldsymbol{\alpha}\| = |\lambda| \, \|\boldsymbol{\alpha}\|$,
(ii) $|(\boldsymbol{\alpha}, \boldsymbol{\beta})| \leq \|\boldsymbol{\alpha}\| \, \|\boldsymbol{\beta}\|$ *(**Cauchy-Schwarz inequality**)*,
(iii) $\|\boldsymbol{\alpha} + \boldsymbol{\beta}\| \leq \|\boldsymbol{\alpha}\| + \|\boldsymbol{\beta}\|$ *(**triangle inequality**)*.

Proof:

(i) $\|\lambda\boldsymbol{\alpha}\| = \sqrt{(\lambda\boldsymbol{\alpha}, \lambda\boldsymbol{\alpha})} = \sqrt{\lambda^2(\boldsymbol{\alpha}, \boldsymbol{\alpha})} = |\lambda| \sqrt{(\boldsymbol{\alpha}, \boldsymbol{\alpha})} = |\lambda| \, \|\boldsymbol{\alpha}\|$.

The proofs of (ii) and (iii) are similar to those of the same conclusions in §4.2.

Applying Cauchy-Schwarz inequality to the inner product space $C[a, b]$, we get the following inequality

$$\left|\int_a^b f(x)g(x)\,\mathrm{d}x\right| \leq \left(\int_a^b f^2(x)\,\mathrm{d}x\right)^{1/2} \left(\int_a^b g^2(x)\,\mathrm{d}x\right)^{1/2}.$$

Having proved Cauchy-Schwarz inequality, we can use an inner product to define the included angle between two vectors.

Definition A.3

For two nonzero vectors $\boldsymbol{\alpha}$, $\boldsymbol{\beta}$ in a real inner product space, we define their included angle as

$$\langle \boldsymbol{\alpha}, \boldsymbol{\beta} \rangle = \arccos \frac{(\boldsymbol{\alpha}, \boldsymbol{\beta})}{\|\boldsymbol{\alpha}\| \, \|\boldsymbol{\beta}\|}.$$

Obviously, two nonzero vectors are perpendicular, if and only if, $(\boldsymbol{\alpha},\ \boldsymbol{\beta})=0$. Since the inner product of the zero vector with any vector is zero, we say that the zero vector is orthogonal with any vector, thus we have the following definition.

Definition A.4

If $(\boldsymbol{\alpha},\ \boldsymbol{\beta})=0$ for two vectors $\boldsymbol{\alpha}$, $\boldsymbol{\beta}$ in a real inner product space V, then $\boldsymbol{\alpha}$ and $\boldsymbol{\beta}$ are said to be **orthogonal**, and it is denoted as $\boldsymbol{\alpha}\perp\boldsymbol{\beta}$.

Example A.3 Orthogonal functions under the standard inner product of $C[-1, 1]$

For the standard inner product in $C[-1,\ 1]$, prove that the functions

$$P_1(x)=x, \quad P_2(x)=\frac{1}{2}(3x^2-1)$$

are orthogonal and calculate their lengths.

Proof:

$$(P_1,\ P_2)=\int_{-1}^{1} P_1(x)P_2(x)\,\mathrm{d}x=\frac{1}{2}\int_{-1}^{1}(3x^3-x)\,\mathrm{d}x=0$$

So, P_1 and P_2 are orthogonal.

The length of P_1 is

$$\|P_1\|=\sqrt{\int_{-1}^{1} P_1^2(x)\,\mathrm{d}x}=\sqrt{\int_{-1}^{1} x^2\,\mathrm{d}x}=\sqrt{\frac{2}{3}}.$$

The length of P_2 is

$$\|P_2\|=\sqrt{\int_{-1}^{1} P_2^2(x)\,\mathrm{d}x}=\sqrt{\frac{2}{5}}.$$

■

A.2 Metric Matrices and Orthonormal Bases

For an n-dimensional real linear space V, in order to determine the inner product $(\boldsymbol{\alpha},\ \boldsymbol{\beta})$, we only need to determine the inner product of a basis.

Let $\{\boldsymbol{\varepsilon}_1,\ \boldsymbol{\varepsilon}_2, \ldots, \boldsymbol{\varepsilon}_n\}$ be a basis of V, and

$$(\boldsymbol{\varepsilon}_i,\ \boldsymbol{\varepsilon}_j)=a_{ij},\ i,\ j=1,\ 2, \ldots, n. \tag{A.3}$$

For any two vectors (elements) in V,

$$\boldsymbol{\alpha} = x_1\boldsymbol{\varepsilon}_1 + x_2\boldsymbol{\varepsilon}_2 + \cdots + x_n\boldsymbol{\varepsilon}_n, \quad \boldsymbol{\beta} = y_1\boldsymbol{\varepsilon}_1 + y_2\boldsymbol{\varepsilon}_2 + \cdots + y_n\boldsymbol{\varepsilon}_n.$$

From the properties of the inner product, we get

$$\begin{aligned}
(\boldsymbol{\alpha}, \boldsymbol{\beta}) &= (x_1\boldsymbol{\varepsilon}_1 + x_2\boldsymbol{\varepsilon}_2 + \cdots + x_n\boldsymbol{\varepsilon}_n, y_1\boldsymbol{\varepsilon}_1 + y_2\boldsymbol{\varepsilon}_2 + \cdots + y_n\boldsymbol{\varepsilon}_n) \\
&= \sum_{i=1}^{n}\sum_{j=1}^{n} (\boldsymbol{\varepsilon}_i, \boldsymbol{\varepsilon}_j)x_i y_j = \sum_{i=1}^{n}\sum_{j=1}^{n} a_{ij}x_i y_j = \sum_{i=1}^{n} x_i \sum_{j=1}^{n} a_{ij}y_j \\
&= \sum_{i=1}^{n} x_i (a_{i1}, a_{i2}, \ldots, a_{in}) \begin{pmatrix} y_1 \\ y_2 \\ \vdots \\ y_n \end{pmatrix} \\
&= (x_1, x_2, \ldots, x_n) \begin{pmatrix} a_{11} & a_{12} & \cdots & a_{1n} \\ a_{21} & a_{22} & \cdots & a_{2n} \\ \vdots & \vdots & & \vdots \\ a_{n1} & a_{n2} & \cdots & a_{nn} \end{pmatrix} \begin{pmatrix} y_1 \\ y_2 \\ \vdots \\ y_n \end{pmatrix} = \boldsymbol{x}^{\mathrm{T}}\boldsymbol{A}\boldsymbol{y}, \qquad \text{(A.4)}
\end{aligned}$$

where $\boldsymbol{x} = (x_1, x_2, \ldots, x_n)^{\mathrm{T}}$ and $\boldsymbol{y} = (y_1, y_2, \ldots, y_n)^{\mathrm{T}}$ are the coordinate vectors of $\boldsymbol{\alpha}, \boldsymbol{\beta}$ respectively under the basis $\{\boldsymbol{\varepsilon}_1, \boldsymbol{\varepsilon}_2, \ldots, \boldsymbol{\varepsilon}_n\}$.

$$\boldsymbol{A} = (a_{ij})_{n\times n} = ((\boldsymbol{\varepsilon}_i, \boldsymbol{\varepsilon}_j))_{n\times n} \qquad \text{(A.5)}$$

is called the metric matrix of an Euclidean space under the basis $\{\boldsymbol{\varepsilon}_1, \boldsymbol{\varepsilon}_2, \ldots, \boldsymbol{\varepsilon}_n\}$, or the metric matrix of $\{\boldsymbol{\varepsilon}_1, \boldsymbol{\varepsilon}_2, \ldots, \boldsymbol{\varepsilon}_n\}$ in short. Since an inner product has symmetric property, then a metric matrix $\boldsymbol{A}$ is real and symmetric. From the above discussions, we can calculate the inner product of any two vectors $\boldsymbol{\alpha}$ and $\boldsymbol{\beta}$ by their coordinate vectors according to $\boldsymbol{x}^{\mathrm{T}}\boldsymbol{A}\boldsymbol{y}$ as in (A.4) when the metric matrix $\boldsymbol{A}$ with respect to a basis is known. Thus, the metric matrix determines the inner product completely.

THEOREM A.2

The two metric matrices of two different bases in an Euclidean space are congruent.

Proof: Let the metric matrix of a basis $\{\boldsymbol{\varepsilon}_1, \boldsymbol{\varepsilon}_2, \ldots, \boldsymbol{\varepsilon}_n\}$ be $\boldsymbol{A}$ and let us formally express $\boldsymbol{A}$ in (A.5) as a product of matrices (it is a matrix form and the vectors may not be in vector space $\mathbb{R}^n$), that is,

$$\boldsymbol{A} = \begin{pmatrix} \boldsymbol{\varepsilon}_1 \\ \boldsymbol{\varepsilon}_2 \\ \vdots \\ \boldsymbol{\varepsilon}_n \end{pmatrix} (\boldsymbol{\varepsilon}_1, \boldsymbol{\varepsilon}_2, \ldots, \boldsymbol{\varepsilon}_n). \qquad ①$$

Here, the product of $\boldsymbol{\varepsilon}_i$ and $\boldsymbol{\varepsilon}_j$ is the inner product of $(\boldsymbol{\varepsilon}_i, \boldsymbol{\varepsilon}_j)$.

Let the metric matrix of another basis $\{\boldsymbol{\eta}_1, \boldsymbol{\eta}_2, \ldots, \boldsymbol{\eta}_n\}$ be

$$\boldsymbol{B} = \begin{pmatrix} \boldsymbol{\eta}_1 \\ \boldsymbol{\eta}_2 \\ \vdots \\ \boldsymbol{\eta}_n \end{pmatrix} (\boldsymbol{\eta}_1, \boldsymbol{\eta}_2, \ldots, \boldsymbol{\eta}_n). \qquad ②$$

Thc transition matrix from basis $\{\boldsymbol{\varepsilon}_1, \boldsymbol{\varepsilon}_2, \ldots, \boldsymbol{\varepsilon}_n\}$ to basis $\{\boldsymbol{\eta}_1, \boldsymbol{\eta}_2, \ldots, \boldsymbol{\eta}_n\}$ is $\boldsymbol{C}$, that is,

$$(\boldsymbol{\eta}_1, \boldsymbol{\eta}_2, \ldots, \boldsymbol{\eta}_n) = (\boldsymbol{\varepsilon}_1, \boldsymbol{\varepsilon}_2, \ldots, \boldsymbol{\varepsilon}_n)\boldsymbol{C}. \qquad ③$$

Substituting ③ into ② yields

$$\boldsymbol{B} = \boldsymbol{C}^{\mathrm{T}} \begin{pmatrix} \boldsymbol{\varepsilon}_1 \\ \boldsymbol{\varepsilon}_2 \\ \vdots \\ \boldsymbol{\varepsilon}_n \end{pmatrix} (\boldsymbol{\varepsilon}_1, \boldsymbol{\varepsilon}_2, \ldots, \boldsymbol{\varepsilon}_n)\boldsymbol{C} = \boldsymbol{C}^{\mathrm{T}}\boldsymbol{A}\boldsymbol{C}. \qquad ④$$

So, $\boldsymbol{A}$ and $\boldsymbol{B}$ are congruent. □

Example A.4 Finding the metric matrix of a given basis and hence the standard inner product

Let $\boldsymbol{\eta}_1 = (1, 0, 0, 0)^{\mathrm{T}}$, $\boldsymbol{\eta}_2 = (1, 1, 0, 0)^{\mathrm{T}}$, $\boldsymbol{\eta}_3 = (1, 1, 1, 0)^{\mathrm{T}}$, and $\boldsymbol{\eta}_4 = (1, 1, 1, 1)^{\mathrm{T}}$. Find the metric matrix of this basis and calculate the standard inner product $(\boldsymbol{\alpha}, \boldsymbol{\beta})$ of

$$\boldsymbol{\alpha} = \boldsymbol{\eta}_1 + \boldsymbol{\eta}_2 + \boldsymbol{\eta}_3 + \boldsymbol{\eta}_4, \quad \boldsymbol{\beta} = \boldsymbol{\eta}_1 - \boldsymbol{\eta}_2 + \boldsymbol{\eta}_3.$$

Solution: Obviously, $\{\boldsymbol{\eta}_1, \boldsymbol{\eta}_2, \boldsymbol{\eta}_3, \boldsymbol{\eta}_4\}$ is a basis of $\mathbb{R}^4$, so the metric matrix under the basis $\{\boldsymbol{\eta}_1, \boldsymbol{\eta}_2, \boldsymbol{\eta}_3, \boldsymbol{\eta}_4\}$ is

$$A = \begin{pmatrix} (\eta_1, \eta_1) & (\eta_1, \eta_2) & (\eta_1, \eta_3) & (\eta_1, \eta_4) \\ (\eta_2, \eta_1) & (\eta_2, \eta_2) & (\eta_2, \eta_3) & (\eta_2, \eta_4) \\ (\eta_3, \eta_1) & (\eta_3, \eta_2) & (\eta_3, \eta_3) & (\eta_3, \eta_4) \\ (\eta_4, \eta_1) & (\eta_4, \eta_2) & (\eta_4, \eta_3) & (\eta_4, \eta_4) \end{pmatrix} = \begin{pmatrix} 1 & 1 & 1 & 1 \\ 1 & 2 & 2 & 2 \\ 1 & 2 & 3 & 3 \\ 1 & 2 & 3 & 4 \end{pmatrix}.$$

Hence, we get

$$(\boldsymbol{\alpha}, \boldsymbol{\beta}) = (1, \ 1, \ 1, \ 1)\, A \begin{pmatrix} 1 \\ -1 \\ 1 \\ 0 \end{pmatrix} = 6.$$

We can also use Theorem A.2 to solve this problem. Let $\{\boldsymbol{\varepsilon}_1, \boldsymbol{\varepsilon}_2, \boldsymbol{\varepsilon}_3, \boldsymbol{\varepsilon}_4\}$ be the natural basis. Then, the transition matrix from $\{\boldsymbol{\varepsilon}_1, \boldsymbol{\varepsilon}_2, \boldsymbol{\varepsilon}_3, \boldsymbol{\varepsilon}_4\}$ to $\{\boldsymbol{\eta}_1, \boldsymbol{\eta}_2, \boldsymbol{\eta}_3, \boldsymbol{\eta}_4\}$ is

$$\boldsymbol{C} = \begin{pmatrix} 1 & 1 & 1 & 1 \\ 0 & 1 & 1 & 1 \\ 0 & 0 & 1 & 1 \\ 0 & 0 & 0 & 1 \end{pmatrix}.$$

We know that the metric matrix of the natural basis is $\boldsymbol{I}$, so the metric matrix of basis $\{\boldsymbol{\eta}_1, \boldsymbol{\eta}_2, \boldsymbol{\eta}_3, \boldsymbol{\eta}_4\}$ is $\boldsymbol{A} = \boldsymbol{C}^{\mathrm{T}}\boldsymbol{I}\boldsymbol{C} = \boldsymbol{C}^{\mathrm{T}}\boldsymbol{C}$. Therefore,

$$\begin{aligned}(\boldsymbol{\alpha}, \boldsymbol{\beta}) &= (\ 1,\ \ 1,\ \ 1,\ \ 1\ )\boldsymbol{A}\begin{pmatrix} 1 \\ -1 \\ 1 \\ 0 \end{pmatrix} \\ &= (\ 1,\ \ 1,\ \ 1,\ \ 1\ )\boldsymbol{C}^{\mathrm{T}}\boldsymbol{C}\begin{pmatrix} 1 \\ -1 \\ 1 \\ 0 \end{pmatrix} = 6.\end{aligned}$$

We can use a simpler and easier method by calculating $(\boldsymbol{\alpha}, \boldsymbol{\beta}) = 4 \times 1 + 2 \times 1 = 6$, which is from $\boldsymbol{\alpha} = \boldsymbol{\eta}_1 + \boldsymbol{\eta}_2 + \boldsymbol{\eta}_3 + \boldsymbol{\eta}_4 = (4,\ 3,\ 2,\ 1)$, $\boldsymbol{\beta} = \boldsymbol{\eta}_1 - \boldsymbol{\eta}_2 + \boldsymbol{\eta}_3 = (1,\ 0,\ 1,\ 0)$. ■

In addition, we have

$$(\boldsymbol{\alpha}, \boldsymbol{\alpha}) = \boldsymbol{x}^{\mathrm{T}}\boldsymbol{A}\boldsymbol{x} > 0,$$

for any nonzero vector $\boldsymbol{\alpha}$ (i.e., the coordinate vector of $\boldsymbol{\alpha}$ under basis $\{\boldsymbol{\varepsilon}_1, \boldsymbol{\varepsilon}_2, \boldsymbol{\varepsilon}_3, \boldsymbol{\varepsilon}_4\}$ is $\boldsymbol{x} \neq \boldsymbol{0}$) according to property (iv) of the inner product (A.6). Therefore, the metric matrix $\boldsymbol{A}$ is positive definite.

We have shown in §6.4 that a positive definite matrix must be congruent to the identity matrix, that is, for a positive definite matrix $\boldsymbol{A}$ there exists an invertible matrix $\boldsymbol{C}$, such that

$$\boldsymbol{C}^T\boldsymbol{A}\boldsymbol{C} = \boldsymbol{I}.$$

Therefore, there exists a basis $\{\boldsymbol{\alpha}_1, \boldsymbol{\alpha}_2, \ldots, \boldsymbol{\alpha}_n\}$ in Euclidean space from Theorem A.2 that satisfies the relation

$$(\boldsymbol{\alpha}_1, \boldsymbol{\alpha}_2, \ldots, \boldsymbol{\alpha}_n) = (\boldsymbol{\varepsilon}_1, \boldsymbol{\varepsilon}_2, \ldots, \boldsymbol{\varepsilon}_n)\boldsymbol{C},$$

and the metric matrix of the basis is the identity matrix, namely,

$$\begin{pmatrix} (\alpha_1, \alpha_1) & (\alpha_1, \alpha_2) & \cdots & (\alpha_1, \alpha_n) \\ (\alpha_2, \alpha_1) & (\alpha_2, \alpha_2) & \cdots & (\alpha_2, \alpha_n) \\ \vdots & \vdots & & \vdots \\ (\alpha_n, \alpha_1) & (\alpha_n, \alpha_2) & \cdots & (\alpha_n, \alpha_n) \end{pmatrix} = \begin{pmatrix} 1 & & & \\ & 1 & & \\ & & \ddots & \\ & & & 1 \end{pmatrix}. \tag{A.7}$$

So,

$$(\alpha_i, \alpha_j) = \begin{cases} 1, j = i, \\ 0, j \neq i, \end{cases} \quad i, j = 1, 2, \ldots, n. \tag{A.8}$$

Definition A.5

Let $\alpha_1, \alpha_2, \ldots, \alpha_n$ be n vectors in an n-dimensional Euclidean space V. If any α_i, α_j satisfy (A.8), then $\{\alpha_1, \alpha_2, \ldots, \alpha_n\}$ is called an **orthonormal basis**.

Since the metric matrix of a standard orthogonal basis is the identity matrix, the inner product of α, β is $(\alpha, \beta) = x^{\mathrm{T}} y$, where x and y are the corresponding coordinate vectors under an orthonormal basis. The previous inner product defined in $\mathbb{R}^n$ is defined under the natural orthonormal basis $\{\varepsilon_1, \varepsilon_2, \ldots, \varepsilon_n\}$. Therefore, we also call it the standard inner product.

From the discussions above, we know that there exists an orthonormal basis in a general Euclidean space. To find an orthonormal basis, we still use the Gram-Schmidt orthogonalization procedure. It is easy for readers to prove that the method is applicable to general Euclidean spaces.

Example A.5 Finding an orthonormal basis of $\mathbb{R}[x]_3$

In $\mathbb{R}[x]_3 = \{a_0 + a_1 x + a_2 x^2 \mid a_0, a_1, a_2 \in \mathbb{R}\}$, define an inner product

$$(f, g) = \int_0^1 f(x) g(x)\, \mathrm{d}x.$$

Find an orthonormal basis of $\mathbb{R}[x]_3$ on interval [0, 1].

Solution: It's easy to verify that $f_0 = 1$, $f_1 = x$, $f_2 = x^2$ is a basis of $\mathbb{R}[x]_3$. One can derive an orthonormal basis of $\mathbb{R}[x]_3$ from this basis by using Gram-Schmidt's orthogonalization.

First, orthogonalize in the following manner:

$$g_0 = f_0 = 1, \quad \|g_0\|^2 = (g_0, g_0) = \int_0^1 1^2\, \mathrm{d}x = 1,$$

$$g_1 = f_1 - \frac{(f_1, g_0)}{(g_0, g_0)} g_0 = x - \frac{1}{1}\left(\int_0^1 x \cdot 1\, \mathrm{d}x\right) \cdot 1 = x - \frac{1}{2},$$

$$\|g_1\|^2=(g_1,\ g_1)=\int_0^1\left(x-\frac{1}{2}\right)^2\mathrm{d}x=\frac{1}{3}\left(x-\frac{1}{2}\right)^3\Big|_0^1=\frac{1}{12},$$

$$g_2=f_2-\frac{(f_2,\ g_1)}{(g_1,\ g_1)}g_1-\frac{(f_2,\ g_0)}{(g_0,\ g_0)}g_0$$

$$=x^2-12\left(x-\frac{1}{2}\right)\int_0^1 x^2\left(x-\frac{1}{2}\right)\mathrm{d}x-\int_0^1 x^2\cdot 1\,\mathrm{d}x$$

$$=x^2-\left(x-\frac{1}{2}\right)-\frac{1}{3}=x^2-x+\frac{1}{6},$$

$$\|g_2\|^2=(g_2,\ g_2)=\int_0^1\left(x^2-x+\frac{1}{6}\right)^2\mathrm{d}x=\frac{1}{180}.$$

Then, normalize them,

$$p_0=\frac{g_0}{\|g_0\|}=1,$$

$$p_1=\frac{g_1}{\|g_1\|}=\sqrt{3}(2x-1),$$

$$p_2=\frac{g_2}{\|g_2\|}=\sqrt{5}(6x^2-6x+1),$$

and p_0, p_1, and p_2 are the elements of an orthonormal basis of $\mathbb{R}[x]_3$. It is a set of orthonormal polynomials of $\mathbb{R}[x]_3$ with respect to the inner product in (A.2). ■

A.3 Inner Product of Complex Vectors, Unitary Space

In this section, we will introduce the inner product in linear spaces over the complex field and unitary spaces.

Definition A.6

Let V be a linear space over the complex field $\mathbb{C}$. A complex number $(\boldsymbol{\alpha},\ \boldsymbol{\beta})$ is defined for any two vectors $\boldsymbol{\alpha}$, $\boldsymbol{\beta}$ in V, which obeys the following axioms:

(i) $(\boldsymbol{\alpha},\ \boldsymbol{\beta})=\overline{(\boldsymbol{\beta},\ \boldsymbol{\alpha})}$;
(ii) $(k\boldsymbol{\alpha},\ \boldsymbol{\beta})=k(\boldsymbol{\alpha},\ \boldsymbol{\beta})$;
(iii) $(\boldsymbol{\alpha}+\boldsymbol{\beta},\boldsymbol{\gamma})=(\boldsymbol{\alpha},\ \boldsymbol{\gamma})+(\boldsymbol{\beta},\ \boldsymbol{\gamma})$;
(iv) $(\boldsymbol{\alpha},\ \boldsymbol{\alpha})\geq 0$, the equality holds, if and only if, $\boldsymbol{\alpha}=0$,

(where $\boldsymbol{\alpha},\ \boldsymbol{\beta},\ \boldsymbol{\gamma} \in V,\ k \in \mathbb{C}$). Then, $(\boldsymbol{\alpha},\ \boldsymbol{\beta})$ is called an inner product of $\boldsymbol{\alpha}$ and $\boldsymbol{\beta}$ of V over the complex field F. The linear space V over the complex field with the above inner product is called a **unitary space**.

It is important to remember that $(\boldsymbol{\alpha},\ k\boldsymbol{\beta}) = \overline{(k\boldsymbol{\beta},\ \boldsymbol{\alpha})} = \overline{k(\boldsymbol{\beta},\ \boldsymbol{\alpha})} = \bar{k}(\boldsymbol{\alpha},\ \boldsymbol{\beta})$

$$(\mathbf{0},\ \boldsymbol{\beta}) = (0\boldsymbol{\alpha},\ \boldsymbol{\beta}) = 0(\boldsymbol{\alpha},\ \boldsymbol{\beta}) = 0.$$

Example A.6 Verification of the axioms of inner product of complex vectors

Let column vectors $\boldsymbol{\alpha} = (a_1,\ a_2, \ldots, a_n)^{\mathrm{T}}$ and $\boldsymbol{\beta} = (b_1,\ b_2, \ldots, b_n)^{\mathrm{T}}$ be two vectors in the n-dimensional complex vector space $\mathbb{C}^n$, and define the operation among $\boldsymbol{\alpha}$ and $\boldsymbol{\beta}$ as,

$$(\boldsymbol{\alpha},\ \boldsymbol{\beta}) = a_1\overline{b_1} + a_2\overline{b_2} + \cdots + a_n\overline{b_n}. \tag{A.9}$$

Then, the operation defined by (A.9) is an inner product.

Solution: Obviously, $(\boldsymbol{\alpha},\ \boldsymbol{\beta}) = \overline{\boldsymbol{\beta}}^{\mathrm{T}}\boldsymbol{\alpha}$. From (A.9), we get

$$\overline{(\boldsymbol{\beta},\ \boldsymbol{\alpha})} = \overline{\overline{a_1}b_1 + \overline{a_2}b_2 + \cdots + \overline{a_n}b_n} = \overline{b_1}a_1 + \overline{b_2}a_2 + \cdots + \overline{b_n}a_n = (\boldsymbol{\alpha}, \boldsymbol{\beta}).$$

Hence, axiom (i) holds.

The verifications of axioms (ii) and (iii) are comparatively easy. Here, we verify axiom (iv). Since

$$(\boldsymbol{\alpha},\ \boldsymbol{\alpha}) = a_1\overline{a_1} + a_2\overline{a_2} + \cdots + a_n\overline{a_n},$$

where $a_i\overline{a_i} = |a_i|^2 \geq 0,\ i = 1,\ 2, \ldots, n$. So, $(\boldsymbol{\alpha},\ \boldsymbol{\alpha}) \geq 0$ and $(\boldsymbol{\alpha},\ \boldsymbol{\alpha}) = 0$, if and only if,

$$|a_i| = 0,\ i = 1,\ 2, \ldots, n, \text{ that is, } \boldsymbol{\alpha} = (a_1,\ a_2, \ldots, a_n)^{\mathrm{T}} = \mathbf{0}. \quad \blacksquare$$

From axiom (iv) above, we can define the length of complex vectors.

Definition A.7

Let $\boldsymbol{\alpha}$ be a vector in the unitary space V, the length of $\boldsymbol{\alpha}$ is defined as

$$\|\boldsymbol{\alpha}\| = \sqrt{(\boldsymbol{\alpha}, \boldsymbol{\alpha})}.$$

Henceforth, we will use $|\cdots|$ to express modulus of a complex number, and $\|\cdots\|$ to express the **length** (usually called **norm**) **of a complex vector**.

THEOREM A.3

For any two vectors $\boldsymbol{\alpha},\ \boldsymbol{\beta} \in \mathbb{C}^n$, we have

(i) $|(\boldsymbol{\alpha},\ \boldsymbol{\beta})| \le \|\boldsymbol{\alpha}\|\ \|\boldsymbol{\beta}\|$ (**Cauchy-Schwarz inequality**),
(ii) $\|\boldsymbol{\alpha}+\boldsymbol{\beta}\| \le \|\boldsymbol{\alpha}\| + \|\boldsymbol{\beta}\|$ (**triangle inequality**).

Proof:

(i) When $\boldsymbol{\beta} = \mathbf{0}$, the Cauchy-Schwarz inequality is true obviously. When $\boldsymbol{\beta} \neq \mathbf{0}$, we have

$$(\boldsymbol{\alpha}+k\boldsymbol{\beta},\ \boldsymbol{\alpha}+k\boldsymbol{\beta}) = (\boldsymbol{\alpha},\ \boldsymbol{\alpha}) + |k|^2(\boldsymbol{\beta},\ \boldsymbol{\beta}) + k(\boldsymbol{\beta},\ \boldsymbol{\alpha}) + \bar{k}(\boldsymbol{\alpha}, \boldsymbol{\beta}).$$

Since $\bar{k}(\boldsymbol{\alpha},\ \boldsymbol{\beta}) = \bar{k}\ \overline{(\boldsymbol{\beta},\ \boldsymbol{\alpha})} = \overline{k\ (\boldsymbol{\beta},\ \boldsymbol{\alpha})}$, the sum of conjugate complex numbers is equal to the sum of their real parts, that is, $z+\bar{z} = 2\text{Re}\ z$. So, the above expression can be expressed as

$$\begin{aligned}(\boldsymbol{\alpha}+k\boldsymbol{\beta},\ \boldsymbol{\alpha}+k\boldsymbol{\beta}) &= (\boldsymbol{\alpha},\ \boldsymbol{\alpha}) + |k|^2\,(\boldsymbol{\beta},\ \boldsymbol{\beta}) + 2\text{Re}\left\{\bar{k}(\boldsymbol{\alpha},\ \boldsymbol{\beta})\right\} \\ &= \|\boldsymbol{\alpha}\|^2 + |k|^2\,\|\boldsymbol{\beta}\|^2 + 2\text{Re}\left\{\bar{k}(\boldsymbol{\alpha},\ \boldsymbol{\beta})\right\} \ge 0. \qquad \text{(A.10)}\end{aligned}$$

Suppose

$$k = -\frac{(\boldsymbol{\alpha},\ \boldsymbol{\beta})}{\|\boldsymbol{\beta}\|^2}, \qquad ①$$

then

$$|k|^2 = k\bar{k} = \frac{|(\boldsymbol{\alpha},\ \boldsymbol{\beta})|^2}{\|\boldsymbol{\beta}\|^4}, \qquad ②$$

$$\bar{k}(\boldsymbol{\alpha},\ \boldsymbol{\beta}) = -\frac{\overline{(\boldsymbol{\alpha}, \boldsymbol{\beta})}}{\|\boldsymbol{\beta}\|^2}(\boldsymbol{\alpha}, \boldsymbol{\beta}) = -\frac{|(\boldsymbol{\alpha},\ \boldsymbol{\beta})|^2}{\|\boldsymbol{\beta}\|^2}. \qquad ③$$

Substituting ② and ③ into (A.10), it follows that

$$\|\boldsymbol{\alpha}\|^2\,\|\boldsymbol{\beta}\|^2 \ge |(\boldsymbol{\alpha},\ \boldsymbol{\beta})|^2,$$

so

$$|(\boldsymbol{\alpha},\ \boldsymbol{\beta})| \le \|\boldsymbol{\alpha}\|\ \|\boldsymbol{\beta}\|.$$

The equality sign holds, if and only if, $\boldsymbol{\alpha}+k\boldsymbol{\beta} = \mathbf{0}$, that is, $\boldsymbol{\alpha}$ and $\boldsymbol{\beta}$ are linearly dependent.

(ii) Let $k=1$ in (A.10). With $\text{Re}\,\{(\boldsymbol{\alpha},\ \boldsymbol{\beta})\} \le |(\boldsymbol{\alpha},\ \boldsymbol{\beta})| \le \|\boldsymbol{\alpha}\|\ \|\boldsymbol{\beta}\|$, we obtain

$$\|\boldsymbol{\alpha}+\boldsymbol{\beta}\|^2 = (\boldsymbol{\alpha}+\boldsymbol{\beta},\ \boldsymbol{\alpha}+\boldsymbol{\beta}) \le \|\boldsymbol{\alpha}\|^2 + \|\boldsymbol{\beta}\|^2 + 2\,\|\boldsymbol{\alpha}\|\ \|\boldsymbol{\beta}\|.$$

So,

$$\|\alpha+\beta\| \le \|\alpha\| + \|\beta\| . \qquad \square$$

By virtue of the Cauchy-Schwarz inequality, a definition of the included angle of nonzero vectors α and β can be given.

Definition A.8

For two nonzero vectors α, β in a unitary space, define the included angle $\langle\alpha, \beta\rangle$ as

$$\langle\alpha,\ \beta\rangle = \arccos\frac{(\alpha,\ \beta)}{\|\alpha\|\ \|\beta\|},\qquad 0 \le \langle\alpha,\ \beta\rangle \le \pi.$$

If $\langle\alpha,\ \beta\rangle = 0$, then α and β are said to be **orthogonal** or **perpendicular to each other** and is denoted as $\alpha \perp \beta$.

When α and β are orthogonal, taking $k=1$ in expression (A.10), we get the Pythagorean theorem

$$\|\alpha+\beta\|^2 = \|\alpha\|^2 + \|\beta\|^2 .$$

With the same argument as in the case of a finite-dimensional Euclidean space, one can show that a finite-dimensional unitary space also has the metric matrices of base and orthonormal bases. An orthonormal basis can be obtained from a basis by Gram-Schmidt orthogonalization. However, we must pay attention to the fact that in this case, the elements of the metric matrix $A = (a_{ij})$ of a basis $\{\varepsilon_1,\ \varepsilon_2, \ldots, \varepsilon_n\}$ satisfy

$$a_{ji} = (\varepsilon_j,\ \varepsilon_i) = \overline{(\varepsilon_i,\ \varepsilon_j)} = \overline{a_{ij}},\ i, j = 1,\ 2, \ldots, n,$$

so

$$\overline{A}^{\mathrm{T}} = A. \tag{A.11}$$

The metric matrix A of a basis in a unitary space is called a **Hermitian matrix**.

In addition, if the coordinate vectors of α and β under the basis $\{\varepsilon_1,\ \varepsilon_2, \ldots, \varepsilon_n\}$ are x and y ($x,\ y \in \mathbb{C}^n$), respectively, then

$$(\alpha,\ \beta) = \overline{y}^{\mathrm{T}} A x, \tag{A.12}$$

$$(\alpha,\ \alpha) = \overline{x}^{\mathrm{T}} A x = (Ax,\ x). \tag{A.13}$$

The expression (A.13) is called a **Hermitian quadratic form** of complex variables $x_1,\ x_2, \ldots, x_n$.

A.4 Unitary Matrices and Hermitian Quadratic Forms

Definition A.9

If an nth-order complex matrix $\boldsymbol{U}$ satisfies

$$\overline{\boldsymbol{U}}^{\mathrm{T}}\boldsymbol{U}=\boldsymbol{I}, \tag{A.14}$$

then $\boldsymbol{U}$ is called a unitary matrix.

A unitary matrix has the following properties (the proof is left to readers as an exercise).

(i) The modulus of the determinant of a unitary matrix is 1, that is, $|\det \boldsymbol{U}|=1$.
(ii) $\boldsymbol{U}^{-1}=\overline{\boldsymbol{U}}^{\mathrm{T}}$, but $\overline{\boldsymbol{U}}^{\mathrm{T}}$ is still a unitary matrix.
(iii) The product of two unitary matrices is still a unitary matrix.
(iv) The set of column vectors and set of row vectors of a unitary matrix are both orthonormal bases of $\mathbb{C}^n$.

Definition A.10

If $\boldsymbol{A}=(a_{ij})$ is a Hermitian matrix of order n ($\overline{\boldsymbol{A}}^{\mathrm{T}}=\boldsymbol{A}$), then

$$\overline{\boldsymbol{x}}^{\mathrm{T}}\boldsymbol{A}\boldsymbol{x}=\sum_{i=1}^{n}\sum_{j=1}^{n}a_{ij}\overline{x_i}x_j \quad (a_{ij}=\overline{a_{ji}}), \tag{A.15}$$

is called a Hermitian quadratic form over $x_1, x_2, \ldots, x_n$.

For any $\boldsymbol{x}\in\mathbb{C}^n$, the Hermitian quadratic form $\overline{\boldsymbol{x}}^{\mathrm{T}}\boldsymbol{A}\boldsymbol{x}$ is a real number, due to the fact that

$$\overline{\overline{\boldsymbol{x}}^{\mathrm{T}}\boldsymbol{A}\boldsymbol{x}}=\overline{(\overline{\boldsymbol{x}}^{\mathrm{T}}\boldsymbol{A}\boldsymbol{x})^{\mathrm{T}}}=\overline{\boldsymbol{x}}^{\mathrm{T}}\overline{\boldsymbol{A}}^{T}\boldsymbol{x}=\overline{\boldsymbol{x}}^{\mathrm{T}}\boldsymbol{A}\boldsymbol{x}.$$

THEOREM A.4

Eigenvalues of a Hermitian matrix **A** *are all real numbers and the eigenvectors belonging to different eigenvalues are orthogonal.*

Proof: Let $\boldsymbol{A}\boldsymbol{x}=\lambda\boldsymbol{x}$ ($\boldsymbol{x}\neq 0$), then

$$\overline{\boldsymbol{x}}^{\mathrm{T}}\boldsymbol{A}\boldsymbol{x}=\lambda\overline{\boldsymbol{x}}^{\mathrm{T}}\boldsymbol{x}=\lambda(\boldsymbol{x},\boldsymbol{x}), \quad \lambda=\frac{\overline{\boldsymbol{x}}^{\mathrm{T}}\boldsymbol{A}\boldsymbol{x}}{(\boldsymbol{x},\ \boldsymbol{x})}.$$

Since $(\boldsymbol{x}, \boldsymbol{x}) > 0$ and $\overline{\boldsymbol{x}}^{\mathrm{T}}\boldsymbol{A}\boldsymbol{x}$ is a real number, so eigenvalue λ is a real number.

Let $\boldsymbol{A}\boldsymbol{x}_1 = \lambda_1\boldsymbol{x}_1$, $\boldsymbol{A}\boldsymbol{x}_2 = \lambda_2\boldsymbol{x}_2$, λ_1 and λ_2 be different eigenvalues (real numbers). Then,

$$\lambda_1\overline{\boldsymbol{x}_2}^{\mathrm{T}}\boldsymbol{x}_1 = \overline{\boldsymbol{x}_2}^{\mathrm{T}}\boldsymbol{A}\boldsymbol{x}_1 = \overline{(\boldsymbol{A}\boldsymbol{x}_2)}^{\mathrm{T}}\boldsymbol{x}_1 = \overline{\lambda_2\boldsymbol{x}_2}^{\mathrm{T}}\boldsymbol{x}_1 = \lambda_2\overline{\boldsymbol{x}_2}^{\mathrm{T}}\boldsymbol{x}_1.$$

Since $\lambda_1 \neq \lambda_2$, so $\overline{\boldsymbol{x}_2}^{\mathrm{T}}\boldsymbol{x}_1 = (\boldsymbol{x}_1, \boldsymbol{x}_2) = 0$, that is, eigenvectors $\boldsymbol{x}_1$ and $\boldsymbol{x}_2$ are orthogonal. □

THEOREM A.5

For any nth-order Hermitian matrix $\boldsymbol{A}$, *there exists a unitary matrix* $\boldsymbol{U}$, *such that*

$$\boldsymbol{U}^{-1}\boldsymbol{A}\boldsymbol{U} = \overline{\boldsymbol{U}}^{\mathrm{T}}\boldsymbol{A}\boldsymbol{U} = \mathrm{diag}(\lambda_1, \lambda_2, \ldots, \lambda_n),$$

where $\lambda_1, \lambda_2, \ldots, \lambda_n$ *are n eigenvalues of* $\boldsymbol{A}$ *and the n column vectors of* $\boldsymbol{U}$ *are orthonormal eigenvectors corresponding to* $\lambda_1, \lambda_2, \ldots, \lambda_n$, *respectively.*

The proof of Theorem A.5 is similar to that of Theorem 5.12 in §5.3. From Theorem A.5, we can immediately obtain the following theorem.

THEOREM A.6

For any Hermitian quadratic form $\overline{\boldsymbol{x}}^{\mathrm{T}}\boldsymbol{A}\boldsymbol{x}$, *there exists a unitary transformation* $\boldsymbol{x} = \boldsymbol{U}\boldsymbol{y}$, *such that*

$$\overline{\boldsymbol{x}}^{\mathrm{T}}\boldsymbol{A}\boldsymbol{x} = \overline{\boldsymbol{y}}^{\mathrm{T}}(\overline{\boldsymbol{U}}^{\mathrm{T}}\boldsymbol{A}\boldsymbol{U})\boldsymbol{y} = \lambda_1\overline{y_1}y_1 + \lambda_2\overline{y_2}y_2 + \cdots + \lambda_n\overline{y_n}y_n.$$

In Chapter 5, we stated that not all matrices are similar to diagonal matrices. With the concept of the unitary matrix, we can prove that every matrix over the complex field is similar to an upper triangular matrix (Theorem A.7). This conclusion is significant in practical applications.

THEOREM A.7

For any real or complex matrix $\boldsymbol{A}$ *of order n, there exists a unitary matrix* $\boldsymbol{U}$, *such that*

$$\boldsymbol{U}^{-1}\boldsymbol{A}\boldsymbol{U} = \begin{pmatrix} \lambda_1 & * & \cdots & * \\ & \lambda_2 & \cdots & * \\ & & \ddots & \vdots \\ & & & \lambda_n \end{pmatrix}, \tag{A.16}$$

where the upper triangular matrix is a complex matrix.

We can prove the theorem by mathematical induction. The proof is basically the same as Theorem A.5.

COROLLARY 1.1

If eigenvalues of a real matrix $\mathbf{A}$ *are all real, then there exists an orthogonal matrix* $\boldsymbol{Q}$*, such that* $\boldsymbol{Q}^{-1}\mathbf{A}\boldsymbol{Q}$ *is an upper triangular matrix.*

Exercises

A.1 Let $\boldsymbol{A}$ be a positive definite matrix of order n and define $(\boldsymbol{\alpha}, \boldsymbol{\beta}) = \boldsymbol{\alpha}^{\mathrm{T}}\boldsymbol{A}\boldsymbol{\beta}$ for any two vectors $\boldsymbol{\alpha} = (x_1, x_2, \ldots, x_n)^{\mathrm{T}}$, $\boldsymbol{\beta} = (y_1, y_2, \ldots, y_n)^{\mathrm{T}}$ in $\mathbb{R}^n$.

(i) Prove that $\mathbb{R}^n$ forms an Euclidean space under this definition [i.e., verify that $(\boldsymbol{\alpha}, \boldsymbol{\beta})$ satisfies the four axioms of inner product].

(ii) Find the metric matrix of $\mathbb{R}^n$ under the natural basis $\{\boldsymbol{\varepsilon}_1, \boldsymbol{\varepsilon}_2, \ldots, \boldsymbol{\varepsilon}_n\}$.

(iii) Provide the Cauchy-Schwarz inequality for this space.

A.2 Let $\boldsymbol{\alpha}, \boldsymbol{\beta} \in \mathbb{R}^3$ and define $(\boldsymbol{\alpha}, \boldsymbol{\beta})$ as follows. Which definition satisfies the axioms of inner product in $\mathbb{R}^3$?

(i) $(\boldsymbol{\alpha}, \boldsymbol{\beta}) = a_1b_1 + a_2b_2 + a_3b_3 + 2a_2b_3 - 2a_3b_2$,

(ii) $(\boldsymbol{\alpha}, \boldsymbol{\beta}) = a_1b_1 + a_2b_2 + a_3b_3 - a_2b_3 - a_3b_2$,

(iii) $(\boldsymbol{\alpha}, \boldsymbol{\beta}) = {a_1}^2{b_1}^2 + {a_2}^2{b_2}^2 + {a_3}^2{b_3}^2$,

(iv) $(\boldsymbol{\alpha}, \boldsymbol{\beta}) = a_1b_1 + a_3b_3$.

A.3 Which one of the following operations defines an inner product in $C[-1, 1]$?

(i) $(f, g) = \int_{-1}^{1} f(x)^2 g(x)^2\,\mathrm{d}x$,

(ii) $(f, g) = \int_{-1}^{1} x f(x) g(x)\,\mathrm{d}x$,

(iii) $(f, g) = \int_{-1}^{1} x^2 f(x) g(x)\,\mathrm{d}x$,

(iv) $(f, g) = -\int_{-1}^{1} x f(x) g(x)\,\mathrm{d}x$,

(v) $(f, g) = \int_{-1}^{1} \mathrm{e}^{-x} f(x) g(x)\,\mathrm{d}x$.

A.4 Let $\{\boldsymbol{\varepsilon}_1, \boldsymbol{\varepsilon}_2, \boldsymbol{\varepsilon}_3, \boldsymbol{\varepsilon}_4, \boldsymbol{\varepsilon}_5\}$ be an orthonormal basis of a 5-dimensional Euclidean space V and $W = L(\boldsymbol{\alpha}_1, \boldsymbol{\alpha}_2, \boldsymbol{\alpha}_3)$, where $\boldsymbol{\alpha}_1 = \boldsymbol{\varepsilon}_1 + \boldsymbol{\varepsilon}_5$, $\boldsymbol{\alpha}_2 = \boldsymbol{\varepsilon}_1 - \boldsymbol{\varepsilon}_2 + \boldsymbol{\varepsilon}_4$, $\boldsymbol{\alpha}_3 = 2\boldsymbol{\varepsilon}_1 + \boldsymbol{\varepsilon}_2 + \boldsymbol{\varepsilon}_3$. Find an orthonormal basis of W.

A.5 Let $\{\boldsymbol{\alpha}_1, \boldsymbol{\alpha}_2, \ldots, \boldsymbol{\alpha}_n\}$ be an orthonormal basis of an n-dimensional Euclidean space V. Prove that

(i) if $\boldsymbol{\alpha} \in V$, such that $(\boldsymbol{\gamma}, \boldsymbol{\alpha}_i) = 0$ for $i = 1, 2, \ldots, n$, then $\boldsymbol{\gamma} = 0$,

(ii) if $\boldsymbol{\gamma}_1, \boldsymbol{\gamma}_2 \in V$, such that $(\boldsymbol{\gamma}_1, \boldsymbol{\alpha}) = (\boldsymbol{\gamma}_2, \boldsymbol{\alpha})$ for any $\boldsymbol{\alpha} \in V$, then $\boldsymbol{\gamma}_1 = \boldsymbol{\gamma}_2$.

A.6 Define an inner product $(f, g) = \int_{-1}^{1} f(x)g(x)\,\mathrm{d}x$ in $\mathbb{R}[x]_4$. Try to find the metric matrix of the basis $\{1, x, x^2, x^3\}$, and construct an orthonormal basis from this basis.

A.7 Prove that $\mathbb{R}[x]$ forms an Euclidean space from the definition

$$(f, g) = \int_0^{+\infty} f(x)g(x)\mathrm{e}^{-x}\,\mathrm{d}x,$$

for any two polynomials $f(x)$ and $g(x)$ in $\mathbb{R}[x]$ (all the polynomials are with real coefficients). Find three orthonormal polynomials from $f_0 = 1$, $f_1 = x$, $f_2 = x^2$ by Gram-Schmidt orthogonalization with the inner product above.

A.8 For any two nth-order complex matrices $\boldsymbol{A}$ and $\boldsymbol{B}$ in $\mathbb{C}^{n\times n}$, define $(\boldsymbol{A}, \boldsymbol{B}) = \mathrm{tr}(\boldsymbol{A}\overline{\boldsymbol{B}}^{\mathrm{T}})$. Verify that it satisfies the four axioms of the inner product.

A.9 Prove that $\boldsymbol{A} = \boldsymbol{B}\overline{\boldsymbol{B}}^{\mathrm{T}}$ is a Hermitian matrix for any complex matrix $\boldsymbol{B}$.

A.10 Prove the four properties of a unitary matrix.

A.11 Let $\boldsymbol{U}$ be an nth-order unitary matrix, $\boldsymbol{x}$ and $\boldsymbol{y}$ be n-dimensional complex column vectors. Prove that $(\boldsymbol{U}\boldsymbol{x}, \boldsymbol{U}\boldsymbol{y}) = (\boldsymbol{x}, \boldsymbol{y})$.

A.12 Prove Theorem A.5.

A.13 Prove Theorem A.7.

A.14 For the following Hermitian matrices, find a unitary matrix $\boldsymbol{U}$ such that $\overline{\boldsymbol{U}}^{\mathrm{T}}\boldsymbol{A}\boldsymbol{U}$ is a diagonal matrix.

(i) $\boldsymbol{A} = \begin{pmatrix} 1 & 2\mathrm{i} \\ -2\mathrm{i} & 1 \end{pmatrix}$,

(ii) $\boldsymbol{A} = \begin{pmatrix} 2 & 3-3\mathrm{i} \\ 3+3\mathrm{i} & 5 \end{pmatrix}$.

A.15 Let $\boldsymbol{P}_1$ and $\boldsymbol{P}_2$ be both mth-order matrices, $\boldsymbol{P} = \begin{pmatrix} \boldsymbol{P}_1 & \boldsymbol{A} \\ \boldsymbol{0} & \boldsymbol{P}_2 \end{pmatrix}$ is a unitary matrix. Prove that $\boldsymbol{A} = \boldsymbol{0}$, $\boldsymbol{P}_1$ and $\boldsymbol{P}_2$ are both unitary matrices.

Appendix

B Jordan Canonical Form (Abstract)

In Chapter 5, we have seen that an nth-order matrix $\boldsymbol{A}$ over the complex field is similar to a diagonal matrix, if and only if, $\boldsymbol{A}$ has n linearly independent eigenvectors. If there are m linearly independent eigenvectors $(m < n)$, we can verify that $\boldsymbol{A}$ must be similar to a Jordan matrix that is made up of Jordan blocks. The definition of a Jordan matrix is as follows.

Definition B.1

A quasi-diagonal matrix

$$\boldsymbol{J} = \begin{pmatrix} \boldsymbol{J}_1 & & & \\ & \boldsymbol{J}_2 & & \\ & & \ddots & \\ & & & \boldsymbol{J}_s \end{pmatrix},$$

where

$$\boldsymbol{J}_i = \begin{pmatrix} \lambda_i & 1 & & \\ & \lambda_i & \ddots & \\ & & \ddots & 1 \\ & & & \lambda_i \end{pmatrix}, \quad i = 1, 2, \ldots, s,$$

is called a **Jordan matrix**, where each $\boldsymbol{J}_i$ is a square matrix called **Jordan block**.

When $J_1 = (\lambda_1)$, $J_2 = (\lambda_2)$, ..., $J_s = (\lambda_s)$ are all Jordan blocks of order 1, then J is a diagonal matrix. So, a diagonal matrix is a special case of Jordan matrix.

An nth-order matrix A over the complex field is similar to a Jordan matrix, that is, there exists an invertible matrix P, such that

$$P^{-1}AP = J = \begin{pmatrix} J_1 & & & \\ & J_2 & & \\ & & \ddots & \\ & & & J_s \end{pmatrix}.$$

Obviously, λ_i in J_i is an eigenvalue of A, but λ_i and λ_j may be the same even when $i \neq j$. However, the column vectors of P are not all of the eigenvectors of A.

The Jordan matrix similar to A is called a **Jordan canonical form** of A.

The theory of Jordan canonical forms is comparatively complicated and we only introduce the main points of the theory (i.e., we do not provide the proofs) and the method of finding a Jordan canonical form of a given matrix.

Definition B.2

If the elements a_{ij} of a matrix $A = (a_{ij})$ are polynomials in λ, then A is called a λ-matrix and denoted as $A(\lambda)$.

For example, a characteristic matrix of A, $\lambda I - A$ is a λ-matrix.

We can also perform elementary operations on a λ-matrix. The three elementary operations are:

(i) multiply some row (column) of the matrix by a nonzero constant,
(ii) add some row (column) multiplied by a polynomial $\varphi(\lambda)$ to another row (column) in the matrix,
(iii) switch two rows (columns) of the matrix.

Definition B.3

If a λ-matrix $A(\lambda)$ can be changed to $B(\lambda)$ by elementary operations, then $A(\lambda)$ and $B(\lambda)$ are said to be equivalent, which is denoted as $A(\lambda) \cong B(\lambda)$.

THEOREM B.1

The characteristic matrix $\boldsymbol{A}(\lambda) = \lambda \boldsymbol{I} - \boldsymbol{A}$ of any nth-order matrix $\boldsymbol{A}$ is equivalent to a diagonal λ-matrix, that is,

$$\boldsymbol{A}(\lambda) = \lambda \boldsymbol{I} - \boldsymbol{A} \cong \begin{pmatrix} d_1(\lambda) & & & \\ & d_2(\lambda) & & \\ & & \ddots & \\ & & & d_n(\lambda) \end{pmatrix} = \boldsymbol{D}(\lambda). \tag{B.1}$$

$$A_k(\lambda) = D_k(\lambda) \ (k = 1, 2, \ldots, n), \tag{B.2}$$

where

(i) *$d_i(\lambda)(i = 1, 2, \ldots, n)$ is a monic polynomial, that is, a polynomial with the coefficient of the highest-degree term being 1,*

(ii) *$d_i(\lambda) | d_{i+1}(\lambda)$ [i.e., $d_{i+1}(\lambda) = q_i(\lambda) d_i(\lambda)$, $q_i(\lambda)$ is also a polynomial of λ] $i = 1, 2, \ldots, n-1$,*

(iii) *$A_k(\lambda)$ and $D_k(\lambda)$ express separately the highest common factor of all the kth-order minors of $\boldsymbol{A}(\lambda)$ and $\boldsymbol{D}(\lambda)$, respectively.*

From the conclusion of Theorem B.1, we get

$$D_k(\lambda) = d_1(\lambda) d_2(\lambda) \cdots d_k(\lambda), \quad k = 1, 2, \ldots, n, \tag{B.3}$$

$$d_1(\lambda) = D_1(\lambda) = A_1(\lambda), \tag{B.4}$$

$$A_k(\lambda) = D_k(\lambda) = D_{k-1}(\lambda) d_k(\lambda) = A_{k-1}(\lambda) d_k(\lambda).$$

So,

$$d_k(\lambda) = \frac{A_k(\lambda)}{A_{k-1}(\lambda)}, \quad k = 2, 3, \ldots, n. \tag{B.5}$$

From the above discussion, it is clear that $d_1(\lambda), d_2(\lambda), \ldots, d_k(\lambda)$ are determined uniquely by $\boldsymbol{A}(\lambda) = \lambda \boldsymbol{I} - \boldsymbol{A}$. They are called the **invariant factors** of $\lambda \boldsymbol{I} - \boldsymbol{A}$ (henceforth, for brevity, we will refer to them as invariant factors of $\boldsymbol{A}$). Since $A_n(\lambda) = |\lambda \boldsymbol{I} - \boldsymbol{A}| = D_n(\lambda)$ is an nth-degree polynomial of λ, therefore the sum of degrees of n invariant factors is equal to n.

Example B.1 Invariant factors of the characteristic matrix

Find invariant factors of the characteristic matrix $\boldsymbol{J}(\lambda)=\lambda \boldsymbol{I}-\boldsymbol{J}$ of the third-order matrix

$$\boldsymbol{J}=\begin{pmatrix} a & 1 & 0 \\ 0 & a & 1 \\ 0 & 0 & a \end{pmatrix}.$$

Solution:
Method 1: Find invariant factors $d_1(\lambda)=A_1(\lambda)$ according to (B.4) and $d_k(\lambda)=\frac{A_k(\lambda)}{A_{k-1}(\lambda)}$ according to (B.5).

First, calculate all the first-, second-, and third-order subdeterminants of

$$\boldsymbol{J}(\lambda)=\lambda \boldsymbol{I}-\boldsymbol{J}=\begin{pmatrix} \lambda-a & -1 & 0 \\ 0 & \lambda-a & -1 \\ 0 & 0 & \lambda-a \end{pmatrix},$$

then it is easy to obtain their highest common factors separately,

$$J_1(\lambda)=1, \quad J_2(\lambda)=1, \quad J_3(\lambda)=(\lambda-a)^3.$$

Thus, we get the invariant factors of $\boldsymbol{J}(\lambda)$,

$$d_1(\lambda)=J_1(\lambda)=1,$$

$$d_2(\lambda)=\frac{J_2(\lambda)}{J_1(\lambda)}=1,$$

$$d_3(\lambda)=\frac{J_3(\lambda)}{J_2(\lambda)}=(\lambda-a)^3.$$

Method 2: Apply elementary operations, transform $\boldsymbol{J}(\lambda)=\lambda \boldsymbol{I}-\boldsymbol{J}$ to form (B.1) (ⓘ indicates the ith row and $[j]$ indicates the jth column as follows).

$$\lambda \boldsymbol{I}-\boldsymbol{J}=\begin{pmatrix} \lambda-a & -1 & 0 \\ 0 & \lambda-a & -1 \\ 0 & 0 & \lambda-a \end{pmatrix} \xrightarrow{[1]+[2]\times(\lambda-a)}$$

$$\begin{pmatrix} 0 & -1 & 0 \\ (\lambda-a)^2 & \lambda-a & -1 \\ 0 & 0 & \lambda-a \end{pmatrix} \xrightarrow[{[1]\rightleftharpoons[2]}]{② + ① \times(\lambda-a)} \begin{pmatrix} -1 & 0 & 0 \\ 0 & (\lambda-a)^2 & -1 \\ 0 & 0 & \lambda-a \end{pmatrix}$$

$$\xrightarrow{[2]+[3]\times(\lambda-a)^2}\begin{pmatrix}-1 & 0 & 0\\ 0 & 0 & -1\\ 0 & (\lambda-a)^3 & \lambda-a\end{pmatrix}$$

$$\xrightarrow[{[2]\rightleftharpoons[3]}]{③+②\times(\lambda-a)}\begin{pmatrix}-1 & 0 & 0\\ 0 & -1 & 0\\ 0 & 0 & (\lambda-a)^3\end{pmatrix}\xrightarrow[②\times(-1)]{①\times(-1)}$$

$$\begin{pmatrix}1 & 0 & 0\\ 0 & 1 & 0\\ 0 & 0 & (\lambda-a)^3\end{pmatrix},$$

so the invariant factors of $\boldsymbol{J}(\lambda)=\lambda\boldsymbol{I}-\boldsymbol{J}$ are 1, 1, $(\lambda-a)^3$. ■

Since an nth-degree polynomial can certainly be factored into the product of n first-degree factors over the complex field, invariant factors of degree ≥ 1 of $\lambda\boldsymbol{I}-\boldsymbol{A}$ can all be factored into the product of several powers of linear factors. These powers of linear factors are called **elementary divisors** of $\lambda\boldsymbol{I}-\boldsymbol{A}$ (or of $\boldsymbol{A}$, in short). However, in the elementary divisors of $\boldsymbol{A}(\lambda)$, the same power of one linear factor may appear repeatedly. For example, an elementary divisor of $\boldsymbol{J}(\lambda)=\lambda\boldsymbol{I}-\boldsymbol{J}$ in Example B.1 is $(\lambda-a)^3$. Another example, when

$$\lambda\boldsymbol{I}-\boldsymbol{A}\cong\begin{pmatrix}1 & & & & \\ & 1 & & & \\ & & 1 & & \\ & & & (\lambda-3)^2 & \\ & & & & (\lambda+1)(\lambda-3)^2\end{pmatrix}, \tag{B.6}$$

elementary divisors of $\boldsymbol{A}$ are $(\lambda-3)^2$, $(\lambda-3)^2$, and $(\lambda+1)$.

THEOREM B.2

$\boldsymbol{A}\sim\boldsymbol{B}$, if and only if, $\lambda\boldsymbol{I}-\boldsymbol{A}\cong\lambda\boldsymbol{I}-\boldsymbol{B}$. (The proof is omitted.)

Since the equivalence relation implies transitivity, a necessary and sufficient condition of $\lambda\boldsymbol{I}-\boldsymbol{A}\cong\lambda\boldsymbol{I}-\boldsymbol{B}$ is that they must have the same list of invariant factors. Since they have the same common factors, they must have the same list of elementary divisors and vice versa. Hence, we have the following theorem.

THEOREM B.3

$A \sim B$, if and only if, $\lambda I - A$ and $\lambda I - B$ have the same list of elementary divisors.

THEOREM B.4

If elementary divisors of the matrix $\lambda I - A$ of an nth-order matrix A are $(\lambda - \lambda_1)^{m_1}, (\lambda - \lambda_2)^{m_2}, \ldots, (\lambda - \lambda_k)^{m_k}$, where $\sum_{i=1}^{k} m_i = n$, then

$$A \sim J = \begin{pmatrix} J_1 & & & \\ & J_2 & & \\ & & \ddots & \\ & & & J_k \end{pmatrix}, \tag{B.7}$$

where

$$J_i = \begin{pmatrix} \lambda_i & 1 & & \\ & \lambda_i & \ddots & \\ & & \ddots & 1 \\ & & & \lambda_i \end{pmatrix}_{m_i\, th\text{-}order}, \quad i = 1, 2, \ldots, k.$$

Proof: From Example B.1, we know that the invariant factors of $\lambda I - J_i$ are $1, \ldots, 1, (\lambda - \lambda_i)^{m_i}$, elementary divisors are $(\lambda - \lambda_i)^{m_i}, i = 1, 2, \ldots, k$. So, $\lambda I - J_1$ and $\lambda I - A$ have the same list of elementary divisors. According to Theorem B.3, A is similar to J. □

Since $\lambda I - A$ has elementary divisors and the elementary divisors are determined by A uniquely, thus the sum of the degrees of elementary divisors is equal to the order of the matrix; therefore, any nth-order matrix is similar to a Jordan matrix over the complex field from Theorem B.4.

Here, we need to explain further. Changing the order of Jordan blocks in a Jordan matrix J does not affect the elementary divisors of $\lambda I - J$. Hence, if the order of Jordan blocks is not considered, the Jordan canonical form J of a matrix A is unique.

From Theorem B.4, we know that A is similar to a diagonal matrix, if and only if, elementary divisors of $\lambda I - A$ are all linear factors.

Example B.2 Jordan canonical form of matrices (I)

Given that

$$\lambda I - A \cong \begin{pmatrix} 1 & & & & \\ & 1 & & & \\ & & 1 & & \\ & & & (\lambda-3)^2 & \\ & & & & (\lambda+1)(\lambda-3)^2 \end{pmatrix},$$

$$\lambda I - B \cong \begin{pmatrix} 1 & & & & \\ & 1 & & & \\ & & 1 & & \\ & & & 1 & \\ & & & & (\lambda+4)(\lambda+3)^4 \end{pmatrix}.$$

Find the Jordan canonical forms of matrices A and B.

Solution: The elementary divisors of $\lambda I - A$ are $(\lambda-3)^2$, $(\lambda-3)^2$, $(\lambda+1)$; the elementary divisors of $\lambda I - B$ are $(\lambda+3)^4$, $(\lambda+4)$. So,

$$A \sim J = \begin{pmatrix} \begin{matrix} 3 & 1 \\ 0 & 3 \end{matrix} & & \\ & \begin{matrix} 3 & 1 \\ 0 & 3 \end{matrix} & \\ & & -1 \end{pmatrix}$$

$$\text{or } A \sim \begin{pmatrix} \begin{matrix} 3 & 1 \\ 0 & 3 \end{matrix} & & \\ & -1 & \\ & & \begin{matrix} 3 & 1 \\ 0 & 3 \end{matrix} \end{pmatrix},$$

$$\text{or } A \sim \begin{pmatrix} -1 & & & & \\ & 3 & 1 & & \\ & 0 & 3 & & \\ & & & 3 & 1 \\ & & & 0 & 3 \end{pmatrix}.$$

$$B \sim J = \begin{pmatrix} -3 & 1 & 0 & 0 & \\ 0 & -3 & 1 & 0 & \\ 0 & 0 & -3 & 1 & \\ 0 & 0 & 0 & -3 & \\ & & & & -4 \end{pmatrix}$$

$$\text{or } B \sim \begin{pmatrix} -4 & & & & \\ & -3 & 1 & 0 & 0 \\ & 0 & -3 & 1 & 0 \\ & 0 & 0 & -3 & 1 \\ & 0 & 0 & 0 & -3 \end{pmatrix}.$$

■

Example B.3 Jordan canonical form of a matrix (II)

Find the Jordan canonical form of

$$A = \begin{pmatrix} 3 & 1 & 0 \\ -4 & -1 & 0 \\ 4 & -8 & -2 \end{pmatrix}.$$

Solution:

$$\lambda I - A = \begin{pmatrix} \lambda-3 & -1 & 0 \\ 4 & \lambda+1 & 0 \\ -4 & 8 & \lambda+2 \end{pmatrix} \cong \begin{pmatrix} 0 & -1 & 0 \\ (\lambda-1)^2 & \lambda+1 & 0 \\ 8\lambda-28 & 8 & \lambda+2 \end{pmatrix}$$

$$\cong \begin{pmatrix} 0 & -1 & 0 \\ (\lambda-1)^2 & 0 & 0 \\ 8\lambda-28 & 0 & \lambda+2 \end{pmatrix} \cong \begin{pmatrix} 1 & 0 & 0 \\ 0 & (\lambda-1)^2 & 0 \\ 0 & 8\lambda-28 & \lambda+2 \end{pmatrix}$$

$$\cong \begin{pmatrix} 1 & 0 & 0 \\ 0 & (\lambda-1)^2 & 0 \\ 0 & -44 & \lambda+2 \end{pmatrix} \cong \begin{pmatrix} 1 & 0 & 0 \\ 0 & 0 & \frac{1}{44}(\lambda+2)(\lambda-1)^2 \\ 0 & -44 & \lambda+2 \end{pmatrix}$$

$$\cong \begin{pmatrix} 1 & 0 & 0 \\ 0 & 1 & 0 \\ 0 & 0 & (\lambda+2)(\lambda-1)^2 \end{pmatrix}.$$

So, elementary divisors of $\lambda I - A$ are $\lambda+2$, $(\lambda-1)^2$, hence the Jordan canonical form of A is

$$\left(\begin{array}{c|cc} -2 & 0 & 0 \\ \hline 0 & 1 & 1 \\ 0 & 0 & 1 \end{array}\right) \text{ or } \left(\begin{array}{cc|c} 1 & 1 & 0 \\ 0 & 1 & 0 \\ \hline 0 & 0 & -2 \end{array}\right).$$

■

Example B.4 Jordan canonical form of a matrix and similarity

Let

$$A = \begin{pmatrix} 1 & 2 & 0 \\ 0 & 2 & 0 \\ -2 & -2 & 1 \end{pmatrix}.$$

Is A similar to a diagonal matrix? If it is not similar to any diagonal matrix, find an invertible matrix P, such that $P^{-1}AP$ is a Jordan canonical form.

Solution:

$$|\lambda \boldsymbol{I}-\boldsymbol{A}|=\begin{vmatrix} \lambda-1 & -2 & 0 \\ 0 & \lambda-2 & 0 \\ 2 & 2 & \lambda-1 \end{vmatrix}=(\lambda-1)^2(\lambda-2),$$

the eigenvalues of $\boldsymbol{A}$ are $\lambda_1=1$ (double), $\lambda_2=2$.

There is only one linearly independent eigenvector of double eigenvalue $\lambda_1=1$, which is

$$\boldsymbol{x}_1=(0,0,1)^{\mathrm{T}},$$

where $\boldsymbol{A}$ is not similar to any diagonal matrix.

To find the Jordan canonical form of $\boldsymbol{A}$, we can use the methods of Examples B.1 or B.3 to obtain

$$\boldsymbol{A}\sim\boldsymbol{J}=\begin{pmatrix} 1 & 1 & 0 \\ 0 & 1 & 0 \\ 0 & 0 & 2 \end{pmatrix},$$

that is, there exists an invertible matrix $\boldsymbol{P}$, such that $\boldsymbol{P}^{-1}\boldsymbol{A}\boldsymbol{P}=\boldsymbol{J}$. Let

$$\boldsymbol{P}=(\boldsymbol{\xi}_1,\boldsymbol{\xi}_2,\boldsymbol{\xi}_3),$$

then

$$\boldsymbol{A}\boldsymbol{P}=\boldsymbol{P}\boldsymbol{J},$$

$$\boldsymbol{A}(\boldsymbol{\xi}_1,\boldsymbol{\xi}_2,\boldsymbol{\xi}_3)=(\boldsymbol{\xi}_1,\boldsymbol{\xi}_2,\boldsymbol{\xi}_3)\begin{pmatrix} 1 & 1 & 0 \\ 0 & 1 & 0 \\ 0 & 0 & 2 \end{pmatrix},$$

hence

$$\boldsymbol{A}\boldsymbol{\xi}_1=\boldsymbol{\xi}_1,\quad \boldsymbol{A}\boldsymbol{\xi}_2=\boldsymbol{\xi}_1+\boldsymbol{\xi}_2,\quad \boldsymbol{A}\boldsymbol{\xi}_3=2\boldsymbol{\xi}_3,$$

where $\boldsymbol{\xi}_1$ is an eigenvector corresponding to $\lambda_1=1$, that is, $\boldsymbol{\xi}_1=x_1=(0,0,1)^{\mathrm{T}}$, and $\boldsymbol{\xi}_3$ is an eigenvector corresponding to $\lambda_2=2$, which is, $\boldsymbol{\xi}_3=(2,1,-6)^{\mathrm{T}}$.

$\boldsymbol{\xi}_2$ is not an eigenvector of $\boldsymbol{A}$, but substituting $\boldsymbol{\xi}_1$ into

$$\boldsymbol{A}\boldsymbol{\xi}_2=\boldsymbol{\xi}_1+\boldsymbol{\xi}_2,$$

that is,

$$(\boldsymbol{A}-\boldsymbol{I})\boldsymbol{\xi}_2=\boldsymbol{\xi}_1,$$

we can obtain

$$\boldsymbol{\xi}_2=(-\frac{1}{2},0,0)^{\mathrm{T}}.$$

So,

$$P = (\xi_1, \xi_2, \xi_3) = \begin{pmatrix} 0 & -\frac{1}{2} & 2 \\ 0 & 0 & 1 \\ 1 & 0 & -6 \end{pmatrix},$$

we obtain

$$P^{-1}AP = J = \begin{pmatrix} 1 & 1 & 0 \\ 0 & 1 & 0 \\ 0 & 0 & 2 \end{pmatrix}.$$

Since the choice of eigenvectors ξ_1 and ξ_3 can be different, so ξ_2 can be different as well. Therefore, P is not unique. ■

From the derivation of P, we know that if the Jordan canonical form of A is made up of s Jordan blocks, then A has s linearly independent eigenvectors and vice versa. However, readers must pay attention to the fact that just by working out s linearly independent eigenvectors of A, we cannot immediately obtain its s Jordan blocks. For example, suppose λ_i is a fourfold eigenvalue of A and there are two linearly independent eigenvectors of A which belong to λ_i. Consequently, there must be two Jordan blocks in Jordan canonical form of A whose leading diagonal element is λ_i, but it may be one of the two cases

$$\left(\begin{array}{cc:cc} \lambda_i & 1 & & \\ 0 & \lambda_i & & \\ \hdashline & & \lambda_i & 1 \\ & & 0 & \lambda_i \end{array}\right) \quad \text{or} \quad \left(\begin{array}{c:ccc} \lambda_i & & & \\ \hdashline & \lambda_i & 1 & \\ & & \lambda_i & 1 \\ & & & \lambda_i \end{array}\right).$$

Of course, for the given A, the Jordan form must be either one of these cases.

Exercises

QB.1–B.8 Find the Jordan canonical forms for these matrices.

B.1 $\begin{pmatrix} 1 & 1 & 0 \\ -4 & 5 & 0 \\ 1 & 0 & 2 \end{pmatrix}$. **B.2** $\begin{pmatrix} 3 & 0 & 8 \\ 3 & -1 & 6 \\ -2 & 0 & -5 \end{pmatrix}$.

B.3 $\begin{pmatrix} 4 & 5 & -2 \\ -2 & -2 & 1 \\ -1 & -1 & 1 \end{pmatrix}$. **B.4** $\begin{pmatrix} 1 & -1 & 2 \\ 3 & -3 & 6 \\ 2 & -2 & 4 \end{pmatrix}$.

B.5 $\begin{pmatrix} 3 & 7 & -3 \\ -2 & -5 & 2 \\ -4 & -10 & 3 \end{pmatrix}$. **B.6** $\begin{pmatrix} -4 & 2 & 10 \\ -4 & 3 & 7 \\ -3 & 1 & 7 \end{pmatrix}$.

B.7 $\begin{pmatrix} 3 & 1 & 0 & 0 \\ -4 & -1 & 0 & 0 \\ 7 & 1 & 2 & 1 \\ -7 & -6 & -1 & 0 \end{pmatrix}$.

B.8 $\begin{pmatrix} 0 & 1 & 0 & \cdots & 0 & 0 \\ 0 & 0 & 1 & \cdots & 0 & 0 \\ \vdots & \vdots & \vdots & \ddots & \vdots & \vdots \\ 0 & 0 & 0 & \cdots & 0 & 1 \\ 1 & 0 & 0 & \cdots & 0 & 0 \end{pmatrix}$.

B.9 For the following Jordan blocks $\boldsymbol{J}_i$, find the kth-degree power $\boldsymbol{J}_i^k$.

(i) $\begin{pmatrix} \lambda & 1 & 0 \\ 0 & \lambda & 1 \\ 0 & 0 & \lambda \end{pmatrix}$,

(ii) $\begin{pmatrix} \lambda & 1 & & \\ & \lambda & \ddots & \\ & & \ddots & 1 \\ & & & \lambda \end{pmatrix}_{m\text{th-order}}$.

B.10 For the matrix $\boldsymbol{A}$ in **QB.1**, find an invertible matrix $\boldsymbol{P}$ such that $\boldsymbol{P}^{-1}\boldsymbol{AP}$ is Jordan canonical form, and find $\boldsymbol{A}^k$.

B.11 Let $\boldsymbol{A}$ be a nilpotent matrix (i.e., there exists a positive integer k such that $\boldsymbol{A}^k = \boldsymbol{0}$), then find $\det(\boldsymbol{A} - 2\boldsymbol{I})$.

Answers to Selected End-of-Chapter Exercises

Chapter 1 Determinants

1.1 0 **1.3** $(a-b)^2$ **1.5** 0 **1.7** 0

1.9 256 **1.11** –8 **1.13** –7 **1.15** 32

1.17 –60 **1.23** $abcd$ **1.25** 0

1.27 $(a_1a_4 - b_1b_4)(a_2a_3 - b_2b_3)$

1.29 $(-1)^{\frac{n(n+1)}{2}} \prod_{k=1}^{n} k!$

1.31 $1, -1, -1, 1$

1.33 $(a+1)^2 = 4b$

1.35 $f(x) = 7 - 5x^2 + 2x^3$

1.37 Expanding it about the first column, obtain a recurrence formula $D_n = xD_{n-1} + a_n$, $D_2 = \begin{vmatrix} x & -1 \\ a_2 & x+a_1 \end{vmatrix}$.

1.39 Apply mathematical induction: It is true for $n = 1$. Assume the conclusion to be true for all determinants whose order is less than n, and obtain a recurrence formula $D_n = 2(\cos\theta)D_{n-1} - D_{n-2}$ from the expansion of the determinant about the last column.

1.41 $(-1)^{\frac{n(n+1)}{2}}(n+1)^{n-1}$

1.43 $(-1)^{\frac{n(n+1)}{2}} \frac{n+1}{2} n^{n-1}$. Add every column to the first column, extract the common factor $\frac{1}{2}n(n+1)$, subtract the former row from the next row in turn starting from the last row, so that the first column is changed into $1, 0, \cdots, 0$, and expand the determinant by the first column; then for the complementary minor of a_{11}, add the first row multiplied by (-1) to the other rows, then add every column to the last column, and use the result of Example 1.2 in §1.1.

1.45 Apply mathematical induction on the order n of the determinant: It is clearly true for $n = 1$. Assume it to be true for order $(n-1)$ and then prove that the conclusion is true for order n, after expanding the nth-order determinant and using differentiation, then invoking the induction hypothesis.

1.47 $\begin{vmatrix} x_1 & y_1 & 1 \\ x_2 & y_2 & 1 \\ x_3 & y_3 & 1 \end{vmatrix} = 0$

1.49 $\left(x - \frac{1}{2}\right)^2 + y^2 + z^2 = \left(\frac{3}{2}\right)^2$.

Center $\left(\frac{1}{2}, 0, 0\right)$; radius $= \frac{3}{2}$.

Chapter 2 Matrices

2.1 $(1, 2, 2, 1)$

2.3 No solution.

2.5 When $p = 1$, it has infinitely many solutions: $(1 - k_1 - k_2, k_1, k_2)$ in which k_1, k_2 are arbitrary constants; it has no solution when $p = -2$; it has the only solution: $\left[-\frac{p+1}{p+2}, \frac{1}{p+2}, \frac{(p+1)^2}{p+2}\right]$ when $p \neq 1$ and $p \neq -2$.

2.7 548, 653, 758, 863, 968.

2.21 $A^8 = I_2$, $A^9 = \begin{pmatrix} 7 & -12 \\ 4 & -7 \end{pmatrix}$, $A^{2n} = I_2$, $A^{2n+1} = \begin{pmatrix} 7 & -12 \\ 4 & -7 \end{pmatrix}$.

2.23 $\begin{pmatrix} \cos n\phi & \sin n\phi \\ -\sin n\phi & \cos n\phi \end{pmatrix}$, $\begin{pmatrix} \cos \frac{n\pi}{2} & \sin \frac{n\pi}{2} \\ -\sin \frac{n\pi}{2} & \cos \frac{n\pi}{2} \end{pmatrix}$.

2.27 $\begin{pmatrix} a & b \\ 0 & a \end{pmatrix}$, a and b are arbitrary constants.

2.39 A^k is still a symmetric matrix, B^k is a symmetric matrix when k is an even number, and an antisymmetric matrix when k is an odd number.

2.41 (ii) $\frac{1}{6}\begin{pmatrix} 6 & 2 \\ -6 & -1 \\ -18 & -5 \end{pmatrix}$ (iii) $\begin{pmatrix} 1 & -3 & 3 \\ 0 & 1 & -2 \end{pmatrix}$

2.43 (i) $|A| \neq 0$ (ii) No

2.45 $A^{-1} = \frac{1}{2}(A - I)$, $(I - A)^{-1} = -\frac{1}{2}A$

2.47 Prove $(A^{-1})^{\mathrm{T}} = A^{-1}$ by using the operation properties and $A^{\mathrm{T}} = A$.

2.51 $\begin{pmatrix} 1 & 0 & 0 & 0 \\ -1 & 1 & 0 & 0 \\ 0 & -1 & 1 & 0 \\ 0 & 0 & -1 & 1 \end{pmatrix}$

2.53 $\begin{pmatrix} 0 & 0 & \cdots & 0 & \frac{1}{a_n} \\ \frac{1}{a_1} & 0 & \cdots & 0 & 0 \\ 0 & \frac{1}{a_2} & \cdots & 0 & 0 \\ \vdots & \vdots & \ddots & \vdots & \vdots \\ 0 & 0 & \cdots & \frac{1}{a_{n-1}} & 0 \end{pmatrix}$

2.55 $\begin{pmatrix} 20 & -15 & 13 \\ -105 & 77 & -58 \\ -152 & 112 & -87 \end{pmatrix}$

***2.57** (i) $\begin{pmatrix} 1 & 0 & 0 \\ 2 & 1 & 0 \\ 0 & \frac{1}{8} & 1 \end{pmatrix}\begin{pmatrix} 1 & -2 & 3 \\ 0 & 8 & -4 \\ 0 & 0 & \frac{3}{2} \end{pmatrix}$

(ii) $\begin{pmatrix} 1 & 0 & 0 & 0 \\ \frac{2}{3} & 1 & 0 & 0 \\ 0 & \frac{6}{7} & 1 & 0 \\ 0 & 0 & \frac{14}{15} & 1 \end{pmatrix}\begin{pmatrix} 3 & 1 & 0 & 0 \\ 0 & \frac{7}{3} & 1 & 0 \\ 0 & 0 & \frac{15}{7} & 1 \\ 0 & 0 & 0 & \frac{31}{15} \end{pmatrix}$

2.59 Partition B by columns into $B = (b_1, b_2, \cdots, b_n)$.

2.61 $A^{-1} = \begin{pmatrix} \mathbf{0} & C^{-1} \\ B^{-1} & \mathbf{0} \end{pmatrix}$

2.63 Apply elementary row operations to the corresponding partitioned matrix, add the first row multiplied on the left by $(-CA^{-1})$ to the second row.

2.65 $\begin{pmatrix} A_{n-1}^{-1}(I + bw^{-1}cA_{n-1}^{-1}) & -A_{n-1}^{-1}bw^{-1} \\ -w^{-1}cA_{n-1}{}^{-1} & w^{-1} \end{pmatrix}$, in which $w = a_{nn} - cA_{n-1}{}^{-1}b$.

2.67 (i) $-\frac{1}{24}$ (ii) -6 (iii) $-\frac{1}{6}$ (iv) $-\frac{1}{6}$ (v) $(-3)^4(-2)^3$

2.69 $|A^*|^4\,|A| = a^{13}$

2.71 -7, because $\alpha\beta^T = \sum_{i=1}^{3}[\alpha^T\beta]_{ii} = -2 + (-2) + (-3)$.

2.73 $|A| = 2$, $B = 3(A - I)^{-1}A = 3[A^{-1}(A - I)]^{-1} = 3(I - \frac{1}{2}A^*)^{-1} = \mathrm{diag}(6, 2, 1)$

2.75 $|I - A| = |A^T - I|\,|A| = (-1)^n\,|I - A|$, so $|I - A| = 0$.

2.77 Using $A^n = 2^{n-1}A$ yields $|kI - A^n| = k^2(k - 2^n)$.

2.79 $A=\left[(C-B)^T\right]^{-1}=\begin{pmatrix}\frac{1}{2} & 0 & 0\\ 0 & \frac{1}{2} & 0\\ 0 & 0 & 1\end{pmatrix}$,

$A^{-1}=(C-B)^T=\text{diag}(2,2,1)$.

2.81 $|AB+I|=1-2a_{23}^2\neq 0,\ a_{23}\neq\dfrac{\pm 1}{\sqrt{2}}$.

2.83 Using $A^*=|A|\,A^{-1}$,

(i) $(A^{-1})^*=\left|A^{-1}\right|A=(A^*)^{-1}$

(ii) $(A^{\mathrm{T}})^*=|A|\,(A^{\mathrm{T}})^{-1}=(|A|\,A^{-1})^{\mathrm{T}}=(A^*)^{\mathrm{T}}$

(iii) $(kA)^*=k^n\,|A|\,(kA)^{-1}=k^{n-1}\,|A|\,A^{-1}=k^{n-1}A^*$

2.85 Suppose $A=(a_{ij})$,

(i) take $E_{ii}=\text{diag}(0,\ldots,0,1,0,\ldots,0)$, get $a_{ij}=0 (i\neq j, i,j=1,2,\ldots,n)$ from $E_{ii}A=AE_{ii} (i=1,2,\ldots,n)$;

(ii) prove $a_{11}=a_{22}=\cdots=a_{nn}$.

2.87 Prove that the sum of the main diagonal elements of $(AB-BA)$ is equal to zero.

2.89

$$\begin{pmatrix}(a-b)^n & C_1^n(a-b)^{n-1} & C_2^n(a-b)^{n-2} & C_3^n(a-b)^{n-3}\\ 0 & (a-b)^n & C_1^n(a-b)^{n-1} & C_2^n(a-b)^{n-2}\\ 0 & 0 & (a-b)^n & C_1^n(a-b)^{n-1}\\ 0 & 0 & 0 & (a-b)^n\end{pmatrix},$$

$$\frac{\begin{pmatrix}1 & -C_1^n(a-b)^{-1} & \left[(C_1^n)^2-C_2^n\right](a-b)^{n-2} & \left[-(C_1^n)^3+2C_2^nC_1^n-C_3^n\right](a-b)^{-3}\\ 0 & 1 & -C_1^n(a-b)^{-1} & \left[(C_1^n)^2-C_2^n\right](a-b)^{-2}\\ 0 & 0 & 1 & -C_1^n(a-b)^{-1}\\ 0 & 0 & 0 & 1\end{pmatrix}}{(a-b)^n}$$

2.91 Apply mathematical induction to the order of the matrix and use the method of partitioning to prove that the conclusion is true for any nth-order matrix.

2.93 $\begin{pmatrix}A^{-1} & -A^{-1}BC^{-1}\\ 0 & C^{-1}\end{pmatrix}$

Chapter 3 Systems of Linear Equations

3.1 $\frac{5}{4}\alpha_1+\frac{1}{4}\alpha_2-\frac{1}{4}\alpha_3-\frac{1}{4}\alpha_4$

3.3 It is linearly dependent.

3.5 When $\alpha=\mathbf{0}$, it is linearly dependent; when $\alpha\neq\mathbf{0}$, it is linearly independent.

3.9 We prove the necessity condition using reduction to absurdity and apply Theorem 3.4.

3.11 We prove this using reduction to absurdity.

3.13 (i) The rank is 3; $\boldsymbol{\alpha}_1, \boldsymbol{\alpha}_2, \boldsymbol{\alpha}_4$; $\boldsymbol{\alpha}_3 = \boldsymbol{\alpha}_1 - 5\boldsymbol{\alpha}_2$.

(ii) The rank is 3; $\boldsymbol{\alpha}_1, \boldsymbol{\alpha}_2, \boldsymbol{\alpha}_4$; $\boldsymbol{\alpha}_3 = 3\boldsymbol{\alpha}_1 + \boldsymbol{\alpha}_2$; $\boldsymbol{\alpha}_5 = \boldsymbol{\alpha}_4 - \boldsymbol{\alpha}_1 - \boldsymbol{\alpha}_2$.

(iii) The rank is 3; $\boldsymbol{\alpha}_1, \boldsymbol{\alpha}_2, \boldsymbol{\alpha}_3$; $\boldsymbol{\alpha}_4 = 5\boldsymbol{\alpha}_2 - 3\boldsymbol{\alpha}_1 - \boldsymbol{\alpha}_3$.

***3.15** (i) Use the definition of rank (Definition 3.6) or reduce it to the normal-form by elementary operations.

(ii) Use the definition of rank (Definition 3.6) as well as the converse proposition to proposition (ii) in **Q3.8**, to prove that it is not true.

3.17 Prove rank$(\boldsymbol{AB}) < m$ using the result from **Q3.16**.

3.19 The rank is 3; $\begin{vmatrix} 1 & 3 & 5 \\ 0 & -1 & -3 \\ 0 & 0 & 4 \end{vmatrix}$.

3.21 The rank is 3; $\begin{vmatrix} -1 & -3 & -2 \\ 3 & 1 & -3 \\ -5 & -6 & 1 \end{vmatrix}$.

***3.25** Utilizing the equivalent normal-form of $\boldsymbol{A}$, that is, $\boldsymbol{PAQ} = (\boldsymbol{I}_m, \boldsymbol{0})$, $\boldsymbol{AQ} = (\boldsymbol{P}^{-1}, \boldsymbol{0})$.

***3.27** Reduce the equivalent normal-form to the sum of r matrices with rank 1, in the second part of the proof, use reduction to absurdity and Property 3.1 in §3.3.

3.29 (i) $(8, 0, 0, -10)^{\mathrm{T}} + k_1(-9, 1, 0, 11)^{\mathrm{T}} + k_2(-4, 0, 1, 5)^{\mathrm{T}}$

(ii) $(-16, 23, 0, 0, 0)^{\mathrm{T}} + k_1(1, -2, 1, 0, 0)^{\mathrm{T}} + k_2(1, -2, 0, 1, 0)^{\mathrm{T}} + k_3(5, -6, 0, 0, 1)^{\mathrm{T}}$

3.31 $\boldsymbol{\varepsilon}_i = (0, \cdots, 0, 1, 0, \cdots, 0)^{\mathrm{T}}$ $(i = 1, 2, \cdots, n)$ is a solution, that is, $\boldsymbol{A\varepsilon}_i = \boldsymbol{0}$, then use block multiplication.

3.37 $(a_1 + a_2 + a_3 + a_4, a_2 + a_3 + a_4, a_3 + a_4, a_4, 0)^{\mathrm{T}} + k(1, 1, 1, 1, 1)^{\mathrm{T}}$

3.39 (c) **3.41**. (e) **3.43**. (c)

3.45 (i) Yes, it can (since $\boldsymbol{\alpha}_2$ and $\boldsymbol{\alpha}_3$ are linearly independent).

(ii) No, it cannot (otherwise, $\boldsymbol{\alpha}_4$ becomes linearly expressible by $\boldsymbol{\alpha}_2, \boldsymbol{\alpha}_3$, which is in contradiction to the linear independence of $\boldsymbol{\alpha}_2, \boldsymbol{\alpha}_3, \boldsymbol{\alpha}_4$).

3.47 $\mathrm{r}(\boldsymbol{B}) = n$ (that is, n column vectors of $\boldsymbol{B}$ are linearly independent) can be deduced from $\mathrm{r}(\boldsymbol{AB}) = n$, since $\mathrm{r}(\boldsymbol{B}) < n$ implies $\mathrm{r}(\boldsymbol{AB}) < n$.

3.49 $a = \dfrac{1}{1-n}$

3.51 $m = 2, n = 4, t = 6$.

3.53 It has no solution. Since $\mathrm{r}(\boldsymbol{A}_1) = n - 1$, $\mathrm{r}(\boldsymbol{A}_1, \boldsymbol{\alpha}_n) = n$.

3.55 $\boldsymbol{L} = \begin{pmatrix} \boldsymbol{I}_k & 0 \\ -\boldsymbol{A}_{12}\boldsymbol{A}_{11}^{-1} & \boldsymbol{I}_{n-k} \end{pmatrix}, \boldsymbol{U} = \begin{pmatrix} \boldsymbol{I}_k & -\boldsymbol{A}_{11}^{-1}\boldsymbol{A}_{12} \\ \boldsymbol{0} & \boldsymbol{I}_{n-k} \end{pmatrix}$.

3.57 We use equivalent normal-forms to yield $\boldsymbol{A} = \boldsymbol{P}^{-1}\begin{pmatrix} \boldsymbol{I}_r & 0 \\ 0 & 0 \end{pmatrix}\boldsymbol{Q}^{-1}$, and then use

$$\begin{pmatrix} \boldsymbol{I}_r & 0 \\ 0 & 0 \end{pmatrix}_{m\times n} = \begin{pmatrix} \boldsymbol{I}_r & 0 \\ 0 & 0 \end{pmatrix}_{m\times n}\begin{pmatrix} \boldsymbol{I}_r & 0 \\ 0 & 0 \end{pmatrix}_{n\times n}.$$

3.59 Condition of necessity: Among all the nonzero coefficients such that $k_1\boldsymbol{\alpha}_1 + k_2\boldsymbol{\alpha}_2 + \ldots + k_r\boldsymbol{\alpha}_r = \boldsymbol{0}$, there must be the "smallest" subscript i such that $k_i \neq 0$ and $k_j = 0$ $(j > i)$; then prove that $1 < i \leq r$ and $\boldsymbol{\alpha}_1, \boldsymbol{\alpha}_2, \ldots, \boldsymbol{\alpha}_{i-1}$ are linearly independent (we can prove this using reduction to absurdity).

3.61 Suppose there are two vectors $\boldsymbol{\alpha}_i, \boldsymbol{\alpha}_j$, we use reduction to absurdity to linearly express the set with the vectors appearing before $\boldsymbol{\alpha}_i, \boldsymbol{\alpha}_j$.

3.63 (i) and (ii): $\mathrm{r}(\boldsymbol{A}) = 1$, that is, all the row vectors are proportional.

3.65 Use the fact that a necessary and sufficient condition under which a nonhomogeneous system of linear equations has a solution is that the rank of the augmented matrix is equal to the rank of the coefficient matrix.

3.67 Use the result in **Q3.34**.

3.69 Use the result of Example 3.14 in §3.4 and Property 3.1 in §3.3.

Chapter 4 Vector Spaces and Linear Transformations

4.1 $\frac{1}{4}(5, 1, -1, -1)^{\mathrm{T}}$

4.3 (i) $\begin{pmatrix} 1 & 0 & 0 & 1 \\ 1 & 1 & 0 & 1 \\ 0 & 1 & 1 & 1 \\ 0 & 0 & 1 & 0 \end{pmatrix}, \begin{pmatrix} 0 \\ -1 \\ 0 \\ 1 \end{pmatrix}.$

(ii) $\begin{pmatrix} 0 & 1 & -1 & 1 \\ -1 & 1 & 0 & 0 \\ 0 & 0 & 0 & 1 \\ 1 & -1 & 1 & -1 \end{pmatrix}, \begin{pmatrix} 1 \\ 3 \\ 1 \\ -1 \end{pmatrix}.$

(iii) $(3, 1, 0, -2)^{\mathrm{T}}$

4.5 (ii) $k_1(-5, 3, 1, 0)^{\mathrm{T}} + k_2(5, -3, 0, 1)^{\mathrm{T}}$

4.9 (i) $\frac{1}{\sqrt{3}}(1, 1, 1)^{\mathrm{T}}, \frac{1}{\sqrt{6}}(-2, 1, 1)^{\mathrm{T}}, \frac{1}{\sqrt{2}}(0, -1, 1)^{\mathrm{T}}; \left(0, -\frac{\sqrt{6}}{2}, \frac{\sqrt{2}}{2}\right).$

(ii) $\frac{1}{\sqrt{3}}(1, -1, 1)^{\mathrm{T}}, \frac{1}{\sqrt{6}}(-1, 1, 2)^{\mathrm{T}}, \frac{1}{\sqrt{2}}(1, 1, 0)^{\mathrm{T}}; \left(\frac{2\sqrt{3}}{3}, -\frac{\sqrt{6}}{3}, 0\right).$

4.11 $-\frac{6}{7}, \frac{2}{7}, -\frac{6}{7}, -\frac{3}{7}, -\frac{6}{7}$ or $-\frac{6}{7}, -\frac{2}{7}, \frac{6}{7}, \frac{3}{7}, -\frac{6}{7}.$

4.13 The proof can be obtained immediately by the definition $A^{\mathrm{T}}A = I$.

4.17 (i) Yes (ii) No (iii) No (iv) No (v) No (vi) Yes (vii) No (viii) Yes

4.19 (i), (iv) and (v) are true while (ii), (iii) and (vi) are false.

4.21 Dimension is 2; orthonormal basis is $\left\{\frac{1}{\sqrt{3}}(-1, 1, 0, 1, 0)^{\mathrm{T}}, \frac{1}{\sqrt{15}}(1, -2, 0, 3, 1)^{\mathrm{T}}\right\}.$

4.23 Dimension is 3; basis is $\{(1, 1, 0, 0)^{\mathrm{T}}, (1, 0, -1, 0)^{\mathrm{T}}, (1, 0, 0, 1)^{\mathrm{T}}\}$; $\{(1, 1, 0, 0)^{\mathrm{T}}, (1, 0, -1, 0)^{\mathrm{T}}, (1, 0, 0, 1)^{\mathrm{T}}, (1, 0, 0, 0)^{\mathrm{T}}\}$.

4.25 We only need to prove that the vector sets $\{\boldsymbol{\alpha}, \boldsymbol{\beta}\}$ and $\{\boldsymbol{\beta}, \boldsymbol{\gamma}\}$ are equivalent, that is, they can be linearly expressed by each other.

4.27 By definition, for instance, to prove $\boldsymbol{\alpha}_2, \boldsymbol{\alpha}_3, \cdots, \boldsymbol{\alpha}_n, \boldsymbol{\alpha}_{n+1}$ are linearly independent, let $k_2\boldsymbol{\alpha}_2 + k_3\boldsymbol{\alpha}_3 + \cdots + k_n\boldsymbol{\alpha}_n + k_{n+1}\boldsymbol{\alpha}_{n+1} = 0$. Substitute the known condition $\boldsymbol{\alpha}_{n+1} = x_1\boldsymbol{\alpha}_1 + x_2\boldsymbol{\alpha}_2 + x_3\boldsymbol{\alpha}_3 + \cdots + x_n\boldsymbol{\alpha}_n$ ($x_1, \cdots, x_n$ are all nonzero numbers) into the equality and get $k_1 = k_2 = \cdots = k_n = k_{n+1} = 0$.

4.29 (ii) $\begin{pmatrix} 0 & 1 & 1 & 1 \\ 1 & 0 & 1 & 1 \\ 1 & 1 & 0 & 1 \\ 1 & 1 & 1 & 0 \end{pmatrix}$

(iii) Old basis: $(0, 1, 2, -3)^{\mathrm{T}}$; New basis: $(0, -1, -2, 3)^{\mathrm{T}}$.

4.33 (i) Let $x_1\boldsymbol{\alpha}_1 + x_2\boldsymbol{\alpha}_2 = x_3\boldsymbol{\beta}_1 + x_4\boldsymbol{\beta}_2$. Solving the corresponding homogeneous system of linear equations in x_1, x_2, x_3, x_4, we have the result $V_1 \cap V_2$, moreover the basis of $V_1 + V_2 = L(\boldsymbol{\alpha}_1, \boldsymbol{\alpha}_2, \boldsymbol{\beta}_1, \boldsymbol{\beta}_2)$ is a maximal linearly independent subset of $\{\boldsymbol{\alpha}_1, \boldsymbol{\alpha}_2, \boldsymbol{\beta}_1, \boldsymbol{\beta}_2\}$.

4.35 Expand the basis $\{\boldsymbol{\alpha}_1, \cdots, \boldsymbol{\alpha}_m\}$ of W_1 to the basis $\{\boldsymbol{\alpha}_1, \cdots, \boldsymbol{\alpha}_m, \boldsymbol{\alpha}_{m+1}, \cdots, \boldsymbol{\alpha}_n\}$ of $\mathbb{R}^n$, take $W_2 = L\{\boldsymbol{\alpha}_m, \cdots, \boldsymbol{\alpha}_n\}$ then $W_1 \oplus W_2 = R^n$.

4.37 $W_1 \perp W_3$ and $W_1 = W_3^{\perp}$, $W_3 \perp W_4$.

4.41 (i) $\begin{pmatrix} 1 & 1 & 0 \\ 1 & -1 & 0 \\ 0 & 0 & 1 \end{pmatrix}$ (ii) $\begin{pmatrix} 0 & 2 & 2 \\ 1 & 0 & -1 \\ 0 & 0 & 1 \end{pmatrix}$

4.43 (i) $\begin{pmatrix} -2 & 9 & 7 \\ -1 & 2 & 1 \\ -1 & 3 & 2 \end{pmatrix}.$

(ii) $(3, 2, -7)^{\mathrm{T}}, (-12, -5, 22)^{\mathrm{T}}, (-9, -3, 15)^{\mathrm{T}}.$

(iii) $(6, -2, 0)^{\mathrm{T}}$

(iv) $(-7, -5, 17)^{\mathrm{T}}$

(v) $(4k-8, 2-2k, 18-9k)^{\mathrm{T}}$, k is an arbitrary constant.

4.45 $\begin{pmatrix} 0 & 1 & -1 & -1 \\ 0 & 0 & 2 & -1 \\ 0 & 0 & 0 & 3 \\ 0 & 0 & 0 & 0 \end{pmatrix}$

4.47 Use definition, let $c_1\xi + c_2\sigma(\xi) + \cdots + c_k\sigma^{k-1}(\xi) = 0$, multiplying two sides of the equality on the left by σ^{k-1} yields $c_1 = 0$, similarly, deduce $c_2 = \cdots = c_k = 0$. (The proof is similar to the proof of **Q3.46**.)

4.49 $\forall \boldsymbol{\alpha} \in R^3$, $\sigma(\boldsymbol{\alpha}) \in L(\boldsymbol{\alpha}_1, \boldsymbol{\alpha}_2)$, so $\sigma(\boldsymbol{\alpha}) = x_1\boldsymbol{\alpha}_1 + x_2\boldsymbol{\alpha}_2 = (x_1 + x_2, 0 + 2x_2, -x_1 + 2x_2)$.

4.51 (i) Use $(\boldsymbol{A}+\boldsymbol{I})(\boldsymbol{I}-\boldsymbol{A}) = (\boldsymbol{I}-\boldsymbol{A})(\boldsymbol{A}+\boldsymbol{I})$.
(ii) Use $\boldsymbol{I} = \boldsymbol{A}^{\mathrm{T}}\boldsymbol{A}$.

4.53 Use reduction to absurdity.

4.57 Let $\xi \in L(\boldsymbol{\alpha}_1, \ldots, \boldsymbol{\alpha}_s) \cap L(\boldsymbol{\beta}_1, \ldots, \boldsymbol{\beta}_t)$, then $\xi = x_1\boldsymbol{\alpha}_1 + \cdots + x_s\boldsymbol{\alpha}_s = x_{s+1}\boldsymbol{\beta}_1 + \cdots + x_{s+t}\boldsymbol{\beta}_t$, that is, $x_1\boldsymbol{\alpha}_1 + \cdots + x_s\boldsymbol{\alpha}_s - x_{s+1}\boldsymbol{\beta}_1 - \cdots - x_{s+t}\boldsymbol{\beta}_t = 0$. The intersection of the two subspaces can be obtained by solving the homogeneous system of linear equations that corresponds to the vector equation.

4.59 There exists $\boldsymbol{\alpha}_1 \notin V_2$, but $\boldsymbol{\alpha}_1 \in V_1$; $\boldsymbol{\alpha}_2 \notin V_1$, but $\boldsymbol{\alpha}_2 \in V_2$, prove that in this case, $\boldsymbol{\alpha}_1 + \boldsymbol{\alpha}_2$ satisfies the condition.

Chapter 5 Eigenvalues and Eigenvectors, Diagonalization of Matrices

For some of the exercises on calculating eigenvectors, we only present a basis of the eigensubspace. Any nonzero vector made out of a linear combination of the eigensubspace basis is also an eigenvector.

5.1 (i) $\dfrac{3}{2} \pm \dfrac{\sqrt{37}}{2}$, $(6, 1 \mp \sqrt{37})^{\mathrm{T}}$.

(ii) 1, $(0, 1, 1)^{\mathrm{T}}$; 2 (double), $(1, 1, 0)^{\mathrm{T}}$.

(iii) 2 (triple), $(1, 1, 0)^{\mathrm{T}}$, $(-1, 0, 1)^{\mathrm{T}}$.

(iv) 1 (quadruple), $(1, 0, 0, 0)^{\mathrm{T}}$,

(v) 1 (triple), $(-1, 1, 1)^{\mathrm{T}}$.

(vi) 1, $(-2, -1, 2)^{\mathrm{T}}$; 4, $(2, -2, 1)^{\mathrm{T}}$; -2, $(1, 2, 2)^{\mathrm{T}}$.

5.3 Use reduction to absurdity, one eigenvector cannot belong to different eigenvalues.

5.5 Use definition.

5.7 Yes, it does.

5.9 $\begin{pmatrix} 2^2(-1)^n - 3\cdot 2^n & 2(-1)^{n+1} + 2^{n+1} \\ 6(-1)^{n+1} - 3\cdot 2^n & 3(-1)^{n+1} + 2^{n+2} \end{pmatrix}$

***5.11** Use $\boldsymbol{A}^{-1}\boldsymbol{A} = \boldsymbol{I}$.

***5.13** $(1, 0, \cdots, 0)^{\mathrm{T}}$

***5.15** We can get it from $|\boldsymbol{I} - \boldsymbol{A}^2| = |\boldsymbol{I} - \boldsymbol{A}|\,|\boldsymbol{I} + \boldsymbol{A}| = 0$.

5.17 It is possible because an upper (lower) triangular matrix with distinct main diagonal entries can have n mutually different eigenvalues.

5.19 A can be diagonalized,

$$\boldsymbol{A} = \begin{pmatrix} 1 & 0 & 0 & 0 \\ 0 & 1 & 0 & 0 \\ -4 & -4 & 1 & 4 \\ 4 & 4 & 0 & 3 \end{pmatrix},$$

$$A^n = \begin{pmatrix} 1 & 0 & 0 & 0 \\ 0 & 1 & 0 & 0 \\ -1+(-3)^n & -1+(-3)^n & 1 & 1-(-3)^n \\ 1-(-3)^n & 1-(-3)^n & 0 & (-3)^n \end{pmatrix}.$$

5.21 (i) $\begin{pmatrix} 21 & -10 \\ 10 & -4 \end{pmatrix}$; $\begin{pmatrix} -43 & 22 \\ -22 & 12 \end{pmatrix}$;

$$\frac{1}{3}\begin{pmatrix} -1+(-2)^{k+2} & 2+(-2)^{k+1} \\ -2-(-2)^{k+1} & 2^2-(-2)^k \end{pmatrix}.$$

(ii) $\begin{pmatrix} 1279 & -640 \\ 640 & -321 \end{pmatrix}$

5.23 $$\begin{pmatrix} \frac{1}{\sqrt{2}} & \frac{1}{\sqrt{6}} & \frac{\sqrt{3}}{6} & \frac{1}{2} \\ -\frac{1}{\sqrt{2}} & \frac{1}{\sqrt{6}} & \frac{\sqrt{3}}{6} & \frac{1}{2} \\ 0 & -\frac{2}{\sqrt{6}} & \frac{\sqrt{3}}{6} & \frac{1}{2} \\ 0 & 0 & -\frac{\sqrt{3}}{2} & \frac{1}{2} \end{pmatrix} \begin{pmatrix} 2 & & & \\ & 2 & & \\ & & 2 & \\ & & & -2 \end{pmatrix}$$

★5.27 $(-1)^n 2^{n-r}$

5.31 Yes, 1 is an eigenvalue.

5.33 Prove that $\boldsymbol{A}(\boldsymbol{B}x) = \lambda_0(\boldsymbol{B}x)$.

5.37 First, prove that $\boldsymbol{A}^2 = 0$, then use dim $R(\boldsymbol{A}) = 1$.

5.39 (i) From $(\lambda\boldsymbol{I} - \boldsymbol{A})\boldsymbol{\xi} = 0$, obtain the eigenvalues corresponding to $\boldsymbol{\xi}$, $\lambda = -1$, $a = -3$, $b = 0$.

(ii) No. From $|\lambda\boldsymbol{I} - \boldsymbol{A}| = (\lambda+1)^3$, we know that -1 is a triple eigenvalue of $\boldsymbol{A}$, then from $\mathrm{r}(-\boldsymbol{I} - \boldsymbol{A}) = 2$, we also know that the eigenvalue $\lambda = -1$ has only one linearly independent eigenvector, so $\boldsymbol{A}$ cannot be similar to a diagonal matrix.

5.41 Since $\boldsymbol{A}$ can be diagonalized, there are two eigenvectors corresponding to $\lambda_1 = 2$ that are *linearly independent*, so that $\mathrm{r}(\lambda_1\boldsymbol{I} - \boldsymbol{A}) = 1$, where we have $x = -y = 2$; the other eigenvalue λ_2 satisfies: $\lambda_2 + 2 + 2 = 1 + 4 + 5$, so that $\lambda_2 = 6$.

$$P = \begin{pmatrix} 1 & 1 & 1 \\ -1 & 0 & -2 \\ 0 & 1 & 3 \end{pmatrix} \quad \Lambda = \begin{pmatrix} 2 & 0 & 0 \\ 0 & 2 & 0 \\ 0 & 0 & 6 \end{pmatrix}.$$

5.43 (c) **5.45** (d)

Chapter 6 Quadratic Forms

★6.7 (i) False. (ii) True.

6.9 $$\begin{pmatrix} \frac{2}{\sqrt{5}} & 0 & 0 & 0 & \frac{1}{\sqrt{5}} \\ \frac{-1}{\sqrt{5}} & 0 & 0 & 0 & \frac{2}{\sqrt{5}} \\ 0 & 0 & 1 & 0 & 0 \\ 0 & \frac{2}{\sqrt{13}} & 0 & \frac{3}{\sqrt{13}} & 0 \\ 0 & \frac{3}{\sqrt{13}} & 0 & -\frac{2}{\sqrt{13}} & 0 \end{pmatrix}, \quad \begin{pmatrix} 5 & & & & \\ & 5 & & & \\ & & 5 & & \\ & & & -8 & \\ & & & & 0 \end{pmatrix}.$$

6.11 (i) ${z_1}^2 - {z_2}^2 - {z_3}^2$; $\begin{cases} x_1 = z_1 + z_2 - z_3, \\ x_2 = z_1 - z_2 - z_3, \\ x_3 = z_3. \end{cases}$

(ii) $y_1{}^2-3y_2{}^2-\frac{8}{3}y_3{}^2$; $\begin{cases} x_1=y_1-y_2-\frac{5}{3}y_3, \\ x_2=y_2-\frac{1}{3}y_3, \\ x_3=y_3. \end{cases}$

(iii) $y_1^2-4y_2^2+9y_3^2-\frac{49}{9}y_4^2$;

$$\begin{cases} x_1=y_1-3y_2-\frac{5}{2}y_3-\frac{4}{9}y_4, \\ x_2=y_2+\frac{3}{2}y_3+\frac{2}{3}y_4, \\ x_3=y_3+\frac{7}{9}y_4, \\ x_4=y_4. \end{cases}$$

***6.15** Transform $\boldsymbol{A}$ into canonical form, construct x_1, x_2, and x_3 from the fact that indices of positive inertia and negative inertia are both nonzero.

6.17 (i) $y_1{}^2-y_2{}^2-y_3{}^2$, $\begin{pmatrix} 1 & 1 & -1 \\ 1 & -1 & -1 \\ 0 & 0 & 1 \end{pmatrix}$;

(ii) $y_1^2-y_2^2-y_3^2$, $\begin{pmatrix} 1 & -\frac{\sqrt{3}}{3} & -\frac{5}{\sqrt{24}} \\ 0 & \frac{\sqrt{3}}{3} & -\frac{1}{\sqrt{24}} \\ 0 & 0 & \frac{3}{\sqrt{24}} \end{pmatrix}$;

(iii) $y_1{}^2+y_2{}^2-y_3{}^2-y_4{}^2$,

$$\begin{pmatrix} 1 & -\frac{5}{6} & -\frac{3}{2} & -\frac{4}{21} \\ 0 & \frac{1}{2} & \frac{1}{2} & \frac{2}{7} \\ 0 & \frac{1}{3} & 0 & \frac{1}{3} \\ 0 & 0 & 0 & \frac{3}{7} \end{pmatrix}.$$

6.21 (i) is positive definite, while (ii) and (iii) are not positive definite.

6.23 $\frac{n+1}{2}y_1{}^2+\frac{1}{2}(y_2{}^2+y_3{}^2+\cdots+y_n{}^2)$, it is positive definite,

$$\boldsymbol{Q}=\begin{pmatrix} \frac{1}{\sqrt{3}} & -\frac{1}{\sqrt{2}} & \frac{1}{\sqrt{6}} \\ \frac{1}{\sqrt{3}} & 0 & -\frac{2}{\sqrt{6}} \\ \frac{1}{\sqrt{3}} & \frac{1}{\sqrt{2}} & \frac{1}{\sqrt{6}} \end{pmatrix}.$$

6.25 (i) $t>2$ (ii) $|t|<\sqrt{\frac{5}{3}}$

6.27 Prove that $x^T\boldsymbol{P}^T\boldsymbol{P}x>0$ ($\forall x\neq 0$) by definition.

6.31 (i) negative definite (ii) not definite (iii) not definite (iv) negative semi-definite

6.39 Consider the eigenvalues of $\boldsymbol{A}+t\boldsymbol{I}$.

6.41 There exists an invertible matrix $\boldsymbol{C}_1$, such that $\boldsymbol{C}_1{}^{\mathrm{T}}\boldsymbol{B}\boldsymbol{C}_1=\boldsymbol{I}$, and $\boldsymbol{C}_1{}^{\mathrm{T}}\boldsymbol{A}\boldsymbol{C}_1$ is a real symmetric matrix.

6.43 Multiply the second row of partitioned matrix on the *LHS* by $(-x^{\mathrm{T}}\boldsymbol{A}^{-1})$, then add it to the first row, and use the fact that $\boldsymbol{A}^{-1}$ is positive definite.

6.45 Use the result of **Q6.44**.

6.47 (i) From the fact that the rank of the matrix $\boldsymbol{A}$ associated with the quadratic form is 2, where $|\boldsymbol{A}|=0$, we get $c=3$.

(ii) The eigenvalues of $\boldsymbol{A}$ are $\lambda_1=4, \lambda_2=9, \lambda_3=0$, by orthogonal transformation $x=\boldsymbol{Q}y$, the equation $x^{\mathrm{T}}\boldsymbol{A}x=1$ can be transformed to $y^{\mathrm{T}}\boldsymbol{Q}^{\mathrm{T}}\boldsymbol{A}\boldsymbol{Q}y=4y_1^2+9y_2^2=1$, which is an elliptic cylindrical surface.

6.49 From the system of linear equations $x_1+a_1x_2=0, x_2+a_2x_3=0, \ldots, x_n+a_nx_1=0$, when $1+(-1)^{n-1}a_1a_2\cdots a_n\neq 0$, the system has only the zero solution. When $1+(-1)^{n-1}a_1a_2\cdots a_n\neq 0$, $f(x_1, x_2, \ldots, x_n)=0$ if $x=(x_1, x_2, \ldots, x_n)^{\mathrm{T}}=0$; $f(x_1, x_2, \ldots, x_n)=x^{\mathrm{T}}\boldsymbol{A}x>0$ $\forall x\neq 0$, and in this case, the quadratic form $f(x_1, x_2, \ldots, x_n)$ is positive definite.

Chapter 7 Applications

7.7 $x_1=(0.3,0.7)^{\mathrm{T}}$, $x_2=(0.45,0.55)^{\mathrm{T}}$, $x_3=(0.581,0.419)^{\mathrm{T}}$, $x_n=(0.6,0.4)^{\mathrm{T}}$, $(n\geq 11)$.

7.9 $\left(\frac{1}{n},\frac{1}{n},\cdots,\frac{1}{n}\right)^{\mathrm{T}}$

7.15 In an invertible matrix $\boldsymbol{P}$, $p_{i3}=0$, $p_{i1}=0$, $(i\neq 3)$, the rest are all 1.

7.17 There will be 2584 pairs at the end of month 18.

7.19 $u_n=\frac{1}{3}(2^n+(-1)^{n-1})$

7.23 (i) Least squares solution: $(\frac{19}{11},-\frac{5}{11})^{\mathrm{T}}$; projection vector: $(\frac{9}{11},\frac{24}{11},-\frac{5}{11})^{\mathrm{T}}$; distance from $\boldsymbol{b}$ to $R(\boldsymbol{A})$: $\frac{2\sqrt{11}}{11}$.

(ii) Least squares solution: $(0,\frac{1}{2},\frac{3}{4})^{\mathrm{T}}$; projection vector: $\frac{1}{4}(-1,-1,-1,3)^{\mathrm{T}}$; distance from $\boldsymbol{b}$ to $R(\boldsymbol{A})$: $\frac{1}{2}$.

7.27 $y=\frac{7}{2}x-\frac{1}{2}$

Appendix A Inner Product Space, Hermitian Quadratic Form

A.1 (ii) A

(iii) $|(\boldsymbol{\alpha},\boldsymbol{\beta})|\leq\sqrt{\boldsymbol{\alpha}^T\boldsymbol{A}\boldsymbol{\alpha}}\sqrt{\boldsymbol{\beta}^T\boldsymbol{A}\boldsymbol{\beta}}$

A.3 (i) does not satisfy (ii) does not satisfy

(iii) satisfy (iv) does not satisfy

(v) satisfy

A.7 Three orthonormal polynomials: 1, $x-1$, $\frac{1}{2\sqrt{2}}(x-2)^2$.

A.15 Prove by definition and properties of partitioned matrix.

Appendix B Jordan Canonical Form (Abstract)

B.1 $\begin{pmatrix}2&0&0\\0&3&1\\0&0&3\end{pmatrix}$

B.3 $\begin{pmatrix}1&1&0\\0&1&1\\0&0&1\end{pmatrix}$

B.5 $\begin{pmatrix}i&0&0\\0&-i&0\\0&0&1\end{pmatrix}$

B.7 $\begin{pmatrix}1&1&0&0\\0&1&1&0\\0&0&1&1\\0&0&0&1\end{pmatrix}$

B.9 (i) $\begin{pmatrix} \lambda^k & k\lambda^{k-1} & \frac{k(k-1)}{2}\lambda^{k-2} \\ 0 & \lambda^k & k\lambda^{k-1} \\ 0 & 0 & \lambda^k \end{pmatrix}$

(ii) When $k < m-1$,

$$\begin{pmatrix} \lambda^k & C_1^k\lambda^{k-1} & C_2^k\lambda^{k-2} & \cdots & C_{k-1}^k\lambda & 1 & 0 & \cdots & 0 \\ & \lambda^k & C_1^k\lambda^{k-1} & \cdots & C_{k-2}^k\lambda^2 & C_{k-1}^k\lambda & 1 & \cdots & 0 \\ & & \lambda^k & \cdots & C_{k-3}^k\lambda^3 & C_{k-2}^k\lambda^2 & C_{k-1}^k\lambda & \cdots & 0 \\ & & & \ddots & & \cdots & & & \cdots \\ & & & & \ddots & & & & \\ & & & & & \ddots & & & \\ & & & & & & \ddots & & \\ & & & & & & & \ddots & \\ & & & & & & & & \lambda^k \end{pmatrix}.$$

When $k = m-1$,

$$\begin{pmatrix} \lambda^k & C_1^k\lambda^{k-1} & \cdots & C_{k-1}^k\lambda & 1 \\ & \lambda^k & \cdots & C_{k-2}^k\lambda^2 & C_{k-1}^k\lambda \\ & & \ddots & & \vdots \\ & & & \lambda^k & C_1^k\lambda^{k-1} \\ & & & & \lambda^k \end{pmatrix}.$$

When $k > m-1$,

$$\begin{pmatrix} \lambda^k & C_1^k\lambda^{k-1} & \cdots & C_{m-2}^k\lambda^{k-m+2} & C_{m-1}^k\lambda^{k-m+1} \\ & \lambda^k & \cdots & C_{m-3}^k\lambda^{k-m+3} & C_{m-2}^k\lambda^{k-m+2} \\ & & \ddots & & \vdots \\ & & & \lambda^k & C_1^k\lambda^{k-1} \\ & & & & \lambda^k \end{pmatrix}.$$

B.11 $(-2)^n$

Index